ELECTROMAGNETIC FIELDS in BIOLOGY and MEDICINE

ELECTROMAGNETIC FIELDS in BIOLOGY and MEDICINE

Marko S. Markov

The fonts used here are not part of the standard font set and may be corrupted here. This book contains information obtained from authentic and highly regarded sources. Reasonable efforts have been made to publish reliable data and information, but the author and publisher cannot assume responsibility for the validity of all materials or the consequences of their use. The authors and publishers have attempted to trace the copyright holders of all material reproduced in this publication and apologize to copyright holders if permission to publish in this form has not been obtained. If any copyright material has not been acknowledged please write and let us know so we may rectify in any future reprint.

Except as permitted under U.S. Copyright Law, no part of this book may be reprinted, reproduced, transmitted, or utilized in any form by any electronic, mechanical, or other means, now known or hereafter invented, including photocopying, microfilming, and recording, or in any information storage or retrieval system, without written permission from the publishers.

For permission to photocopy or use material electronically from this work, please access www.copyright.com (http://www.copyright.com/) or contact the Copyright Clearance Center, Inc. (CCC), 222 Rosewood Drive, Danvers, MA 01923, 978-750-8400. CCC is a not-for-profit organization that provides licenses and registration for a variety of users. For organizations that have been granted a photocopy license by the CCC, a separate system of payment has been arranged.

Trademark Notice: Product or corporate names may be trademarks or registered trademarks, and are used only for identification and explanation without intent to infringe.

Visit the Taylor & Francis Web site at
http://www.taylorandfrancis.com

and the CRC Press Web site at
http://www.crcpress.com

CRC Press
Taylor & Francis Group
Boca Raton London New York

CRC Press is an imprint of the
Taylor & Francis Group, an **informa** business

CRC Press
Taylor & Francis Group
6000 Broken Sound Parkway NW, Suite 300
Boca Raton, FL 33487-2742

First issued in paperback 2017

© 2015 by Taylor & Francis Group, LLC
CRC Press is an imprint of Taylor & Francis Group, an Informa business

No claim to original U.S. Government works

ISBN-13: 978-1-4822-4850-0 (hbk)
ISBN-13: 978-1-138-74903-0 (pbk)

Visit the Taylor & Francis Web site at
http://www.taylorandfrancis.com

and the CRC Press Web site at
http://www.crcpress.com

Contents

To the Reader

Since the beginning of the twenty-first century, the level of electromagnetic fields (EMF) of natural and man-made origin continuously increases. While the natural physical factors remain relatively stable, man-made EMFs increase and affect life on this planet.

The guiding principle in selecting the topics and authors for this book was to present the contemporary state-of-the-art use of EMF in clinical practice for resolving problems that conventional medical practice cannot successfully treat. More importantly, this book offers a biophysical approach to the discussed problems.

It was my idea to make this book a tribute to William Ross Adey in recognition of his historical achievements in and contributions to bioelectromagnetics. For that reason, the first chapter discusses one of the fundamental ideas of Ross Adey—*biological windows*. Fortunately, Carl Blackman, in Chapter 2, gives a historical overview of the experimental work performed in the laboratory of Ross Adey and by a number of other researchers.

I attempted to arrange the table of contents in a systematic way—starting from EMFs' benefits and hazards to their potential to accelerate healing of tendon injuries in humans and animals.

Arthur Pilla, in Chapter 4, guides readers in the pulsed EMF development from signaling to healing, and Dr. Igor Belyaev discusses the biophysical mechanisms for nonthermal microwave effects in Chapter 5.

There are two chapters on *engineering*. I would like to emphasize the chapter by Jim Seal on step-by-step signal design (Chapter 7) is a good approach to understanding the principles of signal and device engineering starting from level zero to completion of the process, approval of the device, and acquiring leads to international agencies.

Chapters 8 through 10 are *biological*, reporting the research on EMF effects on microcirculation and immune and anti-inflammatory responses. This includes chapters by Ohkubo and Okano from Japan, Gapaev from Russia, and Balcavage et al. from the United States, respectively. It is clear from these chapters how broad the interest of the world scientific community is on the effects of EMF. They also present clarifications on the mechanisms of achieving potential health benefits.

Chapters 11 through 13 are *biophysical*, addressing the effects of EMF and infrasound range vibrations on cell hydrations in various tissues as well as applying computational fluid dynamics methods for studying EMF effects in model systems.

The *medical* part of the book (Chapters 16 through 18) covers various applications of EMF, conducted in different clinical studies all over the world. Starting with the *traditional* use in wound healing and cartilage/bone repair and continuing with the use of EMF for pain control and inhabitation of cancer growth, patients' preference has shifted to EMF application in plastic surgery.

Usually, the safety of EMF is discussed from the viewpoint of patients, neglecting the fact that patients are exposed to short (less than an hour) sessions, while the staff in therapeutic facilities work for hours together in environments of mixed electromagnetic signals. The book offers three chapters on *operator safety*. It includes a general chapter on the exposure assessment and risk, as well as the evaluation of the health status of operators after long hours of work in physical therapy offices.

Chapters 23 and 24 discuss the potential of the relatively new methods of electroporation and nanoelectroporation to be used for electrochemotherapy, gene therapy, and nonthermal ablation. Thanks to Dr. Nuccitelli, Dr. Miklavcic, and Dr. Sersa, these novel clinical methods are described in the book.

The book provides room for two *exotic* topics (Chapters 25 and 26): *dirty* electricity and the treatment of tendon injuries in humans and animals.

As I wrote elsewhere, magnetotherapy is part of biomagnetic technology that requires integrated efforts of experts from engineering, physics, biology, and medicine. It is up to readers to judge to what extent these efforts are presented well.

I believe that readers will be satisfied by the large spectrum of problems covered in relation to the interactions of EMFs with living systems and the clinical benefits of such interactions. I also believe that the book will be a helpful tool for scientists and clinicians, as well as for medical and engineering students.

Marko S. Markov

Editor

Marko S. Markov received his BS, MS, and PhD from Sofia University, Bulgaria. He has been professor and chairman of the Department of Biophysics and Radiobiology for more than 22 years. He was invited professor in a number of European and American academic and industry research centers.

Dr. Markov is well recognized as one of the world's best experts in the clinical application of electromagnetic fields. He was invited speaker at 69 international meetings. He has presented 282 papers and short communications at various international meetings. His list of publications includes 192 published papers. He is a member of the editorial boards of *Electromagnetic Biology and Medicine*.

In 1981, Dr. Markov wrote the book *Professions of the Laser*. In 1988, he coedited *Electromagnetic Fields and Biomembranes*, published by Plenum Press, with Dr. Martin Blank. In 2004, he coedited *Bioelectromagnetic Medicine*, published by Marcel Dekker, with Dr. Paul Rosch, the president of the American Institute of Stress. In 2010, he coedited with Damjian Miklavcic and Andrei Pakhomov *Advanced Electroporation Techniques in Biology and Medicine*, published by CRC Press.

Dr. Markov has edited six special issues of *The Environmentalist* and *Electromagnetic Biology and Medicine* following international workshops on biological effects of electromagnetic fields. He holds five international patents for clinical and technological use of EMF, one U.S. patent, and one pending patent at the U.S. patent office.

Dr. Markov has 45 years of experience in basic science research and 38 years of experience in the clinical application of electromagnetic fields for the treatment of bone and soft tissue patholo-gies. His commercial affiliation includes a series on contractual appointments in Bulgaria, and since 1993, he has been the vice president for research and development in three U.S. companies. His responsibilities include analysis, engineering, and improvement of magnetic field–based modalities for the treatment of various injuries and diseases and pain control. The spectrum of applications from the engineering/physics point of view includes permanent magnets (mainly for pain control), pulsed radiofrequency signals at 27.12 MHz for the treatment of pain and edema in superficial soft tissue (FDA label), and innovation in low frequency range for the inhibition of angiogenesis and tumor growth as well as for pain relief.

Dr. Markov is a cofounder of the International Society of Bioelectricity, the European Bioelectromagnetic association (EBEA), the International Society of Biomagnetism, and the International Society of Bioelectromagnetism. In 1987, Dr. Markov was the organizer of the UNESCO project "EMF and Mankind."

Dr. Markov is an honored citizen of Tcherven briag (Bulgaria), Courtea de Ardes (Romania), Newport (Rhode Island). He received the "Man of the Year" award from the International Biographic Center in 1996 and was listed in the "Dictionary of International Biographies" in 1996.

Contributors

I. Antonov
Department of Medical Physics and Biophysics
Sofia Medical University
Sofia, Bulgaria

Sinerik Ayrapetyan*
Life Sciences International Postgraduate
　　Educational Center
UNESCO Chair
Yerevan, Armenia
E-mail: info@biophys.am

Naira Baghdasaryan
Life Sciences International Postgraduate
　　Educational Center
UNESCO Chair
Yerevan, Armenia

Walter X. Balcavage*
Department of Biochemistry and Molecular
　　Biology
School of Medicine
Indiana University
Indianapolis, Indiana
E-mail: wbalcava@iupui.edu

Sedrak Barseghyan
Life Sciences International Postgraduate
　　Educational Center
UNESCO Chair
Yerevan, Armenia

Igor Belyaev*
Laboratory of Radiobiology
Cancer Research Institute
Slovak Academy of Science
Bratislava, Slovakia
and
Laboratory of Radiobiology
General Physics Institute
Russian Academy of Science
Moscow, Russia
E-mail: Igor.Beliaev@savba.sk

Carl Blackman (retired)*
U.S. Environmental Protection Agency
Raleigh, North Carolina
E-mail: carl.blackman@gmail.com

Matteo Cadossi
Rizzoli Orthopaedic Institute
University of Bologna
Bologna, Italy

Ruggero Cadossi*
IGEA Clinical Biophysics
Carpi, Italy
E-mail: r.cadossi@igeamedical.com

Ivan L. Cameron*
Department of Cellular and Structural Biology
The University of Texas Health Science at
　　San Antonio
San Antonio, Texas
E-mail: cameron@uthscsa.edu

Maja Cemazar
Department of Experimental Oncology
Institute of Oncology Ljubljana
Ljubljana, Slovenia
and
Faculty of Health Sciences
University of Primorska
Koper, Slovenia

E. Dzhambazova
Department of Chemistry, Biochemistry,
　　Physiology and Pathophysiology
Sofia University "St. Kliment Ohridski"
Sofia, Bulgaria

Stéphane J.-P. Egot-Lemaire
EMC-Testcenter AG
Regensdorf, Switzerland

---

* E-mail addresses are provided for each lead chapter contributor.

Andrew B. Gapeyev*
Institute of Cell Biophysics
Russian Academy of Sciences
Pushchino, Moscow Region, Russia
E-mail: a_b_g@mail.ru

Sean Hagberg
Rio Grande Neurosciences, LLC
San Francisco, California

W. Elaine Hardman
Department of Biochemistry and Microbiology
Joan C. Edwards School of Medicine
Marshall University
Huntington, West Virginia

Magda Havas*
Environmental and Resource Studies Program
and
Centre for Health Studies
Trent University
Peterborough, Ontario, Canada
E-mail: drmagdahavas@gmail.com

Armenuhi Heqimyan*
Life Sciences International Postgraduate
   Educational Center
UNESCO Chair
Yerevan, Armenia
E-mail: info@biophys.am

Michel Israel*
Department of Public Health
National Centre of Public Health and Analyses
Pleven Medical University
Sofia, Bulgaria
E-mail: michelisrael@abv.bg

M. Ivanova
National Centre of Public Health and Analyses
Sofia, Bulgaria

Jolanta Karpowicz*
Central Institute for Labour Protection
National Research Institute
Warszawa, Poland
E-mail: jokar@ciop.pl

Tadej Kotnik
Faculty of Electrical Engineering
Department of Biomedical Engineering
University of Ljubljana
Ljubljana, Slovenia

Marko S. Markov*
Research International
Williamsville, New York
E-mail: msmarkov@aol.com

Varsik Martirosyan
Life Sciences International Postgraduate
   Educational Center
UNESCO Chair
Yerevan, Armenia

Harvey N. Mayrovitz*
College of Medical Sciences
Nova Southeastern University
Ft. Lauderdale, Florida
E-mail: mayrovit@nova.edu

Yerazik Mikayelyan
Life Sciences International Postgraduate
   Educational Center
UNESCO Chair
Yerevan, Armenia

Damijan Miklavcic*
Faculty of Electrical Engineering
Department of Biomedical Engineering
University of Ljubljana
Ljubljana, Slovenia
E-mail: Damijan.Miklavcic@fe.uni-lj.si

Lilia Narinyan
Life Sciences International Postgraduate
   Educational Center
UNESCO Chair
Yerevan, Armenia

Edwin Nemoto
Department of Neurosurgery
University of New Mexico Health Sciences
   Center
Albuquerque, New Mexico

---

* E-mail addresses are provided for each lead chapter contributor.

Anna Nikoghosyan
Life Sciences International Postgraduate
    Educational Center
UNESCO Chair
Yerevan, Armenia

Richard Nuccitelli*
BioElectroMed Corp.
Burlingame, California
E-mail: rich@bioelectromed.com

Chiyoji Ohkubo*
Japan EMF Information Center
Tokyo, Japan
E-mail: ohkubo@jeic-emf.jp

Hideyuki Okano
Research Center for Frontier Medical
    Engineering
Chiba University
Chiba, Japan

Richard Parker*
CytoWave, LLC
Delray Beach, Florida
E-mail: rfp@cytowave.com

William Pawluk*
Medical Editor
Timonium, Maryland
E-mail: wpaw@comcast.net

Juan Carlos Pena-Philippides
Department of Neurosurgery
University of New Mexico Health Sciences
    Center
Albuquerque, New Mexico

Arthur A. Pilla*
Department of Biomedical Engineering
Columbia University
and
Department of Orthopedics
Icahn School of Medicine at Mount Sinai
New York, New York
E-mail: aa_pilla@yahoo.com

Christine H. Rohde*
Division of Plastic and Reconstructive Surgery
Columbia University Medical Center
NewYork-Presbyterian Hospital
New York, New York
E-mail: chr2111@cumc.columbia.edu

Tamara Roitbak*
Department of Neurosurgery
University of New Mexico Health Sciences
    Center
Albuquerque, New Mexico
E-mail: troitbak@salud.unm.edu

James G. Seal*
Serial Port Engineering Inc.
Boca Raton, Florida
E-mail: seal9523@bellsouth.net

Gregor Sersa
Department of Experimental Oncology
Institute of Oncology Ljubljana
Ljubljana, Slovenia

Stefania Setti
IGEA Clinical Biophysics
Carpi, Italy

T. Shalamanova
National Centre of Public Health and Analyses
Sofia, Bulgaria

Erin M. Taylor
Division of Plastic and Reconstructive Surgery
Columbia University Medical Center
New York, New York

L. Traikov*
Faculty of Medicine
Department of Medical Physics and Biophysics
Sofia Medical University
Sofia, Bulgaria
E-mail: lltraikov@hotmail.com

* Eh-mail addresses are provided for each lead chapter contributor.

Hubert Trzaska*
Electromagnetic Environment Protection Lab
Telecommunications and
    Teleinformatics Department
and
Department of Electronics
Technical University of Wroclaw
Wroclaw, Poland
E-mail: hubert.trzaska@pwr.edu.pl

A. Ushiama
Department of Environmental Health
National Institute of Public Health
Tokyo, Japan

Lyubina Vesselinova*
Clinic of Physical and Rehabilitation Medicine
Military Medical Detachment for Emergency
    Response
Military Medical Academy
Sofia, Bulgaria
E-mail: lyvess.md@gmail.com

Gabi N. Waite
Department of Cellular & Integrative
    Physiology
School of Medicine
Indiana University
Indianapolis, Indiana

V. Zaryabova
National Centre of Public Health and Analyses
Sofia, Bulgaria

---

* E-mail addresses are provided for each lead chapter contributor.

1 Biological Windows
A Tribute to Ross Adey

Marko S. Markov

CONTENTS

1.1 INTRODUCTION

I knew the name Ross Adey from the literature and was amazed by the depth of his understanding of physics, medicine, and biology long before I had the chance to meet him in person.

In 1986, I organized in Bulgaria International School *electromagnetic fields* (*EMFs*) *and biomembranes*, and one of the first invitations was sent to his California office. It was the time when neither Internet nor cellular phones were available. One day I was surprised to receive a telephone call from the United States. On the other end of the telephone line was Ross. He apologized that he knows little about Bulgaria, me, and my research and expressed himself in an amazing way of politeness. I briefly described the scope of the meeting and gave him some names of invited lecturers. Then he said: "Dear Professor Markov, I will be pleased to be at your service during this week."

Months after that conversation, I was lucky to meet this remarkable individual who was absolutely accurate in his one-hour presentation and (as always sitting at the back of the room) very actively participated in the discussion. Later on, I learned that "When Ross speaks, people listen."

This tall gentleman was usually silent, but his comments at meetings always have been exactly on the core of the topic. Sometimes his criticism was very strong. In person, Ross was extremely polite and treated everybody with respect. Interestingly enough, I have heard from three individuals the same sentence "Ross was a master in talking to you in a way that makes you feel yourself great."

With his encyclopedic knowledge in biology, physics, engineering, and medicine, Ross had numerous contributions in the areas of bioelectromagnetics and public health. One should add here his interest in communication with people from all over the world via Ham radio, as well as his interest in various arts. Last but not the least—up to his 72 years—he was running at least three marathons a year.

There was something interesting that happened at this Bulgarian conference that I learned 15 years later. In the 2001 BEMS meeting, I introduced to him my 20-year-young daughter, and Ross immediately established contact with her by telling a story of his arrival for the meeting: "Julia, your father did organize a conference in a Bulgarian national park. Once I arrived in the hotel, I quickly change to my sports suit and went to run in the park. Just 100 meter from the hotel, I heard a voice of a lion. 'What lions could be here?' I continue running but the voice of lion came again. So, I decided for my safety not to provoke an incident."

At that moment, I took the floor: "Ross, it was a hundred year old lion in the small zoo in the park."

1.2   BIOLOGICAL WINDOWS: SOMETHING WAITING TO BE DISCOVERED

It is astonishing that among all his contributions in the field of bioelectromagnetics, Dr. Adey introduced the term *window*. It is now close to 40 years since the period of 1975/1976 when he first proposed this term. But there is something more interesting here: more likely it was the time when the accumulation of knowledge in numerous scientific disciplines, such as physics, biology, and medicine, opened this idea in the air. In a short period of several months, three papers had been published by research teams that do not know each other. Despite the difference in terminology, these papers treated the same subject: resonance mechanisms in interaction of EMFs with living systems. The paper from the Soviet Union (Ukolova et al., 1975) introduced *stages* in these interactions; in Sofia, Bulgaria, Markov et al. (1975) offered the term *resonance*, while in Loma Linda's laboratory of Ross Adey, Bawin and Adey (1976) focused on *window*. While the US publication was on frequency windows, the other two groups actually consider amplitude windows. In other words, the authors found evidence for the existence of specific values of amplitudes and frequencies at which the response of the biological system is more pronounced than in the surrounding amplitude and frequency values.

The fundamental question for bioelectromagnetics is the identification of the biochemical and biophysical conditions under which applied magnetic fields (MFs) could be recognized by cells in order to further modulate cell and tissue functioning. This actually is achieved by identifying and investigating *resonance* and *window* conditions.

For a while, the term *window* was not accepted, even rejected by some researchers. Curiously enough, a decade later, a number of *resonance mechanisms* have been introduced and basically confirmed the principle of *window* approach. It is interesting to know that the *window* hypothesis was proposed a decade before Liboff (1985) proposed the ion cyclotron resonance, followed by Lednev's ion parametric resonance (1991) and further resonance models (Blanchard and Blackman, 1994; Blackman, 2014).

Being familiar with the electron structure of the atom, I followed the same approach in respect of the possibility of biological systems to recognize and respond to applied EMF. Defined in the electron structure of the atom are a ground state and several excited states at which the electron might be transferred by a defined energy. The energy that would be less or more than the energy necessary for transition will bring the electron to an unstable energy level, and the electron will quickly return to the ground state.

Following the same logics, we proposed that living systems exist in an *equilibrium* state in respect to electromagnetic conditions. Evolutionarily, they have developed a mechanism that allows them to respond to exogenous MFs. I would like to underline that small perturbations in electromagnetic background might lead to small modifications in metabolic activity that are transient and disappear when the applied field is gone. However, if the field amplitude is large enough to the level at which biological system is ready to respond, the effect usually lasts longer (Markov, 1984, 1994). It was called *resonance level*, but further I accepted the term introduced by Ross Adey.

Even this is not directly related to *windows*, I would like to point that there is evidence pointing that the effects of MFs are stronger when the biological system is out of equilibrium. Eventually, this could explain the benefit observed when MFs are applied to treat a number of medical problems—thus helping the restoration of the healthy status of the organ and system (Nindl et al., 2002).

Turning back to the *window* hypothesis, we need to point out that all terms in use, for example, *amplitude window*, *frequency window*, and *time window*, might be unified as *biological window* that to certain extent corresponds to the subject of interaction—biological systems.

While in the early days of development of the *window* hypothesis many efforts were applied for studying the range of frequencies or amplitudes that might affect the biological subject, today, the situation is completely different. Very often one may see publications that claim *window* or *resonance* effects when the authors report positive effects of exposure to EMF. The problem here is that such *window* is determined by experiment with a field testing of only one frequency or amplitude. So it makes the claim not reasonable.

During evolution, most of the living organisms developed specific mechanisms for the perception of natural electric and MFs. These mechanisms require specific combinations of physical parameters of the applied fields to be detected by biological systems. In other words, the *windows* are means by which discrete MF/EMFs are detected by living organisms. Depending on the level of structural organization, these mechanisms of detection and response may be seen at different levels: at membrane, cellular, or tissue level. Sometimes, the *windows* function via signal transduction cascade or central nervous system.

Interesting examples might be seen in geotropism when the roots of plants became oriented along the local geomagnetic field or in navigation of birds and fish.

The sensitivity of the biological systems to weak MF has been described elsewhere (Markov et al., 1979, 1984, 1989) mainly in respect to the dependence of bioeffects on the amplitude or frequency of the applied fields. It may be interesting to know that all early publications made links between *windows* and information transfer (Adey, 1981, 1986, 1993a,b; Markov, 1979, 1984).

Very often magnetobiology neglects the three features of the interaction of EMF with living system:

- Transfer of energy
- Transfer of material
- Transfer of information

Let me go back to my own *window* research. My first publication on *windows* was in the book related to biological information transfer. The idea of this paper (even not expressed in today terms) was that one way or another, the interactions of biological systems have informational character, which means that there are discrete levels of amplitude and frequency that are more plausible than others.

One of the problems here is that international committees like ICNIRP (International Commission on Non-Ionizing Radiation Protection) and IEEE (Institute of Electrical and Electronics Engineers) emphasized for decades on the energy transfer. Moreover, they claim that no biological effects are possible without heat transfer.

There two very important obstacles here: threshold and thermal considerations. To start with, we need to point out that EMFs are frequently discussed under the umbrella *radiation* and introducing the category of nonionizing radiation. For that reason, the effects of EMFs are studied in parallel with the effects of ionizing radiation. Well, radiation constitutes energy, and for that reason, the energy interactions with any physical or biological body are connected with damage or heating of the body when the intensity of the radiation is above certain threshold level. Yes, this is correct for ionizing radiation.

For decades, the idea of thermal effects in bioelectromagnetics had been introduced and became the subject of intensive discussions, related to specific absorption rate (SAR) as useful criteria. It is clear that SAR requires a threshold value determination. I am categorically against the thermal approach as the only possible mechanism for EMF interactions. This approach continues for several decades based upon the engineering approach developed by ICNIRP and IEEE committees. They went even further, claiming that "only possible biological mechanism of EMF interaction is thermal" (Cho and D'Andrea, 2003).

However, hundreds of studies and publications reported biological and clinical effects at low-intensity and low-frequency EMFs, as well as at static magnetic fields (SMFs). At these interactions, it is very unlikely, or even impossible, to expect thermal effects, and the threshold level approach is not reasonable. Having in mind that even the simple biological systems are nonlinear, the possible means of interactions might involve an informational transfer.

In a series of experiments designed to study effects of large-range (0–100 mT) SMFs applied to biosystems with different levels of organization (cellular membranes, microorganisms, plants, and animals), a specific maximum of the observed effects was found at 45 mT (Markov, 1991; Markov and Guttler, 1989). Later on, in a very detailed study of the SMF and low-frequency EMF effects on myosin phosphorylation, the 45 mT window and another window at 15 mT were confirmed (Markov, 2005). Later experiments with Ca^{2+} interactions suggested that calcium efflux, as well as calcium/calmodulin models, also could be attributed to *windows*. It would be interesting to note that the myosin phosphorylation data on amplitude windows are based upon calcium interactions that were first investigated by the Ross Adey group in respect of frequency windows. In another chapter in this book, Blackman describes interesting facts about the history of the development of *window* hypothesis.

1.3   LATEST PAPER OF ROSS ADEY

Let us briefly describe the fundamental contribution of Ross Adey in his last paper (Adey, 2004). If one looks only at the title *Potential Therapeutic Applications of Nonthermal Electromagnetic Fields: Ensemble Organization of Cells in Tissue as a Factor in Biological Field Sensing*, he or she will see reflections in all principles in bioelectromagnetics, starting with ensemble organization of cells as a factor in sensing of physical factors in biological systems through the ability of EMFs to initiate biological response to potential therapeutic applications.

The paper starts with a comparison of natural and man-made electromagnetic environments, followed with the description of thermal and nonthermal biological effects. For me, the most fundamental is the third section discussing the sensitivity to nonthermal stimuli in tissue and functional implications. Thus, the role of field intermittency and exposure duration in seeking optimal therapeutic responses is presented. The paper ends with the discussion of the role of cellular ensembles.

It would be fair to say that Ross Adey's paper is the pearl in the crone of his publications and we only have to say "Thank you for your papers, and for your many contributions in bioelectromagnetics."

REFERENCES

Adey WR (1981) Tissue interactions with nonionizing electromagnetic fields. *Physiol Rev* 61: 435–514.
Adey WR (1986) The sequence and energetics of cell membrane transductive coupling in intracellular enzyme systems. *Bioelectrochem Bioenerget* 15: 447–456.
Adey WR (1993a) Electromagnetic technology and the future of bioelectromagnetics. In Blank M (ed.). *Electricity and Magnetism in Biology and Medicine*. Plenum Press, New York, pp. 101–108.
Adey WR (1993b) Electromagnetics in biology and medicine. In Massumoto H (ed.). *Modern Radio Science*. Oxford University Press, London, U.K., pp. 231–249.
Adey WR (2004) Potential therapeutic applications of nonthermal electromagnetic fields: Ensemble organization of cells in tissue as a factor in biological field sensing. In Rosch PJ, Markov M (eds.). *Bioelectromagnetic Medicine*. Marcel Dekker, New York, pp. 1–14.
Bawin SM, Adey WR (1976) Sensitivity of calcium binding in cerebral tissue to weak environmental electric fields at low frequency. *Proc Natl Acad Sci USA* 73: 1999–2003.
Blackman C (2014) A replication and extension of the Adey group's calcium efflux results. In Markov M (ed.). *Electromagnetic Fields in Biology and Medicine*. CRC Press, Boca Raton, FL.
Blanchard JP, Blackman C (1994) Clarification and application of an Ion Parametric Resonance Model for magnetic field interaction with biological systems. *Bioelectromagnetics* 15: 217–231.
Cho CK, D'Andrea JA (2003) Review of the effects of RF fields on various aspects of human health. *Bioelectromagnetics* 24: S5–S6.
Lednev VV (1991) Possible mechanism for the influence of weak magnetic fields on biological systems. *Bioelectromagnetics* 12: 71–75.
Liboff AR (1985) Cyclotron resonance in membrane transport. In Chiabrera A, Nicolini C, Schwan HP (eds.). *Interactions between Electromagnetic Fields and Cells*. Plenum Press, London, U.K., pp. 281–290.
Markov MS (1979) Informational character of magnetic field action on biological systems. In Jensen K, Vassileva Yu (eds.). *Biophysical and Biochemical Information Transfer in Recognition*. Plenum Press, New York, pp. 496–500.

Markov MS (1984) Influence of constant magnetic field on biological systems. In Allen MJ (ed.). *Charge and Field Effects in Biosystems*. Abacus Press, Kent, U.K., pp. 319–329.

Markov MS (1991) Electromagnetic field influence on membranes. In Bender M (ed.). *Interfacial Phenomena in Biological Systems*. Marcel Dekker, New York, pp. 172–191.

Markov MS (1994) Biological effects of extremely low frequency magnetic fields. In Ueno S (ed.). *Biomagnetic Stimulation*. Plenum Press, New York, pp. 91–103.

Markov MS (2005) Biological windows: A tribute to W. Ross Adey. *The Environmentalist* 25(2/3): 67–74.

Markov MS, Guttler JP (1989) Electromagnetic field influence on electroporation of erythrocytes. *Studia Biophys* 130: 211–214.

Markov MS, Goltzev VN, Doltchinkova VR, Michailova D (1989) Modification of the chloroplast delayed fluorescence by electroporation. *Semin. Biophys.* 5: 71–78.

Markov MS, Todorov NG (1984) Electromagnetic field stimulation of some physiological processes. *Stud. Biophys.* 99: 151–156.

Markov MS, Todorov SI, Petrova RP (1979) Influence of constant magnetic field on the absorption spectra in leaves of Zea mays. *Ann. Sofia Univ.* 68: 109–116, book 4.

Markov MS, Todorov SI, Ratcheva MR (1975) Biomagnetic effects of the constant magnetic field action on water and physiological activity. In Jensen K, Vassileva Yu (eds.). *Physical Bases of Biological Information Transfer*. Plenum Press, New York, pp. 441–445.

Nindl G, Johnson MT, Hughes EF, Markov MS (2002) Therapeutic electromagnetic field effects on normal and activated Jurkat cells. In *International Workshop of Biological Effects of Electromagnetic Fields*, Rhodes, Greece, October 7–11, 2002, pp. 167–173.

Ukolova MA, Kvakina EB, Garkavi LX (1975) *Stages of Magnetic Field Action. Problems of Action of Magnetic Field on Biological Systems*. Nauka, Moscow, Russia, pp. 55–710.

2 Replication and Extension of Adey Group's Calcium Efflux Results*

Carl Blackman

CONTENTS

2.1 INTRODUCTION

For most scientists studying the biological effects of radio-frequency (RF) radiation in the early 1970s, the only mechanism of action was heating. The newly commercialized microwave oven had been brought on the market to make kitchen duties easier and it succeeded. Then many scientists attended the New York Academy of Sciences Meeting on February 12–15, 1974, entitled Conference on the Biological Effects of Non-Ionizing Radiation. Some attendees were forever changed by the presentations made by Ross Adey's group from the UCLA Brain Research Institute. Dr. Suzanne Bawin presented results involving the release of calcium ions (using Ca-45 as a tracer and as a surrogate measure of GABA release), which occurred when 147 MHz carrier waves were sinusoidally amplitude modulated (AM) at frequencies associated with EEG recordings. Bawin used a chick brain tissue preparation and found a power density window well below thermalizing levels, within which enhanced calcium ion efflux occurred and, most importantly, a modulation frequency window centered at 16 Hz, with no effect below 6 Hz and above 20 Hz and no modulation present [1]. Following this meeting, three EPA scientists (including this author) were given approval to visit Adey's laboratory at UCLA to get a better sense of the probability that these results were not artifacts. Adey's group members were very open and provided the background research findings to explain why they had even conceived of the experiment. We concluded that the results appeared genuine and that opinion was augmented by Adey's laboratory request that we try to replicate the results.

Initial attempts to replicate the calcium efflux work over a 6–8-month period were unsuccessful, as was an attempt to extend the research to mammalian cells in culture. The scientist attempting the replication assumed substantial administrative duties, so I became the replacement to continue the attempt to replicate Bawin's original findings.

---

* *Disclaimer*: The opinions expressed in this article are solely those of the author, and not necessarily those of the US Environmental Protection Agency.

2.2 RADIO-FREQUENCY RADIATION EXPERIMENTS

The breakthrough occurred after 9 months of extensive work, which included a number of phone conversations with Bawin. The week before my technician and I were scheduled to visit Bawin in her laboratory, we had our first success. This time, our visit was very productive because we discussed problems we both saw and options to resolve them. One big problem was the rinsing of the tissues to remove loosely attached Ca-45 ions. The results of the first replication of Bawin's report—showing a power density and an amplitude modulation window, centered on 16 Hz, in the same regions reported by Bawin—were published in 1979 [2]. The research reported in the following descriptions used 16 Hz amplitude modulation of RF carrier waves unless explicitly stated.

While we were in Bawin's lab, we also discussed what other steps needed to be improved and decided that we should test different buffers, in place of bicarbonate in the original recipe, to stabilize pH, because in the bicarbonate buffer, the pH changed, dropping from 7.4 to 6.6 during the 20 min treatment period, irrespective of the presence of the electromagnetic fields (EMFs) (data not published).

We also worked very closely with a statistician in our laboratory at the EPA to optimize our procedures for maximum statistical power and with an electrical engineer from Duke University to investigate some of the unexpected EMF dependencies we were observing.

Our statistician encouraged us to increase the statistical power of his analysis by changing from the Bawin protocol, which exposed 10 chick forebrains at a time in 2 racks of 5, placed in either side of a TEM exposure chamber and repeating the exposure for 4 replicates, to exposing only 4 chick forebrains at a time in the same 2 racks (thereby separating the samples by double the original separation distance) and repeating the exposure to accumulate 8 replicates. We immediately observed a narrowing of each power density window of effect [3]. To investigate this result, which we thought might reveal an artifact in the procedure, we started collaborating with William Joines [3]. He was able to model and calculate the electric fields at the tissue surface for each power density and separation distance we had used. The interpretation of his work is that exposed samples also reradiate the fields and the more closely spaced samples were receiving a broader range of field strength than the more widely spaced samples [4]. Joines' calculation method and analysis also allowed us to reconcile the lack of agreement, using similar AM fields, when we tried to align power density windows at different carrier frequencies, namely, 50, 147, and 450 MHz [5]. By converting the exposures from power density to electric field values at the surface of the forebrain hemisphere, the results were found to align [4,6,7]. Finally, we tested the gross physiological state of the brain tissue just after removal from the organism, and it showed no electrical activity (unpublished results), but the same tissues were as metabolically active (determined by oxygen utilization measured by a Clark oxygen electrode), just after removal from the animal as it was at the end of the treatment period [7].

We also tried to improve the physiological medium used to bathe the chick forebrain halves during calcium labeling and during the EMF exposure to determine if we could get larger changes in efflux from the EMF exposures. First, we tried different buffers, including HEPES, and although they held the pH closer to the 7.4 set point, we never detected a radiation-induced change in calcium efflux [7]. The issue of HEPES buffer negating EMF-induced effects arose in a completely different experimental context many years later. A Swedish group reported in a meeting presentation that they were unable to detect the inhibitory influences of specific combinations of ac and dc magnetic fields on neurite outgrowth from PC-12 cells stimulated with nerve growth factor that we had published [8] and references therein. We noted that the medium they used to grow and treat their PC-12 cells contained HEPES buffer, unlike the more traditional medium customarily used to grow the cells. We subsequently presented results demonstrating that when increasing molarities of HEPES in the physiological range were added to our medium, the EMF-induced inhibitory effect was progressively attenuated (Blackman et al. 1998, *BEMS 20 Annual Meeting*,

June 7–11, St. Petersburg, FL; these results are unpublished). The lesson from these results is to be certain a replication is truly a replication.

We also tried to increase the biological response in the chick forebrain experiments by raising and lowering the molarities of potassium and independently of calcium from the values in the physiological medium, 5.6 mM KCl and 2.16 mM $CaCl_2$, respectively. We observed an effect window for KCl, where 2.8 or 8.4 mM suppressed the field-induced response, and for $CaCl_2$, 1.06 and 4.24 mM suppressed the response, while 3.18 mM did not. Seeing no improvement in the response to EMF exposure, we abandoned this approach (data not published). During these early investigations, we noticed that temperature change occurred during the washing step and during the exposure period, and we monitored and took steps to stabilize this critical parameter during washing and during the 20 min exposure period [9].

As Adey's group and our group published more reports, questions were raised in a review paper [10] that stimulated us to repeat our 50 MHz, 16 Hz AM, exposures [3] under improved conditions of temperature control. We replicated the two power density windows we initially reported, thereby removing any doubt about the response, and found four more response windows [11]. These six intensity windows exhibited interesting dependencies that served as a basis for speculation on the initial transduction step(s) and the likely amplification processes that led to the observable expression, calcium ion efflux. Subsequent study by Thompson and colleagues [12] applied techniques developed for cooperative oxygen atom attachment to hemoglobin to our data and found that the intensity response windowing reported for 50 and for 147 MHz radiation, AM at 16 Hz, could be fit with few exceptions to a cooperative model of calcium ion efflux from cell membranes.

The data from Adey's group and from our laboratory stimulated others to examine the response of other biological systems to AM radiation. Dutta and colleagues [13] used cultures of human (IMR-32) and hybrid mouse Chinese hamster (NG-108-15) neuroblastoma cells and showed that specific windows of calcium ion efflux were present in two SAR (a measure of absorbed energy per weight) regions, 0.05 and 0.75–1.0 W/kg, for 915 MHz fields AM at 16 Hz. No changes in field-induced calcium ion efflux were detected at SAR values lower, 0.01 W/kg, between 0.075, 0.1, and 0.5 W/kg, or above those values, 2 and 5 W/kg. When unmodulated 915 MHz fields were used at 0.05 W/kg, there was no effect on calcium ion release as expected, but surprisingly, when unmodulated fields at 1.0 W/kg were used, there was a statistically significant increase in calcium release. This critical result indicated that at high enough SAR values, it is possible for unmodulated fields to produce effects usually reserved for RF carrier waves with specific AM frequencies.

Dutta et al. [14] showed intensity- and modulation-specific frequency-induced changes in calcium ion release when these same two cell lines were exposed to 16 Hz AM, 147 MHz radiation. Dutta et al. [15] used 16 Hz AM, 147 MHz radiation exposures and showed an intensity window at 0.01, 0.02, and 0.05 W/kg SAR, but not at 0.001, 0.005, or 0.1 W/kg, for enhanced acetylcholine activity in the NG-108-15 cells, thus demonstrating that a cholinergic neurotransmitter system could be responsive to this radiation in a low SAR region.

Further, Dutta et al. [16] also found that enolase, a cytoplasmic enzyme that was contained in a plasmid in *Escherichia coli*, could be altered by 147 MHz fields, AM at 16 Hz, and by the ELF fields at 16 and 60 Hz. This result demonstrated that it was not just membrane properties that could be affected by the exposures.

Another research group, Schwartz et al., tested the influence of AM 240 MHz RF on calcium ion efflux using isolated but intact frog heart stimulated to beat at 0.5 Hz [17]. They reported enhanced calcium ion release only from the intact frog heart (bathed in Ringer's solution with only bicarbonate present as a buffer) that occurred at specific intensities and modulation frequencies, in a manner similar to that found with calcium ion release from the brain tissue. However, Schwartz and Mealing [18] were unsuccessful under similar RF conditions (1 GHz, AM at both 0.5 or 16 Hz) in observing altered calcium ion release when they examined the response of a more traditional atrial strip preparation. This negative result might have been expected because the atrial strips were bathed in Ringer's solution that was buffered with 10 mM HEPES to maintain proper pH. Blackman et al. [7] had reported that the presence of HEPES buffer prevented radiation-induced calcium ion release from brain tissue preparations.

There were no established theories that would explain this biological behavior that was responsive to windowed biological responses to modulation frequency and to intensity; certainly, there were no reasonable explanations involving thermalizing actions of the radiation. Thus, these results arose from a discovery process, that is, letting the biological response identify interactions of the fields with biological entities, without a specific theoretical mechanism of action that would cause/predict the observed effects. Published research near the end of the last century driven by the discovery process provided strong empirical information and revealed clues to and requirements for the development of mechanistic understanding of the underlying processes. However, there are a number of parameters that need to be monitored and controlled to perform a faithful replication of earlier published results, some perhaps still not known; in many cases, researchers new to this research area do not take the time to understand the parameter space that needs to be controlled. For example, O'Connor et al. [19] recently used state-of-the-art apparatus in a very-well-designed experiment to test whether GSM RF fields could alter cellular calcium ion movement. Unfortunately, the experimental design required the presence of HEPES buffer during the exposure period, but that buffer had been shown previously to inhibit AM RF field-induced changes [7]. Thus, like Schwartz and Mealing [18], this ignorance of historical detail likely caused an otherwise elegant experiment to become a minor footnote. One lesson from this circumstance is to encourage investigators new to this research area to contact the original authors or their contemporaries to get advice and details that are essential for genuine replication/extension experiments and to avoid pitfalls. If historical details of discovery experiments requiring control of many parameters are ignored, there is much lower probability that replication experiments can be properly performed and usefully integrated into the historical database.

The need to pay attention to experimental details also applies to review articles. For example, two summaries of earlier literature have missed critical components that could have served to more completely define essential parameters that need to be controlled and tested. For example, Juutilainen and de Seze [20] reviewed the history of AM RF field effects, but it missed a critical report. Specifically, Schwartz and Mealing's [18] paper, included in the review, reported no change in the release of calcium ions when frog-heart-derived atrial strips were exposed to 1 GHz RF, using 0.5 or 16 Hz AM (see note earlier about the presence of HEPES buffer potentially inhibiting calcium ion release due to AM RF). However, an earlier publication by Schwartz et al. [17], in the same journal and cited in [18], examined calcium ion release from intact, beating frog heart (in Ringer's solution with only bicarbonate present as a buffer) exposed to AM 240 MHz RF and reported two intensity windows at 0.15 and 0.30 W/kg, but only at 16 Hz, not at 5 Hz AM. The omission of this earlier publication prevented acknowledgment of the authors' expansion of effect targets to a broader biological tableau for response to AM RF fields. Further, a report by Hoyto et al. [21] demonstrated that at the lowest power density examined, 6 W/kg, an effect occurred in the presence and in the absence of the amplitude modulation, and thus, it was incorrectly concluded that modulation was not an important operational factor. The fundamental problem was that a review of the previous literature was inadequate because it had not included Dutta et al.'s paper [13] that reported an intensity window of effects at 0.05 W/kg only when AM is present, no effects between 0.075 and 0.5 W/kg, and an effect at 1.0 W/kg when AM is both present and absent. Thus, it can be argued that Hoyto et al. [21] had actually replicated the Dutta et al. results [13] at 1.0 W/kg but should have tested lower power density before erroneously concluding that true modulation-specific results could not be found with their biological system.

The fundamental lesson is if the details of discovery experiments, such as the need to control and use experimental parameters properly, are ignored, there is a greatly reduced probability that attempts at replication experiments will be successful.

2.3   ELF ELECTROMAGNETIC FIELD EXPERIMENTS

Adey's research group also conducted experiments using chick forebrain tissues to test the effects of ELF electric fields generated between two parallel metal plates. Bawin and Adey [22] reported statistically significant, altered calcium ion release at 6 and 16 Hz at 10 and 56 V/m but no changes at 1 or 32 Hz or at 5 or 100 V/m at 6 or 16 Hz. These results clearly demonstrated an intensity and a frequency window for altered calcium ion efflux. This paper showed that the modulation frequencies that were so critical to the success of the AM RF exposures could also produce similar effects directly.

Blackman et al. [9] showed that 16 Hz ELF EMF, generated in a TEM cell (which produced orthogonal electric and magnetic components), could cause enhanced release of calcium ions in two regions, between 5 and 7.5 and between 35 and 50 Vpp/m, that were clearly separated by a no-effect region between 10 and 30 Vpp/m, as well as by no-effect regions below and above the two effect regions. Neither 1 nor 30 Hz fields at the maximum effect intensity, 40 Vpp/m, had any significant effect on the endpoint. This result essentially independently replicated Bawin and Adey's finding [22].

Research expanded using ELF exposures alone, because it was easier to control the intensity output than with the AM RF generation apparatus and because it simplified the experimental conditions to allow for easier interpretation of the results. Initially, Blackman et al. [23] examined the influence of frequencies from 1 to 120 Hz and intensities from 25 to 65 Vpp/m to search for patterns between the two parameters that might lead to mechanistic hypotheses to be tested. The results revealed patches or *islands* of effects in a *sea* of no effects, and the data were sufficient to develop a hypothesis that the dc field in the earth might be involved in the assignment of responsive frequencies. This hypothesis was tested by Blackman et al. [24], and it was found to be the basis for the frequency assignments. This report has served as the basis for a number of other studies with other biological systems. Other scientists, including Liboff and Lednev, have refined and extended this original finding.

Subsequently, Blackman et al. [25] reported on the frequency dependence for calcium ion efflux every 15 out to 510 Hz, which also provided information to hypothesize three speculative initial transduction models consistent with the data. One of the models—based on highly significant effects every 30 Hz from 15 to 315 Hz, with the exception of 165 Hz, where 15 and 315 Hz responses were shown to be sensitive to changes in the dc field flux density—is generally consistent with a transition metal complex in magnetic resonance, but no more information is available to speculate further.

Another of the models [25], based on a positive effect at 405 Hz, involved a hypothetical field interaction with carbon-13, which was predicted to have a peak resonance at the dc field value in the experiment of 406.8 Hz. Carbon-13 is a naturally occurring isotope that composes 1.1% of all carbon atoms in the body, so it is certainly within the greater membrane of the chick forebrain surface that was thought to be involved in the release process. Plans were made to follow up on this possibility by enriching the isotope in membranes using cell culture studies, but the experiment could not be brought to fruition.

The importance of the alignment between the dc and the ac magnetic field for the calcium efflux assay was first tested by Blackman et al. [26]. They found that the ac and dc magnetic field components had to be orthogonal for the exposure to affect calcium ion efflux; if the field components were parallel, there was no effect. This result seemed to rule out a cyclotron resonance phenomenon and directed the search for a magnetic resonance-type phenomenon.

This frequency and intensity dependence observed in Blackman et al. [23] caused another question to appear, namely, would the environment in which the eggs were incubated (device energized at 60 Hz) have any impact on the response of the brain tissues in the calcium efflux assay? We devised a way to incubate the eggs for the 21-day period, exposing some eggs to 60 Hz electric fields and other eggs to 50 Hz electric fields at 10 Vrms/m in air [27], and assayed them for response to conditions that caused statistically significant changes in calcium efflux [28].

We were only able to perform one-half of the intended experiment, but we were able to demonstrate that for eggs exposed to 60 Hz electric fields during incubation, the brain tissues exhibited a significant response to 50 Hz fields but not to 60 Hz fields, in agreement with the results from commercially incubated eggs [23]. In contrast, the calcium ion efflux from brains from chicks exposed during incubation to 50 Hz fields was not affected by either 50 or 60 Hz fields. This result indicates that some imprinting may have occurred during embryogenesis and organogenesis, thus altering responses to specific EMF treatments [29]. It is obvious from these results that there may be inability to replicate biological responses under identical treatment conditions if there are differences in the electric power frequency or other EMF conditions that exist in the two laboratories. This is particularly of concern for cell culture incubators, where high magnetic and electric fields can occur both from the electric power and also from control systems that reduce temperature fluctuations.

One issue that arose early in these studies was the inability to carefully control the temperature of the forebrains during rinsing and during EMF treatment. We eventually were able to get the laboratory room temperature stabilized through all seasons of the year. In Blackman et al. [30], we reported the calcium release response changes induced by 16 Hz, 14.1 Vrms/m and 64 nTrms following the standard 20 min exposure period. When temperatures in the brain tissue samples were ascending (similar to all previous experiments, see Blackman et al. [9]), there were field-induced effects at 35°C–37°C, but not at 38°C or 39°C. When the temperatures were descending during EMF exposure, there were no changes due to EMF exposure at any temperature from 35°C to 38°C. When temperature of the samples was held stable, there were significant calcium efflux effects from EMF exposure observed for 36°C and 37°C, but no significant changes were observed for samples held at 35°C or 38°C. The temperature range at which EMF-induced effects occurred is thought to be the normal range of the brain tissue occurring in the animal and may reflect the involvement of a normal physiological process, perhaps components at a phase transition, where some lipid components or molecules embedded therein would be biologically active. The concept of sensitivity in a cell component held at a phase transition, where long-range molecular order is altered and kinetics are changed, may be consistent with the influence of EMF exposures that have been reported earlier. For example, under different EMF exposure conditions, reaction rates of isolated enzyme-membrane preparations only exhibited EMF sensitivity at breaks in the Arrhenius plots as a function of temperature where phase transitions occur [31–35].

2.4 CONCLUSIONS

These results arose from discovery processes, that is, letting the biological response identify interactions of the fields with biological entities, without a specific theoretical mechanism of action that would cause/predict the observed effects.

The sum total of the experiments we performed to replicate and then extend the calcium ion efflux results reported by Adey's group of scientists led to many unexpected revelations. We had no theoretical mechanism of action that was specific enough to guide us to particular tests. We were essentially doing *discovery* research, that is, letting the biological systems identify conditions that would allow biological interactions with EMFs. The results do highlight the complex nature of EMFs and their interaction with biological systems, which should not be taken lightly. There may be certain EMF and biological circumstances that could enhance the propensity for detrimental effects and other circumstances where they might find utility in medical treatment. At the very least, they may lead to further understanding of underlying process that, while unknown, is essential for proper homeostasis.

The following table is a catalog of exposure and treatment issues that need to be addressed when attempts at replication or extension of previous, positive-effect studies are constructively undertaken.

Conditions That Need to be Monitored and Reported in Research Reports

EMF parameters:

Radio-frequency radiation

Power density, SAR, field strength of radio-frequency fields (carrier waves), maximum versus average power

Modulation, amplitude, frequency, pulsed:

Wave form, frequency, repetition rate, percent modulation, pulse width and height

Low-frequency fields

Electric fields (E) and magnetic fields (H): average and peak intensities, relative orientations

Static fields (dc) and oscillating (ac): waveform and characteristics, relative orientations of ac and dc fields

Intensity and frequency combinations

Biological parameters:

In vivo versus in vitro: life stage

Genetic sensitivities, for example, mutations and susceptibilities, critical and otherwise

Prior exposure history: EMF and chemical

Buffering components, in vitro preparations

Sources of subjects, descriptions and characterizations, sensitivities

REFERENCES

1. S.M. Bawin, L.K. Kaczmarek, W.R. Adey, Effects of modulated vhf fields on the central nervous system. *Annals of the New York Academy of Sciences* 247 (1975) 74–81.
2. C.F. Blackman, J.A. Elder, C.M. Weil, S.G. Benane, D.C. Eichinger, D.E. House, Induction of calcium ion efflux from brain tissue by radio-frequency radiation: Effects of modulation-frequency and field strength. *Radio Science* 14 (6S) (1979) 93–98.
3. C.F. Blackman, S.G. Benane, W.T. Joines, M.A. Hollis, D.E. House, Calcium-ion efflux from brain tissue: Power-density versus internal field-intensity dependencies at 50-MHz RF radiation. *Bioelectromagnetics* 1 (3) (1980) 277–283.
4. W.T. Joines, C.F. Blackman, Power density, field intensity, and carrier frequency determinants of RF-energy-induced calcium-ion efflux from brain tissue. *Bioelectromagnetics* 1 (3) (1980) 271–275.
5. A.R. Sheppard, S.M. Bawin, W.R. Adey, Models of long-range order in cerebral macromolecules: Effects of sub-ELF and of modulated VHF and UHF fields. *Radio Science* 14 (6S) (1979) 141–145.
6. W.T. Joines, C.F. Blackman, Equalizing the electric field intensity within chick brain immersed in buffer solution at different carrier frequencies. *Bioelectromagnetics* 2 (4) (1981) 411–413.
7. C.F. Blackman, W.T. Joines, J.A. Elder, Calcium-ion efflux in brain tissue by radiofrequency radiation, in: K.H. Illinger (ed.). *Biological Effects of Nonionizing Radiation*, Vol. 157. American Chemical Society, Washington, DC, 1981, pp. 299–314.
8. C.F. Blackman, J.P. Blanchard, S.G. Benane, D.E. House, The ion parametric resonance model predicts magnetic field parameters that affect nerve cells. *FASEB Journal* 9 (7) (1995) 547–551.
9. C.F. Blackman, S.G. Benane, L.S. Kinney, W.T. Joines, D.E. House, Effects of ELF fields on calcium-ion efflux from brain tissue in vitro. *Radiation Research* 92 (3) (1982) 510–520.
10. R.D. Myers, D.H. Ross, Radiation and brain calcium: A review and critique. *Neuroscience & Biobehavioral Reviews* 5 (4) (1981) 503–543.
11. C.F. Blackman, L.S. Kinney, D.E. House, W.T. Joines, Multiple power-density windows and their possible origin. *Bioelectromagnetics* 10 (2) (1989) 115–128.
12. C.J. Thompson, Y.S. Yang, V. Anderson, A.W. Wood, A cooperative model for Ca(++) efflux windowing from cell membranes exposed to electromagnetic radiation. *Bioelectromagnetics* 21 (6) (2000) 455–464.
13. S.K. Dutta, A. Subramoniam, B. Ghosh, R. Parshad, Microwave radiation-induced calcium ion efflux from human neuroblastoma cells in culture. *Bioelectromagnetics* 5 (1) (1984) 71–78.

14. S.K. Dutta, B. Ghosh, C.F. Blackman, Radiofrequency radiation-induced calcium ion efflux enhancement from human and other neuroblastoma cells in culture. *Bioelectromagnetics* 10 (1989) 197–202.

15. S.K. Dutta, K. Das, B. Ghosh, C.F. Blackman, Dose dependence of acetylcholinesterase activity in neuroblastoma cells exposed to modulated radio-frequency electromagnetic radiation. *Bioelectromagnetics* 13 (4) (1992) 317–322.

16. S.K. Dutta, M. Verma, C.F. Blackman, Frequency-dependent alterations in enolase activity in *Escherichia coli* caused by exposure to electric and magnetic fields. *Bioelectromagnetics* 15 (5) (1994) 377–383.

17. J.L. Schwartz, D.E. House, G.A. Mealing, Exposure of frog hearts to CW or amplitude-modulated VHF fields: Selective efflux of calcium ions at 16 Hz. *Bioelectromagnetics* 11 (4) (1990) 349–358.

18. J.L. Schwartz, G.A. Mealing, Calcium-ion movement and contractility in atrial strips of frog heart are not affected by low-frequency-modulated, 1 GHz electromagnetic radiation. *Bioelectromagnetics* 14 (6) (1993) 521–533.

19. R.P. O'Connor, S.D. Madison, P. Leveque, H.L. Roderick, M.D. Bootman, Exposure to GSM RF fields does not affect calcium homeostasis in human endothelial cells, rat pheocromocytoma cells or rat hippocampal neurons. *PLoS One* 5 (7) (2010) e11828.

20. J. Juutilainen, R. de Seze, Biological effects of amplitude-modulated radiofrequency radiation. *Scandinavian Journal of Work Environment & Health* 24 (4) (1998) 245–254.

21. A. Hoyto, J. Juutilainen, J. Naarala, Ornithine decarboxylase activity of L929 cells after exposure to continuous wave or 50 Hz modulated radiofrequency radiation—A replication study. *Bioelectromagnetics* 28 (7) (2007) 501–508.

22. S.M. Bawin, W.R. Adey, Sensitivity of calcium binding in cerebral tissue to weak environmental electric fields oscillating at low frequency. *Proceedings of the National Academy of Sciences of the United States of America* 73 (6) (1976) 1999–2003.

23. C.F. Blackman, S.G. Benane, D.E. House, W.T. Joines, Effects of ELF (1–120 Hz) and modulated (50 Hz) RF fields on the efflux of calcium ions from brain tissue in vitro. *Bioelectromagnetics* 6 (1) (1985) 1–11.

24. C.F. Blackman, S.G. Benane, J.R. Rabinowitz, D.E. House, W.T. Joines, A role for the magnetic field in the radiation-induced efflux of calcium ions from brain tissue in vitro. *Bioelectromagnetics* 6 (4) (1985) 327–337.

25. C.F. Blackman, S.G. Benane, D.J. Elliott, D.E. House, M.M. Pollock, Influence of electromagnetic fields on the efflux of calcium ions from brain tissue in vitro: A three-model analysis consistent with the frequency response up to 510 Hz. *Bioelectromagnetics* 9 (3) (1988) 215–227.

26. C.F. Blackman, S.G. Benane, D.E. House, D.J. Elliott, Importance of alignment between local DC magnetic field and an oscillating magnetic field in responses of brain tissue in vitro and in vivo. *Bioelectromagnetics* 11 (2) (1990) 159–167.

27. W.T. Joines, C.F. Blackman, R.J. Spiegel, Specific absorption rate in electrically coupled biological samples between metal plates. *Bioelectromagnetics* 7 (2) (1986) 163–176.

28. C.F. Blackman, D.E. House, S.G. Benane, W.T. Joines, R.J. Spiegel, Effect of ambient levels of power-line-frequency electric fields on a developing vertebrate. *Bioelectromagnetics* 9 (2) (1988) 129–140.

29. C.F. Blackman, Can EMF exposure during development leave an imprint later in life? *Electromagnetic Biology and Medicine* 25 (4) (2006) 217–225.

30. C.F. Blackman, S.G. Benane, D.E. House, The influence of temperature during electric- and magnetic-field-induced alteration of calcium-ion release from in vitro brain tissue. *Bioelectromagnetics* 12 (3) (1991) 173–182.

31. R.B. Olcerst, S. Belman, M. Eisenbud, W.W. Mumford, J.R. Rabinowitz, The increased passive efflux of sodium and rubidium from rabbit erythrocytes by microwave radiation. *Radiation Research* 82 (2) (1980) 244–256.

32. G.A. Fisher, G.C. Li, G.M. Hahn, Modification of the thermal response by D_2O. I. Cell survival and the temperature shift. *Radiation Research* 92 (3) (1982) 530–540.

33. S.F. Cleary, F. Garber, L.M. Liu, Effects of X-band microwave exposure on rabbit erythrocytes. *Bioelectromagnetics* 3 (4) (1982) 453–466.

34. R.P. Liburdy, A. Penn, Microwave bioeffects in the erythrocyte are temperature and pO_2 dependent: Cation permeability and protein shedding occur at the membrane phase transition. *Bioelectromagnetics* 5 (2) (1984) 283–291.

35. J.W. Allis, B.L. Sinha-Robinson, Temperature-specific inhibition of human red cell Na+/K+ ATpase by 2,450-MHz microwave radiation. *Bioelectromagnetics* 8 (2) (1987) 203–212.

3 Benefit and Hazard of Electromagnetic Fields

Marko S. Markov

CONTENTS

3.1 HAZARD AND BENEFIT: BRIEF

Natural electromagnetic fields (EMFs) and especially magnetic fields (MFs) have been factors that accompanied the development and evolution of the planet Earth and life on the planet. However, the role of natural and man-made MFs and EMFs in the origin and evolution of life and the effects of contemporary MFs/EMFs on human life are still not well investigated and understood. Both the hazard and benefit of these fields are subjects of various studies and review papers, and very often the conclusion is that "this is a controversial issue." I should partially agree with the statement, but the problem is not well defined. The general controversy is in the duration of exposure. The hazard from MFs/EMFs in natural and man-made conditions is due to continuous exposure to these physical factors. The most important issue here is that these fields act in long periods of time. Contrary, the benefits in clinical use are results of short sessions of exposure to known MFs/EMFs in controlled conditions. It is now well understood that these therapeutic fields are orders of magnitude stronger than natural fields.

Contemporary science has enough evidence that life originated and developed in the presence of a number of physical factors with terrestrial and space origin including MFs and EMFs. One of the most important factors is the geomagnetic field, formed during the geological evolution of the planet. In addition, the Earth has been exposed to the influence of ionizing and nonionizing radiation with space origin.

During the last 120 years, the biosphere has been exposed to increasing number and variety of EMFs related to industrial methods for generating electricity and further innovations in technology, communication, transportation, home equipment, and education.

To discuss the potential hazard of MFs/EMFs, one should have reliable information about geological structure and demographics of giving regions, as well as thorough investigation of the medical conditions under discussion. Epidemiology itself cannot be anything else but a tool in analysis, and the emphasis should be on biology and medicine. At the same time, proper dosimetry of EMFs is even more important, and the dosimetry should be done by professionals in biophysics and magnetobiology.

Nearly half a century ago, brilliant Soviet magnetobiologist Yuri Kholodov wrote the book *Man in the Magnetic Web* (Kholodov, 1976). Long before the occurrence of mobile communications, Kholodov pointed out that the entire biosphere is immersed in the ocean of the electromagnetic waves. Note: at that time, there was no Internet nor mobile communications. But humans slowly became immersed in millions and millions of EMFs from radio and TV emitters.

At the opposite site is the magnetotherapy. Starting immediately after World War II, various methods of EMF therapy entered the contemporary medicine.

It would be fair to say that today's western medicine is based mainly on the achievements of chemistry, which have been further utilized and expanded by the pharmaceutical industry. Unfortunately, nearly all pharmaceuticals affect not only the target tissues but the entire organism, and in many cases, it causes adverse effects. In contrast, magnetotherapy provides noninvasive, safe, and easily applied methods to directly treat the site of injury, the source of pain, and inflammation.

The MF therapy is often a subject of publications that categorically affirm or reject the possibility of MFs to cause health effects. The authors of many papers use the word *controversial* when they speak about magnetobiology and magnetotherapy. It should be understood that magnetotherapy is not a controversial issue. The problems occur when general claims are made by scientists or clinicians that the therapy works or does not work when only one MF is applied for the treatment of a specific problem.

To properly evaluate potential benefit of MF therapy, one needs to apply basic science and clinical medicine approaches. This idea will be further developed in the chapter, but here, I would like to stress on several points. First, each EMF contains electric and magnetic components. For low- and medium-frequency fields, electric component is shielded by the body (physical or biological) and transferred into electric current over the body surface. The MF component is capable to penetrate into the tissue. Interestingly enough, the electromagnetic properties of living tissues are similar to those of air. This fact allows to perform 3-D evaluation of a given MF and to predict what MF parameters are at a specific distance from the generating systems and thus to predict what the MF in the target tissue would be.

It is also important to know that different MFs applied for the treatment of different medical problems eventually might cause different results. Therefore, the maximal beneficial effect could be achieved after proper diagnosis, obtaining knowledge of the reasons for medical problem, and close cooperation with engineers and biophysicists.

3.2 ROLE OF MAGNETIC/ELECTROMAGNETIC FIELDS IN THE ORIGIN AND EVOLUTION OF LIFE

The development of physics, geology, and cosmology provides substantial evidence that the universe is a dynamic system that among other characteristics possesses MFs. The Earth, therefore, is exposed to the influence of various MFs with space origin. It is well accepted that the Earth possesses own MF formed during the geological evolution of the planet. Unfortunately, despite the development of physics and geology, reliable information about the values of the MFs/EMFs during the evolution of the planet and of the life in the biosphere is missing and eventually will never be known.

Probably for that reason, all attempts to recreate origination of the first living cell had failed. Biologists know that the presence of water, carbohydrates and amino acids, ultraviolet light, and lightning is necessary to mimic *the birth* of the first living cell. However, they did not know what MF at that time was.

The magnetobiology has evidence that when living creature is placed in an environment that is shielded from ambient MF, some changes in the organisms are observed. When microorganisms confined in μ-metal cylinder are missing usual geomagnetic field, the cells are searching for adaptation to newly created conditions and find this way through the mutation in their genetic apparatus (Pavlovich, 1975).

It is time to say that contemporary science now has convincing evidence that the geomagnetic field serves as a protector against space ionizing radiation and MFs reaching the atmosphere, thereby protecting the biosphere and all forms of life. The Earth's magnetosphere is determined by the Earth's MF, as well as by the solar and interplanetary MFs. In the magnetosphere, a mix of ozone molecules, free ions, and electrons from the Earth's ionosphere is maintained by electromagnetic forces with terrestrial origin.

Usually, the role of space factors in the geological development of our planet is discussed by different groups of scientists. To completely understand the role of space factors and especially the role of the solar MF, serious attention should be paid to the little known research of the Russian scientist Leonid Chizevsky. Being an expert in heliophysics, biophysics, space biology, cosmobiology, and geobiology, studying the impact of cosmic physical factors on the processes in living nature, Chizevsky found a relationship between solar activity cycles and many phenomena in the biosphere. He demonstrated that the physical fields of the Earth and its surroundings should be taken into account as being among the main factors influencing the state of the biosphere. He claimed that variations of solar activity and dependent geomagnetic oscillations have impact on any type of life. Chizevsky proposed that biosphere history is shaped by the 11-year cycles in the Sun's activity that triggered solar magnetism and further geomagnetic storms manifesting in power shortages, plane crashes, epidemics, grasshopper infestations, upheavals, and revolts (Chizhevsky, 1976).

3.3 CONTEMPORARY EMF CONDITIONS AND HAZARD WITH WI-FI COMMUNICATIONS

3.3.1 HAZARD FROM EMF

Very often, the news media discuss how dangerous EMF/MF might be for human and environmental health, especially in relation to cancer initiation. The hazard should be considered in respect to the continuous exposure to EMFs in the workplace and/or occupational conditions, while at the same time, short, controlled exposure to specific EMFs makes possible therapeutic benefit.

The hazard issue in the western scientific community has been discussed beginning with the power-line EMFs and continuing with wireless communications.

During the last three decades, more and more voices have been raised that it might be certain detrimental effects of EMFs on the biosphere and human life. Very often, the news media discuss how dangerous EMF/MF might be for human and environmental health, especially in relation to cancer initiation. In the United States, the hazard issue has been discussed since the middle of the 1980s, beginning with the power-line EMFs and continuing after 2000 with wireless communications.

The fast development of satellite communications, followed by wireless communications and recently Wi-Fi technology, dramatically changes the electromagnetic environment. The entire biosphere is exposed to continuous action of complex and unknown (by sources, amplitudes, frequencies) EMFs. We usually neglect this complex that includes radio and TV transmissions, satellite signals, mobile phones, and base stations and wireless communications (Markov and Grigoriev, 2013).

By 2010 in the United States, 285 million mobile phone subscribers have been registered (for a little bit more than 300 million inhabitants). In Brazil, the number of mobile phones is larger than the entire population of the country. It is estimated that in the world, more than 6 billion mobile phones are in use for approximately 7.5 billion people living on this planet.

Having this in mind, the evaluation and prediction of the potential adverse effects from using wireless communications (any mobile device, including), especially by children, becomes a question of crucial importance. The twenty-first century is marked with exponentially increasing development of technologies that provide wireless communications. To the pollution of the atmosphere with radio and TV signals, not only satellite communications but also any varieties of the Wi-Fi networks are added.

It is not well known fact that a cellular telephone delivers a power density of radio-frequency radiation, which is about 2 billion times greater than similar fields that occur naturally in the environment. Since the mobile phones are designed to operate at the side of the user's head, a large part of the transmitted energy is radiated directly into that person's brain. Therefore, the absorbed energy potentially could cause dangerous and damaging biological effects within the brain. The small cellular telephones effectively deposit large amounts of energy into small areas of the user's head and brain.

The evaluation and prediction of the potential adverse effects from using wireless communications, especially for children, becomes a question of crucial importance.

I will use the word *controversial* once again. The hazard from high-frequency EMFs used in the twenty-first-century communication is frequently represented as *controversial*, and it is absolutely incorrect. It is not a controversial issue; it is a conflict of interest of industry on one side and of humans and the environment on the other.

It is remarkable that the International Agency of Research on Cancer (IARC) in the summer of 2011 classified radio-frequency EMF as a possible carcinogen. Immediately after the publication of this qualification, IARC was the target for the complaints from industry and engineering committees like the International Commission on Non-Ionizing Radiation Protection (ICNIRP) and Institute of Electrical and Electronics Engineers (IEEE). Speaking of the potential hazard of Wi-Fi technologies, one should not forget that it includes not only mobile phones but also more importantly all forms of emitters and distributors of Wi-Fi signals, mainly antennas, base stations, and satellites. In many public locations, own systems are introduced in order to facilitate work performance. Well, this might be understood. However, why are Wi-Fi communications secured in subway tunnels? When I was writing this chapter, the US news media are discussing the possibility of allowing use of mobile devices in aircraft during flight. It obviously requires high and oriented power to which all passengers in the trains and planes are exposed, just to make the users of mobile devices comfortable. It is forgotten that in the conditions of confined subway tunnels and planes, to use mobile signal significant increase of the delivered power of the signal is needed.

It is my belief that a special attention should be paid to the potential harm the twenty-first-century society can cause to children. As it was already explained, to the action of all means of electronic communications are exposed all living organisms, including elderly people and newly born babies. There are realities of our life, and it is now impossible to protect children from the *cocktails* of electromagnetic radiation the twenty-first century offers. Unfortunately, children are probably the most aggressive part of population in the use of wireless devices—starting from toys to planshets to smartphones (Markov, 2012).

The cell phone users are seriously misled. When buying a cell phone, they assume that this radiation-emitting device was tested for safety in human health. This is wrong: cell phones (never mind who is the manufacturer or which generation they are) have not been tested. For the first time in human history, radiation-emitting devices were placed directly to human head and exposed human brain to microwave radiation.

The industry executives, engineers, and dosimetry experts continuously misrepresent the issues related to potential hazard of cell phone radiation. The engineering committees of ICNIRP and

IEEE continue to mislead scientific and medical communities, as well as the general public, that the only meaningful effects of EMF are thermal.

Let me refer to Robert Kane and his 1995 book *Cellular Telephone Russian Roulette*: "Never in human history has ever been such a practice as we now encounter with the marketing and distributing of products to the human biological system by an industry with foreknowledge of those effects" (Kane, 1995).

It is known that the human head is a complex structure of many different tissue types. Each of these types—skin, bone, fat, cerebrospinal fluid, brain, dura, etc.—absorbs RF energy in its own way. In addition, even having the same structural components, different heads vary in volume and shape. This is even more correct when one considers children's head. Besides this difference, more important is the fact that a child's brain is in process of development and any exposure to RF is more detrimental for children than for adults.

Every evaluation of the *hazard* as well as every standard for the permissible level of exposure should be done following the precautionary principle: if we do not know that a given food, drink, medication, physical factor, or chemical factor is safe, we should treat it as potentially hazardous.

One of the main problems in evaluation potential hazard of nonionizing radiation is the attempt to treat it in parallel with ionizing radiation. In my opinion, it is an incorrect approach since these are two completely different physical factors. While ionizing radiation basically ionized the tissues and develops thermal effects based upon existing of threshold levels, the nonionizing radiation acts differently.

What do we actually know about nonionizing radiation? Basically *nothing*. We have no complete knowledge of even the simplest and longer studied behavior of natural MFs, electric fields, and EMFs. In an engineering community, a lot of efforts and funds have been spent to prove the thermal effects of EMF while neglecting the nonthermal effects. We should point that biological systems are nonlinear systems and plenty of biophysical studies demonstrated that nonthermal effects prevail in laboratory and clinical settings (Markov, 2007). The nonthermal character of interactions between EMF and living systems has been discussed elsewhere (Markov, 2006).

3.3.2 ENDOGENOUS AND EXOGENOUS MAGNETIC FIELDS

The centuries of development of natural sciences provide enough evidence to claim: "Life is an electromagnetic event." Contemporary biology knows that all physiological processes are performed with the movement of electric charges, ions, and dipoles within the cell interior, through the plasma membranes and in communication between different cells. The movement of charges generates electric currents, and as a result, MFs could originate within living tissues. These MFs are commonly classified as endogenous fields.

On the other hand, every MF generated outside the biological system is an exogenous field. Usually, these are fields connected with the physical means of generation. It includes solar and terrestrial MFs, as well as EMFs generated by industrial and communication systems. However, some fields with biological origin may fall in this category. For example, the MF created by one organ in the human body would be endogenous to another organ.

The necessity to distinguish endogenous and exogenous MFs is also related to the use of two terms: biomagnetism and magnetobiology, which mistakenly are used as synonyms. The semantics of the words, however, should suggest the difference. Biomagnetism is an area of science that deals with MFs generated by biological systems, while magnetobiology studies the effects of exogenous MFs when applied to biological objects.

3.3.3 BENEFITS OF CLINICAL APPLICATION OF EMF

MFs and EMFs slowly become recognized by western medicine as a real physical modality that promises healing of various health problems even when conventional medicine has failed. The human history keeps numerous evidences that MFs have been used to treat some maladies.

Ancient physicians in China, Japan, and Europe successfully applied natural magnetic materials in their daily practice. One of the earliest scientific records could be found in the book *De Magnete*, written in 1600 by William Gilbert, a nature philosopher known in physics and mathematics and as a personal physician of the British Queen (Gilbert, 1600).

The contemporary magnetotherapy began immediately after World War II by introducing both MFs and EMFs generated by various wave shapes of the supplying systems. Starting in Japan, this modality quickly moved to Europe, first in Romania and the former Soviet Union. By 1985, nearly all European countries had designed and manufactured their own magnetotherapeutic systems. The first book on magnetotherapy was published by Nencho Todorov in Bulgaria in 1992, summarizing the experience of utilizing specific EMF for the treatment of 2700 patients having 33 different pathologies (Todorov, 1982).

The application of MFs for treating specific medical problems such as arthritis, fracture unification, chronic pain, wound healing, insomnia, and headache has steadily increased during the last decades. In contrary to pharmacotherapy, when the medication is spread along the entire body (which needs doses of sometimes three orders of magnitude stronger than needed), magnetotherapy provides noninvasive, easy-to-apply methods for a defined period of time.

There is a large body of basic science and clinical evidence that time-varying MFs can modulate molecular, cellular, and tissue functions in a physiologically and clinically significant manner, most recently summarized in several books and review articles (Shupak, 2003; Rosch and Markov, 2004; Barnes and Greenebaum, 2007; Markov, 2007; Lin, 2011).

3.3.4 MECHANISMS OF DETECTION AND RESPONSE TO EMF

The fundamental question for engineers, scientists, and clinicians is to identify the biochemical and biophysical conditions under which applied MFs could be recognized by cells in order to further modulate cell and tissue functioning. It is also important for the scientific and medical communities to comprehend that different MFs applied to different tissues could cause different effects. Nevertheless, hundreds of studies had been performed in search of one unique mechanism of action of MF on living systems. It would be fair to say that these efforts are determined to fail. Why? It is for the same reason that during the millions of years of evolution of life, enormous number of different living systems originated. It is difficult to believe that the same response will be seen at bone and soft tissue, at elephant and butterfly, and at microorganism and buffalo. Biology knows that the geographical and climate conditions created genetic and physiological differences in the organisms from the same species.

The problem of mechanisms of interactions might be discussed from different points of view: engineering and physics, biology, and medicine. More plausible is to follow the signal-transduction cascade that postulates that in any biological system, the modifications that may occur as a result of the influence of the applied MF on structures such as cellular membrane or specific proteins, conformational changes, and/or charge redistribution could be initiated and by signal-transduction mechanism can be spread over the cell or tissue.

In a different chapter of this book, the biological mechanism of response to EMF is discussed in respect of *window* and *resonance* approaches; therefore, it will not be discussed here (Markov, 2014).

It should be taken into account that EMF interactions with living tissues are going on in three channels:

1. Transport of energy
2. Transport of material
3. Transport of information

At different stages and types of interactions, one or another of these pathways might prevail. In general, all these transports exist simultaneously. What is the most important is that the EMF interaction with living tissues occurs with the transport of charges that results in electric current and MF generation. To accurately evaluate the possibility of detecting and utilizing exogenous EMF, it is necessary to involve biophysics approach. Why biophysics approach could be plausible for detection of the exogenous EMF? First of all, because biological systems are nonlinear systems, biophysics is the only science that integrates biological and physical approaches. Secondly, the biophysics is the science that has investigated the action of nonionizing radiation in a different way from the action of the ionizing radiation, for example, threshold or window interactions. In a recent paper, Belyaev (2015) emphasizes that the biological systems are not only nonlinear systems but also nonequilibrium systems.

3.3.5 DOES THRESHOLD EXIST?

EMFs are frequently discussed under the umbrella *radiation* in the category of nonionizing radiation. For that reason, the research of effects of EMF is going in parallel with studying the effects of ionizing radiation. As basic physics teaches, radiation constitutes energy, and for that reason, the energy interactions with any physical or biological body are connected with damage or heating of the body when the intensity of the radiation is above certain threshold level. Yes, this is correct for ionizing radiation because the effects of ionizing radiation are energy based. For decades, the same approach has been applied in nonionizing research. For example, the idea of thermal effects in bioelectromagnetics had been introduced and became the subject of intensive discussions, related to specific absorption rate (SAR) as useful criteria. It is clear that SAR requires a threshold value determination.

However, better part of publications referred to SAR in incorrect manner, neglecting the word *absorption* rate. It is illogical to state "the object was exposed to such SAR." There is no physical reason to consider exposure as mirror image to absorption.

Hundreds of studies and publications reported biological and clinical effects of low-intensity and low-frequency EMFs, as well as of static MFs. At these interactions, it is very unlikely, or even impossible, to expect thermal effects, and the threshold level approach is not reasonable. An important point in these considerations is that even when exposed to static MFs, a relative movement of charged particles (electrons and ions) in MF occurs and the MF initiated effects that depend on the specific parameters of the applied field, but not at threshold. Having in mind that even the simplest biological systems are nonlinear systems, the possible means of interactions might involve an informational transfer.

It is my opinion that the engineering attempts to consider EMF interactions with living systems from the view point that *only* thermal effects are meaningful is, simply said, inaccurate. Much more evidence is available that these interactions are nonthermal (Markov, 2006).

3.4 MAGNETOTHERAPY

As was already mentioned, the use of MFs for treatment of medical problems has a long history. However, there are still a lot of problems related to implementation of this approach.

What should magnetotherapy be, and how should it be developed? Magnetotherapy is a part of bioelectromagnetic technology and therefore requires rigorous interdisciplinary research efforts and coordinated, educational programs. Magnetotherapy cannot be developed without the joint efforts of physicists, engineers, biologists, and physicians. An important role will be played by medical practitioners, including physical and occupational therapists, who routinely use physical modalities, while the scientists need to create dosimetry and methodology for magnetotherapy. It should be noted that MF stimulation requires as precise dosage as any other therapy. However, *dosage* in magnetotherapy is more complicated because it requires identifying a number of physical parameters that characterize the MF-generating system. It is important to establish the proper target for magnetotherapy. For example, it has been shown that to stimulate coagulation, one combination of parameters of applied field is required, whereas stimulation of anticoagulation requires another field configuration (Todorov, 1982; Markov and Todorov, 1984). In other words, "different magnetic fields produce different effects in different biotargets under differing conditions of exposure" (Markov, 2000).

An evaluation of the efficacy of these modalities should be based on recognition of the clinical problem, identification of the physiological responses, and a critical review of the reported basic science and clinical data (which include patient status). Any magnetic stimulation starts with identification of the MF parameters needed for the desired target tissue. The ability of MFs to modulate biological processes is determined firstly by the physiological state of the injured tissue, which establishes whether or not a physiologically relevant response can be achieved, and secondly by achieving effective dosimetry of the applied MFs at the target site.

Why in this section the emphasis is on MF, not EMFs? The reason is that for static and low-frequency EMFs, the only important value is the MF strength. In case of the low-frequency EMF, the electric field component does not penetrate any physical and biological body, but is transferred in electric current over the body surface (Markov, 2000).

In general, EMF therapeutic modalities can be categorized in the following groups:

- Static/permanent MFs
- Low-frequency sine waves
- Pulsed EMFs (PEMFs)
- Pulsed radio-frequency fields (PRFs)
- Transcranial magnetic/electric stimulation
- Millimeter waves
- Electroporation and nanoelectroporation

Permanent MFs can be created by various permanent magnets as well as by passing direct current (dc) through a coil.

Low-frequency sine wave EMFs mostly utilize a 60 Hz (in United States and Canada) and a 50 Hz (in Europe and Asia) frequency used in power lines.

PEMFs are usually low-frequency fields with very specific wave shapes and amplitude. The variety of commercially available PEMF devices makes it difficult to compare the physical and engineering characteristics of devices, and it is the main obstacle in the analysis of the biological and clinical effects of those devices.

PRFs utilize the frequency of 27.12 MHz in two modifications: in continuous mode, it usually produces deep heat, while in pulsed (nonthermal) mode, it is used for soft tissue stimulation.

More recently, *millimeter waves* (having very high frequency of 30–100 GHz) have been used in the treatment of a number of diseases, especially in the countries of the former Soviet Union:

Transcranial magnetic stimulation represents stimulation of selected portion of the brain by applying very short magnetic pulses of up to 8 T.

Electroporation and nanoelectroporation can be achieved by applying high-frequency and extremely high-frequency pulses that allow reaching cellular interior.

Magnetic stimulation provides beneficial and reproducible healing effects even when other methods have failed. However, there is a lack of uniformity among medical practitioners with respect to stimulation, the parameters of the applied fields, and lack of defined biophysical mechanism capable of explaining the observed bioeffects. Therefore, a systematic study of MF action on biological systems has to consider the following important parameters:

- Type of field
- Intensity of induction
- Gradient (dB/dt)
- Vector (dB/dx)

- Frequency
- Pulse shape
- Component (electric or magnetic)
- Localization
- Time of exposure
- Depth of penetration

A common problem when comparing the effects of magnetic devices is that each manufacturer uses their own system of characterizing the product and in most cases, the stated MF strength is different than that obtained in direct measurement. My own experience with distributors of magnetothera-peutic systems suggests that the description of the field parameters in the promotional and technical materials is overstated when indicating the MF strength. Sometimes it is done because of inaccurate evaluation of the parameters, but in many cases, the reason is to show that the field is very strong. Unfortunately, in magnetotherapy, *more does not necessarily mean better.*

What could be done to avoid the complications that occur with inaccurate description of the device parameters? First of all, make sure that the parameters of device are as accurate as possible— requesting confirmation from the manufacturer regarding the methods of evaluation of the param-eters. Secondly, search for opportunity that you make own calibration of the device.

Let me distinguish two types of dosimetry: physical and biophysical. The *physical dosimetry* describes the parameters of the generating system of the specific device. In most cases, these param-eters are calculated by basic engineering principles and most often without real measurements. The *biophysical dosimetry* is always based on measurements. It is important for biology and medicine to have knowledge for the exact MF at the target site. In most cases, the target is located at certain dis-tance from the applicator. One significant advantage here is the fact that for static and low-frequency MFs (as most of the therapeutic devices are), the magnetic and dielectric properties of the biological tissues are similar to the properties of air. This allows biophysical dosimetry to be performed in air and basically to create 3-D structure of any signal in use in laboratory or clinical settings. Thus, we would know the MF parameters at the target site and eventually select the optimal parameters for treatment of specific medical problem.

Very often, magnetotherapy is a subject of discussion about the accuracy of claims regard-ing the efficacy of the applied MF/EMF's effects. There is a serious reason for this *controversial* statements—the claims of authors of scientific and clinical studies that MF *does* or *does not* cause biological response. In most cases, such statements are based upon studying one only signal. The correct statement is, "Our study shows that *this* magnetic field affects *that* medical problem." It is not very probable that the same MF would cause the same effects in wound healing and bone unification.

3.5 MAGNETIC AND ELECTROMAGNETIC STIMULATION

Several decades of clinical application of various EMFs have clearly demonstrated the potential benefit of the use of selected MFs for the treatment of various medical problems. The success of magnetotherapy depends on the proper diagnosis and selection of physical parameters of applied fields. To be more precise, *any therapy that involves MFs/EMFs should mandatorily include the knowledge of the exact parameters of the field at the target site.* Not what is the MF at generator, but what is the field at the target.

Several double-blind studies published in the late 1990s have demonstrated the potential of a static MF to provide significant pain relief. Valbona et al.'s study (1997) showed that an SMF of 300–500 G decreases the pain score in postpolio patients as 76% vs. 19% in placebo group, and Colbert's study used mattresses that utilize ceramic permanent magnets with surface strength of about 1000 G (1999). The estimated field strength on the body surface in Colbert's study was in the range 300–500 G, depending on the body mass of the patient. This MF helps patients suffering from fibromyalgia and improves the status of the patients in the real treatment group with more than 30%.

Soft tissue and bone/cartilage systems have been successfully treated with the most notable being treatments for problems related to muscular-skeleton system (Bassett et al., 1974; Markov and Todorov, 1984; Detlavs, 1987; Bassett, 1994; Adey, 2004; Markov, 2004; Pilla, 2007). Magnetotherapeutic systems have been successfully applied to treat vascular, immune, and endocrine systems (Todorov, 1982; Rosch and Markov, 2004; Barnes and Greenebaum, 2007; Sandra, 2013).

A literature survey indicates that many electric and magnetic modalities have been developed to heal nonunion fractures and wounds (Bassett, 1989; Markov et al., 1992; Vodovnik and Karba, 1992; Markov and Pilla, 1995; Seaborne et al., 1996; Mayrovitz, 2015). The noninvasive EMF most often employed in the United States for soft tissue applications is short-wave pulsed radio frequency (PRF), based on the continuous 27.12 MHz sinusoidal diathermy signals and used for decades for deep tissue heating. The pulsed version of this signal was originally reported to elicit a nonthermal biological effect by Ginsburg (1934). PRF MFs have reduced posttraumatic and postoperative pain and edema in soft tissues and have some applications to wound healing, burn treatment, ankle sprains, hand injuries, and nerve regeneration. Pulsed radio-frequency MF treated pressure sores in patients resulting in significant reduction (up to 47%) in the mean sore area after 2 weeks of treatment, while the mean duration of pressure sores (before treatment) was 13.5 weeks (Seaborne et al., 1996).

In addition to accelerated wound healing, MF modalities have been shown to significantly increase local blood flow in the stimulated area improving the status of the ischemic tissue (Ohkubo and Okano, 2004, 2015). Magnetic and electric stimulation has been associated with increased collagen deposition, enhanced ion transport, amino acid uptake, fibroblast migration, ATP, and protein synthesis, including a significant increase in the rate of protein and DNA synthesis after stimulation of human fibroblasts in tissue culture. One area of interest, mainly in basic science, is the effect of EMF and MF on cell proliferation. Most cells normally differentiate to a specific morphology and function. In pathological conditions, cell proliferation is usually suppressed (in conditions of chronic wounds) or enhanced (in the case of neoplastic growth). MF stimulation of the skin fibroblast resulting in significant increase in collagen secretion and protein concentration has been reported, and these results suggest a favorable alteration in the proliferative and migratory capacity of epithelial and connective tissue cells involved in tissue regeneration and repair.

Over the past two decades, several methods for therapy of the peripheral vascular system using static MFs have been developed. The clinical outcome of this therapy includes analysis of hemodynamics, microcirculation, transcapillary phenomena, and morphological and cytochemical characteristics of blood components, including lymphocytes, erythrocytes, leukocytes, and thrombocytes. The therapeutic efficiency depends on the status of the patient (age, general health, gender) as well as on the disease stage. There is also a distinct relationship between specific diseases and the MF parameters that initiate optimal response (Mayrovitz, 2004, 2015).

Improved blood perfusion in the magnetically stimulated tissue has been an assumed mechanism for the stimulatory effects on the regenerative processes. These clinical observations, along with the findings that blood flow and metabolic activity increase after long-term muscle stimulation, motivated a series of studies of the effects of MFs on different health problems (Ohkubo and Okano, 2004; Mayrovitz, 2015).

3.6 ABOUT REPLICATION AND DOUBLE-BLIND STUDIES

These two issues are not linked; however, they deserve a special attention, from both scientific and clinical view points. I cannot recall any real *replication* study. The reasons for this could be seen in different ways. First, scientists always want to do original studies, not to replicate somebody else's designed protocol. It might be seen in the chapter of the book by Blackman (2015) that from a biochemical point of view, it is important to have actually not only the same source of chemicals but the same production batch (even chemicals from the most respected suppliers). On the other hand, magnetic/electromagnetic conditions in different laboratories vary, and in most cases, the background EMFs are not measured and described in publications.

Colbert et al. (2007) performed an analysis of 56 double-blind clinical studies to find that only two accurately reported all 10 of the selected parameters for analysis. One should not be surprised that the same therapeutic device is reported to cause different effects in different clinical settings. The questions are as follows: "Are these devices exactly the same?" Are they manufactured by the same company? Did you measure the parameters of the device; are you sure that the target receives the same magnetic stimulation? In order to quantify MF exposure at the target tissue, the target site must be clearly named, the distance from MF-generating system to the target must be defined, and the dosing parameters must be precisely reported.

I always worry when I hear *double-blind clinical trials*. Such type of clinical trial assumes that half of the individuals will be placed in placebo/control group. This *golden standard* in the twenty-first-century medicine is basically inhumane. Who gave the right of the investigator to place a large number of individuals in a group that will not receive real treatment? This is not an experiment with animals. The investigator actually denies the care for the medical problem or at least postpones it.

What is the goal of double-blind studies? The goal is to prevent bias in the study. It is so easy that this bias is prevented with placing an intermediate stage or person between the person who performs the exposure and the person who analyzes the data. Especially with chronic medical problems, patients might have own control and the clinical trial would effectively analyze the success of applied therapy. Moreover, that statistics applied in the clinical trials does not care for the improvement of a specific patient, but for average numbers for the group.

3.7 THE FUTURE

As it was shown in the first part of this chapter, in the twenty-first century, there has been an exponentially increasing use of electromagnetic polluters in the sense of exposing humans and the entire biosphere to various EMFs, with no control and no interest from the side of the WHO and public health offices. At this moment, nobody knows what the potential effects of these polluters are. Even the IARC classification that the microwaves used in cellular communications are potentially hazardous addresses only one small fraction of the entire spectrum. If the public health system abandons this field, the industry is here. With more and more items.

Here are just two examples from recent development. The *smart meters* became installed on the outside walls of each house with the purpose to provide minute-by-minute information about the use of electricity by various home utilities and electronics. Effectively, industry installs small radiostation on each house. What will happen with the inhabitants of the house—industry does not care. What will happen with the biosphere as a result of billions of transmitters—nobody cares.

Google launched *smart glasses* in May 2014. The owners of this quite expensive toy will be able to use Field Trip Instructor, Word Lens Translator, Google Now, and Google Maps. The owner will be able to participate in conversation and even conference calls, to make videos, etc.

A problem that is not new, but escaped from the horizon of scientists and public health experts, is on electric vehicles. These types of cars need electric power to charge the high-capacity batteries as well as this battery will be a driving force for the vehicle. Did somebody analyze the EMFs inside the vehicle?

There is something even more troublesome: it was reported that there are now already developed *magnetic bacteria* that could be incorporated in probiotic bacteria and be used as a tool for diagnostics and therapy of diseases of internal organs.

It is time now to ring the bell. This probably will not be enough. Scientific and medical communities must stand opposing uncontrolled introduction of toys and tools that utilize EMFs.

It is also time that the aforementioned communities increase their efforts in further development of the background of electromagnetic biology and medicine. We know now that EMFs are effective in the treatment of soft tissue, tendon, and orthopedic problems, in anti-inflammatory and immune systems. More efforts are needed in clarifying biological mechanisms of interactions of EMF with living tissues, in designing new therapeutic modalities.

Analyzing the reported biological and clinical data obtained with devices and signals in use for MF therapy, one might conclude that some types of signals are more promising for the future development of the MF therapy. It appears that semi–sine waves are more effective compared to continuous sine waves. This approach is based on rectification of the continuous sinusoidal signal. Let me note that the actual shape of the rectified signal significantly differs from the ideal semi–sine wave drawing in textbooks.

There have been reported two different approaches for the utilization of these signals. One relies on constructing an elliptical or spherical coil, which could be moved around the patient body (Williams et al., 2001), and the other relies on the application of the MF on the upper or lower limbs, assuming that the results appear following systemic effects when the benefit is obtained at sites distant from the site of application (Ericsson et al., 2004).

There is another type of application of EMF, which became very popular recently. It is the distribution of mattresses and pads that have incorporated different coils and are sold on the Internet mainly with promises to cure everything. Even a brief look on this type of *devices* shows that manufacturers do not know or at least do not provide adequate information about the MF delivered to the patient's body or body parts. Could these items help for some health problems? Eventually, yes. It appears that it might be useful for body relaxation, for wellness purposes, and for treatment of muscle fatigue. In a double-blind study of using *magnetic mattresses*, Colbert et al. (1999) demonstrated that static MF significantly (with 33%) improves the status of patients with fibromyalgia.

It is reasonable to expect that the advantages of powerful computer technologies should be used in designing new magnetotherapeutic devices. At first, it should be the computerized control of the signal and maintenance of the parameters of the signal during the whole treatment session. This has been implemented already in a large number of therapeutic systems. Next is the inclusion of user-friendly software packages with prerecorded programs, as well as with the ability to modify programs depending on the patient's needs. With appropriate sensors, the feedback information could be recorded and used during the course of therapy. Third, the computer technology provides the opportunity to store the data for the treatment of individuals in a large database and further analyze the cohort of data for a particular study or disease.

Finally, let me say that nearly all therapeutic devices are engineered without consideration of the biological properties of the target tissue—only by the intuition of the design engineers. Therefore, these empirically designed devices require both physical and biophysical dosimetries. While the physical dosimetry is dealing with the physical characteristics of the generating system, the biophysical dosimetry requires exact knowledge for the MF at the target site.

In a recently published paper, Parker and Markov propose an analytical approach in designing the therapeutic signal. This approach requires knowledge about the spontaneous MF generated by normal and injured tissues. Such signals could be obtained by using SQUID magnetometers and appropriate elaboration of the recorded data and designing the signal that might be therapeutically effective (Parker and Markov, 2011). One step-by-step approach in the design of a therapeutic device is presented in this book (Seal, 2014).

Much work remains to be done in designing both the technology and methodology of the application of magnetotherapeutic devices. The proper diagnosis of the medical problem and the understanding of the biophysical mechanisms of EMF interactions with injured/diseased tissues are the first two steps to be implemented in choosing the type of PEMF stimulation. Further, the design of the appropriate treatment protocol and the choice of clinical outcomes might facilitate the success of the therapy.

REFERENCES

Adey WR (2004) Potential therapeutic applications of nonthermal electromagnetic fields: Ensemble organization of cells in tissue as a factor in biological field sensing. In: Rosch PJ, Markov MS (eds.), *Bioelectromagnetic Medicine*. Marcel Dekker, New York, pp. 1–14.

Barnes F, Greenebaum B (eds.) (2007) *Handbook of Biological Effects of Electromagnetic Fields*, 3rd edn. CRC Press, Boca Raton, FL.

Bassett CAL (1989) Fundamental and practical aspects of therapeutic uses of pulsed electromagnetic fields (PEMFs). *Critical Review Biomedical Engineering* 17:451–529.

Bassett CAL (1994) Therapeutic uses of electric and magnetic fields in orthopedics. In: Karpenter D and Ayrapetyan S (eds.), *Biological Effects of Electric and Magnetic Fields*. Academic Press, San Diego, CA, pp. 13–18.

Bassett CAL, Pawluk RJ, Pilla AA (1974) Acceleration of fracture repair by electromagnetic fields. *Annals of New York Academic Science* 238:242–262.

Belyaev I (2015) Biophysical mechanisms for nonthermal microwave effects. In: Markov MS (ed.), *Electromagnetic Fields in Biology and Medicine*. CRC Press, Boca Raton, FL, pp. 49–68.

Blackman C (2015) Replication and extension of the Adey group's calcium efflux results. In: Markov MS (ed.), *Electromagnetic Fields in Biology and Medicine*. CRC Press, Boca Raton, FL, pp. 7–14.

Cadossi R, Cadossi M, Setti S (2014) Physical regulation in cartilage and bone repair In: Markov MS (ed.), *Electromagnetic Fields in Band Medicine*. CRC Press, Boca Raton, FL.

Chizhevsky LI (1976) *The Terrestrial Echo of Solar Storms*. Nauka, Moscow (in Russian).

Colbert AP, Markov MS, Banerji M, Pilla AA (1999) Magnetic mattress pad use in patients with fibromyalgia: A Randomized double-blind pilot study. *Journal of Back and Musculoskeletal Rehabilitation* 13(1):19–31.

Colbert AP, Wahbeh H, Markov M et al. (2007) Static magnetic field therapy: A critical review of treatment parameters. *Complementary and Alternative Medicine* doi:10.1093/ecam/nem131.

Detlavs I (1987) *Electromagnetic Therapy in Traumas and Diseases of the Support-Motor Apparatus*. RMI, Riga, Latvia.

Ericsson AD, Hazlewood CF, Markov MS (2004) Specific Biochemical changes in circulating lymphocytes following acute ablation of symptoms in Reflex Sympathetic Dystrophy (RSD): A pilot study. In: Kostarakis P (ed.), *Proceedings of Third International Workshop on Biological Effects of EMF*. Kos, Greece, October 4–8, pp. 683–688.

Gilbert W (1600) *De Magnete* (written in Latin) Translated and published by Dower publication, London, U.K., p. 368.

Ginsburg AJ (1934) Ultrashort radio waves as a therapeutic agent. *Medical Records* 19:1–8.

Kane R (1995) *Cellular Phones: Russian Roulette*. Vantage Press Inc., New York, p. 241.

Kholodov YA (1976) *Man in Magnetic Web*. Nauka, Moscow (in Russian).

Lin J (ed.) (2011) *Electromagnetic fields in Biological Systems*. CRC Press, Boca Raton, FL.

Markov MS (1995) Electric current and electromagnetic field effects on soft tissue: Implications for wound healing. *Wounds* 7(3):94–110.

Markov MS (2000) Dosimetry of magnetic fields in the radiofrequency range. In: Klauenberg BJ and Miklavcic D (eds.), *Radio Frequency Radiation Dosimetry*. Kluwer Academic Press, New York, pp. 239–245.

Markov MS (2002) Can magnetic and electromagnetic fields be used for pain relief? *American Pain Society Bulletin* 12(1):3–7.

Markov MS (2004) Magnetic and electromagnetic field therapy: Basic principles of application for pain relief. In: Rosch PJ, Markov MS (eds.), *Bioelectromagnetic Medicine*. Marcel Dekker, New York, pp. 251–264.

Markov MS (2006) Thermal versus nonthermal mechanisms of interactions between electromagnetic fields and biological systems. In: Ayrapetyan SN and Markov M (eds.), *Bioelectromagnetics: Current Concepts*. Springer, Dordrecht, the Netherlands, pp. 1–16.

Markov MS (2007) Pulsed electromagnetic field therapy: History, state of the art and future. *The Environmentalist* 27:465–475.

Markov MS (2012) Impact of physical factors on the society and environment. *The Environmentalist* 32(2):121–130.

Markov MS, Grigoriev YG (2013) WiFi technology—An uncontrolled experiment on human health. *Electromagnetic Biology and Medicine* 32(2):121.

Markov MS, Pilla AA (1995) Electromagnetic field stimulation of soft tissue: Pulsed radiofrequency treatment of post-operative pain and edema. *Wounds* 7(4):143–151.

Markov MS, Todorov NG (1984) Electromagnetic field stimulation of some physiological processes. *Studia Biophysica* 99:151–156.

Markov MS, Wang S, Pilla AA (1992) Effects of weak low frequency and DC magnetic fields on myosin phosphorylation in a cell-free preparation. *Bioelectrochemistry and Bioenergetics* 30:119–125.

Mayrovitz H (2004) Electromagnetic linkage in soft tissue wound healing. In: Rosch PJ, Markov MS (eds.), *Bioelectromagnetic Medicine*. Marcel Dekker, New York, pp. 461–484.

Mayrovitz H (2015) Electromagnetic fields for soft tissue wound healing. In: Markov MS (ed.), *Electromagnetic Fields in Biology and Medicine*. CRC Press, Boca Raton, FL, pp. 231–252.

Ohkubo C, Okano H (2004) Static magnetic fields and microcirculation. In: Rosch PJ, Markov MS (eds.), *Bioelectromagnetic Medicine*. Marcel Dekker, New York, pp. 563–591.

Ohkubo C, Okano H (2015) Magnetic field influences on the microcirculation. In: Markov MS (ed.), *Electromagnetic Fields in Biology and Medicine*. CRC Press, Boca Raton, FL, pp. 103–128.

Parker R, Markov M (2011) An analytical design technique for magnetic field therapy devices. *The Environmentalist* 31(2):155–160.

Pavlovich SA (1975) *Shielding Magnetic Fields May Cause Mutations in Microorganisms*. Nauka, Moscow (in Russian).

Pilla AA (2007) Mechanisms and therapeutic applications of time-varying and static magnetic fields. In: Barnes F, Greenebaum B (eds.), *Handbook of Biological Effects of Electromagnetic Fields*, 3rd edn. CRC Press, Boca Raton, FL, pp. 351–411.

Rosch PJ, Markov MS (eds.) (2004) *Bioelectromagnetic Medicine*. Marcel Dekker, New York.

Sandra S (2013) *Magnetoterapia*. Kiadja, Budapest 343pp. (in Hungarian).

Seaborne D, Quirion-DeGirardi C, Rousseau M (1996) The treatment of pressure sores using pulsed electromagnetic energy (PEME). *Physiotherapy Canada* 48:131–137.

Shupak N (2003) Therapeutic uses of pulsed magnetic-field exposure: A review. *Radio Science Bulletin* 307:9–32.

Todorov N (1982) *Magnetotherapy*. Meditzina i Physcultura Publishing House, Sofia, Bulgaria.

Valbona C, Hazlewood C, Jurida G (1997) Response of pain to static magnetic fields in postpolio patients: A double blind pilot study. *Archives of Physical Medicine and Rehabilitation* 78:1200–1203.

Vodovnik L, Karba R (1992) Treatment of chronic wounds by means of electric and electromagnetic fields. *Medical and Biological Engineering and Computing* 30:257–266.

Williams CD, Markov MS, Hardman WE, Cameron IL (2001) Therapeutic electromagnetic field effects on angiogenesis and tumor growth. *Anticancer Research* 21:3887–3892.

4 Pulsed Electromagnetic Fields
From Signaling to Healing

Arthur A. Pilla

CONTENTS

4.1 INTRODUCTION

Pulsed electromagnetic fields (PEMFs), from extremely low frequency (ELF) to pulsed radio frequency (PRF), have been successfully employed as adjunctive therapy for the treatment of delayed, nonunion, and fresh fractures and fresh and chronic wounds. Recent increased understanding of the mechanism of action of PEMF signals has permitted technologic advances allowing the development of PRF devices that are portable and disposable; can be incorporated into dressings, supports, and casts; and can be used over clothing. This broadens the use of nonpharmacological, noninvasive PEMF therapy to the treatment of acute and chronic pain and inflammation.

The vast majority of PEMF signals employed clinically do not directly cause a physiologically significant temperature rise in the target cell/tissue area. Furthermore, many experimental and clinical studies show there are preferential responses to the waveform configuration that cannot be explained on the basis of power or energy transfer alone. It was proposed more than 40 years ago that PEMF signals could be configured to be bioeffective by matching the amplitude spectrum of the in situ electric field to the kinetics of ion binding (Pilla, 1972, 1974a,b, 1976). Models have been developed that show that nonthermal PEMF could be configured to modulate electrochemical processes at the electrified interfaces of macromolecules by assessing its detectability versus background voltage fluctuations in a specific ion binding pathway (Pilla et al., 1994, 1999; Pilla, 2006). This has allowed the a priori configuration of nonthermal pulse-modulated radio-frequency (RF) signals that have been demonstrated to modulate calmodulin (CaM)-dependent nitric oxide (NO) signaling in many biological systems (Pilla et al., 2011; Pilla, 2012, 2013).

Given the aforementioned points, PEMF therapy is rapidly becoming a standard part of surgical care, and new, more significant, clinical applications for osteoarthritis, brain and cardiac ischemia, and traumatic brain injury are in the pipeline. This study reviews recent evidence that suggests that CaM-dependent NO signaling, which modulates the biochemical cascades of living cells and tissues employed in response to external challenges, is an essential PEMF transduction pathway. Cellular, animal, and clinical results are presented that provide support for this mechanism of action of PEMF and that may provide a unifying explanation for the reported effects of a variety of PEMF signal configurations on tissue repair, angiogenesis, pain, and inflammation.

4.2 PEMF AS A FIRST MESSENGER

The electrochemical information transfer (ECM) model (Pilla, 1974b, 2006) proposed that weak nonthermal PEMF could be configured to modulate voltage-dependent electrochemical processes at electrified aqueous cell and molecular interfaces. Use of the ECM model for PEMF signal configuration required knowledge of the initial kinetics of ion binding in the target pathway from which physiologically meaningful equivalent electric circuits could be developed. The model provided a means for configuring nonthermal EMF signals to couple to electrochemical kinetics to modulate target processes such as ion binding and transmembrane transport. Application of the ECM model led to the a priori configuration of bone growth stimulator (BGS) signals that are now part of the standard armamentarium of orthopedic practice worldwide for the treatment of recalcitrant bone fractures (Basset et al., 1977; Fontanesi et al., 1983; Aaron et al., 2004). Signal-to-thermal-noise ratio (SNR) analysis was later added to the ECM model to enable more precise signal configurations that could be effective in specific ion binding pathways (Pilla et al., 1994).

The proposed PEMF signal transduction pathway begins with voltage-dependent Ca^{2+} binding to CaM, which is driven by increases in cytosolic Ca^{2+} concentrations in response to chemical and/or physical challenges (including PEMF itself) at the cellular level. Ca/CaM binding has been well characterized, with a binding time constant reported to be in the range of 1–10 ms (Blumenthal and Stull, 1982), whereas the dissociation of Ca^{2+} from CaM requires the better part of a second (Daff, 2003). Thus, Ca/CaM binding is kinetically asymmetrical, that is, the rate of binding exceeds the rate of dissociation by several orders of magnitude ($k_{on} \gg k_{off}$), driving the reaction in the forward direction according to the concentration and voltage dependence of Ca^{2+} binding. The asymmetry in Ca/CaM binding kinetics provides an opportunity to configure any PEMF waveform to induce an electric field that can produce a net increase in the population of bound Ca^{2+} (activated CaM) (Pilla et al., 1999; Pilla, 2006). However, this is only possible if pulse duration or carrier period is significantly shorter than bound Ca^{2+} lifetime (Pilla et al., 2011). Thus, Ca^{2+} binds to CaM when the voltage at the binding site increases; however, Ca^{2+} does not immediately dissociate from CaM when the voltage decreases as the waveform decays or changes polarity, because Ca^{2+} bound in the initial phase of the waveform is sequestered for the better part of a second during which activated CaM activates its target enzyme. Thus, the Ca/CaM signaling pathway can exhibit rectifier-like properties for any PEMF signal because ion binding kinetics are asymmetrical, not because there is a nonlinearity in electrical response to RF signals (Kowalczuk et al., 2010). What follows will demonstrate how a nonthermal PEMF signal can be configured to optimally modulate Ca^{2+} binding to CaM using the ECM model by assessing its electrical detectability in the binding pathway.

4.2.1 PEMF Signal Configuration

Initial configuration of a PEMF signal that can modulate CaM-dependent enzyme activity starts with an analysis of the kinetic equations describing the two-step Ca/CaM/enzyme binding process using the ECM model (Pilla et al., 1999; Pilla, 2006). Thus,

$$Ca^{2+} + CaM \underset{k_{off}}{\overset{k_{on}}{\rightleftarrows}} CaM^* + Enz \underset{k'_{off}}{\overset{k'_{on}}{\rightleftarrows}} Enz^* \qquad (4.1)$$

where CaM* is the activated CaM (Ca^{2+} bound) and Enz* is the activated enzyme target (CaM* bound). In all cases, the forward reaction of Ca^{2+} binding is significantly faster than Ca^{2+} dissociation ($k_{on} \gg k_{off}$). It is also important to note that CaM*/enzyme binding is significantly slower than Ca/CaM binding ($k_{on} \gg k'_{on}$). Under these conditions, Ca^{2+} binding per se can be treated as a linear system for the small changes in voltage at the binding site produced by nonthermal PEMF. Thus, the change in bound Ca^{2+} concentration with time, $\Delta Ca(s)$, may be written as

$$\Delta Ca(s) = \frac{k_{on}}{sCa^0} \left[-\Delta Ca(s) + \kappa E(s) + \Delta Enz * (s) \right] \qquad (4.2)$$

where s is the real-valued Laplace transform frequency, E(s) the induced voltage from PEMF, κ the voltage dependence of Ca^{2+} binding, and Ca^0 the initial concentration of bound Ca^{2+} (homeostasis, too small to activate CaM). Note that the Laplace transform is positive-valued at all frequencies and inherently accounts for the kinetic asymmetry of Ca/CaM binding (Pilla, 2006).

The net increase in Ca^{2+} bound to CaM from PEMF is proportional to the binding current, $I_A(s)$, which, in turn, is proportional to $\Delta Ca(s)$ from Equation 4.2. From this, as shown in detail elsewhere (Pilla, 2006), binding impedance, $Z_A(s)$, may be written as

$$Z_A(s) = \frac{E(s)}{I_A(s)} = \frac{1}{q_e \kappa} \left[\frac{1 + sCa^0}{\dfrac{k_{on}}{sCa^0}} \right] \qquad (4.3)$$

Equation 4.3 shows that the kinetics of Ca^{2+} binding to CaM may be represented by a series R_A–C_A electrical equivalent circuit, where R_A is the equivalent resistance of binding, inversely proportional to k_{on}, and C_A is the equivalent capacitance of binding, proportional to bound Ca^{2+}. The time constant for Ca^{2+} binding is, thus, $\tau_A = R_A C_A$, which is proportional, as expected, to $1/k_{on}$. To evaluate SNR for the Ca/CaM target, the quantity of interest is the effective voltage, $E_b(s)$, induced across C_A, evaluated for simple ion binding (Equation 4.3) using standard circuit analysis techniques (Cheng, 1959):

$$E_b(s) = \frac{\left(\dfrac{1}{sC_A} \right) E(s)}{\left(R_A^2 + \left(\dfrac{1}{sC_A} \right)^2 \right)^{1/2}} \qquad (4.4)$$

Equation 4.4 clearly shows that $E_b(s)$ is dependent upon E(s) for any waveform, for example, rectangular, sinusoidal, arbitrary, or chaotic. In every case, only that portion of the applied electric field that appears at the Ca/CaM binding pathway, $E_b(s)$, can increase surface charge (bound Ca^{2+}). Thus, as long as it is nonthermal, the total applied energy in E(s) is not the dose metric. Rather, the frequency spectrum of $E_b(s)$, for any E(s), is taken as a measure of biological reactivity. The calculation of SNR has been described in detail elsewhere (Pilla et al., 1994; Pilla, 2006). Briefly, thermal voltage noise in the binding pathway may be evaluated via (DeFelice, 1981)

$$RMS_{noise} = \left[4kT \int_{\omega_1}^{\omega_2} Re\left[Z_A(\omega) \right] d\omega \right]^{1/2} \qquad (4.5)$$

where RMS_{noise} is the root mean square of the thermal voltage noise spectral density across C_A; Re is the real part of the total binding impedance, Z_A; $\omega = 2\pi f$; and the limits of integration (ω_1, ω_2) are determined by the band pass of binding, typically 10^{-2}–10^7 rad/s. SNR is evaluated using

$$SNR = \frac{E_b(s)}{RMS_{noise}} \qquad (4.6)$$

The ECM model with SNR analysis can be applied to any ion binding process, provided it is kinetically asymmetrical with $k_{on} \gg k_{off}$, as well as to any nonthermal weak PEMF signal configuration (Pilla, 2006; Pilla et al., 2011). The model has proven to be very useful for the a posteriori analysis of many different PEMF signals particularly for studies in which the response was either marginal or did not exist.

4.2.2 ECM MODEL VERIFICATION

An example of the use of Equation 4.6 to compare the expected efficacy of PRF signals with different pulse modulations, assuming Ca/CaM binding as the transduction pathway, is shown in Figure 4.1a. As may be seen, the rate of Ca^{2+} binding to CaM is expected to be increased identically using either a 2 ms burst of 27.12 MHz repeating at 1 burst/s with 5 uT peak amplitude (signal A) or a 0.065 ms burst of 27.12 MHz repeating at 1 burst/s with 200 uT peak amplitude (signal B). However, the 0.065 ms burst signal at 5 uT peak amplitude (signal C) is not expected to be effective. This was tested on CaM-dependent myosin light chain kinase (MLCK) in a cell-free enzyme assay for myosin light chain (MLC) phosphorylation (Markov et al., 1994; Pilla, 2006). MLC is a contractile protein of physiological importance in muscle and blood and lymph vessel tone. As may be seen in Figure 4.1b, MLC phosphorylation was increased twofold versus control after one 5 min exposure for signals A and B, whereas there was no significant difference versus shams for signal C, just as predicted by the ECM model.

The cell-free MLC assay was also employed to assess the effect of burst duration of the 27.12 MHz RF carrier, keeping burst repetition and amplitude constant (Pilla, 2006). These results are shown

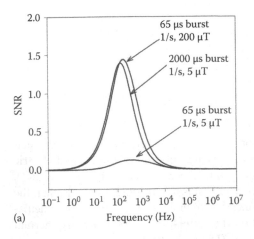

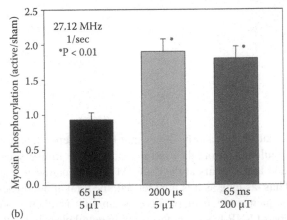

(a)

(b)

FIGURE 4.1 ECM model applied to configure RF PEMF. (a) SNR analysis in Ca/CaM binding pathway shows 2000 µs burst at 5 µT and 65 µs burst at 200 µT should be equally effective versus 65 µs at 5 µT that was predicted ineffective. (b) Five-minute PEMF exposure on MLCK cell-free assay verified ECM prediction. (Adapted from Pilla, A.A., Mechanisms and therapeutic applications of time varying and static magnetic fields, in *Biological and Medical Aspects of Electromagnetic Fields*, F. Barnes and B. Greenebaum, eds., CRC Press, Boca Raton FL, 2006, p. 351.)

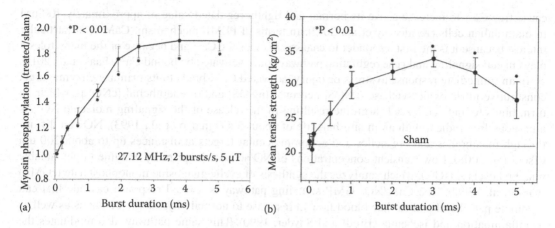

FIGURE 4.2 Pulse modulation of RF PEMF impacts outcome. (a) MLC phosphorylation is increased nearly twofold as burst duration of 27.12 MHz RF PEMF signal increases. Maximum effect occurs at 2–3 ms beyond which saturation occurs. (Adapted from Pilla, A.A., Mechanisms and therapeutic applications of time varying and static magnetic fields, in *Biological and Medical Aspects of Electromagnetic Fields*, F. Barnes and B. Greenebaum, eds., CRC Press, Boca Raton FL, 2006, p. 351.) (b) Wound strength in a rat model similarly increases to 2–3 ms; however, in contrast to the cell-free enzyme assay, tensile strength does not saturate, but peaks as burst duration increases. (Adapted from Strauch, B. et al., *Plast. Reconstr. Surg.*, 120, 425, 2007.)

in Figure 4.2a wherein the increase in bound Ca^{2+} appears to reach saturation as burst duration approaches 2–3 ms. Further verification of the ECM model was obtained from a study of the effect of PRF on wound healing in the rat (Strauch et al., 2007). The study was designed as a prospective, placebo-controlled, blinded trial in which rats were treated for 30 min twice daily with PRF signals with identical carrier frequency (21.12 MHz), burst repetition rate (2 bursts/s), and amplitude (5 μT), but with progressively increasing burst duration. The results are shown in Figure 4.2b, wherein it may be seen that, as for the MLCK assay, peak effect occurred at a burst duration of between 2 and 3 ms. However, instead of a saturation effect, wound strength was lower as burst duration increased. Thus, in a complete tissue, there appears to be a window within which maximum acceleration of healing occurs. The possible reasons for this behavior in living cells and tissue and its implications for the clinical applications of this PRF signal will be considered in detail in the following.

These and other studies suggested an RF signal configuration consisting of a 27.12 MHz carrier pulse modulated with a 2 ms burst repeating at 2 bursts/s, having an amplitude of 4 ± 1 μT. Unless otherwise indicated, this signal configuration was employed in the studies reviewed here. Single-exposure times varied between 15 and 30 min, except where indicated.

4.3 NITRIC OXIDE SIGNALING

PEMF signals with a vast range of waveform parameters have been reported to reduce pain and inflammation (Ross and Harrison, 2013b) and enhance healing (Pilla, 2006). Using the ECM model, a common unifying mechanism has been proposed, which involves Ca^{2+}-dependent NO signaling, to quantify the relation between signal parameters and bioeffect (Pilla et al., 2011). Intracellular calcium ions play an important role in the signal transduction pathways; a cell utilizes to respond to external challenges, for example, cell growth and division, apoptosis, metabolism, synaptic transmission, and gene expression (Bootman et al., 2001; Mellstrom et al., 2008). Regulation of cytosolic Ca^{2+} concentration is orchestrated by an elaborate system of pumps, channels, and binding proteins found both in the plasma membrane and on intracellular organelles such as the endoplasmic reticulum (Harzheim et al., 2010). High-affinity proteins (e.g., CaM, troponin) mediate the multiple physiological responses regulated by changes in intracellular Ca^{2+} concentrations produced by challenges that cause free Ca^{2+} to increase above its normal and tightly regulated value of approximately 100 nM in mammalian cells (Konieczny et al., 2012). In terms of PEMF mechanism, CaM is of particular interest because it is the first responder to changes in cytosolic Ca^{2+} and because of the many roles it plays in cell signaling and gene regulation pathways once activated by bound Ca^{2+} (Faas et al., 2011).

In an immediate response to stress or injury, activated CaM binds to its primary enzyme target constitutive nitric oxide synthase (cNOS, neuronal [nNOS] and/or endothelial [eNOS]), which, in turn, binds to and catalyzes L-arginine resulting in the release of the signaling molecule NO. As a gaseous free radical with an in situ half-life of about 5 s (Ignarro et al., 1993), NO can diffuse through membranes and organelles and act on molecular targets at distances up to about 200 μm (Tsoukias, 2008). Low transient concentrations of NO activate its primary enzyme target, soluble guanylyl cyclase (sGC), which catalyzes the synthesis of cyclic guanosine monophosphate (cGMP) (Cho et al., 1992). The CaM/NO/cGMP signaling pathway is a rapid response cascade that can modulate peripheral and cardiac blood flow in response to normal physiologic demands as well as to inflammation and ischemia (Bredt and Snyder, 1990). This same pathway also modulates the release of cytokines, such as interleukin-1beta (IL-1β), which is proinflammatory (Ren and Torres, 2009), and growth factors such as basic fibroblast growth factor (FGF-2) and vascular endothelial growth factor (VEGF), which are important for angiogenesis, a necessary component of tissue repair (Werner and Grose, 2003).

Following a challenge such as a bone fracture, surgical incision, or other musculoskeletal injury, repair commences with an inflammatory stage during which the proinflammatory cytokines, such as IL-1β, are released from macrophages and neutrophils that rapidly migrate to the injury site. IL-1β upregulates inducible nitric oxide synthase (iNOS), which is not Ca^{2+} dependent and therefore not modulated directly by PEMF. Large sustained amounts of NO that are produced by iNOS in the wound bed (Lee et al., 2001) are proinflammatory and can lead to increased cyclooxygenase-2 (COX-2) and prostaglandins (PGE). These processes are a natural and necessary component of healing but are often unnecessarily prolonged, which can lead to increased pain and delayed or abnormal healing (Broughton et al., 2006). The natural anti-inflammatory regulation produced by CaM/NO/cGMP signaling attenuates IL-1β levels and downregulates iNOS (Palmi and Meini, 2002).

4.3.1 PEMF MODULATES NO SIGNALING

PEMF that produces sufficient SNR in the Ca/CaM binding pathway causes a burst of additional NO to be released from a cell, which has already been challenged. In addition to causing cytosolic Ca^{2+} to increase, inflammatory challenges, injury, and temperature all cause proinflammatory cytokines to be released from macrophages and neutrophils, the first cellular responders. This upregulates iNOS, resulting in the release of large sustained amounts of NO, which is proinflammatory. A recent study specifically examined the effect of PRF on CaM-dependent NO signaling in real time. Dopaminergic cells (MN9D) in phosphate buffer were challenged acutely with a nontoxic concentration of lipopolysaccharide (LPS), which causes an immediate increase in cytosolic Ca^{2+}. As reviewed earlier, any free cytosolic Ca^{2+} above approximately 100 nM instantly activates CaM, which, in turn, instantly activates cNOS. The result is immediate NO production. PRF was applied during LPS challenge. The results (Figure 4.3a) showed that PRF approximately tripled NO release from challenged MN9D cells within seconds, as measured electrochemically in real time using a NO-selective membrane electrode specially designed for use in cell cultures (Pilla, 2012). It was also verified in this study that cells challenged with phosphate buffer alone did not respond to this PRF signal, as shown in Figure 4.3b. This indicates that PEMF can act as an anti-inflammatory only when intracellular Ca^{2+} increases and provides one explanation for the lack of adverse effects reported in all clinical studies that employ this PRF signal.

Other studies have confirmed that this PRF signal can augment CaM-dependent NO and cGMP release from human umbilical vein endothelial cell (HUVEC) and fibroblast cultures

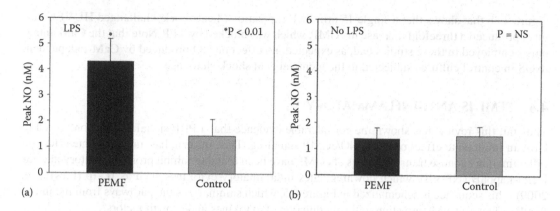

FIGURE 4.3 Effect of PEMF on neuronal cells challenged acutely with LPS (a) or phosphate buffer (b). NO was measured electrochemically with a NO-selective membrane electrode in real time during challenge. PEMF augmented CaM-dependent NO by nearly threefold when cells were challenged with LPS. In contrast, there was no PEMF effect on NO release for cells challenged with phosphate buffer. (Adapted from Pilla, A.A., *Biochem. Biophys. Res. Commun.*, 426, 330, 2012.)

(Pilla et al., 2011; Pilla, 2013). These studies additionally showed that a PRF effect could be obtained only if the cells were challenged (e.g., temperature shock) sufficiently to cause increases in cytosolic Ca^{2+} large enough to satisfy the Ca/CaM dependence of cNOS (Bredt and Snyder, 1990). Direct measurement of cytosolic Ca^{2+} binding to CaM under physiological conditions in living cells or tissue has not yet been successfully performed under PEMF exposure, primarily because of the submicromolar concentrations of Ca^{2+} involved. However, the CaM antagonists N-(6-Aminohexyl)-5-chloro-1-naphthalenesulfonamide hydrochloride (W-7) and trifluoroperazine (TFP) were able to block the PEMF effect on additional NO release, supporting CaM activation, which only occurs if Ca^{2+} binds to CaM, as a principle PEMF target for the modulation of tissue repair. A typical example of the effect of PRF on NO release in heat shock–challenged fibroblast cultures is summarized in Figure 4.4a, which shows a single 15 min PRF exposure produced a nearly twofold increase in NO, which was blocked with W-7. Another example can be seen in

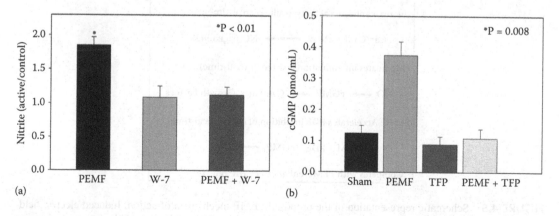

FIGURE 4.4 Effect of PEMF on CaM/NO/cGMP signaling in fibroblast and HUVEC cultures. (a) A single 15 min exposure of RF PEMF in fibroblast cultures increased NO by twofold, which could be blocked by the CaM antagonist W-7. (Adapted from Pilla, A.A. et al., *Biochem. Biophys. Acta.*, 1810, 1236, 2011.) (b) Similar PEMF exposure in HUVEC cultures increased cGMP by threefold, which, in this case, was blocked by the CaM antagonist TFP. These results are consistent with CaM/NO/cGMP as the transduction pathway for PEMF bioeffects. (Adapted from Pilla, A.A., *Electromagn. Biol. Med.*, 32, 123, 2013.)

Figure 4.4b that shows that a single 15 min PRF exposure of heat shock–challenged HUVEC cultures produced a threefold increase in cGMP, which was blocked by TFP. Note that the CaM antagonists employed in these studies had, as expected, no effect on NO produced by CaM-independent iNOS in control cultures subjected to the identical heat shock challenge.

4.4 PEMF IS ANTI-INFLAMMATORY

Thus far, this review has shown the considerable evidence that a PRF signal can be configured to have an immediate effect on CaM/NO/cGMP signaling. This, in turn, has an immediate effect on inflammation because transient bursts of cGMP have been found to inhibit proinflammatory nuclear factor-kappaB (NF-κB), which downregulates inflammatory cytokines such as IL-1β (Lawrence, 2009). This sequence is schematized in Figure 4.5, which summarizes the pathways from the initial PEMF effect on CaM activation to the modulation of cytokines and growth factors.

To illustrate, RF PEMF has been reported to modulate inflammatory cytokines in fibroblasts and keratinocytes (Moffett et al., 2012). Cell cultures were placed at room temperature (heat shock challenge) and exposed PRF for 15 min. Cytokine expression was assayed 2 h after PEMF exposure. The results in Figure 4.6 show that PRF downregulated the proinflammatory cytokine IL-1β and upregulated the anti-inflammatory cytokines IL-5, IL-6, and IL-10, consistent with an anti-inflammatory PEMF effect via modulation of CaM/NO/cGMP signaling.

PEMF has been reported to downregulate iNOS at the mRNA and protein levels in monocytes (Reale et al., 2006) and proinflammatory cytokines in human keratinocytes (Vianale et al., 2008). Weak electric fields partially reversed the decrease in the production of extracellular matrix caused by exogenous IL-1β in full-thickness articular cartilage explants from osteoarthritic adult human knee joints (Brighton et al., 2008). PRF reduced IL-1β in cerebrospinal fluid 6 h after posttraumatic brain injury in a rat model (Rasouli et al., 2012). PEMF downregulated IL-1β and upregulated IL-10 in a mouse cerebral ischemia model (Pena-Philippides et al., 2014b) and upregulated IL-10 within 7 days in a chronic inflammation model in the mouse (Pena-Philippides et al., 2014a).

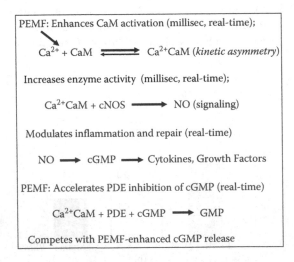

FIGURE 4.5 Schematic representation of the proposed PEMF mechanism of action. Induced electric field from the PEMF signal enhances CaM activation, which enhances cNOS activation. This, in turn, enhances CaM-dependent NO release, which enhances cGMP release. This reduces inflammatory cytokines and increases anti-inflammatory cytokines. As inflammation is reduced, CaM/NO/cGMP signaling modulates growth factor release, allowing tissue repair to be accelerated. Also shown is PEMF acceleration of PDE regulation of cGMP, which places limits on PEMF dosing.

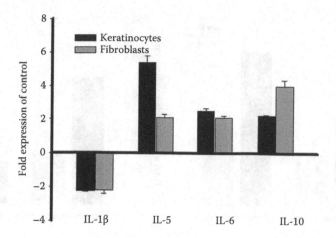

FIGURE 4.6 Effect of RF PEMF on inflammatory cytokine panel in fibroblasts and keratinocytes. A single 15 min exposure of cultures placed at room temperature (heat shock) produced several-fold changes in inflammatory cytokines. PEMF downregulated the proinflammatory cytokine IL-1β and upregulated the anti-inflammatory cytokines IL-5, IL-6, and IL-10. These results are consistent with CaM/NO/cGMP signaling as the PEMF transduction pathway. (Adapted from Moffett, J. et al., *J. Pain. Res.*, 12, 347, 2012.)

In the clinical setting, PRF has been shown to enhance the management of postoperative pain and inflammation. Several double-blind, placebo-controlled, randomized studies, including breast reduction (Rohde et al., 2010), breast augmentation (Hedén and Pilla, 2008; Rawe et al., 2012), and autologous flap breast reconstruction (Rohde et al., 2012), have reported that nonthermal PRF fields significantly accelerate postoperative pain and inflammation reduction and, concomitantly, reduce postoperative narcotic requirements. Two of these studies examined the effect of PRF on levels of IL-1β in wound exudates, as well as wound exudate volume in the first 24 h postoperatively (Rohde et al., 2010, 2012). Figure 4.7a shows that IL-1β in wound exudates of active patients in both studies was 50% of that for sham patients at 6 h postoperative. Figure 4.7b shows wound exudate volume at 12 h postoperative was approximately twofold higher in sham versus active patients in both studies. Reductions in IL-1β and wound exudate volume are consistent with a PEMF effect on inflammation via CaM/NO/cGMP signaling.

An important result of PEMF modulation of CaM/NO/cGMP signaling is the modulation of nociception (Cury et al., 2011), perhaps since NO signaling modulates the sensitivity of opioid receptors (Chen et al., 2010). Clinically, NO enhances the actions of narcotics for postoperative analgesia (Lauretti et al., 1999), which may play a role in the reported effects of RF PEMF signals on reduction of postoperative narcotic usage (Rohde et al., 2010, 2012, 2014; Rawe et al., 2012). CaM-dependent NO also reduces IL-1β, which in turn reduces PDE and COX-2. Thus, addition of an NO donor to NSAIDs and aspirin enhances analgesia (Velazquez et al., 2005; Borhade et al., 2012). Modulation of the endogenous opioid pathway by physical modalities has previously been described for ELF magnetic fields (Kavaliers and Ossenkopp, 1991; Prato et al., 1995) and transcutaneous electrical nerve stimulation (Sluka et al., 1999). It has recently been reported that PRF signals increase endogenous opioid precursors in human epidermal keratinocytes and dermal fibroblasts both at the mRNA and protein levels (Moffett et al., 2012). The same PRF signal was also used in a rat pain behavior model, wherein pain reduction was, in part, related to PRF modulation of the release of β-endorphin (Moffett et al., 2010, 2011). Thus, PEMF can reduce inflammation and potentially enhance the action of narcotics, NSAIDS, and aspirin, all of which reduce patient morbidity.

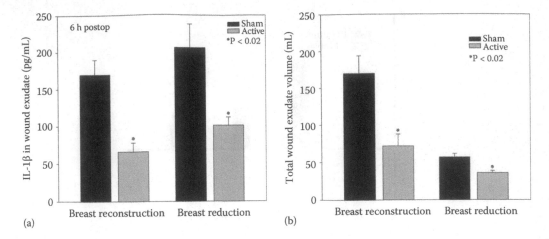

FIGURE 4.7 Effect of RF PEMF on inflammation in two double-blinded, randomized, and placebo-controlled clinical studies. (a) PEMF reduced IL-1β in wound exudates, which was approximately twofold higher in sham versus active patients at 6 h postoperative in both breast reconstruction and breast reduction studies. (b) Total wound exudate volume was approximately twofold higher in sham versus active patients at 12 h postoperative in both studies. These results are consistent with a PEMF effect on inflammation via CaM/NO/cGMP signaling. (Adapted from Rohde, C. et al., *Plast. Reconstr. Surg.*, 125, 1620, 2010; Rohde, C. et al., *Plast. Reconstr. Surg.*, 130, 91, 2012.)

4.4.1 CLINICAL TEST OF ECM MODEL

In the clinical arena, the ECM/SNR model can be used a posteriori to analyze the effectiveness of different configurations of PRF signals assuming the transduction pathway is CaM-dependent NO signaling. This was performed for two independent double-blind, placebo-controlled, and randomized studies that assessed the effect of PRF on postoperative pain relief and narcotic requirements in breast augmentation patients (Hedén and Pilla, 2008; Rawe et al., 2012). One study used PRF signal A consisting of a 2 ms burst of 27.12 MHz repeating at 2 bursts/s and inducing a peak electric field of 5 ± 2 V/m (predicted effective by ECM). The second study used PRF signal B consisting of a 0.1 ms burst of 27.12 MHz repeating at 1000 bursts/s and inducing a peak electric field of 0.4 ± 0.1 V/m. Both studies reported accelerated postoperative pain reduction. However, sham patients in the first study (signal A) had 2.7-fold more pain at POD2 compared to active patients. In contrast, sham patients in the second study (signal B) had only 1.3-fold more pain than active patients at POD2. Clearly, pain reduction in patients treated with signal A was substantially faster than that for patients treated with signal B. Relative effectiveness of both signals is shown in Figure 4.8, wherein it may be seen that dosimetry for signal A, in terms of SNR at Ca/CaM binding sites, is significantly larger than that for signal B. Clearly, signal A would be expected to produce a burst of NO significantly higher than that produced by signal B that could lead to a larger anti-inflammatory effect.

4.5 PEMF CAN MODULATE ANGIOGENESIS

As tissue responds to injury, PEMF can rapidly modulate the relaxation of the smooth muscles controlling blood and lymph vessel tone through the CaM/NO/cGMP cascade (McKay et al., 2007). PEMF also enhances growth factor release through the same cascade in endothelial cells to modulate angiogenesis. PEMF modulation of eNOS activity may, therefore, be a useful strategy to augment angiogenesis for tissue repair and possibly other conditions that require vascular plasticity, such as ischemia (Cooke, 2003). An early study showed that PEMF augmented the creation of

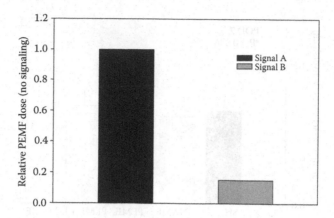

FIGURE 4.8 Effect of pulse modulation parameters of RF PEMF on clinical efficacy for postoperative pain reduction. Relative PEMF dose is measured assuming the transduction mechanism is the modulation of CaM/NO/cGMP signaling. A posteriori analysis of SNR for each signal reveals that signal B is expected to enhance CaM activation by 20% versus 100% for signal A. Clinical data confirm sham patients in signal A study had 2.7-fold more pain at POD2 compared to active patients, compared to only 1.3-fold more pain in the signal B study. (Adapted from Hedén, P. and Pilla, A.A., *Aesthetic. Plast. Surg.*, 32, 660, 2008; Rawe, I.M. et al., *Aesthetic Plast. Surg.*, 36, 458, 2012.)

tubular, vessel-like structures from endothelial cells in culture in the presence of growth factors (Yen-Patton et al., 1998). Another study confirmed a sevenfold increase in endothelial cell tubule formation in vitro (Tepper et al., 2004). Quantification of angiogenic proteins demonstrated a five-fold increase in FGF-2, suggesting that PEMF modulates angiogenesis by increasing FGF-2 production. This same study also reported PEMF increased vascular in-growth more than twofold when applied to an implanted Matrigel plug in mice, with a concomitant increase in FGF-2, similar to that observed in vitro. PEMF significantly increased neovascularization and wound repair in normal mice, and particularly in diabetic mice, through an endogenous increase in FGF-2, which could be eliminated by using a FGF-2 inhibitor (Callaghan et al., 2008). Similarly, a PRF signal of the type used clinically for wound repair was reported to significantly accelerate vascular sprouting from an arterial loop transferred from the hind limb to the groin in a rat model (Roland et al., 2000). This study was extended to examine free flap survival on the newly produced vascular bed (Weber et al., 2005). Results showed 95% survival of PRF-treated flaps compared to 11% survival in the sham-treated flaps, suggesting a significant clinical application for PRF signals in reconstructive surgery. Another study (Delle Monache et al., 2008) reported that PEMF increased the degree of endothelial cell proliferation and tubule formation and accelerated the process of wound repair, suggesting a mechanism based upon a PEMF effect on VEGF receptors. In the clinical setting, PRF has been reported to enhance fresh and chronic wound repair (Kloth et al., 1999; Kloth and Pilla, 2010; Strauch et al., 2007; Guo et al., 2012).

A recent study evaluated the effect of PRF on cardiac angiogenesis in a reproducible thermal myocardial injury model. The injury was created in the region of the distal aspect of the left anterior descending artery at the base of the heart in a blinded rat model (Patel et al., 2006; Strauch et al., 2006, 2009; Pilla, 2013). PRF exposure was 30 min twice daily for 3, 7, 14, or 21 days. Sham animals were identically exposed, but received no PRF signal. A separate group of animals treated for 7 days received L-nitroso-arginine methyl ester (L-NAME), a general NOS inhibitor, in their drinking water. Upon sacrifice, myocardial tissue specimens were stained with CD-31, and the number of new blood vessels was counted on histological sections at the interface between normal and necrotic muscle at each time point by three independent blinded histologists. The results showed mean new vessel count was not significantly increased by PEMF at day 3, but was significantly increased at

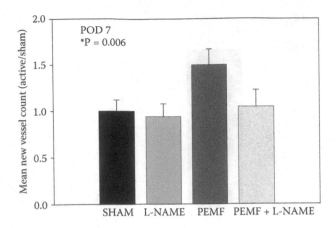

FIGURE 4.9   Effect of RF PEMF on angiogenesis in a thermal myocardial injury model. The results show that PEMF significantly increased new vessel growth and that L-NAME, a general NOS inhibitor administered in drinking water, blocked the PEMF effect. These results suggest PEMF was acting through the CaM/NO/cGMP pathway since blockage occurred far upstream of the production of growth factors necessary for angiogenesis. (Adapted from Pilla, A.A., *Electromagn. Biol. Med.*, 32, 123, 2013.)

day 7 (+50%, P = 0.006), day 14 (+67%, P = 0.004), and day 21 (+99%, P < 0.001). The results shown in Figure 4.9 for day 7 indicate L-NAME completely blocked the PEMF effect on angiogenesis, suggesting the transduction pathway for the PRF effect on angiogenesis in this study involved CaM-dependent NO signaling.

4.6   PEMF, NO SIGNALING, AND PHOSPHODIESTERASE

It will be recalled that PRF effects in living systems appear to occur within a window defined by a combination of pulse modulation parameters and amplitude of induced electric field (see Figure 4.2b). It will also be recalled that activated CaM can activate a number of enzymes, including MLCK, cNOS, calmodulin-dependent protein kinase II (CaMPKII), and some phosphodiesterase (PDE) isoforms. It is the latter that appears to play a large role in the effectiveness of PEMF to enhance NO signaling. This is because the dynamics of Ca^{2+}-dependent NO release are tightly regulated through a negative feedback mechanism that involves PDE isoenzymes (Mo et al., 2004; Batchelor et al., 2010). These studies show the rate of Ca^{2+}-dependent NO production in challenged tissue will not only be modulated by PEMF but also by PDE activity, which, itself, is also Ca^{2+}-dependent. In essence, the augmentation of cGMP and therefore the modulation of cytokines and growth factors that follows from the augmentation of NO by PEMF (see Figure 4.5) can be curtailed by PDE that converts cGMP to GMP, which is inactive. Thus, PEMF can enhance both the activity of cNOS, which increases NO release, and the activity of PDE, which decreases NO release in tissue by inhibiting cGMP. Thus, if PEMF increases NO too rapidly, PDE isoenzyme activity may also be enhanced sufficiently to reduce or block the effects of PEMF on NO signaling.

Thus, the predictions of the ECM model for signal configuration now need to take into account CaM/NO/cGMP/PDE signaling dynamics. The PRF signal considered here produces an instantaneous burst of NO during each signal burst, in real time (Pilla, 2012). The amount of increased NO is directly dependent upon induced electric field amplitude, burst duration, repetition rate, and total exposure time. Each of these parameters will modify the accumulated increase of NO in tissue produced by PEMF. As NO increases so does cGMP, and this dynamic is regulated by a CaM-dependent PDE isoform. It has been suggested that the extreme sensitivity of NO receptors is due to the low affinity of PDE to cGMP, meaning NO activates sGC faster than PDE inhibits cGMP (Batchelor et al., 2010). However, PEMF also enhances PDE activation, suggesting that the natural CaM/NO/cGMP dynamic can be disrupted if the PEMF signal enhances NO too rapidly. As PEMF-enhanced NO production increases, PDE activation also increases, and the dynamic can favor cGMP inhibition (see Figure 4.5). This may be the explanation for the existence of peak responses to PRF in cellular and animal studies (Pilla, 2006; Strauch et al., 2007; Pilla et al., 2011). It may also explain recent clinical results in which, for an identical PRF configuration, a treatment regimen, which produced NO at fourfold the rate of a known effective signal configuration for postoperative pain relief, was ineffective (Taylor et al., 2014).

This modification of the ECM model may be tested in other studies. Two recent publications show that PEMF effects depend upon signal configuration. In the first, a PEMF effect on breast cancer cell apoptosis was significant when the same waveform, applied for the same exposure time, was repeated at 20 Hz but not at 50 Hz (Crocetti et al., 2013). The second study showed that PEMF significantly reduced the expression of inflammatory markers, tumor necrosis factor (TNF), and NF-κB in challenged macrophages when the same waveform, applied for the same exposure time, was repeated at 5 Hz, but not at 15 or 30 Hz (Ross and Harrison, 2013a). In both studies, Ca^{2+}-dependent NO signaling modulates the expressions of these inflammatory markers (Ha et al., 2003; Yurdagul et al., 2013), suggesting the effect of increased repetition rate is consistent with increased production of NO at a rate high enough for negative feedback from PDE isoforms to predominate, thus blocking the PEMF effect on Ca^{2+}-dependent NO signaling.

4.7 DISCUSSION AND CONCLUSIONS

There is now sufficient evidence to support the proposal that CaM-dependent NO signaling is a primary PEMF transduction pathway. There is abundant recent data showing PEMF effects on NO, cGMP, and inflammatory cytokines at the cellular, animal, and clinical levels, as well as evidence for enhanced wound repair and tissue healing. Many studies have suggested NO plays a role in biological responses to weak nonthermal EMF without proposing a transduction mechanism. Perhaps the first was a report that a burst-type RF PEMF increased NO and cGMP in a rat cerebellum cell-free supernatant (Miura et al., 1993). This study determined that L-NAME and EDTA could inhibit the effect, suggesting CaM-dependent NO release was involved. Interestingly, when a continuous frequency waveform was applied, there were no effects on NO, confirming that bioeffects require matching amplitude spectrum of the signal to the kinetics of the proposed ion binding target (Pilla, 2006; Pilla et al., 2011). Several studies have reported that NO mediates the effect of PEMF on osteoblast proliferation and differentiation (Diniz et al., 2002; Cheng et al., 2011) and chondrocyte proliferation (Fitzsimmons et al., 2008).

In this review, emphasis was placed on the evidence that a PRF signal could be configured a priori to modulate CaM/NO/cGMP signaling. However, the predictions of the ECM model apply to any PEMF signal. For example, a study on the effect of PEMF on bone repair in a rat fibular osteotomy model compared the original BGS signal (Pilla, 2006) with a newer pulse-burst waveform having much shorter pulse duration (Midura et al., 2005). This study showed the original BGS waveform accelerated bone repair by producing a twofold increase in callus volume and a twofold increase in the rate of hard callus formation at 2 weeks, followed by a twofold increase in callus stiffness at 5 weeks. In contrast, the new pulse burst waveform had no significant effect on any of these parameters in this model. Application of the ECM model, assuming a CaM/NO/cGMP signaling target, provides an explanation for the differing PEMF effects. Although both PEMF signals were repetitive asymmetrical pulse bursts, they had large differences in the amplitude of their respective frequency components in the Ca/CaM binding pathway. The new pulse burst induced an electric field consisting of approximately a 1 ms burst of relatively short, asymmetrical pulses (4/12 μs) having a peak amplitude of 0.05 V/m, repeating at approximately 1 Hz. In contrast, the induced electric field from the original BGS signal consisted of a 5 ms burst of significantly longer asymmetrical pulses (200/20 μs) having significantly higher peak amplitude (2 V/m) and repetition rate (15 Hz). Comparison of SNR in the Ca/CaM binding pathway for these EMF signals shows that the SNR frequency spectrum of SNR for the new pulse burst signal does not effectively couple to the Ca/CaM binding pathway. Thus, application of the ECM model, assuming Ca/CaM/cGMP as the transduction pathway, would have predicted this signal ineffective for bone repair, as was indeed reported. It is of interest to note that peak SNR in the Ca/CaM binding pathway for the original BGS signal is approximately fivefold lower than that for the PRF signal considered in this review (Pilla et al., 2011). This large difference in dosimetry may provide one explanation for the generally observed slow response to BGS signals in clinical applications.

It is also possible to use the ECM model to predict whether nonthermal bioeffects may be produced from the RF signals emitted by cellular phones using the ECM model. Indeed, the asymmetrical kinetics of Ca/CaM binding may also be a potential signaling target for a GSM waveform. Evaluation of SNR in the Ca/CaM target pathway indicates that SNR for a single 577 μs pulse of a 1800 MHz GSM signal at 20 V/m falls within the same frequency range and has a peak value similar to that for a single 2000 μs pulse of a 27.12 MHz RF signal at 12 V/m, typical measured field strengths within the biological target for these signals (Pilla and Muehsam, 2010). Thus, GSM signals may have nonthermal bioeffects, including therapeutic effects, as suggested by recent reports showing that long-term exposure to a GSM signal protects against and reverses cognitive impairment in a mouse model of Alzheimer's disease (Arendash et al., 2010, 2012; Dragicevic et al., 2011). Also, it has been suggested that inflammation, mediated by both IL-1β and iNOS, will enhance the deposition of β-amyloid (Chiarini et al., 2006). According to the transduction mechanism proposed here, a GSM signal would be expected to downregulate both factors, which may prevent or reverse the effects of Alzheimer's disease and any other neurodegenerative disease with an inflammatory component. It is to be noted that the mechanism proposed here is based upon kinetic asymmetry in ion binding kinetics and does not depend upon nonlinearity in the electrical response of the target to RF signals (Kowalczuk et al., 2010).

While the results presented here support an electric field effect on CaM signaling via the asymmetrical voltage-dependent kinetics of Ca^{2+} binding to CaM (k_{on}), it is important to emphasize that the slow dissociation kinetics of Ca/CaM (k_{off}) can be responsive to weak dc and combined dc/ac magnetic (B) fields. Indeed, it has been reported that weak B fields can affect NO signaling (Palmi et al., 2002; Reale et al., 2006). Of the many models proposed to explain the bioeffects of weak magnetic fields, those involving modulation of the bound trajectory of a charged ion by classic Lorentz force (Chiabrera et al., 1993; Edmonds, 1993; Muehsam and Pilla, 2009a,b) are most relevant to asymmetrical ion binding kinetics, suggesting that weak B-field effects on the trajectory of the ion within the binding site itself could affect reactivity. One interpretation of this is that weak dc and certain combinations of weak ac/dc magnetic fields could enhance or inhibit the exit of the target ion from the binding site, thereby accelerating or inhibiting the overall reaction rate by manipulating dissociation kinetics (k_{off}), even in the presence of thermal noise. This has been tested with success for CaM-dependent myosin phosphorylation (Muehsam and Pilla, 2009b) suggesting that analysis of B-field effects on bound ion trajectories could be used to explain and predict weak magnetic field effects on CaM-dependent NO signaling via modulation of Ca/CaM dissociation kinetics.

Taken together, the results reviewed here strongly support that PEMF signals can be configured to instantaneously modulate NO release in injured tissue. At the cellular level, PEMF-mediated NO signaling could be the common transduction mechanism in studies that report up- and downregulation of anti- and proinflammatory genes (Brighton et al., 2008; Moffett et al., 2012; Pena-Philippides et al., 2014a,b) and modulation of adenosine pathways (De Mattei et al., 2009; Vincenzi et al., 2013; Adravanti et al., 2014). At the clinical level, there is strong support that PEMF-mediated signaling is anti-inflammatory and can enhance angiogenesis. NO release via cNOS is dynamic (Batchelor et al., 2010) and closely linked to the negative feedback provided by PDE inhibition of cGMP (Mo et al., 2004). This causes ensuing dynamic consequences in the rates of up- or downregulation of growth factors and cytokines. For example, iNOS activity in the inflammatory stage of healing can be rapidly downregulated by inhibition of NF-kB in a negative feedback mechanism (Chang et al., 2004).

The proposed transduction mechanism is consistent with the hypothesis that a nonthermal PEMF signal can be configured a priori to act as a first messenger in CaM-dependent signaling pathways that include NO and cyclic nucleotides relevant to tissue growth, repair, and maintenance. Of the multitude of intracellular Ca^{2+} buffers, CaM is an important early responder because CaM-dependent cytokine and growth factor release orchestrates rapid cellular and tissue response to physical and mechanical challenges. Ca^{2+} binding to CaM is voltage dependent, and its kinetic asymmetry ($k_{on} \gg k_{off}$) yields rectifier-like properties, allowing the use of an electrochemical model to configure effective EMF signals to couple efficiently with this pathway. The ECM model is potentially powerful enough to unify the observations of many groups and may offer a means to explain both the wide range of reported bioeffects as well as the many equivocal reports from PEMF studies. The predictions of the proposed model open a host of significant possibilities for configuration of nonthermal EMF signals for clinical and wellness applications that can reach far beyond fracture repair and wound healing. Active studies are underway to assess the utility of the known anti-inflammatory activity of PRF for the treatment of traumatic brain injury, neurodegenerative diseases, cognitive disorders, degenerative joint disease, neural regeneration, and cardiac and cerebral ischemia.

REFERENCES

Aaron RK, Ciombor DM, Simon BJ. Treatment of nonunions with electric and electromagnetic fields. *Clin Orthop.* 2004;419:21–29.

Adravanti P, Nicoletti S, Setti S, Ampollini A, de Girolamo L. Effect of pulsed electromagnetic field therapy in patients undergoing total knee arthroplasty: A randomised controlled trial. *Int Orthop.* 2014;38(2):397–403.

Arendash GW, Mori T, Dorsey M, Gonzalez R, Tajiri N, Borlongan C. Electromagnetic treatment to old Alzheimer's mice reverses β-amyloid deposition, modifies cerebral blood flow, and provides selected cognitive benefit. *PLoS ONE.* 2012;7(4):e35751.

Arendash GW, Sanchez-Ramos J, Mori T et al. Electromagnetic field treatment protects against and reverses cognitive impairment in Alzheimer's disease mice. *J Alzheimers Dis.* 2010;19:191–210.

Basset CAL, Pilla AA, Pawluk R. A non-surgical salvage of surgically-resistant pseudoarthroses and nonunions by pulsing electromagnetic fields. *Clin Orthop.* 1977;124:117–131.

Batchelor AM, Bartus K, Reynell C, Constantinou S, Halvey EJ, Held KF, Dostmann WR, Vernon J, Garthwaite J. Exquisite sensitivity to subsecond, picomolar nitric oxide transients conferred on cells by guanylyl cyclase-coupled receptors. *Proc Natl Acad Sci USA* 2010;107:22060–22065.

Blumenthal DK, Stull JT. Effects of pH, ionic strength, and temperature on activation by calmodulin and catalytic activity of myosin light chain kinase. *Biochemistry* 1982;21:2386–2391.

Bootman MD, Lipp P, Berridge MJ. The organisation and functions of local Ca(2+) signals. *J Cell Sci.* 2001;114:2213–2222.

Borhade N, Pathan AR, Halder S et al. NO-NSAIDs. Part 3: Nitric oxide-releasing prodrugs of non-steroidal anti-inflammatory drugs. *Chem Pharm Bull (Tokyo)* 2012;60:465–481.

Bredt DS, Snyder SH. Isolation of nitric oxide synthetase, a calmodulin-requiring enzyme. *Proc Natl Acad Sci USA* 1990;87:682–685.

Brighton CT, Wang W, Clark CC. The effect of electrical fields on gene and protein expression in human osteoarthritic cartilage explants. *J Bone Joint Surg Am.* 2008;90:833–848.

Broughton G 2nd, Janis JE, Attinger CE. Wound healing: An overview. *Plast Reconstr Surg.* 2006; 117(7 Suppl):1e-S–32e-S.

Callaghan MJ, Chang EI, Seiser N, Aarabi S, Ghali S, Kinnucan ER, Simon BJ, Gurtner GC. Pulsed electromagnetic fields accelerate normal and diabetic wound healing by increasing endogenous FGF-2 release. *Plast Reconstr Surg.* 2008;121:130–141.

Chang K, Lee SJ, Cheong I, Billiar TR, Chung HT, Han JA, Kwon YG, Ha KS, Kim YM. Nitric oxide suppresses inducible nitric oxide synthase expression by inhibiting post-translational modification of IkappaB. *Exp Mol Med.* 2004;36:311–324.

Chen Y, Boettger MK, Reif A et al. Nitric oxide synthase modulates CFA-induced thermal hyperalgesia through cytokine regulation in mice. *Mol Pain* 2010;6:13–17.

Cheng DK. *Analysis of Linear Systems*. Addison-Wesley, London, U.K., 1959.

Cheng G, Zhai Y, Chen K, Zhou J, Han G, Zhu R, Ming L, Song P, Wang J. Sinusoidal electromagnetic field stimulates rat osteoblast differentiation and maturation via activation of NO-cGMP-PKG pathway. *Nitric Oxide* 2011;25:316–325.

Chiabrera A, Bianco B, Moggia E. Effect of lifetimes on ligand binding modeled by the density operator. *Bioelectrochem Bioenerg*. 1993;30:35–42.

Chiarini A, Dal Pra I, Whitfield JF, Armato U. The killing of neurons by beta-amyloid peptides, prions, and pro-inflammatory cytokines. *Ital J Anat Embryol*. 2006;111:221–246.

Cho HJ, Xie QW, Calaycay J, Mumford RA, Swiderek KM, Lee TD, Nathan C. Calmodulin is a subunit of nitric oxide synthase from macrophages. *J Exp Med*. 1992;176:599–604.

Cooke JP. NO and angiogenesis. *Atheroscler Suppl*. 2003;4:53–60.

Crocetti S, Beyer C, Schade G, Egli M, Fröhlich J, Franco-Obregón A. Low intensity and frequency pulsed electromagnetic fields selectively impair breast cancer cell viability. *PLoS ONE*. 2013;8(9):e72944.

Cury Y, Picolo G, Gutierrez VP et al. Pain and analgesia. The dual effect of nitric oxide in the nociceptive system. *Nitric Oxide* 2011;25:243–254.

Daff S. Calmodulin-dependent regulation of mammalian nitric oxide synthase. *Biochem Soc Trans*. 2003; 31:502–505.

DeFelice LJ. *Introduction to Membrane Noise*. Plenum, New York, 1981, pp. 243–245.

Delle Monache S, Alessandro R, Iorio R, Gualtieri G, Colonna R. Extremely low frequency electromagnetic fields (ELF-EMFs) induce in vitro angiogenesis process in human endothelial cells. *Bioelectromagnetics* 2008;29:640–648.

De Mattei M, Varani K, Masieri FF, Pellati A, Ongaro A, Fini M, Cadossi R, Vincenzi F, Borea PA, Caruso A. Adenosine analogs and electromagnetic fields inhibit prostaglandin E2 release in bovine synovial fibroblasts. *Osteoarthritis Cart*. 2009;17:252–262.

Diniz P, Soejima K, Ito G. Nitric oxide mediates the effects of pulsed electromagnetic field stimulation on the osteoblast proliferation and differentiation. *Nitric Oxide* 2002;7:18–23.

Dragicevic N, Bradshaw PC, Mamcarz M, Lin X, Wang L, Cao C, Arendash GW. Long-term electromagnetic field treatment enhances brain mitochondrial function of both Alzheimer's transgenic mice and normal mice: A mechanism for electromagnetic field-induced cognitive benefit? *Neuroscience* 2011;185:135–149.

Edmonds DT. Larmor Precession as a mechanism for the detection of static and alternating magnetic fields. *Bioelectrochem Bioenerg*. 1993;30:3–12.

Faas GC, Raghavachari S, Lisman JE, Mody I. Calmodulin as a direct detector of Ca^{2+} signals. *Nat Neurosci*. 2011;14:301–304.

Fitzsimmons RJ, Gordon SL, Kronberg J, Ganey T, Pilla AA. A pulsing electric field (PEF) increases human chondrocyte proliferation through a transduction pathway involving nitric oxide signaling. *J Orthop Res*. 2008;26:854–859.

Fontanesi G, Giancecchi F, Rotini R, Cadossi R. Treatment of delayed union and pseudarthrosis by low frequency pulsing electromagnetic stimulation. Study of 35 cases. *Ital J Orthop Traumatol*. 1983;9:305–318.

Guo L, Kubat NJ, Nelson TR, Isenberg RA. Meta-analysis of clinical efficacy of pulsed radio frequency energy treatment. *Ann Surg*. 2012;255:457–467.

Ha KS, Kim KM, Kwon YG et al. Nitric oxide prevents 6-hydroxydopamine-induced apoptosis in PC12 cells through cGMP-dependent PI3 kinase/Akt activation. *FASEB J*. 2003;17:1036–1047.

Hedén P, Pilla AA. Effects of pulsed electromagnetic fields on postoperative pain: A double-blind randomized pilot study in breast augmentation patients. *Aesthetic Plast Surg*. 2008;32:660–666.

Ignarro LJ, Fukuto JM, Griscavage JM, Rogers NE, Byrns RE. Oxidation of nitric oxide in aqueous solution to nitrite but not nitrate: Comparison with enzymatically formed nitric oxide from L-arginine. *Proc Natl Acad Sci USA* 1993;90:8103–8107.

Kavaliers M, Ossenkopp KP Opioid systems and magnetic field effects in the land snail, Cepaea nemoralis. *Bio Bull*. 1991;180:301–309.

Kloth LC, Berman JE, Sutton CH, Jeutter DC, Pilla AA, Epner ME. Effect of pulsed radio frequency stimulation on wound healing: A double-blind pilot clinical study. In *Electricity and Magnetism in Biology and Medicine*, Bersani F (ed.). Plenum, New York, 1999, pp. 875–878.

Kloth LC, Pilla AA. Electromagnetic stimulation for wound repair. In *Wound Healing: Evidence Based Management*, 4th edn., McCulloch JM, Kloth LC (eds.). Davis, Philadelphia, PA, 2010, pp. 514–544.

Konieczny V, Keebler MV, Taylor CW. Spatial organization of intracellular Ca2+ signals. *Semin Cell Dev Biol.* 2012;23:172–180.

Kowalczuk C, Yarwood G, Blackwell R et al. Absence of nonlinear responses in cells and tissues exposed to RF energy at mobile phone frequencies using a doubly resonant cavity. *Bioelectromagnetics* 2010;31:556–565.

Lauretti GR, de Oliveira R, Reis MP et al. Transdermal nitroglycerine enhances spinal sufentanil postoperative analgesia following orthopedic surgery. *Anesthesiology* 1999;90(3):734–739.

Lawrence T. The nuclear factor NF-kappaB pathway in inflammation. *Cold Spring Harb Perspect Biol.* 2009;1(6):a001651.

Lee RH, Efron D, Tantry U, Barbul A. Nitric oxide in the healing wound: A time-course study. *J Surg Res.* 2001;101:104–108.

Markov MS, Muehsam DJ, Pilla AA. Modulation of cell-free myosin phosphorylation with pulsed radio frequency electromagnetic fields. In *Charge and Field Effects in Biosystems 4*, Allen MJ, Cleary SF, Sowers AE (eds.). World Scientific, Hackensack, NJ, 1994, pp. 274–288.

McKay JC, Prato FS, Thomas AW. A literature review: The effects of magnetic field exposure on blood flow and blood vessels in the microvasculature. *Bioelectromagnetics* 2007;28:81–98.

Mellstrom B, Savignac M, Gomez-Villafuertes R et al. Ca²⁺-operated transcriptional networks: Molecular mechanisms and in vivo models. *Physiol Rev.* 2008;88:421–449.

Midura RJ, Ibiwoye MO, Powell KA, Sakai Y, Doehring T, Grabiner MD, Patterson TE, Zborowski M, Wolfman A. Pulsed electromagnetic field treatments enhance the healing of fibular osteotomies. *J Orthop Res.* 2005;23:1035–1046.

Miura M, Takayama K, Okada J. Increase in nitric oxide and cyclic GMP of rat cerebellum by radio frequency burst-type electromagnetic field radiation. *J Physiol.* 1993;461:513–524.

Mo E, Amin H, Bianco IH, Garthwaite J. Kinetics of a cellular nitric oxide/cGMP/phosphodiesterase-5 pathway. *J Biol Chem.* 2004;279:26149–26158.

Moffett J, Fray LM, Kubat NJ. Activation of endogenous opioid gene expression in human keratinocytes and fibroblasts by pulsed radiofrequency energy fields. *J Pain Res.* 2012;12:347–357.

Moffett J, Griffin NE, Ritz MC et al. Pulsed radio frequency energy field treatment of cells in culture results in increased expression of genes involved in the inflammation phase of lower extremity diabetic wound healing. *J Diabetic Foot Complications* 2010;2:57–64.

Moffett J, Kubat NJ, Griffin NE et al. Pulsed radio frequency energy field treatment of cells in culture results in increased expression of genes involved in angiogenesis and tissue remodeling during wound healing. *J Diabetic Foot Complications* 2011;3:30–39.

Muehsam DJ, Pilla AA. A Lorentz model for weak magnetic field bioeffects: Part I—Thermal noise is an essential component of AC/DC effects on bound ion trajectory. *Bioelectromagnetics* 2009a; 30:462–475.

Muehsam DJ, Pilla AA. A lorentz model for weak magnetic field bioeffects: Part II—Secondary transduction mechanisms and measures of reactivity. *Bioelectromagnetics* 2009b;30:476–488.

Palmi M, Meini A. Role of the nitric oxide/cyclic GMP/Ca²⁺ signaling pathway in the pyrogenic effect of interleukin-1beta. *Mol Neurobiol.* 2002;25:133–147.

Patel MK, Factor SM, Wang J, Jana S, Strauch B. Limited myocardial muscle necrosis model allowing for evaluation of angiogenic treatment modalities. *J Reconstr Microsurg.* 2006;22:611–615.

Pena-Philippides JC, Hagberg S, Nemoto E, Roitbak T. Effect of pulsed electromagnetic field (PEMF) on LPS-induced chronic inflammation in mice. In *Pulsed Electromagnetic Fields*, Markov MS (ed.), CRC Press, Boca Raton, FL, 2015, pp. 165–172 (this volume).

Pena-Philippides JC, Yang Y, Bragina O, Hagberg S, Nemoto E, Roitbak T. Effect of pulsed electromagnetic field (PEMF) on infarct size and inflammation after cerebral ischemia in mice. *Transl Stroke Res.* 2014b;5:491–500.

Pilla AA. Electrochemical information and energy transfer in vivo. *Proceedings 7th IECEC*, Washington, DC, 1972, pp. 761–764.

Pilla AA. Electrochemical information transfer at living cell membranes. *Ann NY Acad Sci.* 1974a;238:149–170.

Pilla AA. Mechanisms of electrochemical phenomena in tissue growth and repair. *Bioelectrochem Bioenerg.* 1974b;1:227–243.

Pilla AA. On the possibility of an electrochemical trigger for biological growth and repair processes. *Bioelectrochem Bioenerg.* 1976;3:370–373.

Pilla AA. Mechanisms and therapeutic applications of time varying and static magnetic fields. In *Biological and Medical Aspects of Electromagnetic Fields*, Barnes F and Greenebaum B (eds.). CRC Press, Boca Raton FL, 2006, pp. 351–411.

Pilla AA. Electromagnetic fields instantaneously modulate nitric oxide signaling in challenged biological systems. *Biochem Biophys Res Commun*. 2012;426:330–333.

Pilla AA. Nonthermal electromagnetic fields: From first messenger to therapeutic applications. *Electromagn Biol Med*. 2013;32:123–136.

Pilla AA, Fitzsimmons R., Muehsam DJ, Rohde C, Wu JK, Casper D. Electromagnetic fields as first messenger in biological signaling: Application to calmodulin-dependent signaling in tissue repair. *Biochem Biophys Acta* 2011;1810:1236–1245.

Pilla AA, Muehsam DJ. Non-thermal bioeffects from radio frequency signals: Evidence from basic and clinical studies, and a proposed mechanism. *Proceedings, 32nd Annual Meeting, Bioelectromagnetics Society*, Frederick, MD, June 2010.

Pilla AA, Muehsam DJ, Markov MS, Sisken BF. EMF signals and ion/ligand binding kinetics: Prediction of bioeffective waveform parameters. *Bioelectrochem Bioenerg*. 1999;48:27–34.

Pilla AA, Nasser PR, Kaufman JJ. Gap junction impedance, tissue dielectrics and thermal noise limits for electromagnetic field bioeffects. *Bioelectrochem Bioenerg*. 1994;35:63–69.

Prato FS, Carson JJ, Ossenkopp KP et al. Possible mechanisms by which extremely low frequency magnetic fields affect opioid function. *FASEB J*. 1995;9:807–814.

Rasouli J, Lekhraj R, White NM, Flamm ES, Pilla AA, Strauch B, Casper D. Attenuation of interleukin-1beta by pulsed electromagnetic fields after traumatic brain injury. *Neurosci Lett*. 2012;519:4–8.

Rawe IM, Lowenstein A, Barcelo CR, Genecov DG. Control of postoperative pain with a wearable continuously operating pulsed radiofrequency energy device: A preliminary study. *Aesthetic Plast Surg*. 2012;36:458–463.

Reale M, De Lutiis MA, Patruno A, Speranza L, Felaco M, Grilli A, Macrì MA, Comani S, Conti P, Di Luzio S. Modulation of MCP-1 and iNOS by 50-Hz sinusoidal electromagnetic field. *Nitric Oxide* 2006;15:50–57.

Ren K, Torres R. Role of interleukin-1beta during pain and inflammation. *Brain Res Rev*. 2009;60:57–64.

Rohde C, Chiang A, Adipoju O, Casper D, Pilla AA. Effects of pulsed electromagnetic fields on IL-1β and post operative pain: A double-blind, placebo-controlled pilot study in breast reduction patients. *Plast Reconstr Surg*. 2010;125:1620–1629.

Rohde C., Hardy K, Asherman J, Taylor E, Pilla AA, PEMF therapy rapidly reduces post-operative pain in TRAM flap patients. *Plast Reconstr Surg*. 2012;130(5S-1):91–92.

Rohde C, Taylor E, Pilla A. Pulsed electromagnetic fields accelerate reduction of post-operative pain and inflammation: Application to plastic and reconstructive surgical procedures. 2014, this volume.

Roland D, Ferder MS, Kothuru R, Faierman T, Strauch B. Effects of pulsed magnetic energy on a microsurgically transferred vessel. *Plast Reconstr Surg*. 2000;105:1371–1374.

Ross CL, Harrison BS. Effect of pulsed electromagnetic field on inflammatory pathway markers in RAW 264.7 murine macrophages. *J Inflamm Res*. 2013a;6:45–51.

Ross CL, Harrison BS. The use of magnetic field for the reduction of inflammation: A review of the history and therapeutic results. *Altern Ther Health Med*. 2013b;19(2):47–54.

Sluka KA, Deacon M, Stibal A et al. Spinal blockade of opioid receptors prevents the analgesia produced by TENS in arthritic rats. *J Pharmacol Exp Ther*. 1999;289:840–846.

Strauch B, Herman C, Dabb R, Ignarro LJ, Pilla AA. Evidence-based use of pulsed electromagnetic field therapy in clinical plastic surgery. *Aesthet Surg J*. 2009;29:135–143.

Strauch B, Patel MK, Navarro A, Berdischevsky M, Pilla AA. Pulsed magnetic fields accelerate wound repair in a cutaneous wound model in the rat. *Plast Reconstr Surg*. 2007;120:425–430.

Strauch B, Patel MK, Rosen D, Casper D, Pilla AA. Pulsed magnetic fields increase angiogenesis in a rat myocardial ischemia model. *Proceedings, 28th Annual Meeting*, Bioelectromagnetics Society, Frederick, MD, June 2006.

Taylor E, Hardy K, Alonso A, Pilla A, Rohde C. Pulsed electromagnetic field (PEMF) dosing regimen impacts pain control in breast reduction patients. *J Surg Res*. 2014 [Epub ahead of print].

Tepper OM, Callaghan MJ, Chang EI et al. Electromagnetic fields increase in vitro and in vivo angiogenesis through endothelial release of FGF-2. *FASEB J*. 2004;18:1231–1233.

Tsoukias NM. Nitric oxide bioavailability in the microcirculation: Insights from mathematical models. *Microcirculation* 2008;15:813–834.

Velazquez C, Praveen Rao PN, Knaus EE. Novel nonsteroidal antiinflammatory drugs possessing a nitric oxide donor diazen-1-ium-1,2-diolate moiety: Design, synthesis, biological evaluation, and nitric oxide release studies. *J Med Chem*. 2005;48:4061–4067.

Vianale G, Reale M, Amerio P, Stefanachi M, Di Luzio S, Muraro R. Extremely low frequency electromagnetic field enhances human keratinocyte cell growth and decreases proinflammatory chemokine production. *Br J Dermatol*. 2008;158:1189–1196.

Vincenzi F, Targa M, Corciulo C, Gessi S, Merighi S, Setti S, Cadossi R, Goldring MB, Borea PA, Varani K. Pulsed electromagnetic fields increased the anti-inflammatory effect of A_2A and A_3 adenosine receptors in human T/C-28a2 chondrocytes and hFOB 1.19 osteoblasts. *PLoS ONE* 2013;8(5):e65561.

Weber RV, Navarro A, Wu JK, Yu HL, Strauch B. Pulsed magnetic fields applied to a transferred arterial loop support the rat groin composite flap. *Plast Reconstr Surg.* 2005;114:1185–1189.

Werner S, Grose R. Regulation of wound healing by growth factors and cytokines. *Physiol Rev.* 2003;83:835–870.

Yen-Patton GP, Patton WF, Beer DM et al. Endothelial cell response to pulsed electromagnetic fields: Stimulation of growth rate and angiogenesis in vitro. *J Cell Physiol.* 1998;134:37–39.

Yurdagul A Jr, Chen J, Funk SD, Albert P, Kevil CG, Orr AW. Altered nitric oxide production mediates matrix-specific PAK2 and NF-κB activation by flow. *Mol Biol Cell.* 2013;24:398–408.

5 Biophysical Mechanisms for Nonthermal Microwave Effects

Igor Belyaev

CONTENTS

5.1 THERMAL AND NT BIOLOGICAL EFFECTS OF MWs

Exposures to radio-frequency (RF, 3 kHz–300 GHz) electromagnetic radiation or microwaves (MWs, 300 MHz–300 GHz) vary in many parameters: intensity (incident flux power density [PD], specific absorption rate [SAR]), wavelength/frequency, near field/far field, polarization (linear, circular), continuous wave (CW) and pulsed fields (which include variables such as pulse repetition rate, pulse width or duty cycle, pulse shape, and pulse to average power), modulation (amplitude, frequency, phase, complex), static magnetic field (SMF) and electromagnetic stray fields at the place of exposure, overall duration and intermittence of exposure (continuous, interrupted), and acute and chronic exposures. With increased absorption of energy, the so-called thermal effects of MWs are usually observed that deal with MW-induced heating. SAR is a main determinate for thermal MW effects. Several other physical parameters of exposure including carrier frequency, modulation, polarization, and coherence time of exposure have been reported to be of importance for the so-called nonthermal (NT) biological effects that are induced by MW at intensities well below measurable heating. The literature on the NT MW effects is very broad. There are four lines of evidence for the NT MW effects: (1) altered cellular responses in laboratory in vitro studies and results of chronic exposures in vivo studies (Cook et al., 2006; Grigoriev et al., 2003; Huss et al., 2007; Lai, 2005), (2) results of medical application of NT MW in the former Soviet Union countries (Betskii et al., 2000; Devyatkov et al., 1994; Pakhomov and Murphy, 2000; Sit'ko, 1989), (3) hypersensitivity to electromagnetic fields (EMFs) (Hagstrom et al., 2013; Havas and Marrongelle, 2013), and (4) epidemiological studies suggesting increased cancer risks for mobile phone users (Hardell et al., 2013a,b). It should be noted that along with detrimental effects, beneficial effects of NT MW have also been reported (Arendash et al., 2010; Dragicevic et al., 2011) supporting long-standing notion that interplay of physical MW parameters and biological parameters of the exposed object defines the value of the NT MW effect (Sit'ko, 1989). Majority of scientific community in this field is moving to better characterize and discover mechanisms for NT MW effects. International IARC expert panel has concluded: "Although it has been argued that RF radiation cannot induce physiological effects at exposure intensities that do not cause an increase in tissue temperature, it is likely that not all mechanisms of interaction between weak RF-EMF (with the various signal modulations used in wireless communications) and biological structures have been discovered or fully characterized." This paper is not intended to be a comprehensive review of the literature on the NT MW effects. In this review, we will describe recent developments in physical mechanisms.

5.2   MOST IMPORTANT PHYSICAL VARIABLES FOR THE NT MW EFFECTS

It is widely accepted that the NT MW effects depend on a variety of physical parameters. Most representative so far IARC expert review panel on RF carcinogenicity has stated that "the reproducibility of reported effects may be influenced by exposure characteristics (including SAR or power density (PD), duration of exposure, carrier frequency, type of modulation, polarization, continuous versus intermittent exposures, pulsed-field variables, and background electromagnetic environment" (IARC, 2013). These exposure characteristics were recently described in detail (Belyaev, 2010a) and are briefly summarized in the following.

5.2.1   Carrier Frequency

Resonance response of leaving cells to NT MW was reported. Slight changes in carrier frequency about 2–4 MHz resulted in disappearance of NT MW effects because of high quality of resonance-like responses (Belyaev et al., 1992a, 1996; Shcheglov et al., 1997a). Significant narrowing in resonance response with decreasing PD has been found when studying the growth rate in yeast cells (Grundler, 1992) and chromatin conformation in thymocytes of rats (Belyaev and Kravchenko, 1994). In Grundler's study, the half-with of resonance decreased from 16 MHz to 4 MHz as PD decreased from 10^{-2} W/cm^2 to 5 pW/cm^2 (Grundler, 1992). Small change in carrier frequency by 10 MHz has reproducibly resulted in cell-type-dependent appearance (915 MHz) or disappearance (905 MHz) in effects of GSM mobile phone on DNA repair foci in human cells (Belyaev et al., 2009; Markova et al., 2005, 2010).

5.2.2   Nonlinearity: Sigmoid Dependencies and Power Windows

One of the earliest observations of a threshold in response to NT MW was published by Frey (1967). In this study, a threshold of 30 µW/cm^2 was found by Frey to evoke brain stem responses to RF in cats (Frey, 1967). This value was four orders of magnitude lower than intensities needed to cause internal body temperature increase. Devyatkov and colleagues have found that a wide variety of the NT MW effects in vitro and in vivo display sigmoid dependence on intensity above certain intensity thresholds (Devyatkov, 1973).

In their pioneering study on blood–brain barrier (BBB) permeability, Oscar and Hawkins exposed rats to MW at 1.3 GHz and analyzed BBB permeability by measuring uptake of several neutral polar substances in certain areas of the brain (Oscar and Hawkins, 1977). A single, 20 min exposure to CW MW increased the uptake of D-mannitol at average PD of less than 3 mW/cm². Increased permeability was observed both immediately and 4 h after exposure, but not 24 h after exposure. After an initial rise at 0.01 mW/cm², the permeability of cerebral vessels to saccharides decreased with increasing MW power at 1 mW/cm². Thus, the effects of MW were observed within the power window of 0.01–0.4 mW/cm². The findings on *power windows* for BBB permeability have been subsequently corroborated by the group of Persson and Salford (Persson et al., 1997; Salford et al., 1994). In their recent study, the effects of GSM MW on the permeability of the BBB and signs of neuronal damage in rats were investigated using a GSM programmable mobile phone in the 900 MHz band (Eberhardt et al., 2008). The rats were exposed for 2 h at an SAR of 0.12, 1.2, 12, or 120 mW/kg. Albumin extravagation and also its uptake into neurons increased after 14 days. The occurrence of dark neurons in the rat brains increased later, after 28 days. Both effects were already seen at 0.12 mW/kg with only a slight increase, if any, at higher SAR values.

The data obtained in experiments with *Escherichia coli* cells and rat thymocytes provided new evidence for sigmoid PD dependence and suggested that, similar to the effects of extremely low frequency (ELF) EMF, MW effects may be observed within specific *intensity windows* (Belyaev et al., 1992c, 1996; Belyaev and Kravchenko, 1994; Shcheglov et al., 1997b). The most striking example of the sigmoid PD dependence was found at the resonance frequency of 51.755 GHz (Belyaev et al., 1996). When exposing *E. coli* cells at the cell density of 4×10^8 cell/mL, the effect reached saturation at the PD of 10^{-18}–10^{-17} W/cm² and did not change up to PD of 10^{-3} W/cm². This suggested that the PD dependence of MW effects at specific resonance frequencies might have intensity threshold just above the background level.

The dependence of the effect on the exposure level is usually not linear, and the reduction in the effect is much slower than expected according to the decreasing SAR (Suhhova et al., 2013).

5.2.3 Coherence Time

MW exposure of L929 fibroblasts was performed by the Litovitz's group (Litovitz et al., 1993). MW at 915 MHz modulated at 55, 60, or 65 Hz approximately doubled ornithine decarboxylase (ODC) activity after 8 h. Switching the modulation frequency from 55 to 65 Hz at coherence times of 1.0 s or less abolished enhancement, while times of 10 s or longer provided full enhancement. These results suggested that the MW coherence effects are remarkably similar to those of ELF magnetic fields observed by the same authors (Litovitz et al., 1997b).

5.2.4 Modulation

Significant numbers of in vitro and in vivo studies from diverse research groups demonstrate that the NT RF effects depend upon modulation (Blackman, 1984; Blackman et al., 1980; Byus et al., 1984, 1988; d'Ambrosio et al., 2002; Dutta et al., 1984, 1989; Gapeev et al., 1997; Huber et al., 2002, 2005; Lin-Liu and Adey, 1982; Litovitz et al., 1997a; Markkanen et al., 2004; Penafiel et al., 1997; Persson et al., 1997; Veyret et al., 1991). Comprehensive reviews on the role of modulation in appearance of the NT RF are available (Belyaev, 2010a; Blackman, 2009). Recent studies provided new evidence for dependence of the NT RF effects on modulation (Lustenberger et al., 2013; Schmid et al., 2012; Valbonesi et al., 2014).

5.2.5 POLARIZATION

Our research group have consistently reported data showing that the NT MW effects depend on polarization (Alipov et al., 1993; Belyaev et al., 1992a–c, 1993a,b; Belyaev and Kravchenko, 1994; Shcheglov et al., 1997a; Ushakov et al., 2006, 1999). Usually, only one of the two possible circular polarizations, left-handed or right-handed, was effective at each frequency window/resonance. The sign of effective circular polarization (left or right) alternated between frequency windows. The difference between effects of right and left polarizations cannot be explained by heating or by the mechanism of the so-called *hot spots* due to unequal SAR distribution. The varying effects of differently polarized MW and the inversion of effective circular polarization between resonances and after irradiation of cells with X-rays provided strong evidence for the NT mechanisms of MW effects. The data by others supported our findings on the role of polarization in the NT MW effects (Polevik, 2013; Shckorbatov et al., 2009, 2010).

5.2.6 ELECTROMAGNETIC NOISE

Litovitz and colleagues found that ELF magnetic noise inhibited the effects of MW on ODC in L929 cells (Litovitz et al., 1997a). Following studies confirmed that NT MW effects may depend on electromagnetic noise (Burch et al., 2002; Di Carlo et al., 2002; Lai, 2004; Lai and Singh, 2005; Litovitz et al., 1997a; Sun et al., 2013).

5.2.7 STATIC MAGNETIC FIELD

Dependence of MW effects on SMF during exposure has been described (Belyaev et al., 1994a; Gapeev et al., 1997, 1999; Ushakov et al., 2006). SMF is considered as one of the physical variables, which may be important for the NT MW effect (IARC, 2013).

5.2.8 E AND H FIELDS

It is particularly interesting to know whether the NT biological effects of MW are induced by the electric (E) or the magnetic (H) field component of EMF. In fact, separating E from H could be potentially important for evaluating the mechanisms underlying the effects of RF radiation on biological systems. Recent studies have indicated that the E component of the EMF may be more significant than H component in NT MW effects (Schrader et al., 2011).

5.2.9 NEAR AND FAR FIELDS

MW sources close to the body surface as in case of exposure to MP produce near-field exposures. The emitted field is magnetically coupled directly from the antenna into the tissues. The most agreed-upon definition for the near field admits that the near field is less than one wavelength (λ) from the antenna. In the near fields, the E and H fields are not necessarily perpendicular. They are often more nonpropagating in nature and are therefore called fringing fields or induction fields. At increasing distances from the source, the human body progressively takes on properties of a radio antenna, with absorption of radiated energy determined by physical dimensions of the trunk and limbs. This is a far-field exposure, defined as fully developed at 10–30 wavelengths from the source. These fields are approximately spherical waves that can in turn be approximated in a limited region of space by plane waves.

In dependence on type and location of transmitter, exposure may occur in far or near field. Kamenetskii et al. have shown that field structures with a local coupling between the time-varying E and H fields differing from the E–H coupling in regular-propagating free-space electromagnetic waves in a source-free subwavelength region of MW fields can exist (Kamenetskii et al., 2013).

To distinguish such field structures from regular EMF structures, they termed them magnetoelectric (ME) fields. They show existence of sources of MW ME near fields—the *ME particles*. These particles can be represented by small quasi-2D ferrite disks with magnetic-dipolar-oscillation spectra. The near fields originating from such particles are characterized by topologically distinctive power-flow vortices, nonzero helicity, and a torsion degree of freedom. The authors provided a theoretical analysis of properties, including chirality and helicity, of the E and H fields inside and outside of a ferrite particle with magnetic-dipolar-oscillation spectra resulting in the appearance of MW ME near fields. Based on the obtained properties of the ME near fields, the authors suggested possibilities for effective MW sensing of natural and artificial chiral structures.

In specially designed experiments, Gapeev et al. have compared effects in near and far field (Gapeev et al., 1996). These authors exposed mouse peritoneal neutrophils to RF using different types of antenna and wide-band coupling with the object both in near-field and far-field zones. They found that low-intensity RF in the near-field zone modifies the activity of peritoneal neutrophils in a frequency-dependent manner. The RF exposure inhibited luminol-dependent chemiluminescence of neutrophils activated by opsonized zymosan. This effect was not found in the far-field zone.

5.3 BIOLOGICAL SYSTEMS ARE NONLINEAR AND NONEQUILIBRIUM OPEN THERMODYNAMIC SYSTEMS

All objects, whether living or nonliving, are continuously generating EMF due to the thermal agitation of their charged particles. The EMF spectrum that is generated is described by Planck's law for the ideal case of a black body in thermal equilibrium. Thermally generated EMF has a random, noncoherent character. Historically, biology has been steeped in the biochemistry of equilibrium thermodynamics. Heating and heat exchange have been viewed as measures of essential processes in the brain and other living tissues, and intrinsic thermal energy has been seen as concept setting an immutable threshold for external MW stimulation (Adair, 2003). From many studies on the NT MW biological effects, it is clear that heating is not the basis of a broad spectrum of biological phenomena incompatible with this concept. They are consistent with processes in nonequilibrium thermodynamics. There is an emerging notion that the physical mechanisms of the NT MW effects must be based on physics of nonequilibrium and nonlinear systems (Binhi, 2002; Binhi and Rubin, 2007; Bischof, 2003; Brizhik et al., 2009a; Frey, 1974; Frohlich, 1968; Grundler et al., 1992; Kaiser, 1995; Scott, 1999; Srobar, 2009b). Physically, living biological systems are not at equilibrium (they have different energy levels than their surroundings) and are open thermodynamic systems (they can exchange energy and matter with the surroundings) (Trevors and Masson, 2011). Such systems may locally decrease entropy (increase order). Spontaneous nonlinear oscillations occur in biological systems (Kruse and Julicher, 2005; Nakahata et al., 2006). Since living systems are not in thermal equilibrium, their electromagnetic (or generally, vibration) spectrum may also deviate from the thermal spectra given by Planck's law (Cifra et al., 2011). It is generally accepted that physical theories on interactions between MW at low intensities and biological tissues must account for the facts that biological systems do not exist at equilibrium, that the dynamic nature of these systems is controlled by enzyme-mediated reactions, and that primary effects may be amplified by nonlinear biological processes (IARC, 2013).

5.4 κT PROBLEM

A number of subcellular organelles (membrane, mitotic spindle, nucleus, DNA-domain) or even a single molecule (protein, DNA) was considered as the target of interaction with NT MW. Regardless of the target, one of the major problems in explaining NT MW effects is seen in the fact that the quantum of energy of an EMF with a frequency lower than a few THz in most cases is less than the average energy of thermal noise (kT constraint), providing limits for the effects of a single quantum (Adair, 2003). However, the kT problem is based on assumption that primary absorption occurs by the atomic or molecular target under thermal equilibrium conditions in a single-quantum process. This assumption is not scientifically justified. In particular, besides atomic/molecular targets, relatively large particles with almost macroscopic magnetic moment, the so-called magnetite (Binhi, 2002), or charged macromolecular DNA–protein macrocomplexes such as nucleoids and nuclei (Matronchik and Belyaev, 2008) can be a sensitive target for NT MW effects. Moreover, even cell ensembles may be responsible for primary interaction with MW (Belyaev et al., 1994a; Shcheglov et al., 1997a). The interaction with EMF can be of multiple-quantum character and may develop in the absence of thermal equilibrium (Binhi and Rubin, 2007; Panagopoulos et al., 2000, 2002).

The kT problem has recently been challenged by Binhi and Rubin (2007). These authors stress that the notion of a *kT constraint* originates from statistical physics and is only applicable to systems near thermal equilibrium. In such systems, NT MW cannot change the mean energy of cellular structures or, more precisely, the vibration energy of their molecules stored in their degrees of freedom. Degrees of freedom characterize the ways in which a molecule or structure can move (vibrate, rotate, etc.). The energy absorbed by a degree of freedom from MW at certain frequency cannot be *stored* or accumulated at this frequency if this degree of freedom is coupled to other degrees strongly. In that case, the energy at this frequency will be redistributed to other degrees of freedom (frequencies) and will dissipate very rapidly. However, biological systems are thermody-namically far away from equilibrium, and some of their degrees of freedom are weakly coupled to others or to the surrounding heat bath. Therefore, thermalization time (time needed to redistribute energy into other degrees of freedom) may be significantly higher as compared to systems in ther-mal equilibrium. Thus, it is not surprising that NT MW can induce a significant change in energy in some degrees of freedom before dissipation or redistribution of the energy. In accordance with this concept, recent study has revealed an amazing feature of the Fröhlich's systems, namely, that with increasing total pumping, the incoming energy is deposited almost entirely in the lowest-order (fundamental) mode, even though its modal pumping rate becomes less than that of the higher-order modes (Srobar, 2009a).

The kT problem can also be addressed in the frames of stochastic resonance as a possible process involved in EMF nonlinear interactions with biosystems. In stochastic resonance, the sensitivity of a system to a weak periodic signal is actually increased when the optimal level of random noise is added (Hanggi, 2002; McDonnell and Abbott, 2009).

Given classical considerations of electromagnetic waves, another question with respect to NT MW effects has to do with the coupling of energy. The absorption of EMF energy is inefficient when the receiving antenna (molecules, cells, and their ensembles) is considerably smaller than the EMF wavelength. However, coupling and energy transfer to subcellular structures may become greater if there is resonance interaction of NT MW with vibration modes of the cellular structures (Adair, 2002). As already mentioned earlier, cell ensembles may be responsible for primary interaction with MW (Belyaev et al., 1994a; Shcheglov et al., 1997a). Coupling of EMF in the MW region to vibra-tions of cellular structures is consistent with the fundamentals of Fröhlich's theory that predicts resonant interaction of biomolecular structures with external EMFs (Frohlich, 1980). Theoretical analysis of available experimental data on the MW effects at superlow intensities concluded that these effects must be considered using a quantum-mechanical approach (Belyaev et al., 1994a; Binhi, 2002). In quantum-mechanical consideration, the vector potential that interacts with the wave function altering its phase appears to be the coupling pathway (Trukhan and Anosov, 2003, 2007).

5.5 FRÖHLICH'S THEORY

A fundamental theory suggested by Fröhlich postulated that biological systems exhibit coherent longitudinal vibrations of electrically polar structures (Frohlich, 1968). In the Fröhlich's theory, when the energy supply exceeds a critical level, the polar structure enters a condition in which a steady state of nonlinear vibration is reached and energy is stored in a highly ordered fashion.

While the Fröhlich's theory has been criticized, often by using unjustified assumptions such as that biological system is in thermal equilibrium, it has never been dismissed because it does not contradict the basic physical principles. Fröhlich's condensation of oscillators in vibration mode of lowest frequency is usually compared with Bose–Einstein condensation, superconductivity, lasing, and other unique phenomena involving macroscopic quantum coherence (Reimers et al., 2009). Fröhlich's condensates were classified into three types: weak condensates in which profound effects on chemical kinetics are possible, strong condensates in which an extremely large amount of energy is channeled into one vibrational mode, and coherent condensates in which this energy is placed in a single quantum state (Reimers et al., 2009). According to Reimers et al., from those three types, only weak condensates may have profound effects on chemical and enzyme kinetics and may be produced from biochemical energy or from RF, MW, or terahertz radiation (Reimers et al., 2009). The Fröhlich theory has predicted the existence of selective resonant EMF interactions with biosystems (Frohlich, 1970, 1975). In line with Fröhlich theory, multiple experimental studies have shown dependence of NT MW effects on frequency (Belyaev, 2010a). As far as the frequency is concerned, Fröhlich's original estimate was in the vicinity of 10^{11} Hz, but considering the diversity of oscillating species, both lower and higher frequencies may also be expected. It has been shown that energy can be condensed into Fröhlich's fundamental mode responsible for resonance interaction at the frequency of 0.1–1 GHz (Srobar, 2009b). Such condensation of metabolic energy into fundamental mode results in a significant deviation from thermal equilibrium (Srobar, 2009b).

The results of studies with different cell types indicated that narrowing of the resonance window upon decrease in PD is one of the general regularities in cell response to NT MW (Belyaev et al., 1996; Belyaev and Kravchenko, 1994; Grundler et al., 1992; Shcheglov et al., 1997a). This regularity suggests that many coupled oscillators are involved nonlinearly in the biological effects or the response of living cells to MW as has been predicted by Fröhlich (1968).

While the original Fröhlich model considered excitation in membranes, it is not mechanistically limited to any particular cellular structure. Basically, Fröhlich's condensates may form in any biological structure consisting of electrically polar oscillators, such as microtubules (important elements of the cytoskeleton), actin filaments, or DNA. Pokorny considered oscillations in microtubules that are electrically polar structures with extraordinary elastic deformability at low stress and with energy supply from hydrolysis of guanosine triphosphate (GTP) to guanosine diphosphate (GDP) (Pokorný, 2004). At least a part of the energy supplied from hydrolysis can excite vibrations. The author's calculation shows that some forms of such energy are not thermalized but are instead condensed into a pattern of oscillations. In this context, the structure of the cell's cytoskeleton, which is based on microtubules, is stated to satisfy the basic requirements for an oscillating electric field. Author considers an ionic charge layer or cylindrical envelope around the microtubule. Near the surfaces of microtubules, there may be layers of ordered water as much as 200 nm thick. The effects of the ion layer and bound water would be to reduce viscosity and thus promote the survival of excitations within the microtubule. Losses to viscosity in the inner cavity of the microtubule might be particularly low, as it is thought that all the water in this area could be ordered. Srobar has further considered damping in the Fröhlich's theory (Srobar, 2009b).

5.6 SPIN STATES, RADICAL PAIR MECHANISM

Even small EMF effects on radical concentration could potentially affect multiple biological functions (IARC, 2013). Effects of MW on spin-dependent recombination of radicals via the free radical pair mechanism (RPM) have been established in multiple studies and have recently been reviewed (Belyaev, 2010a; Georgiou, 2010). The reported effects include increased production of reactive oxygen species, enhancement of oxidative stress–related metabolic processes, an increase in DNA single-strand breaks, increased lipid peroxidation, and alterations in the activities of enzymes associated with antioxidative defense. Furthermore, many of the changes observed in RF-exposed cells were prevented by pretreatment with antioxidants and radical scavengers (Belyaev, 2010a;

IARC, 2013). Recent review has summarized studies on EMF exposure and oxidative stress in brain (Consales et al., 2012). While the data from different studies should be compared with care in view of variation in physical and biological parameters, most part of collected data have shown effects of ELF and RF EMF on oxidative stress in brain (Consales et al., 2012).

Keilmann has proposed a theory based on molecular spin (Keilmann, 1986). Spin is a kind of intrinsic degree of freedom of molecules and particles, which, in simple words, characterizes rotation of particles around their own axis. Spin is weakly coupled to other degrees of freedom and thus can be nonthermally populated. In other words, NT MW can influence the population of spin states. Triplet and radical molecules are the special target of this theory since they have nonzero spin. Hence, if these molecules belong to a reaction chain where the reactivity is dependent on spin, EMF can influence the reaction rate. This triplet mechanism may also account for frequency dependence of the NT MW effects (resonance dependency on EMF frequency) affecting biochemical reactions involving radicals.

Woodward et al. calculated magnetic RF (10–150 MHz) resonances among the electron–nuclear spin states of the radical pair, which produces a change in the yield of the product formed by recombination of singlet radical pairs (Woodward et al., 2001). If the radical pairs are sufficiently long-lived (>100 ns), a weak RF magnetic field can enhance the singlet ↔ triplet interconversion and so significantly alter the fractions of radical pairs that react via the singlet and triplet channels. This can take place even though the applied field may be much smaller than the hyperfine couplings, and all magnetic interactions are much weaker than the thermal energy per molecule, kT (Woodward et al., 2001). If spin relaxation and diffusive separation of the radicals are slow enough, effects approaching this size might even be seen for fields as weak as the Earth's (Woodward et al., 2001). In general, larger RF effects are expected if the diffusion of radicals is restricted, for example, in more viscous cell nuclei (Brocklehurst and McLauchlan, 1996). The effect of the RF field on recombination of radicals depends strongly on its frequency and orientation of ambient SMF (Henbest et al., 2004). In dependence on Zeeman and hyperfine interactions, complex response to weak static and RF fields can be anticipated in which the number and position of the resonances and the selection rules for RF-induced transitions are determined by the combined effects of the two interactions (Henbest et al., 2004). Radical pair effects were usually studied using organic radicals in isotropic solution with hyperfine interactions smaller than ~80 MHz. However, this frequency range might be considerably extended using radicals with much larger hyperfine interactions, for example, when the unpaired electron is centered on a phosphorus atom, for which resonant effects might be seen at frequencies up to ~1 GHz (Stass et al., 2000). The frequency range of resonant behavior can be also extended under conditions of restricted molecular rotation when anisotropy of the hyperfine interactions may become important (Stass et al., 2000).

Radicals were considered as the part of a system described by nonlinear equations of biochemical kinetics with bifurcations (Grundler et al., 1992). Structured organization of radicals in biological systems may have an important role for estimation of power levels and comparison with sensitivity of chemical reactions in vitro. The best known example suggesting this structural organization in biology is magnetic compass of birds, which is believed to be based on RPM and sensitive to weak magnetic fields (Ritz et al., 2009).

The long-lived radicals with lifetime of hours have been described in leaving cells (Feldermann et al., 2004; Gudkov et al., 2010; Hafer et al., 2008; Hausser, 1960; Held, 1988; Kumagai et al., 2003; Warren and Mayer, 2008). Because RF effects on conversation to triplet depend on the lifetime of radicals, long-lived radicals may underlie the NT RF effects.

5.7   ELECTROSOLITON

The other theory of EMF generation relates to the electrosoliton (Brizhik et al., 2003, 2009a,b). The soliton is a self-reinforcing solitary wave (a wave packet or pulse) that maintains its shape while it propagates. The electrosoliton is the electrical counterpart of a soliton. Electrosolitons can be viewed as moving charges that provide transport of a charge in biological systems and can be considered an important part of EMF generation in the MW frequency region. In their theoretical analysis, the authors showed that the spectrum of biological effects of EMF can be divided into two major bands. The lower-frequency band is connected with an intense form of EMF energy absorption and consequent emission of sound waves of solitons. In contrast, the higher-frequency bands induce soliton transitions to delocalized states and thus are able to destroy the soliton and disrupt the transfer of energy and information.

5.8 MW HEARING

A well-characterized physical mechanism of MW effects is the so-called MW hearing or Frey effect discovered by Alan Frey (1961). Human beings can *hear* MW energy if MWs are specifically modulated. MW hearing depends on several physical parameters and individual traits. Among physical parameters, it is mostly dependent on shape of pulses and energy in one pulse. If the PD at peak value was relatively high, 267 mW/cm^2 at the duty cycle of 0.0015 and a frequency of 1.3 GHz fitting into the range of mobile phone carrier frequencies, threshold value expressed in units of average incident PD was 0.4 mW/cm^2 (Frey, 1962). At first, MW hearing was repeatedly dismissed as an artifact until it was demonstrated in rats in a carefully controlled study by King et al. (Justesen, 1975; King et al., 1971). Frey and Messenger (Frey and Messenger, 1973) demonstrated and Guy et al. (1975) confirmed that an MW pulse with a slow rise time is ineffective in producing an auditory response; only if the rise time is short, resulting in effect in a square wave with respect to the leading edge of the envelope of radiated RF energy, does the auditory response occur. Thus, the rate of change (the first derivative) of the wave form of the pulse is a critical factor in perception. Given a thermodynamic interpretation, it would follow that information can be encoded in the energy and *communicated* to the *listener*. Dependence of MW hearing effects on type and parameters of modulation has been theoretically described and experimentally confirmed (Elder and Chou, 2003). MW hearing occurs in the frequency range from 2.4 MHz to 10 GHz (Elder and Chou, 2003). The threshold for RF hearing of pulsed 2450 MHz fields was related to an energy density of 40 mJ/cm^2 per pulse or energy absorption per pulse of 16 mJ/g, regardless of the peak power of the pulse or the pulse width (less than 32 ms); calculations showed that each pulse at this energy density would increase tissue temperature by about 5×10^{-6}°C (Elder and Chou, 2003). Evidently, any pulsed MWs will generate acoustic transients in brain tissue, which may or may not be audible.

5.9 PLASMA MEMBRANE AND IONS

The cellular membrane has long been considered as a target for interaction with MW (Adey, 1981, 1999; Desai et al., 2009; Frohlich, 1968). Kaiser suggested coupling of EMF to intracellular Ca^2 oscillations as a possible mechanism (Kaiser, 1995). In his model, nonlinear oscillators were considered, which manifested specific phenomena including synchronization, sub- and superharmonic resonances, and sensitivity of biosystems to MW at various frequencies and intensities. A cell membrane–related theory for MW effects has been proposed by Devyatkov and his coworkers (Devyatkov, 1973; Devyatkov et al., 1994). These scientists considered deformations and asymmetries of polar cellular membrane as a mechanism for generation of acoustoelectric waves whose electric component depends on deviation from the healthy state. A normal (healthy) physiological state was characterized by the lowest sensitivity to MW.

It should be stressed that uncoupled membrane channel in the presence of thermal noise and a general *white noise* intensity (in thermodynamic equilibrium) cannot be reckoned as a relevant target for NT MW effects (Astumian et al., 1995). Instead, the response of coupled nonlinear oscillators should be considered. In recent experimental study, response of neurons to applied electric field resulted to changes in somatic potential of 70 µV, below membrane potential noise levels for neurons, demonstrating that emergent properties of neuronal networks can be more sensitive than measurable effects in single neurons (Deans et al., 2007). A theoretical model by Panagopoulos et al. has analyzed effects of ELF fields and ELF-modulated RF on collective behavior of intracellular ions that affects membrane gating (Panagopoulos et al., 2000, 2002; Panagopoulos and Margaritis, 2010).

5.10 DNA, CHROMATIN, NUCLEI

The number of credible possible sources for electromagnetic signals in cells includes DNA (Liboff, 2012). The experimental data by Belyaev and coworkers provided strong evidence that DNA is a target for NT MW effects: (1) changes in supercoiling of DNA loops and helicity of DNA in leaving cells induced by DNA intercalator ethidium bromide and ionizing radiation correlated with changes in response of these cells to differently polarized NT MW (Alipov et al., 1993; Belyaev et al., 1992a–c; Ushakov et al., 1999); (2) direct prove was obtained by exposing cells with genetically increased length of genome and comparing the obtained resonance responses at differently polarized MW with theoretical predictions (Belyaev et al., 1993a). Possibility that DNA may be a target for MW effects was questioned in study by Prohovsky (Prohofsky, 2004). This study analyzed vibration modes in DNA double helix considered as stretched-out molecule under conditions of thermal equilibrium, in Debye approximation for solids. However, this is an unjustified assumption for modeling of DNA in a leaving cell for the fundamental reasons already discussed earlier. The DNA of each human cell would reach about 2 m in length if it would be stretched out, while it is condensed mostly by proteins, RNA, and ions in chromatin by a factor of about 1000 inside the nucleus of about 5 μm in diameter. Chromatin is organized at several levels of organization in discrete units, one of which represents charged domain of DNA supercoiling (DNA loop or chromatin domain). The net charge of nucleus/DNA loop and chromatin condensation ratio is dynamically changed because of dynamic interactions of DNA with other molecules and ions displaying a spectrum of natural oscillations—acoustic, mechanic, and electromagnetic (Belyaev, 2010b; Belyaev et al., 1993a; Matronchik and Belyaev, 2008). Associations of multiple chromatin binding proteins with DNA are transient and last for only a few seconds (Pliss et al., 2010). Emerging evidence indicates the presence of natural oscillations in charged chromatin (Belyaev, 2010b; Binhi et al., 2001; Matronchik and Belyaev, 2008). Natural rotation of entire nucleus in leaving cells including neurons was found (Brosig et al., 2010; Ji et al., 2007; Lang et al., 2010; Levy and Holzbaur, 2008; Park and De Boni, 1991). These effects provide clear evidence for structured dynamic organization of DNA in leaving cells and provide further basis for NT mechanism of EMF effects.

Importantly, local concentration of macromolecules in human nuclei is very high and may reach 160, 210, and 14 mg/mL for proteins, RNA, and DNA, correspondingly (Pliss et al., 2010). This concentration fits to those at which macromolecules crystallize in solutions. Whether chromatin in nuclei is organized as a liquid crystal remains to be investigated (Leforestier and Livolant, 1997). However, it is clear that DNA in the living cell cannot be considered as an aqueous solution of DNA molecules in thermodynamic equilibrium.

Recent study by Pliss et al. provided clear evidence for asymmetric oscillatory movement of the chromatin domains in human cells over short time periods (<1 s) (Pliss et al., 2013). These oscillations were energy independent and characteristic for early, mid-, and late S-phase of cell cycle. In contrast, fluorescent beads of similar dimensions (100 nM) as chromatin domains displayed random Brownian motion following microinjection into the cell nucleus. Slow nonuniform rotation of nuclei in different cell types has been documented using time-lapse microscopy (Gerashchenko et al., 2009; Ji et al., 2007; Levy and Holzbaur, 2008).

Matronchik and Belyaev have considered the effects of ELF and MW on slowly rotating chromatin domain/nucleus in the frame of their model of phase modulation (Matronchik and Belyaev, 2008). In particular, frequency dependency (of resonance type) of the ELF/MW effects and their dependence on SMF have been considered by modeling slow nonuniform rotation of the charged DNA-domain/nucleoid under combined effects of ELF/MW and SMF. This model suggested that the MW exposure results in slow nonuniform rotation of the nucleoid with angular speed that depends on Larmor frequency. The model predicts also that the NT effects of MW are dependent on carrier frequency and SMF in the place of MW exposure.

An important finding for evaluation of mechanisms is dependence of MW resonance spectra on PD during exposure (Shcheglov et al., 1997a). These data on rearrangement of resonance response of cells to MW in dependence on its PD have been interpreted in the framework of the model of electron-conformational interactions (Belyaev et al., 1996).

The Ising model has been applied for primary interaction of MW with DNA (Arinichev et al., 1993). Arinichev et al. explained resonance absorption and effect of MW by the collective long-relaxation optical vibrations within the same frequency range. The primary structure of double-stranded DNA was considered as three tied chains: two sugar–phosphate chains and one chain of complimentary nucleotides linked by hydrogen bonds. One sugar–phosphate chain was considered a 1D lattice whose unit cell consists of two compact molecular groups, that is, deoxyribose (D) and a residue of phosphoric acid (P). Given the polarity of this 1D lattice's unit cell, the optical branch in the law of dispersion corresponds to polar phonons. Radically new features in the spectrum of optical vibrations of the sugar–phosphate chain appeared when the influence of the nucleotide chain was taken into account. This chain was interpreted as a 1D order–disorder ferroelectric-like system due to the double-wall potential of hydrogen bonds between the complimentary nucleotides. The regulatory proteins and ions are bonded to certain DNA sequences in the processes of transcription, replication, repair, and recombination, producing the polarization of DNA and the ordering of proton tunneling. This kind of ordering was interpreted as a local order–disorder phase transition. To describe this phase transition, the solution for the Ising model in the transverse field was obtained with the self-consistent field method. Because of the obtained relationship between vibrations of the sugar–phosphate chain and the collective modes in the chain of complimentary nucleotides, the external MW may affect the local phase transition in hydrogen bonds of nucleotide pairs (partial unwinding of the DNA) and therefore, the process of DNA–protein interactions. The partially unwound DNA regions may be cleaved by endonucleases, resulting in DNA damage. The suggested model explained the different efficiency of right- and left-handed polarized MW (Belyaev et al., 1992a–c, 1993a,b, 1994a, 1996; Belyaev and Kravchenko, 1994; Ushakov et al., 1999). Recent experimental study by Vishnu et al. (Vishnu et al., 2011) confirmed the Arinichev's theoretical model (Arinichev et al., 1993). Vishnu et al. investigated the effect of mobile phone radiation on DNA by using fluorescence technique. Absorption spectra showed increased absorption of DNA after exposure to radiation from mobile phones with different SAR values up to 1.4 W/kg and MW frequencies, 900 MHz–3.2 GHz, which characterizes unwinding of the DNA double strand. Fluorescence intensity of dye-doped DNA solution was reduced suggesting that absorbed energy is used in unwinding the double strand of DNA after exposure to MW.

5.11 MW/ELF MECHANISMS

Some physical mechanisms consider effects ELF and MW in frame of the same physical models (Binhi, 2002; Chiabrera et al., 1991, 2000; Matronchik et al., 1996; Matronchik and Belyaev, 2005, 2008; Panagopoulos et al., 2002).

Binhi's mechanism of quantum interference considers effects of magnetic field on distribution of probabilities of coordinate of charged particle or, in other words, on wave function of charged particle (Binhi, 2003). The effect is based on shift in the phase of wave function and appearance of nonlinear interference. This mechanism has especially interesting application for rotating charged particles like ions in the pockets of rotating proteins. Based on quantum-mechanical calculations stemming from this mechanism, ELF and MW can result, under specific parameters of these fields such as intensity and frequency, in a shift in kinetics of ion binding into the protein pockets, affecting activity of this protein and related biochemistry (Binhi, 2003). Importantly, effective conditions of exposure include SMF at the place of exposure similar to the model of Matronchik and Belyaev, which is also based on combined EMF effects on phase (Belyaev et al., 1994b; Matronchik et al., 1996; Matronchik and Belyaev, 2008).

A recently suggested model considers effects of magnetic vector potential $\mathbf{A}$ (Trukhan and Anosov, 2003, 2007). Magnetic vector potential, which was initially introduced as value for calculation of magnetic field, $\mathbf{B}$ ($\mathbf{B} = \text{rot}\mathbf{A}$), was considered by analogy with the well-known Aharonov–Bohm and Josephson's effects as primary mechanism underlying the effect on charge transport (electrons in macromolecules or protons in water clusters). The effect is based on shift in the wave functions of charges and their interference. Importantly, the effect may appear at very low $\mathbf{A}$, about 3×10^{-5} T m ($\mathbf{B} \sim 50$ μT). In contrast to $\mathbf{B}$, which decreases by $\sim 1/R^3$ with a distance from elementary magnetic dipole, $\mathbf{A}$ decreases as $\sim 1/R^2$, which may provide better interaction between cells in response to EMF. This interaction between cells in response to weak EMF was experimentally established in repeated experiments both with MW (Belyaev et al., 1994a, 1996; Shcheglov et al., 2002, 1997a) and ELF exposure (Belyaev et al., 1995, 1998).

5.12 CONCLUSIONS

Significant progress has been achieved in understanding the mechanisms for NT biological effects of MWs. However, this understanding is not comprehensive. It is generally accepted that more than one physical theory may describe the same phenomena (compare, e.g., Debye model of phonons in a box and Einstein model of quantum harmonic oscillators for solids). Thus, a variety of physical mechanisms may explain interaction of biosystems with NT MW and its important characteristics such as dependence on frequency, polarization, and modulation.

ACKNOWLEDGMENTS

Financial support from the National Scholarship Programme of the Slovak Republic and the Russian Foundation for Basic Research is gratefully acknowledged.

LIST OF ABBREVIATIONS

EMF   electromagnetic field
ELF   extremely low frequency
MW   microwaves
PD   power flux density
RF   radiofrequency
SAR   specific absorption rate
SMF   static magnetic field

REFERENCES

Adair RK. 2002. Vibrational resonances in biological systems at microwave frequencies. *Biophys J* 82(3):1147–1152.

Adair RK. 2003. Biophysical limits on athermal effects of RF and microwave radiation. *Bioelectromagnetics* 24(1):39–48.

Adey WR. 1981. Tissue interactions with nonionizing electromagnetic fields. *Physiol Rev* 61(2):435–514.

Adey WR. 1999. Cell and molecular biology associated with radiation fields of mobile telephones. In: Stone WR, Ueno S (eds.) *Review of Radio Science, 1996–1999*. Oxford, U.K.: Oxford University Press, pp. 845–872.

Alipov YD, Belyaev IY, Kravchenko VG, Polunin VA, Shcheglov VS. 1993. Experimental justification for generality of resonant response of prokaryotic and eukaryotic cells to MM waves of super-low intensity. *Phys Alive* 1(1):72–80.

Arendash GW, Sanchez-Ramos J, Mori T, Mamcarz M, Lin X, Runfeldt M, Wang L et al. 2010. Electromagnetic field treatment protects against and reverses cognitive impairment in Alzheimer's disease mice. *J Alzheimers Dis* 19(1):191–210.

Arinichev AD, Belyaev IY, Samedov VV, Sit'ko SP. 1993. The physical model of determining the electromagnetic characteristic frequencies of living cells by DNA structure. *Second International Scientific Meeting "Microwaves in Medicine"*. Rome, Italy: "La Sapienza" University of Rome, pp. 305–307.

Astumian RD, Weaver JC, Adair RK. 1995. Rectification and signal averaging of weak electric fields by biological cells. *Proc Natl Acad Sci USA* 92(9):3740–3743.

Belyaev I. 2010a. Dependence of non-thermal biological effects of microwaves on physical and biological variables: Implications for reproducibility and safety standards. In: Giuliani L, Soffritti M (eds.) *European Journal of Oncology—Library Non-Thermal Effects and Mechanisms of Interaction between Electromagnetic Fields and Living Matter. An ICEMS Monograph*. Bologna, Italy: Ramazzini Institute, http://www.icems.eu/papers.htm?f =/c/a/2009/12/15/MNHJ1B49KH.DTL, pp. 187–218.

Belyaev IY. 2010b. Radiation-induced DNA repair foci: Spatio-temporal aspects of formation, application for assessment of radiosensitivity and biological dosimetry. *Mut Res* 704(1–3):132–141.

Belyaev IY, Alipov YD, Matronchik AY. 1998. Cell density dependent response of *E. coli* cells to weak ELF magnetic fields. *Bioelectromagnetics* 19(5):300–309.

Belyaev IY, Alipov YD, Matronchik AY, Radko SP. 1995. Cooperativity in *E. coli* cell response to resonance effect of weak extremely low frequency electromagnetic field. *Bioelectrochem Bioenerg* 37(2):85–90.

Belyaev IY, Alipov YD, Polunin VA, Shcheglov VS. 1993a. Evidence for dependence of resonant-frequency of millimeter-wave interaction with *Escherichia-coli* Kl2 cells on haploid genome length. *Electro Magnetobiol* 12(1):39–49.

Belyaev IY, Alipov YD, Shcheglov VS. 1992a. Chromosome DNA as a target of resonant interaction between *Escherichia-coli*-cells and low intensity millimeter waves. *Electro Magnetobiol* 11(2):97–108.

Belyaev IY, Alipov YD, Shcheglov VS, Polunin VA, Aizenberg OA. 1994a. Cooperative response of *Escherichia-coli*-cells to the resonance effect of millimeter waves at super low-intensity. *Electro Magnetobiol* 13(1):53–66.

Belyaev IY, Markova E, Hillert L, Malmgren LOG, Persson BRR. 2009. Microwaves from UMTS/GSM mobile phones induce long-lasting inhibition of 53BP1/g-H2AX DNA repair foci in human lymphocytes. *Bioelectromagnetics* 30(2):129–141.

Belyaev IY, Matronchik AY, Alipov YD. 1994b. Effect of weak static and alternating magnetic fields on the genome conformational state of *E. coli* cells: Evidence for the model of modulation of high frequency oscillations. In: Allen MJ (ed.) *Charge and Field Effects in Biosystems*. Singapore: World Scientific Publish. Co. PTE Ltd., pp. 174–184.

Belyaev IY, Shcheglov VS, Alipov YD. 1992b. Existence of selection rules on helicity during discrete transitions of the genome conformational state of *E. coli* cells exposed to low-level millimetre radiation. *Bioelectrochem Bioenerg* 27(3):405–411.

Belyaev IY, Shcheglov VS, Alipov YD. 1992c. Selection rules on helicity during discrete transitions of the genome conformational state in intact and x-rayed cells of *E. coli* in millimeter range of electromagnetic field. In: Allen MJ, Cleary SF, Sowers AE, Shillady DD (eds.) *Charge and Field Effects in Biosystems*. Basel, Switzerland: Birkhauser, pp. 115–126.

Belyaev IY, Shcheglov VS, Alipov YD, Polunin VA. 1996. Resonance effect of millimeter waves in the power range from 10(−19) to 3×10(−3) W/cm² on *Escherichia coli* cells at different concentrations. *Bioelectromagnetics* 17(4):312–321.

Belyaev IY, Shcheglov VS, Alipov YD, Radko SP. 1993b. Regularities of separate and combined effects of circularly polarized millimeter waves on *E. coli* cells at different phases of culture growth. *Bioelectrochem Bioenerg* 31(1):49–63.

Belyaev SY, Kravchenko VG. 1994. Resonance effect of low-intensity millimeter waves on the chromatin conformational state of rat thymocytes. *Z Naturforsch [C]* 49(5–6):352–358.

Betskii OV, Devyatkov ND, Kislov VV. 2000. Low intensity millimeter waves in medicine and biology. *Crit Rev Biomed Eng* 28(1–2):247–268.

Binhi VN. 2002. *Magnetobiology: Underlying Physical Problems*. San Diego, CA: Academic Press, 473p.

Binhi VN. 2003. Biological effects of non-thermal electromagnetic fields. The problem of understanding and social consequences. In: Binhi VN (ed.) *Physics of Interaction between Leaving Objects and Environment*. Moscow, Russia: MILTA, pp. 43–69.

Binhi VN, Alipov YD, Belyaev IY. 2001. Effect of static magnetic field on *E. coli* cells and individual rotations of ion-protein complexes. *Bioelectromagnetics* 22(2):79–86.

Binhi VN, Rubin AB. 2007. Magnetobiology: The kT paradox and possible solutions. *Electromagn Biol Med* 26(1):45–62.

Bischof M. 2003. Introduction to integrative biophysics. In: Popp F-A, Beloussov LV (eds.). *Integrative Biophysics*. Dordrecht, the Netherlands: Kluwer Academic Publishers, pp. 1–115.

Blackman C. 2009. Cell phone radiation: Evidence from ELF and RF studies supporting more inclusive risk identification and assessment. *Pathophysiology* 16(2–3):205–216.

Blackman CF. 1984. Sub-chapter 5.7.5 Biological effects of low frequency modulation of RF radiation. In: Elder JA, Cahill DF (eds.) *Biological Effects of Radiofrequency Radiation*: EPA-600/8-83-026F, pp. 5-88–5-92.

Blackman CF, Benane SG, Elder JA, House DE, Lampe JA, Faulk JM. 1980. Induction of calcium-ion efflux from brain tissue by radiofrequency radiation: Effect of sample number and modulation frequency on the power-density window. *Bioelectromagnetics* 1(1):35–43.

Brizhik L, Eremko A, Piette B, Zakrzewski W. 2009a. Effects of periodic electromagnetic field on charge transport in macromolecules. *Electromagn Biol Med* 28(1):15–27.

Brizhik L, Musumeci F, Scordino A, Tedesco M, Triglia A. 2003. Nonlinear dependence of the delayed luminescence yield on the intensity of irradiation in the framework of a correlated soliton model. *Phys Rev E Stat Nonlin Soft Matter Phys* 67(2 Pt. 1):021902.

Brizhik LS, Del Giudice E, Popp FA, Maric-Oehler W, Schlebusch KP. 2009b. On the dynamics of self-organization in living organisms. *Electromagn Biol Med* 28(1):28–40.

Brocklehurst B, McLauchlan KA. 1996. Free radical mechanism for the effects of environmental electromagnetic fields on biological systems. *Int J Rad Biol* 69(1):3–24.

Brosig M, Ferralli J, Gelman L, Chiquet M, Chiquet-Ehrismann R. 2010. Interfering with the connection between the nucleus and the cytoskeleton affects nuclear rotation, mechanotransduction and myogenesis. *Int J Biochem Cell Biol* 42(10):1717–1728.

Burch JB, Reif JS, Noonan CW, Ichinose T, Bachand AM, Koleber TL, Yost MG. 2002. Melatonin metabolite excretion among cellular telephone users. *Int J Rad Biol* 78(11):1029–1036.

Byus CV, Kartun K, Pieper S, Adey WR. 1988. Increased ornithine decarboxylase activity in cultured cells exposed to low energy modulated microwave fields and phorbol ester tumor promoters. *Cancer Res* 48(15):4222–4226.

Byus CV, Lundak RL, Fletcher RM, Adey WR. 1984. Alterations in protein kinase activity following exposure of cultured human lymphocytes to modulated microwave fields. *Bioelectromagnetics* 5(3):341–351.

Chiabrera A, Bianco B, Caufman JJ, Pilla AA. 1991. Quantum dynamics of ions in molecular crevices under electromagnetic exposure. In: Brighton CT, Pollack SR (eds.) *Electromagnetics in Medicine and Biology*. San Francisco, CA: San Francisco Press, pp. 21–26.

Chiabrera A, Bianco B, Moggia E, Kaufman JJ. 2000. Zeeman-Stark modeling of the RF EMF interaction with ligand binding. *Bioelectromagnetics* 21(4):312–324.

Cifra M, Fields JZ, Farhadi A. 2011. Electromagnetic cellular interactions. *Prog Biophys Mol Biol* 105(3):223–246.

Consales C, Merla C, Marino C, Benassi B. 2012. Electromagnetic fields, oxidative stress, and neurodegeneration. *Int J Cell Biol* 2012:683897.

Cook CM, Saucier DM, Thomas AW, Prato FS. 2006. Exposure to ELF magnetic and ELF-modulated radiofrequency fields: The time course of physiological and cognitive effects observed in recent studies (2001–2005). *Bioelectromagnetics* 27(8):613–627.

d'Ambrosio G, Massa R, Scarfi MR, Zeni O. 2002. Cytogenetic damage in human lymphocytes following GMSK phase modulated microwave exposure. *Bioelectromagnetics* 23(1):7–13.

Deans JK, Powell AD, Jefferys JG. 2007. Sensitivity of coherent oscillations in rat hippocampus to AC electric fields. *J Physiol* 583(Pt 2):555–565.

Desai NR, Kesari KK, Agarwal A. 2009. Pathophysiology of cell phone radiation: Oxidative stress and carcinogenesis with focus on male reproductive system. *Reprod Biol Endocrinol* 7(1):114.

Devyatkov ND. 1973. Influence of electromagnetic radiation of millimeter range on biological objects (in Russian). *Usp Fiz Nauk* 116:453–454.

Devyatkov ND, Golant MB, Betskij OV. 1994. *Peculiarities of Usage of Millimeter Waves in Biology and Medicine (in Russian)*. Moscow, Russia: IRE RAN, 164p.

Di Carlo A, White N, Guo F, Garrett P, Litovitz T. 2002. Chronic electromagnetic field exposure decreases HSP70 levels and lowers cytoprotection. *J Cell Biochem* 84(3):447–454.

Dragicevic N, Bradshaw PC, Mamcarz M, Lin X, Wang L, Cao C, Arendash GW. 2011. Long-term electromagnetic field treatment enhances brain mitochondrial function of both Alzheimer's transgenic mice and normal mice: A mechanism for electromagnetic field-induced cognitive benefit? *Neuroscience* 185: 135–149.

Dutta SK, Ghosh B, Blackman CF. 1989. Radiofrequency radiation-induced calcium ion efflux enhancement from human and other neuroblastoma cells in culture. *Bioelectromagnetics* 10(2):197–202.

Dutta SK, Subramoniam A, Ghosh B, Parshad R. 1984. Microwave radiation-induced calcium ion efflux from human neuroblastoma cells in culture. *Bioelectromagnetics* 5(1):71–78.

Eberhardt JL, Persson BR, Brun AE, Salford LG, Malmgren LO. 2008. Blood-brain barrier permeability and nerve cell damage in rat brain 14 and 28 days after exposure to microwaves from GSM mobile phones. *Electromagn Biol Med* 27(3):215–229.

Elder JA, Chou CK. 2003. Auditory response to pulsed radiofrequency energy. *Bioelectromagnetics* (Suppl 6):S162–S173.

Feldermann A, Coote ML, Stenzel MH, Davis TP, Barner-Kowollik C. 2004. Consistent experimental and theoretical evidence for long-lived intermediate radicals in living free radical polymerization. *J Am Chem Soc* 126(48):15915–15923.

Frey AH. 1961. Auditory system response to radio frequency energy. *Aerospace Med* 32:1140–1142.

Frey AH. 1962. Human auditory system response to modulated electromagnetic energy. *J Appl Physiol* 17:689–692.

Frey AH. 1967. Brain stem evoked responses associated with low-intensity pulsed UHF energy. *J Appl Physiol* 23(6):984–988.

Frey AH. 1974. Differential biologic effects of pulsed and continuous electromagnetic fields and mechanisms of effect. *Ann NY Acad Sci* 238:273–279.

Frey AH, Messenger R, Jr. 1973. Human perception of illumination with pulsed ultrahigh frequency electromagnetic energy. *Science* 181:356–358.

Frohlich H. 1968. Long-range coherence and energy storage in biological systems. *Int J Quantum Chem* 2:641–652.

Frohlich H. 1970. Long range coherence and the action of enzymes. *Nature* 228(5276):1093.

Frohlich H. 1975. The extraordinary dielectric properties of biological materials and the action of enzymes. *Proc Natl Acad Sci USA* 72(11):4211–4215.

Frohlich H. 1980. The biological effects of microwaves and related questions. In: Marton L, Marton C (eds.) *Advances in Electronics and Electron Physics.* New York, Academic Press, pp. 85–152.

Gapeev AB, Iakushina VS, Chemeris NK, Fesenko EE. 1997. Modulated extremely high frequency electromagnetic radiation of low intensity activates or inhibits respiratory burst in neutrophils depending on modulation frequency (in Russian). *Biofizika* 42(5):1125–1134.

Gapeev AB, Iakushina VS, Chemeris NK, Fesenko EE. 1999. Dependence of EHF EMF effects on the value of the static magnetic field. *Dokl Akad Nauk* 369(3):404–407.

Gapeev AB, Safronova VG, Chemeris NK, Fesenko EE. 1996. Modification of the activity of murine peritoneal neutrophils upon exposure to millimeter waves at close and far distances from the emitter. *Biofizika* 41(1):205–219.

Georgiou CD. 2010. Oxidative stress-induced biological damage by low-level EMFs: Mechanism of free radical pair electron spin- polarization and biochemical amplification. In: Giuliani L, Soffritti M (eds.) *European Journal of Oncology—Library Non-Thermal Effects and Mechanisms of Interaction between Electromagnetic Fields and Living Matter. An ICEMS Monograph.* Bologna, Italy: Ramazzini Institute, pp. 63–113.

Gerashchenko MV, Chernoivanenko IS, Moldaver MV, Minin AA. 2009. Dynein is a motor for nuclear rotation while vimentin IFs is a "brake". *Cell Biol Int* 33(10):1057–1064.

Grigoriev YG, Stepanov VS, Nikitina VN, Rubtcova NB, Shafirkin AV, Vasin AL. 2003. ISTC Report. Biological effects of radiofrequency electromagnetic fields and the radiation guidelines. Results of experiments performed in Russia/Soviet Union. Moscow, Russia: Institute of Biophysics, Ministry of Health, Russian Federation.

Grundler W. 1992. Intensity- and frequency-dependent effects of microwaves on cell growth rates. *Bioelectrochem Bioenerg* 27:361–365.

Grundler W, Kaiser F, Keilmann F, Walleczek J. 1992. Mechanisms of electromagnetic interaction with cellular systems. *Naturwissenschaften* 79(12):551–559.

Gudkov SV, Garmash SA, Shtarkman IN, Chernikov AV, Karp OE, Bruskov VI. 2010. Long-lived protein radicals induced by x-ray irradiation are the source of reactive oxygen species in aqueous medium. *Dokl Biochem Biophys* 430:1–4.

Guy AW, Chou CK, Lin JC, Christensen D. 1975. Microwave induced acoustic effects in mammalian auditory systems and physical materials. *Ann NY Acad Sci* 247:194–218.

Hafer K, Konishi T, Schiestl RH. 2008. Radiation-induced long-lived extracellular radicals do not contribute to measurement of intracellular reactive oxygen species using the dichlorofluorescein method. *Radiat Res* 169(4):469–473.

Hagstrom M, Auranen J, Ekman R. 2013. Electromagnetic hypersensitive Finns: Symptoms, perceived sources and treatments, a questionnaire study. *Pathophysiol Off J Int Soc Pathophysiol* 20(2):117–122.

Hanggi P. 2002. Stochastic resonance in biology. How noise can enhance detection of weak signals and help improve biological information processing. *Chemphyschem* 3(3):285–290.

Hardell L, Carlberg M, Hansson Mild K. 2013a. Use of mobile phones and cordless phones is associated with increased risk for glioma and acoustic neuroma. *Pathophysiology* 20(2):85–110.

Hardell L, Carlberg M, Soderqvist F, Mild KH. 2013b. Pooled analysis of case-control studies on acoustic neuroma diagnosed 1997–2003 and 2007–2009 and use of mobile and cordless phones. *Int J Oncol* 43(4):1036–1044.

Hausser KH. 1960. Detection and identification of long-lived radicals. *Radiat Res* (Suppl 2):480–96.

Havas M, Marrongelle J. 2013. Replication of heart rate variability provocation study with 2.4-GHz cordless phone confirms original findings. *Electromagn Biol Med* 32(2):253–266.

Held KD. 1988. Models for thiol protection of DNA in cells. *Pharmacol Ther* 39(1–3):123–131.

Henbest KB, Kukura P, Rodgers CT, Hore PJ, Timmel CR. 2004. Radio frequency magnetic field effects on a radical recombination reaction: A diagnostic test for the radical pair mechanism. *J Am Chem Soc* 126(26):8102–8103.

Huber R, Treyer V, Borbely AA, Schuderer J, Gottselig JM, Landolt HP, Werth E et al. 2002. Electromagnetic fields, such as those from mobile phones, alter regional cerebral blood flow and sleep and waking EEG. *J Sleep Res* 11(4):289–295.

Huber R, Treyer V, Schuderer J, Berthold T, Buck A, Kuster N, Landolt HP, Achermann P. 2005. Exposure to pulse-modulated radio frequency electromagnetic fields affects regional cerebral blood flow. *Eur J Neurosci* 21(4):1000–1006.

Huss A, Egger M, Hug K, Huwiler-Muntener K, Roosli M. 2007. Source of funding and results of studies of health effects of mobile phone use: Systematic review of experimental studies. *Environ Health Perspect* 115(1):1–4.

IARC. 2013. *IARC Monographs on the Evaluation of Carcinogenic Risks to Humans. Non-Ionizing Radiation, Part 2: Radiofrequency Electromagnetic Fields.* Lyon, France: IARC Press, pp. 1–406.

Ji JY, Lee RT, Vergnes L, Fong LG, Stewart CL, Reue K, Young SG, Zhang Q, Shanahan CM, Lammerding J. 2007. Cell nuclei spin in the absence of lamin b1. *J Biol Chem* 282(27):20015–20026.

Justesen DR. 1975. Microwaves and behavior. *Am Psychol* 392:1–11.

Kaiser F. 1995. Coherent oscillations—Their role in the interaction of weak EMF-fields with cellular systems. *Neural Netw World* 5:751–762.

Kamenetskii EO, Joffe R, Shavit R. 2013. Microwave magnetoelectric fields and their role in the matter-field interaction. *Phys Rev E Stat Nonlin Soft Matter Phys* 87(2):1–31.

Keilmann F. 1986. Triplet-selective chemistry—A possible cause of biological microwave sensitivity. *Zeitschrift Fur Naturforschung C J Biosci* 41(7–8):795–798.

King NW, Justesen DR, Clarke RL. 1971. Behavioral sensitivity to microwave radiation. *Science* 172:398–401.

Kruse K, Julicher F. 2005. Oscillations in cell biology. *Curr Opin Cell Biol* 17(1):20–26.

Kumagai J, Masui K, Itagaki Y, Shiotani M, Kodama S, Watanabe M, Miyazaki T. 2003. Long-lived mutagenic radicals induced in mammalian cells by ionizing radiation are mainly localized to proteins. *Radiat Res* 160(1):95–102.

Lai H. 2004. Interaction of microwaves and a temporally incoherent magnetic field on spatial learning in the rat. *Physiol Behav* 82(5):785–789.

Lai H. 2005. Biological effects of radiofrequency electromagnetic field. In: Wnek GE, Bowlin GL (eds.) *Encyclopedia of Biomaterials and Biomedical Engineering*. New York: Marcel Decker, pp. 1–8.

Lai H, Singh NP. 2005. Interaction of microwaves and a temporally incoherent magnetic field on single and double DNA strand breaks in rat brain cells. *Electromagn Biol Med* 24(1):23–29.

Lang C, Grava S, van den Hoorn T, Trimble R, Philippsen P, Jaspersen SL. 2010. Mobility, microtubule nucleation and structure of microtubule-organizing centers in multinucleated hyphae of Ashbya gossypii. *Mol Biol Cell* 21(1):18–28.

Leforestier A, Livolant F. 1997. Liquid crystalline ordering of nucleosome core particles under macromolecular crowding conditions: Evidence for a discotic columnar hexagonal phase. *Biophys J* 73(4):1771–1776.

Levy JR, Holzbaur EL. 2008. Dynein drives nuclear rotation during forward progression of motile fibroblasts. *J Cell Sci* 121(Pt 19):3187–3195.

Liboff AR. 2012. Electromagnetic vaccination. *Med Hypotheses* 79(3):331–333.

Lin-Liu S, Adey WR. 1982. Low frequency amplitude modulated microwave fields change calcium efflux rates from synaptosomes. *Bioelectromagnetics* 3(3):309–322.

Litovitz TA, Krause D, Penafiel M, Elson EC, Mullins JM. 1993. The role of coherence time in the effect of microwaves on ornithine decarboxylase activity. *Bioelectromagnetics* 14(5):395–403.

Litovitz TA, Penafiel LM, Farrel JM, Krause D, Meister R, Mullins JM. 1997a. Bioeffects induced by exposure to microwaves are mitigated by superposition of ELF noise. *Bioelectromagnetics* 18(6):422–430.

Litovitz TA, Penafiel M, Krause D, Zhang D, Mullins JM. 1997b. The role of temporal sensing in bioelectromagnetic effects. *Bioelectromagnetics* 18(5):388–395.

Lustenberger C, Murbach M, Durr R, Schmid MR, Kuster N, Achermann P, Huber R. 2013. Stimulation of the brain with radiofrequency electromagnetic field pulses affects sleep-dependent performance improvement. *Brain Stimul* 6(5):805–811.

Markkanen A, Penttinen P, Naarala J, Pelkonen J, Sihvonen AP, Juutilainen J. 2004. Apoptosis induced by ultraviolet radiation is enhanced by amplitude modulated radiofrequency radiation in mutant yeast cells. *Bioelectromagnetics* 25(2):127–133.

Markova E, Hillert L, Malmgren L, Persson BR, Belyaev IY. 2005. Microwaves from GSM mobile telephones affect 53BP1 and gamma-H2AX Foci in human lymphocytes from hypersensitive and healthy persons. *Environ Health Perspect* 113(9):1172–1177.

Markova E, Malmgren LOG, Belyaev IY. 2010. Microwaves from mobile phones inhibit 53BP1 focus formation in human stem cells more strongly than in differentiated cells: Possible mechanistic link to cancer risk. *Environ Health Perspect* 118(3):394–399.

Matronchik AI, Alipov ED, Beliaev II. 1996. A model of phase modulation of high frequency nucleoid oscillations in reactions of *E. coli* cells to weak static and low-frequency magnetic fields (in Russian). *Biofizika* 41(3):642–649.

Matronchik AY, Belyaev IY. 2005. Model of slow nonuniform rotation of the charged DNA domain for effects of microwaves, static and alternating magnetic fields on conformation of nucleoid in living cells. In: Pokorny J (ed.) *Fröhlich Centenary International Symposium "Coherence and Electromagnetic Fields in Biological Systems (CEFBIOS-2005)".* Prague, Czech Republic: Institute of Radio Engineering and Electronics, Academy of Sciences of the Czech Republic, pp. 63–64.

Matronchik AY, Belyaev IY. 2008. Mechanism for combined action of microwaves and static magnetic field: Slow non uniform rotation of charged nucleoid. *Electromagn Biol Med* 27(4):340–354.

McDonnell MD, Abbott D. 2009. What is stochastic resonance? Definitions, misconceptions, debates, and its relevance to biology. *PLoS Comput Biol* 5(5):e1000348.

Nakahata Y, Akashi M, Trcka D, Yasuda A, Takumi T. 2006. The in vitro real-time oscillation monitoring system identifies potential entrainment factors for circadian clocks. *BMC Mol Biol* 7:5.

Oscar KJ, Hawkins TD. 1977. Microwave alteration of the blood-brain barrier system of rats. *Brain Res* 126(2):281–293.

Pakhomov AG, Murphy MB. 2000. Comprehensive review of the research on biological effects of pulsed radiofrequency radiation in Russia and the former Soviet Union. In: Lin JC (ed.) *Advances in Electromagnetic Fields in Living System.* New York: Kluwer Academic/Plenum Publishers, pp. 265–290.

Panagopoulos DJ, Karabarbounis A, Margaritis LH. 2002. Mechanism for action of electromagnetic fields on cells. *Biochem Biophys Res Commun* 298(1):95–102.

Panagopoulos DJ, Margaritis LH. 2010. The identification of an intensity 'window' on the bioeffects of mobile telephony radiation. *Int J Radiat Biol* 86(5):358–366.

Panagopoulos DJ, Messini N, Karabarbounis A, Philippetis AL, Margaritis LH. 2000. A mechanism for action of oscillating electric fields on cells. *Biochem Biophys Res Commun* 272(3):634–640.

Park PC, De Boni U. 1991. Dynamics of nucleolar fusion in neuronal interphase nuclei in vitro: Association with nuclear rotation. *Exp Cell Res* 197(2):213–221.

Penafiel LM, Litovitz T, Krause D, Desta A, Mullins JM. 1997. Role of modulation on the effect of microwaves on ornithine decarboxylase activity in L929 cells. *Bioelectromagnetics* 18(2):132–141.

Persson BRR, Salford LG, Brun A. 1997. Blood-Brain Barrier permeability in rats exposed to electromagnetic fields used in wireless communication. *Wireless Networks* 3:455–461.

Pliss A, Kuzmin AN, Kachynski AV, Prasad PN. 2010. Nonlinear optical imaging and raman microspectrometry of the cell nucleus throughout the cell cycle. *Biophys J* 99(10):3483–3491.

Pliss A, Malyavantham KS, Bhattacharya S, Berezney R. 2013. Chromatin dynamics in living cells: Identification of oscillatory motion. *J Cell Physiol* 228(3):609–616.

Pokorný J. 2004. Excitation of vibrations in microtubules in living cells. *Bioelectrochemistry* 63(1–2):321–326.

Polevik ND. 2013. Effect of decimeter polarized electromagnetic radiation on germinating capacity of seeds. *Biophysics* (Russian Federation) 58(4):543–548.

Prohofsky EW. 2004. RF absorption involving biological macromolecules. *Bioelectromagnetics* 25(6):441–451.

Reimers JR, McKemmish LK, McKenzie RH, Mark AE, Hush NS. 2009. Weak, strong, and coherent regimes of Fröhlich condensation and their applications to terahertz medicine and quantum consciousness. *Proc Nat Acad Sci USA* 106(11):4219–4224.

Ritz T, Wiltschko R, Hore PJ, Rodgers CT, Stapput K, Thalau P, Timmel CR, Wiltschko W. 2009. Magnetic compass of birds is based on a molecule with optimal directional sensitivity. *Biophys J* 96(8):3451–3457.

Salford LG, Brun A, Sturesson K, Eberhardt JL, Persson BR. 1994. Permeability of the blood-brain barrier induced by 915 MHz electromagnetic radiation, continuous wave and modulated at 8, 16, 50, and 200 Hz. *Microsc Res Tech* 27(6):535–542.

Schmid MR, Murbach M, Lustenberger C, Maire M, Kuster N, Achermann P, Loughran SP. 2012. Sleep EEG alterations: Effects of pulsed magnetic fields versus pulse-modulated radio frequency electromagnetic fields. *J Sleep Res* 21(6):620–629.

Schrader T, Kleine-Ostmann T, Munter K, Jastrow C, Schmid E. 2011. Spindle disturbances in human-hamster hybrid (A(L)) cells induced by the electrical component of the mobile communication frequency range signal. *Bioelectromagnetics* 32(4):291–301.

Scott A. 1999. *Nonlinear Science: Emergence and Dynamics of Coherent Structures*. Oxford, U.K.: Oxford University Press.

Shcheglov VS, Alipov ED, Belyaev IY. 2002. Cell-to-cell communication in response of *E. coli* cells at different phases of growth to low-intensity microwaves. *Biochim Biophys Acta* 1572(1):101–106.

Shcheglov VS, Belyaev IY, Alipov YD, Ushakov VL. 1997a. Power-dependent rearrangement in the spectrum of resonance effect of millimeter waves on the genome conformational state of *Escherichia coli* cells. *Electro Magnetobiol* 16(1):69–82.

Shcheglov VS, Belyaev IY, Ushakov VL, Alipov YD. 1997b. Power-dependent rearrangement in the spectrum of resonance effect of millimeter waves on the genome conformational state of *E. coli* cells. *Electro Magnetobiol* 16(1):69–82.

Shckorbatov YG, Pasiuga VN, Goncharuk EI, Petrenko TP, Grabina VA, Kolchigin NN, Ivanchenko DD, Bykov VN, Dumin OM. 2010. Effects of differently polarized microwave radiation on the microscopic structure of the nuclei in human fibroblasts. *J Zhejiang Univ Sci B* 11(10):801–805.

Shckorbatov YG, Pasiuga VN, Kolchigin NN, Grabina VA, Batrakov DO, Kalashnikov VV, Ivanchenko DD, Bykov VN. 2009. The influence of differently polarised microwave radiation on chromatin in human cells. *Int J Radiat Biol* 85(4):322–329.

Sit'ko SP (ed.). 1989. *The First All-Union Symposium with International Participation "Use of Millimeter Electromagnetic Radiation in Medicine"*. Kiev, Ukraine: TRC Otklik, 298p.

Srobar F. 2009a. Occupation-dependent access to metabolic energy in Frohlich systems. *Electromagn Biol Med* 28(2):194–200.

Srobar F. 2009b. Role of non-linear interactions by the energy condensation in frohlich systems. *Neural Netw World* 19(4):361–368.

Stass DV, Woodward JR, Timmel CR, Hore PJ, McLauchlan KA. 2000. Radiofrequency magnetic field effects on chemical reaction yields. *Chem Phys Lett* 329(1–2):15–22.

Suhhova A, Bachmann M, Karai D, Lass J, Hinrikus H. 2013. Effect of microwave radiation on human EEG at two different levels of exposure. *Bioelectromagnetics* 34(4):264–274.

Sun WJ, Shen XY, Lu DB, Lu DQ, Chiang H. 2013. Superposition of an incoherent magnetic field inhibited EGF receptor clustering and phosphorylation induced by a 1.8 GHz pulse-modulated radiofrequency radiation. *Int J Rad Biol* 89(5):378–383.

Trevors JT, Masson L. 2011. Quantum microbiology. *Curr Issues Mol Biol* 13(2):43–49.

Trukhan EM, Anosov VN. 2003. Vector potential and biological activity of weak electromagnetic fields. In: Binhi VN (ed.) *Physics of Interaction between Leaving Objects and Environment*. Moscow, Russia: MILTA, pp. 71–86.

Trukhan EM, Anosov VN. 2007. Vector potential as a channel of informational effect on living objects. *Biofizika* 52(2):376–381.

Ushakov VL, Alipov ED, Shcheglov VS, Beliaev I. 2006. Peculiarities of non-thermal effects of microwaves in the frequency range of 51–52 GHz on *E. coli* cells. *Radiats Biol Radioecol* 46(6):719–728.

Ushakov VL, Shcheglov VS, Belyaev IY, Harms-Ringdahl M. 1999. Combined effects of circularly polarized microwaves and ethidium bromide on *E-coli* cells. *Electro Magnetobiol* 18(3):233–242.

Valbonesi P, Franzellitti S, Bersani F, Contin A, Fabbri E. 2014. Effects of the exposure to intermittent 1.8 GHz radio frequency electromagnetic fields on HSP70 expression and MAPK signaling pathways in PC12 cells. *Int J Radiat Biol* 90(5):382–391.

Veyret B, Bouthet C, Deschaux P, de Seze R, Geffard M, Joussot-Dubien J, le Diraison M, Moreau JM, Caristan A. 1991. Antibody responses of mice exposed to low-power microwaves under combined, pulse-and-amplitude modulation. *Bioelectromagnetics* 12(1):47–56.

Vishnu K, Nithyaja B, Pradeep C, Sujith R, Mohanan P, Nampoori VPN. 2011. Studies on the effect of mobile phone radiation on DNA using laser induced fluorescence technique. *Laser Physics* 21(11):1945–1949.

Warren JJ, Mayer JM. 2008. Surprisingly long-lived ascorbyl radicals in acetonitrile: Concerted proton-electron transfer reactions and thermochemistry. *J Am Chem Soc* 130(24):7546–7547.

Woodward JR, Timmel CR, McLauchlan KA, Hore PJ. 2001. Radio frequency magnetic field effects on electron-hole recombination. *Phys Rev Lett* 87(7):077602.

6 Engineering Problems in Bioelectromagnetics

Hubert Trzaska

CONTENTS

6.1 INTRODUCTION

Electromagnetic field (EMF) standards (primary standards) are one of the least accurate standards, as compared to those of other physical magnitudes. Their uncertainty (class), in the best cases, is around several percents, while, for instance, frequency standards represent a class well below 10^{-10}. In order to illustrate achievable classes of the EMF standards, selected results of international EMF standards comparison are shown in Figure 6.1. The comparison was headed by the National Institute of Standards and Technology (NIST) in Boulder—the most competent institution in the field.

A direct comparison of the standards is impossible. The comparison was performed using an EMF probe, which was calibrated in several labs, and then, the results were set up and published by the NIST (Kanda et al., 2000). The approach could lead to many errors and mistakes; however, the level of compared standards well illustrates the uncertainty of even the most accurate standards. By the way, the figure makes it possible to add a comment with regard to the estimation accuracy of a standard. The standard of the Technical University of Wroclaw (TUW) represents a similar accuracy level as the others. However, an optimistic, statistically supported analysis of the accuracy has led to its disagreement with the estimated mean value of the field intensity.

An accuracy of a primary EMF standard is estimated in almost ideal conditions, that is, EMF distribution in the standard is not affected by any material media close or within it, and EMF disturbances due to a calibrated device are assumed as omittable. The accuracy estimations usually take into account only the accuracy of the excitation measurement and the homogeneity of the EMF at the point (area) of concern. In contrary, in exposure systems (secondary standards), widely applied in studies related to electromagnetic compatibility (EMC) and especially in bioelectromagnetics, the accuracy is remarkably reduced by the presence of an object under test (OUT). Several sources for the measuring errors and ways to limit them are presented in this chapter.

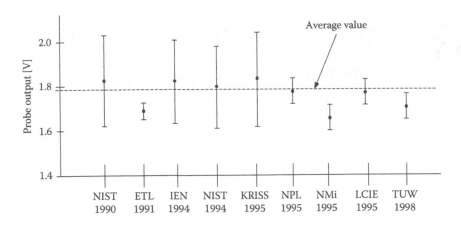

FIGURE 6.1   Results of EMF standards comparison at 100 MHz.

6.2   TEM CELL AS AN EXPOSURE SYSTEM

The TEM cell, due to its advantages, is widely applied as an exposure system in a variety of experiments. The advantages include simple construction, ease of calculation, frequency-independent field intensity inside the cell, and relative strong fields with low-power excitation. However, these advantages may be well limited when an OUT is immersed in the cell; the limitation is a function of the OUT size and its electrical properties. Two effects of OUT are discussed in the following.

6.2.1   MATCHING THE CELL

A schematic diagram of an exposure system with a TEM cell is shown in Figure 6.2.

The cell is fed from a source of output impedance Z_g through a connection cable of similar wave impedance, and it is loaded by matching the load of impedance Z_1. The wave impedance of the empty cell (Z_c) is identical, which assures full matching of the system (say primary standard conditions). As a result of an OUT present inside the cell, the wave impedance of the cell may be different in any of its cross sections. In the simplest case, the impedance is uniform and equals Z_c'. Due to the difference in the impedances, reflections exist at the input (Γ_1) and output (Γ_2) of the cell. These reflections lead to standing waves within the cell; that is, the field intensity is different in any cross section. A relation between an incident wave at the input of the cell V_1^+ and the resultant voltage at its end V_2 may be presented in the following form:

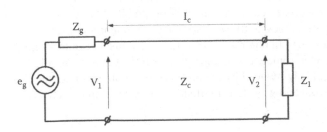

FIGURE 6.2   Schematic diagram of a TEM cell.

TABLE 6.1

Estimated Relations V_2/V_1^+ versus Γ_2

Γ_2	V_2/V_1^+
1	0–2
0.8	0.2–1.8
0.6	0.4–1.6
0.4	0.6–1.4
0.2	0.8–1.2
0	1

$$|V_2| = |V_1^+|\sqrt{1+|\Gamma_2|^2 + 2|\Gamma_2|\cos(2\beta l + \phi)} \tag{6.1}$$

where

β is the propagation constant ($\beta = 2\pi/\lambda$, where λ is the wavelength)
ϕ is the phase angle

The results of the estimations as a function of Γ_2 are given in Table 6.1.

The presented estimation needs a comment: They are performed with the most primitive assumption that the reflections exist only at the end of the cell ($\Gamma_1 = 0$). It excludes, the presence of multiple reflections and much more complex structures of the field within the cell. Propagation constant (β) was assumed to be real, while it is usually complex ($\gamma = \alpha + j\beta$). In order to find maximal differences in the voltages, the length of the cell was assumed to be resonant [cos $(\beta l + \phi) = \pm 1$]. It allowed finding values majorizing any other. However, it must be remembered that as a result of the cell loading with a lossy medium, the wavelength in the cell and electrical length of the cell are unknown. The example presented may suggest that in the experiments a cell of any length may be applied. This is not true. Practice shows that problems with the field uniformity, even within an empty cell, start to create problems for $l > \lambda/2$, what to say about the uniformity in a loaded cell of length exceeding λ (cases presented in the literature). It may be noticed that even for relatively small reflections ($\Gamma_2 = 0.2$), the difference in the field intensity reaches $\pm 20\%$, while the difference in the absorbed power in an OUT (proportional to the square of E) is twice as much. Although it is possible to match a cell, at both its ends, it does not affect the field distribution inside it.

It is suggested that in any experiment, the homogeneity of the field is experimentally checked in both the empty cell and the loaded one. Usually, because of the relatively large sizes of probes and meters available on the market, the application of meters is limited. For this purpose, a small-size probe may be constructed, a schematic diagram of which is shown in Figure 6.3.

The probe contains (say) a 1 cm dipole antenna loaded with a Ge or a Schottky barrier diode (leads of the diode may form the dipole), RC low-pass filter R $\approx$ 10 kΩ and C $\approx$ 1 nF, and a transparent lead (T_l) to a DC millivoltmeter. The lead should be of the length 10–15 cm and may be assembled using 10×100 kΩ resistors in each arm. The end of the lead is blocked with a $C_b \approx 1$ nF capacitor and then connected via a link or screened cable to an indicator. The probe may be calibrated using a cell and, which plays the role of an EMF meter. However, its main role is to assure the possibility to check the homogeneity of the field within an exposure system, and due to this, relative values are of concern. The aforementioned values are valid for frequencies above 30 MHz. For lower frequencies, R and C should be increased.

The probe may be useful for checking any exposure system. However, a warning here is necessary: The majority of DC meters (not to mention many types of EMF meters) are susceptible

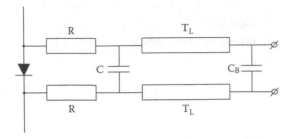

FIGURE 6.3   Schematic diagram of an E-field probe.

to EMF, while only voltage from the probe should be indicated. In order to test the system, it is enough to cross the diode and observe indications. They should not appear if the set is insensitive to EMF. Otherwise, any possibility of field penetration into the device should be found and eliminated (e.g., screening, low-band-pass filters).

6.2.2   MUTUAL COUPLINGS

The phenomenon of mutual couplings is well known from antenna theory and practice. The coupling of an antenna with a material medium affects its input impedance. Although it is not known what is meant by *an input impedance* of an OUT, the final effect is similar: the absorption of the EMF energy in the OUT depends upon the surrounding material objects. In the case of a TEM cell, the nature of the couplings is twofold:

1. Couplings between the OUT and the walls of the cell or, more accurate, couplings between the OUT and the infinite number of its mirror reflections in the cell walls
2. Couplings between OUTs, if more than one is simultaneously exposed in the cell

In order to illustrate the coupling's role, a simple example was analyzed. Two mice are placed between two metallic plates (Figure 6.4a): one mouse in the center and the other between the former and the wall of the cell. For estimation purposes, they were replaced by homogeneous, semiconducting cubes sized 5 cm (Figure 6.4b). Electric parameters of the cubes were assumed typical to living tissues ($\sigma = 0.84$ S/m and $\varepsilon_r = 80$), with a frequency of 100 MHz and E = const = 1 V/m. The results of the calculated absorber power (P) versus distance (d) between the cell walls are presented in Figure 6.5.

The results of these simple estimations confirm the existence and role of the couplings with no regard to any assumed simplifications and accuracy of the procedures applied.

The phenomenon may be neglected only in the case where a single object is exposed and its energy absorption is estimated as the energy balance at the input and the output of the cell. In any other case, the EMF energy absorption in exposed objects is a function of the number of the objects,

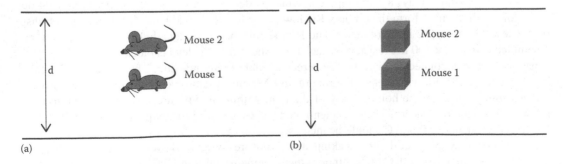

(a)   (b)

FIGURE 6.4   Two mice in a TEM cell (a) and their models (b).

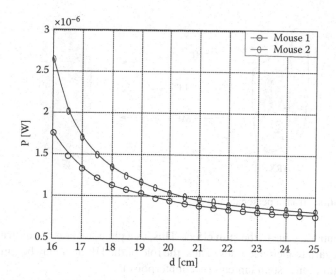

FIGURE 6.5 Results of the absorbed power estimations.

their mutual location, and their distance from the walls of the cell, and, as a result, it is different in any object. While in the cell many objects are simultaneously exposed, the difference may well exceed an order of magnitude.

Couplings may be used in the checking of the probes discussed. When the probe is moved across the walls of the cell, an increase in indications will be observed. It does not mean that the field intensity increases close to the walls. It is due to the increase in the probe's antenna input capacitance and, as a result, the increase in the sensitivity of the device. A similar effect may be observed while changing distance d and keeping E=const.

Only two factors, limiting exposure estimations accuracy and assumed as the most affecting, are briefly summarized. However, it is easy to see what scale of errors may be of concern here. The question remains as to what difference of exposure exists between real energy absorption and (as it is usually done) the one estimated by way of the measurement of energy absorbed in the cell divided by the number of OUTs.

There is a possibility to remarkably limit these errors at the expense of the number and size of the exposed OUTs and the size of the cell. The latter should be possibly small in relation to the shortest wavelength (see Section 6.2.1), while the former requires appropriate separation between exposed objects and between the objects and the walls of the cell (see Section 6.2.2).

6.3 SEVERAL OTHER EXAMPLES OF UNCERTAINTY SOURCES

6.3.1 COMPLETE EXPOSURE

Let us illustrate the problem of complete exposure on the ground of mobile communications. Energy radiated by a handheld transceiver (TRX, for instance, a cellular phone) may be divided into three frequency ranges (Figure 6.6):

- HF frequency at which the system works
- LF H-field due to current flowing from a power source to a power amplifier
- MF fields generated by control part of the device

HF fields, usually appropriately modulated, are applied in the majority of studies in bioelectromagnetics. It should be remembered that the HF EMF is of limited range due to radiation specificity

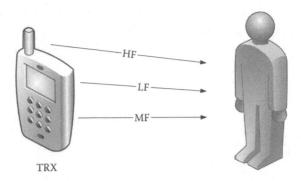

FIGURE 6.6   Radiations from a personal radio.

(near field) and attenuation due to the skin effect. This, as a phenomenon, is confirmed by any model studies—with no regard to their accuracy. Thus, the role of the HF field may be limited only to areas close to a device radiation system (antenna + counterpoise).

Two additional comments can be added to conclude the discussion of this phenomenon:

1. The casing of a device plays the role of the counterpoise, which is sufficient at frequencies above 500 MHz. At lower frequencies, the role of the counterpoise is partly played by a body of an operator as well. It results in a hand and/or mouth HF current. However, this has not been taken into account during any studies.
2. It is proposed in the literature that envelope detection may exist due to the nonlinear properties of cell walls. The author tried to measure the detection at the macroscopic level, unfortunately, without results; however, it does not mean that the phenomenon does not exist. If envelope detection does exist, the detected envelope of the carrier wave may freely propagate through the entire body of the operator.

The LF H-field, due to Biot–Savart's law, is identical to its frequency spectrum and the time dependency to the envelope of the radiated signal. However, an LF signal generated at any point on the body, occurs on the entire body. In order to illustrate the phenomenon, *transmittance* measured through the bodies of two persons (left) is presented in Figure 6.7; the measurements were performed in a *set*, as shown on the right in the figure.

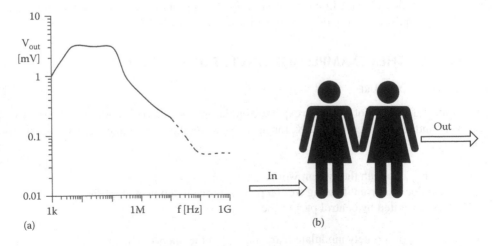

FIGURE 6.7   Results of (a) transmittance measurements and (b) measuring *set.*

A voltage (1 V) was applied to a foot of the first person through two electrodes distanced 10 cm apart. An output voltage was measured on the hand of the second person, using similar electrodes. The measurements were rather of a qualitative character (especially at higher frequencies), but they well illustrate the phenomenon.

MF noises generated by processing part of a device have a wideband character. Measurements show that the spectrum is the most intensive below several hundred of kilohertz, but it is measurable until the megahertz range. Bioeffects caused by similar spectra were intensively carried out before due to radiations of video display terminals. The display terminals were closed when CRT tubes were replaced by flat screens, and their radiation was considered nonhazardous. However, in the case of portable terminals, a direct coupling exists between a device (close to the body of an operator) and the body. As a result, his or her body is well measurable within the whole spectrum.

Two comments on this study are as follows:

1. Measurements performed by the author, on his own body and bodies of several friends, have confirmed that frequencies below several megahertz propagate freely through a body or through bodies of several persons forming a chain with their hands (see Figure 6.7).
2. LF and MF signals are widely used in electro- and magnetotherapy. Thus, it may be supposed that they present a kind of biologic activity. Neglecting their role and limiting investigations only to HF may lead, according to the literature, to disagreement between laboratory studies and epidemiological ones.

Several solutions of exposure systems that assure a possibility of complex exposure were published previously. Similar solutions may be proposed by any electronic engineer; moreover, his or her proposal would take into account local requirements and possibilities. One of the author's proposals to indicate several problems of technical nature is shown in Figure 6.8. The problems are as follows:

- An HF generator is applied to a TRX instead. Usually, a TRX cannot work with full output power in long-term experiments without damage. The generator assures stability and necessary modulation.
- The HF envelope may be taken by its detection (as shown in Figure 6.8) or directly from the modulating part of the generator (if appropriate, output exists in the generator).
- MF signals are taken directly from a TRX. Here, an output power of the device is of secondary importance; moreover, it may work as a receiver or as a transmitter. In both cases, the activity of the processing part is different. Its working system may be selected using a base station simulator. In order to protect the system against external interference, the TRX is placed in a screened chamber (for instance, in a TEM cell).

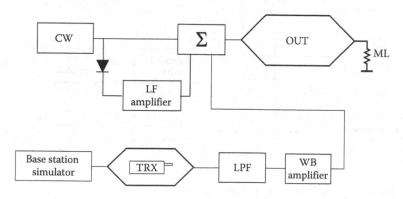

FIGURE 6.8 Block diagram of a proposed exposure system.

6.3.2 Behavior Limitations

The experiments performed in vivo are usually twofold:

1. The animals are kept stationary.
2. They can move in a system generating linearly polarized EMF.

Both the approaches can produce specific errors. The former may be acceptable while intensive exposure is applied, and the exposure effect may be assumed as dominating. In low-level effect studies, it is supposed that EMF effects may be dominated by a stress due to the animal's immobilization. In the latter case, the role of the exposure system may be played by a capacitor, a TEM cell, an antenna, or similar radiator. In all these solutions, EMF is polarized linearly. Maximal EMF energy absorption requires a parallelism of the E-field vector and the main axis of the exposed object. The method of estimation when the object is moving is still unknown.

In order to ensure the moving animal's identical exposure, with no regard to its spatial position, an exposure system that would assure three spatial components of EMF was proposed. The three spatial components would suggest a spherical polarization. Unfortunately, such a polarization is a physical fiction. A quasispherical polarization is proposed instead. The concept is based upon a circularly polarized EMF generation, in which the polarization plane rotates in space. One of possible solutions is shown in Figure 6.9. It allows generating three currents I_x, I_y, and I_z, which can be described as follows:

$$I_x = A \cos \Omega t$$

$$I_y = B \sin \Omega t \sin \omega t \qquad (6.2)$$

$$I_z = C \sin \Omega t \cos \omega t$$

where
   A, B, and C are the amplitudes
   Ω is the carrier wave frequency
   ω is the polarization plane rotation frequency

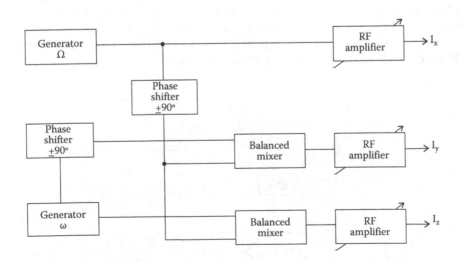

FIGURE 6.9 Block diagram of a three orthogonal current generator.

The currents excite three pairs of mutually perpendicular plates (plate capacitors), three coaxially placed and mutually perpendicular loops, or similar radiation systems. The set allows generating a quasispherical or quasielliptical polarized field with the rotation of the polarization plane. This is the simplest solution. An extension of the system would allow having any EMF polarization, from linear to quasispherical, with any spatial locations of the field vectors and frequencies and directions of their rotations.

6.4 FINAL COMMENTS

Only selected sources of errors, specific to studies in bioelectromagnetics, are presented, and their role in accuracy reduction of an experiment is discussed. Several other comments and suggestions in the field may be found in the literature (e.g., Grudzinski and Trzaska, 2013). Attention was focused upon an exposure system containing a TEM cell, as the most popular tool. The deliberations allow the following conclusions:

- There are cases in which the uncertainty of the exposure may exceed an order of magnitude.
- In these cases, the nature of the experiment is rather qualitative than quantitative.
- Thus, a statistical approach to the experiment results leads to a conclusion similar to the known joke about conjugal infidelity.
- It seems that, in these cases, a case study approach should be obligatory—to try to find maximal and minimal effects and attempt to explain reasons of the differences (including that in exposure).
- An incomplete analysis of the planned experiment conditions may lead to disagreement between the experiment results and effects observed in reality, for instance, complex exposure in the case of mobile telephony, the role of an animal behavior limitation during the experiment, and others.

Some more general comments are as follows:

- The author has never let himself to formulate any suggestions as regards to the medical side of biomedical experiments. Regardless of this they should be planned and supervised, and their results interpreted by medical doctors or/and biologists. However, a participation and supervision of people experienced in electromagnetics and their decisions and responsibility for the technical side of the experiments must be accepted.
- Every single experiment can present errors. The chapter briefly presents only several sources of errors of technical nature. Similar errors exist on the biomedical side, in calculation procedures, codes applied, and others, for instance, the accuracy of biomedical analyses, selection of experimental material, etc.
- With no regard to the errors, the results of experiments, in the majority of cases, are presented only using statistical analysis, not taking into account an uncertainty of the experiment or existing errors. Apart from the aforementioned comment to the statistics used, it is impossible to repeat such experiments in other labs or to verify them.
- This may be the cause of disagreement between the results of experiments performed in *identical* conditions in different labs.
- The aforementioned comments may be seen as being pessimistic, but this is not true. The goal of this study was to focus attention upon the existing limitations in experiment accuracy and possibilities to improve it. Correctly planned experiments, performed in acceptable conditions, have to lead to results repeatable in any lab, within the frame of estimated accuracy.

REFERENCES

Grudzinski E., Trzaska H. EMF standards and exposure systems. *SciTech* 2013.
Kanda M., Camell D., de Vreede J.P.M., Achkar J., Alexander M., Borsero M., Yaima H., Chang N.S., Trzaska H. International comparison GT/RF 86-1 electric field strengths: 27 MHz to 10 GHz. *IEEE Trans* EMC-42 nr 2/2000, 190–205.

7 Signal Design
Step by Step

James G. Seal

CONTENTS

7.1 INTRODUCTION

Contemporary magnetotherapy started developing immediately after World War II with the introduction of various magnetic and electromagnetic field generating systems. As was pointed out elsewhere (Markov, 2007; Parker and Markov, 2011), most, if not all, of the signals were designed through the intuition and skill of the engineer—not by taking into account the interaction of the signals with living tissues but by applying signals that were easily available at the time. This chapter intends to reverse this procedure, by first establishing via research the signals needed and then using engineering (applied physics) to develop a device to perform the task.

The process used to design and develop a medical device is regulated and well documented (Food and Drug Administration CGMP; ISO 13485). Indeed, in 1990, the FDA started a revision of CGMP to use the International Organization for Standardization (ISO) 9001:1994 "Quality Systems—Model for Quality Assurance in Design, Development, Production,

Installation, and Servicing." These standards are available on the Internet. As well, there are websites that one will find an appropriate consultant who can guide you through the regulations and procedures.

This chapter is not intended to provide sufficient information to guide the reader through the complete process of designing and marketing a medical device. It only intends to identify salient features of the processes when a pulsed electromagnetic field (PEMF) generating device is being developed.

The primary overall requirement for any medical device is to ultimately treat the patient with the services claimed. The regulatory approval to market the device usually comes as a result of following FDA GMP and ISO 13485 procedures.

In general, quality procedures stress the concept of *conformance to requirements* (Crosby, 2001). There has been a lot written on this subject over the last 50 years. Most of the material has been centered on the manufacturing procedures used in industry. In this chapter, *conformance to requirements* is defined as

1. State very accurately what you are going to do
2. Do what you say

Of course, there is always a possibility that your first attempt might not conform to all of the requirements. Therefore, there is no reason why you cannot go back to the design, change parts of it, and iterate your way to success. This concept is the foundation of design quality systems.

Therefore, the most important part of the whole design process is to define the requirements.

7.2   DEFINE SIGNAL NEEDED WITHIN THE TISSUE

7.2.1   DEFINE THE TARGET AND SIGNAL

The most important requirement for any PEMF generating device is the definition of the field within the tissue (Markov, 2007). In order to fulfill this requirement, it is necessary to go back to some basic principles as well as to make some general assumptions.

The target tissue should recognize the applied signal. That is, one cannot define the signal unless work has already been done, via research or experimentation, to establish the signal characteristics.

The scope of these definitions will be limited to time-varying (ac) magnetic fields only. The rate of change of the fields is limited to an equivalent bandwidth of 20 kHz. This limitation is used since it represents a majority of the modalities used in the science and clinical application. Other higher-frequency systems that create pulsed radio-frequency signals can be effectively characterized by the envelope of the pulses used. It is assumed that the carrier (i.e., ISM band at 27.125 MHz) is of such a high frequency that the tissue does not actually respond to each cycle of the carrier but to only the presence and absence of the magnetic field. Therefore, that signal is characterized by the pulse wave shape only.

In order to define the field, the target area must be defined. This must be done with intimate knowledge of the physiology of the individual being treated. It is beyond the scope of this chapter to discuss how this location is selected (see Figure 7.1).

The magnetic field can be represented as a vector. All vectors need a reference plane. Figure 7.1 shows a reference 3D origin. For the purposes of this chapter, this will be the reference origin.

Below the skin, there will be a target region, specified in three dimensions, where the required magnetic field must be created. For the example in this document, assume the target region is a cube, which is 1.25 cm on a side, and the upper edge is 1.25 cm below the skin surface (Figure 7.2).

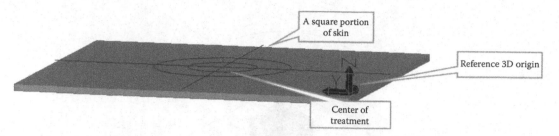

FIGURE 7.1 Patch of skin with location to be treated.

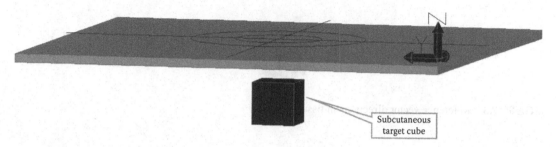

FIGURE 7.2 Target tissue region under the skin.

The magnetic field, being a vector, can be located with reference to the origin by x, y, z offsets, and the direction of the north to south poles can be represented by X, Y, Z direction. For example, if the north pole was pointed up from the skin, then the vector would have a direction of 0, 0, 1, respectively.

The three main components of the field are the strength of the field, as measured in Tesla (generally the term gauss is also popular where 1 T is 10,000 G); the direction of the field, as measured in 3D as a vector; and the rise time (or the shape) of the signal.

An applicator is a generic term used to describe the portion of the system that delivers the magnetic field to the tissue. While it is possible to build applicators with a static and dynamic field component, this chapter will discuss only the dynamic component. The static and dynamic components add as vectors. If there were a static component, then the engineering parameters would be listed as the average field strength across the target tissue measured in Tesla and the vector direction of the static field X, Y, Z (X, Y, Z format).

And for a dynamic field component, the engineering parameters would be listed as the average field strength across the target in Tesla (root mean square [RMS]) and the vector direction of the flux lines within the target X, Y, Z (X, Y, Z format).

Assuming the field is normal to the skin, the average field strength across the target can be determined by averaging the field impinging on the upper edge of the target cube with the field impinging on the lower edge of the target cube (see Figure 7.3).

The orientation and motion of the field vectors may have a significant impact on the currents induced into tissue. In the case of mutual induction of highly conductive materials (e.g., copper), field orientation is very important. Living tissue is significantly less conductive, and the conductivity varies within the tissue (Roth, 2000). Membranes have impedances different from interstitial fluids. Therefore, an ac magnetic field will induce different currents into different types of tissue (Markov, 2007). It is beyond the scope of this chapter to determine the effects of different orientations of the magnetic field, other than to ensure that the direction of the field vector is included in the requirements for the medical device.

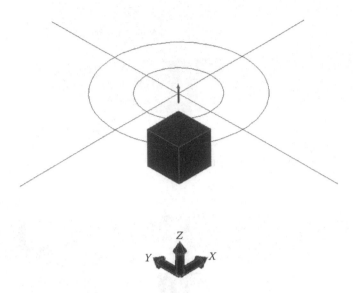

FIGURE 7.3 Reference vector directions in tissue sample.

Since the magnetic field is a vector, the signal (or time-varying amplitude) of the magnetic field can become a very complex requirement. The ac magnetic field can not only have an amplitude component that is changing over time but the direction of the vector can also be changing in direction over time.

In general, the vector direction of the magnetic field has not been well characterized in the literature. Indeed, most applicators can only deliver a linear (unidirectional) field pattern. In the case of these applicators, the *signal* merely influences the amplitude and the polarity of the magnetic field.

Some research has been accomplished when the tissue is located in a Petri dish. In these cases, there is no mention of the direction of the magnetic vector. It is difficult to define the magnetic vector with respect to the cell being studied since the cell moves around in the Petri dish medium during the delivery of the dose. It may be possible that the cells being studied react to the magnetic stimuli differently depending on the magnetic vector orientation with respect to the cell membrane. In order to determine this, the cells may have to be held in a lattice. If this phenomenon is studied, and it is determined that cells do respond differently depending on the magnetic field orientation, it would advance the science tremendously.

However, it is possible to create more complex applicators. If designed, these newer applicators use multiphased signals to control (within certain limits) the direction of the magnetic vector as well as the amplitude of the field.

During the time interval when the signal is changing the amplitude and direction of current in these applicators, the effect on the magnetic field is to vary the amplitude and direction of the magnetic vector. At any one point in space, the vector will only take on one of two possible directions. The direction of the magnetic vector is directly related to the direction of current in the applicator. It would be up to the requirements document to specify only the magnetic vector direction.

Figure 7.3 shows a possible reference vector direction for generating magnetic field requirements. In effect, the vector shown is a 0, 0, 1 field.

The next part of the requirements is to specify the signal. In the simplest case, the signal is the time-varying component of the amplitude of the magnetic field. This can be represented as a sine wave (Figure 7.4).

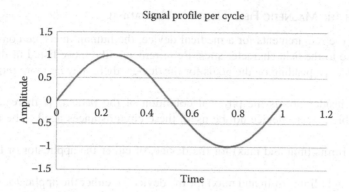

FIGURE 7.4 Sine wave profile.

The frequency and the duty cycle of the sine wave must be specified. The frequency is specified in Hz, and the duty cycle is specified in %ON of the measurement interval. For example, a treatment may require that the magnetic field be amplitude-modulated at 76 Hz with a duty cycle of 50% over a 1 s interval. In this case, the signal would look like as follows (Figure 7.5):

In this example, the engineering parameters would be

Fundamental frequency	76 Hz
Fundamental wave shape	Sine (sine, square, arbitrary)
Duty cycle	50%
Interval	0.25 s

It is also generally assumed that if the designer can create the desired signal in the electronics, then the magnetic field will be identical in wave shape. This assumption is true for simple sine waves but may not be true in the exact sense for more complex wave shapes. The designer must take into account the characteristics of the coil itself. There may be natural resonances and coil impedances that slightly alter the actual shape of the amplitude of the magnetic field. However, for a first approximation, the shape of the driving signal is the same as the shape of the magnetic field.

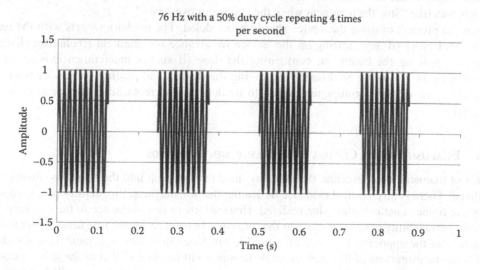

FIGURE 7.5 Sine wave with duty cycle.

7.2.2 ESTABLISH THE MAGNETIC FIELD DELIVERY MECHANISM

When specifying the requirements for a medical device, the human interface considerations rank next in importance to the field characteristics. Requirements like those listed in the following are suggested requisites. Depending on the goals for the device, there may be more requirements to be defined.

Of foremost consideration is the physical condition of the patient. If there are situations in which it is desirable that the device not be used, these circumstances need to be identified in the requirements:

Size: The size limits (min and max) for the device, of either the applicator or the controller or both.

Weight: The weight limits (min and max) for the device, of either the applicator or the controller or both.

Shape: Certain devices may be designed for use in certain locations on the body. The limitations for the shapes, surface curves, etc., are to be specified.

Fastening: The device may need to be fastened to the skin. If this is the case, then the fastening methods need to be specified in terms of what can and what cannot be done. It is good practice to develop delivery mechanisms that are convenient to the patient, as this promotes increased patient compliance. Questions such as "How is the magnetic field disrupted by minor movements of the applicator device?" should be asked, and the requirements should be defined.

Controls: How the operator turns on, sets up, and turns off the device is to be defined.

Sterility: Methods and procedures envisioned to sterilize, as well as sterility requirements, should be specified.

Power: The power source and the amount of power consumed need to be documented. If the device is portable, the battery charge times and operational times need to be specified.

Operational temperatures: Since the device will be in contact with or close to body parts, device surface temperature limits are to be specified.

Warning: If it is envisioned that the device may develop negative side effects when brought near to other metal or magnetic fields, then it is reasonable to require that the device be able to detect the incorrect environments and warn the operator appropriately.

Safety features and controls: It is reasonable to require that the device has safety measures incorporated into the design. These measures may include a forced shutdown of the device when potential safety issues are detected. Safety issues can be resolved by adding to the design requirement features like "Stop the treatment when the applicator overheats."

Then the process of *using* the device should be modeled. The modeling starts with the *opening the packaging* (if any), setting up the device in advance of treatment (including charging batteries), applying the treatment, confirming the dose (if such a mechanism exists, see the discussion in the following on dosage), sterilizing the device, and placing the device back into storage. Any nonreusable component needs to be defined and replaced in preparation for the next session.

7.2.3 ESTABLISH DOSAGE CRITERIA AND MEASUREMENT METHODS

In most of magnetic field research, the magnetic field density deep into the issue has always been calculated. There is no empirical evidence indicating that the magnetic flux density projected to be within the tissue is actually the value realized. However, there is no evidence to the contrary. The magnetic permeability of tissue is close to that of water. In most cases, the magnetic flux density at the surface of the applicator has been measured in air. Since the relative permeability of water is 0.9999, any measurement of flux density made in water will be close to that of the same measurement made in air. However, if more accurate measurements are needed, then it will be necessary to calculate (or model) the transition in permeability between the applicator and the tissue. If this cannot be done, it may be necessary to build a phantom with the magnetic probe mounted inside the solution. Then a more accurate simulation of the tissue will support a more accurate measurement of the magnetic flux density at the target.

In the literature, the concept of dosage has many variables. In general, the *dose* will consist of a magnetic field of a known strength for a known period of time. Auxiliary parameters are the frequency, duty cycle, wave shape, dB/dt (a parameter that is the first derivative of the wave shape), and the harmonic content of the induced field. Some articles describe a power density (in W/cm³) over time as a dose. There appears to be no standard for this measurement.

This is a complex issue. Different tissues react to the magnetic field differently. The direction of the magnetic vector may play in important role in the effective dose.

For practical reasons, the simplest form of a dose would be to integrate over time the instantaneous magnetic field that the tissue experiences.

In this model, the dose is frequency independent. There is a reason for this. The research in the bioeffect of the dynamic magnetic fields has clearly indicated that the bioeffect is related to frequency. There are a few *frequency and level windows* that will produce more effect than a random choice. The nonlinear effects of these windows cannot be easily modeled. Therefore, to be able to compare one dose with another, this model suggests that the dose is described with four parameters: amplitude, time, primary frequency, and bandwidth (harmonics).

The dose (for a specific tissue and bioeffect) is established by the following equation:

$$\text{Dose} = \left(A \times T \times DC\right) \times E\left(f\right) \times F(bw), \tag{7.1}$$

where

Dose is in units of Tesla hours or *Thours*. Alternate units may be gauss hours or *Ghours*

A is the flux density (T)

T is the time (h)

DC is the duty cycle (%). It is defined as (time signal is on/total time) $\times 100$

f is the frequency of the fundamental signal (Hz)

$E(f)$ is the dose efficiency factor (unitless). This function relates the bioeffect to the fundamental frequency. It must be established for every kind of bioeffect and tissue. For example, reduction of edema may be the intended goal. Soft tissue may be the target. Then as data become available for this modality, the curve for $E(f)$ can be established

bw is the bandwidth (Hz). This is the bandwidth generated by the modulation (or wave shape) of the signal. This value is derived as an attempt to include a factor for the rise and fall time of the magnetic field or dB/dt. While it is not well understood how the rise and fall times of the signal contribute to bioeffect, it would be prudent to include a parameter for them. Bandwidth is the difference from the fundamental frequency to the highest frequency at 26 dB down from the fundamental

$F(bw)$ is the dose multiplier (unitless). This function relates the bioeffect to the bandwidth of the signal. It must be established for every kind of bioeffect and tissue

In general, it is expected that a higher flux density and a longer exposure time at a selected frequency and signal will produce an increase in bioeffect. This may not always be true. More flux density may not always mean more bioeffect. Therefore, the model takes this into account by requiring that the $E(f)$ curve has defined limits. There must be an upper and lower limit of the flux density before a particular curve can be used.

At this time, the functions $E(f)$ and $F(bw)$ are not known. However, the model attempts to develop a standard method for establishing dosage. Experimentation and clinical trials are needed to validate the values used in $E(f)$ and $F(bw)$.

7.2.3.1 How to Measure the Field?

The measurement of the characteristics of the signal can be done with a simple Hall effect sensor. An example of this kind of sensor is as follows (Figure 7.6).

There are many devices on the market that can measure magnetic flux density. Some can measure only dc (static) fields. Others can measure time-varying ac fields. The ac types usually specify the lower and upper frequencies the device is calibrated for. These devices may be difficult to use since they do not discriminate which frequency they are measuring. There may be significant 50/60 cycle fields in the vicinity of the applicator being measured. These fields are being generated by current in nearby power wiring. The applicator can be generating fields at the correct frequency, but the power company fields may overcome the applicator fields. The block diagram for a *gauss meter*, which can discriminate by frequency, is shown in Figure 7.7.

The sensor produces a signal directly proportional to the vector component of the magnetic field near the sensor tip. It is necessary to place the tip orthogonal to the direction of the maximum field strength if you want to read the maximum swing in the field. By using a sampling system, it is possible to digitize the signal from the probe and integrate the area under the time waveform. After calibration of the voltage, the results will be in Tesla.

A block diagram of a typical magnetic field meter capable of measuring the ac component of the magnetic field at a specific frequency is as follows (Figure 7.7).

Hall effect sensors are available as individual devices. Most of the devices on the market are non-linear in nature and are suitable for use in magnetic switches. Other devices, like A1362LKTTN-T from Allegro MicroSystems, LLC, or AH49E from BCD Semiconductor Manufacturing Limited can be designed into custom probes. This is by no means a complete list of available Hall effect sensors.

In order to establish the dB/dt component of the dose model, it is necessary to establish the bandwidth the signal is generating. To do this, the signal from the Hall effect sensor can be delivered to a digital scope that has built-in fast Fourier transform (FFT) capability.

The following spectral graph shows a continuous sine wave signal of 25 Hz. The selection of 25 Hz was arbitrary and is only for demonstration of the measurement process.

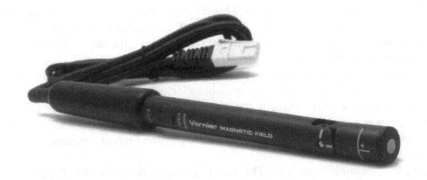

FIGURE 7.6   Off-the-shelf Hall effect sensor. (Courtesy of Vernier Software and Technology, Beaverton, OR.)

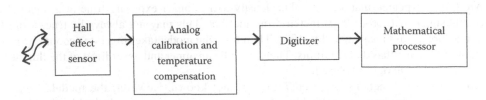

FIGURE 7.7   Typical dc/ac gauss meter.

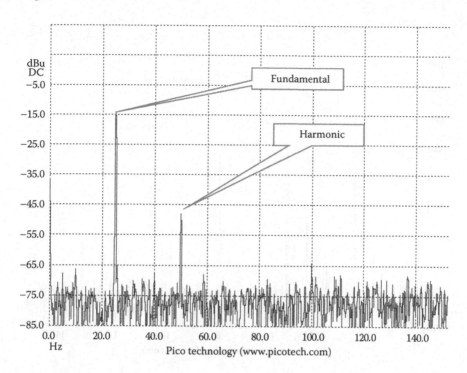

FIGURE 7.8 Spectral view of a 25 Hz sine wave.

The purpose is to determine the maximum bandwidth of the signal being generated. This is needed for the dB/dt portion of the dose model. In this example, all harmonics greater than 26 dB down from the fundamental will be discarded. In the case of Figure 7.8, the fundamental at 25 Hz peaks at −15 dBu, and there are no harmonics less than 26 dB down. Therefore, for the purposes of the dose model, this signal has zero bandwidth.

The bandwidth definition is derived from the definition of *occupied bandwidth*, which is a definition used in communication systems. By setting the level of the harmonics to be 26 dB down from the fundamental, the bandwidth measurement will ensure that 99% of the harmonic energy in the signal is included.

In the case of Figure 7.9, the square wave signal creates harmonics out to 125 Hz or the ninth harmonic. For the purposes of the model, the value of $E(bw)$ is established using the value of 125 Hz.

This bandwidth measurement is directly related to the rise time of the magnetic field. The rise time is generally the inverse of the bandwidth. Therefore, in this example, the rise time would be 1/125 or 8 ms. Work has shown that the rise time has impact on the bioeffect (Pilla, 1999).

7.3   CREATIVE STAGE: CREATING THE DESIGN OF THE DEVICE

Once all of the requirements have been identified and quantified, it is now possible to start the design process.

The process needs to start from the applicator and work backward to the generator.

The design order may look like the following:

1. Establish the requirements.
2. Establish the technology needed to create the magnetic fields that satisfy the requirements. (The coils that generate the magnetic field are sometimes called the *applicator*.)

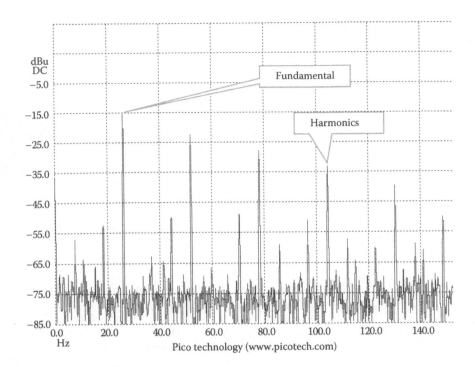

FIGURE 7.9 Harmonics in square wave.

3. Establish what packaging the applicator will need and how it is to be applied.
4. Based on the applicator's coils, establish what signals will be needed.
5. Based on the signals, establish the generator design.
6. Package the generator.
7. Review the design for *conformance to requirements*.

Doing it this way will help ensure that all of the human interface requirements are met early on in the process.

Therefore, the first step is to determine if the requirements can be met by a single coil and signal-generating device.

7.3.1 Select the Coils

Selecting the coils requires going through a list of decisions. Knowing both the required magnetic field density at the required distance from the applicator and the required fundamental frequency is sufficient to allow the designer to start calculating (or modeling) simple coils. There are two ways to estimate the magnetic field density from a coil. One way is to calculate the density based on the coil current, number of turns, and diameter of the coil. The other method is to enter the coil data into a magnetic modeling application and solve for the field density.

There is a selection of magnetic simulation applications on the market. They range in price quite dramatically. The author has discovered a free application called *FEMM 4.2* by David Meeker. This is available at the author's email address (dmeeker@ieee.org). He may be consulted at this website.

This application, FEMM 4.2, can be used to model the coil, or combination of coils, in order to estimate the magnetic field strength within the target area. An example of a single coil follows.

For this example, assume that the requirements are

Target average field strength	10 G
Target range from skin minimum	1.25 cm
Target range from skin maximum	2.5 cm
Target surface region	1.25 × 1.25 cm (Therefore, the target region is 1.25 × 1.25 × 1.25 cm³, which lies 1.25 cm below the skin.)
Target polarization	Linear
Field modulation	Amplitude only

Assume that we will start with a single coil of 5 cm diameter. Surrounding the coil will be only air. We have chosen this diameter rather loosely. The diameter of the coil is at least larger than the target area.

We need to select how many turns of wire the coil needs, what gauge of wire to use, and how much current is needed to meet the requirements. This will be done in an iterative approach.

In order to visualize the coil, Figure 7.10 shows a coil placed above the skin. The magnetic modeling application used is a 2D simulator only. Therefore, the sectional plane shown is how the coil has been cut for simulation purposes.

To begin, start with 200 turns of magnet wire, say AWG30. Enter these data into the simulator program. Assume 1 A of current is flowing through the coil. Create the mesh and run the simulation. FEMM 4.2 will calculate the magnetic flux lines shown earlier. It is then possible to calculate the magnetic flux density by creating a boundary at various distances from the coil (Figure 7.11).

Then the application will solve for the flux density. The units are in Tesla. The flux density at the center of the coil is ~0.0035 mT or 35 G. When the measurement is made close to the coil, the flux density close to the coil is higher than the flux density at the center of the coil (see Figure 7.12).

However, Figure 7.13 shows the flux density profile at 2.5 cm into the tissue. The peak value is 17 G, and it is located at the center of the coil. For radius of values less than the radius of the coil, it is reasonable to average the scalar values of the upper and lower flux densities. In this example (Figures 7.12 and 7.13), the scalar average would be approximately 26 G.

This is not a particularly accurate description of the flux density. In order to better characterize the field, vectors should be used (see Figure 7.14).

This allows you to visualize the direction of the magnetic field at any location with respect to the coil. From Figure 7.14, it is clear that the vector direction of the magnetic field is very different

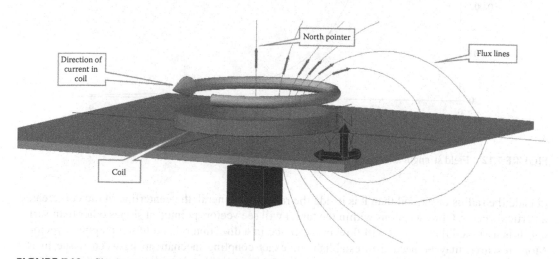

FIGURE 7.10 Single air coil placed over the skin.

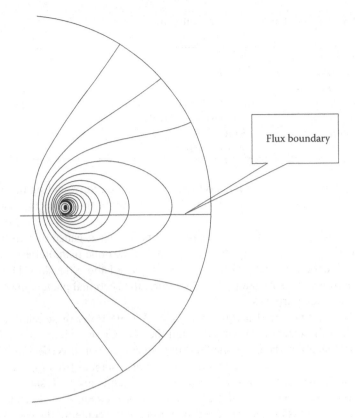

FIGURE 7.11 Selection of an example boundary for flux density calculations.

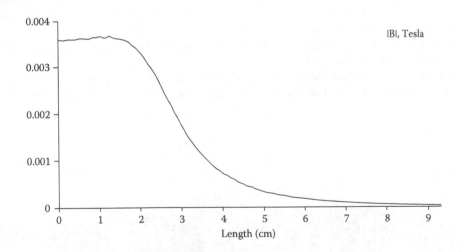

FIGURE 7.12 Field strength at 1.25 cm under the skin.

outside the radius of the coil than it is inside the radius. In general, the centerline of the coil creates a vertical vector. Other locations within the target will see vectors pointed at slopes other than vertical. It is assumed that current will flow in the tissue in a direction related to the magnetic vectors. More research may be needed to establish the exact coupling mechanism between tissue membranes and the magnetic field when the direction of the vectors is taken into account (Figure 7.14).

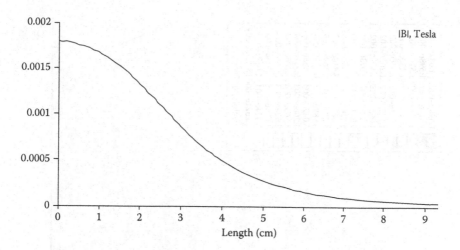

FIGURE 7.13 Flux density at 2.5 cm under the skin.

7.3.2 Based on Selected Coils, the Signal Needed to Drive the Coils Can Be Calculated

The earlier model used a current of 1 A, and it calculated that the average field density is 26 G. It is a straightforward process to scale the current based on the requirements for the field. In this example, the requirement is 0.003 T or 30 G (RMS). Scaling the current up to meet the requirement suggests that the current in the coil will be 1.15 A RMS.

The inductance of the coil may not have a large impact on the coil impedance. However, it is always a good idea to calculate the coil inductance (Figure 7.16).

Inductance of a multilayered air coil (Wheeler 1928)

$$L = \frac{31.6 \cdot r_1^2 \cdot n^2}{6 \cdot r_1 + 9 \cdot l + 10 \cdot (r_2 - r_1)} \tag{7.2}$$

where
 L is the inductance in microhenries
 r_1 is the insider coil radius in meters
 r_2 is the outside coil radius in meters
 l is the coil length in meters
 n is the number of turns total

Based on the aforementioned formula, the coil will be ~4 mH. The coil will be energized with a sine wave of 76 Hz. Using Equation 7.1, the reactive component of the impedance is

$$R = 2 * \text{pi} * H * F \tag{7.3}$$

where
 R is the reactive impedance (Ω)
 H is the inductance = 0.004 H
 F is the frequency = 76 Hz
 pi = 3.14159

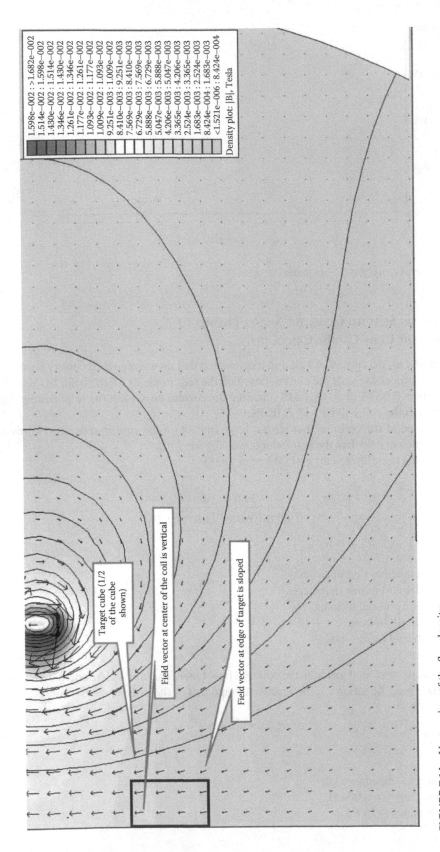

1.598e−002 : >1.682e−002
1.514e−002 : 1.598e−002
1.430e−002 : 1.514e−002
1.346e−002 : 1.430e−002
1.261e−002 : 1.346e−002
1.177e−002 : 1.261e−002
1.093e−002 : 1.177e−002
1.009e−002 : 1.093e−002
9.251e−003 : 1.009e−002
8.410e−003 : 9.251e−003
7.569e−003 : 8.410e−003
6.729e−003 : 7.569e−003
5.888e−003 : 6.729e−003
5.047e−003 : 5.888e−003
4.206e−003 : 5.047e−003
3.365e−003 : 4.206e−003
2.524e−003 : 3.365e−003
1.683e−003 : 2.524e−003
8.424e−004 : 1.683e−003
<1.521e−006 : 8.424e−004

Density plot: |B|, Tesla

Target cube (1/2 of the cube shown)

Field vector at center of the coil is vertical

Field vector at edge of target is sloped

FIGURE 7.14 Vector view of the flux density.

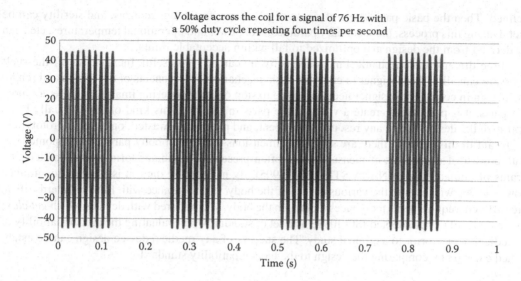

FIGURE 7.15   Example drive voltage for simple coil.

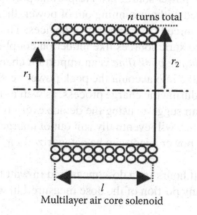

FIGURE 7.16   Inductance of a multilayered air coil (Wheeler's formula; Wheeler 1928) (where turns = 200, radius = 2.5 cm, depth = 0.25 cm, and length = 0.25 cm).

Then the reactive impedance will be ~1.9 Ω. This impedance is small relative to the resistance, and its value will be neglected.

The circuitry needed to drive these signals into the coil will vary. The operational environment will put pressure to select specific designs. For example, if batteries are used as a power source, then getting such a high voltage on the output signal will require either a dc–dc converter to generate the high voltage or output transformers. If the generator uses house power, then an appropriate power supply can be selected early in the design cycle.

7.4   DESIGN THE PACKAGING

Packaging can make or break an otherwise good electrical design. This is where the user interface becomes the priority in the design decision making.

From Section 7.2.2, it can be seen that most of the requirements are associated with who and how the device will be used. Of foremost importance, the physical condition of the patient needs to be defined. Then the basic parameters of size, weight, shape, fastening, controls, and sterility can be met. During this process, the secondary parameters of power and operational temperatures, etc., can be derived from the design and optimized to fall within acceptable limits.

Three-dimensional mechanical modeling software can be very useful. Inexpensive off-the-shelf software can allow the designer to enter the basic mechanical components of the design and render the design in color. The designer needs not have to stop only at rendering images. If there are moving parts, it is possible to create a video of the parts in motion. This kind of effort usually helps optimize the design before any resources are spent, and potentially wasted, on making prototypes.

To aid in this process, there are standard dimensions for human body parts. Anthropometrical data consist of collections of measurements, often presented in tabular format or annotated diagrams of human figures (NASA-STD-3000, 1995). By using these data, it is possible to simulate how a device will fit onto the various curves of the body. In accordance with FDA standards, there are different requirements for devices that touch the body, as compared with devices which *get close* to the body. FDA has adopted ISO 10993 as a set of standards for evaluating the biocompatibility of a medical device prior to a clinical study. The success of any medical device design can be established early on by comparing the design to the biocompatibility standards.

7.4.1 SELECT THE POWER SOURCE

As part of the design process, the power source is an important part of the system. Ultimately, if the magnetic dose cannot be delivered without running out of power, then the device will not have met its requirements. Power can be sourced from an array of devices. These can range from the conventional wall outlet, to batteries, to exotic sources like induction coupling.

In the case of batteries, the *charge back time* is an important characteristic. The selection of the value of this parameter should take into account the peak power needed to charge the batteries and its loading of the power sources during the charge process, as well as the typical lifestyle of the user. For example, if the treatment plan suggests using the device every 6 h, and the charge back time is longer than that interval, the device will eventually run out of charge.

Regardless of the source, the power needs can be categorized into two parameters:

1. Total power needed in watt hours for a dose measured in watt hours
2. Peak power need during any portion of the dose measured in watts

Note: Along with the power consideration comes the waste heat issue. If the device creates heat in either the generator electronics or the applicators, then this heat must be conducted out of the device. There are limits to the allowable temperature increase that devices can cause to the patient. There are also safety issues when hot surfaces exist on any part of the device. Therefore, the design needs to be as efficient as possible so that these issues can be avoided.

7.4.2 SIMULATE THE HUMAN INTERACTION PROCEDURES

There is a significant amount of literature on the human interface with devices. These designs usually need a lot of thought on how they will be used. The misuse of a device can expose a fault that eventually leads to the downfall of the device in the marketplace. Therefore, simulating the device usage and/or creating a mock-up of the device can be very helpful in determining if the design has unintended characteristics.

Many designs that require human interaction fail because the control panel is too complicated. Others fail because the panels are too simple and cannot address the specific needs of the patient. Also, the status indicators may be confusing or difficult to see.

In the case of the applicator, the method of placing it at the treatment site may also require a novel concept. How the applicator is held in place and how the cable travels to the generator need to be established for each proposed application. If the system has no cables, or if other methods to create the magnetic fields are employed, there will always be a need to *dry run* how each treatment is to be accomplished.

7.5   DEVELOP THE COST OF THE MODEL

Along with the development of the mechanical model, the cost to manufacture the device should be simultaneously developed. The cost data are considered to be an input to the iterative optimization of the design. The cost involves the following:

- Basic cost of purchased parts
- Cost of specialized parts
- Cost to inventory the parts
- Cost to assemble
- Cost to test subassemblies
- Cost to test the finished assembly
- Cost to document the device history
- Cost for after-market support

As the cost model becomes more complex, it may also contain the following:

- Cost of borrowing money
- Cost to market the device
- Cost to distribute the device, including the cost of demo units
- Cost of training distributors, etc.
- Cost of warranty support, which involves a whole new set of costs associated with warranty repair activities

This chapter will not be able to discuss any of the complex cost parameters. However, when designing any device, there are some general good guidelines a designer should keep in mind. These are the following:

- *Reduce the total number of parts*: The reduction of the number of parts in a product is probably the best opportunity for reducing manufacturing costs. Fewer parts imply fewer purchases, less inventory, less handling, less processing time, less development time, less equipment, less engineering time, less assembly difficulty, less service inspection, and less testing, etc. In general, fewer parts reduce the level of intensity of all activities related to the product during its entire life. Some approaches to part-count reduction are based on the use of one-piece structures and selection of manufacturing processes such as injection molding, extrusion, precision castings, and powder metallurgy, among others.
- *Use standard components*: Standard components are less expensive than custom-made items. The high availability of these components reduces product lead times. Also, their reliability factors are well ascertained.
- *Design parts for multiuse*: In a manufacturing firm, different products can share parts that have been designed for multiuse. These parts can have the same or different functions when used in different products. In order to do this, it is necessary to identify the parts that are suitable for multiuse. For example, all the parts (purchased or made) can be sorted into two groups: the first containing all the parts that are used commonly in all products. Then, part families are created by defining categories of similar parts in each group. The goal is to minimize the number of categories, the variations within the categories, and the number of design features within each variation. The result is a set of standard part families from

which multiuse parts are created. After organizing all the parts into part families, the manufacturing processes are standardized for each part family. The production of a specific part belonging to a given part family would follow the manufacturing routing that has been set up for its family, skipping the operations that are not required for it. Furthermore, in design changes to existing products and especially in new product designs, the standard multiuse components should be used.

- *Design for ease of fabrication*: Select the optimum combination between the material and fabrication process to minimize the overall manufacturing cost. In general, final operations such as painting, polishing, and machine finishing should be avoided. Excessive tolerance, surface-finish requirements, and so on are commonly found problems that result in higher than necessary production cost.

- *Avoid separate fasteners*: The use of fasteners increases the cost of manufacturing a part due to the handling and feeding operations that must be performed. Besides the high cost of the equipment required for them, these operations are not 100% successful, so they contribute to reducing the overall manufacturing efficiency. In general, fasteners should be avoided and replaced, for example, by using tabs or snap fits. If fasteners have to be used, then some guides should be followed for selecting them. Minimize the number, size, and variation used; also, utilize standard components whenever possible. Avoid screws that are too long, or too short, separate washers, tapped holes, and round heads and flatheads (not good for vacuum pickup). Self-tapping and chamfered screws are preferred because they improve placement success. Screws with vertical side heads should be selected for vacuum pickup.

- *Minimize assembly directions*: All parts should be assembled from one direction. If possible, the best way to add parts is from above, in a vertical direction, parallel to the gravitational direction (downward). In this way, the effects of gravity help the assembly process, contrary to having to compensate for its effect when other directions are chosen.

- *Maximize compliance*: Errors can occur during insertion operations due to variations in part dimensions or in the accuracy of the positioning device used. This faulty behavior can cause damage to the part and/or to the equipment. For this reason, it is necessary to include compliance in the part design and in the assembly process. Examples of part built-in compliance features include tapers or chamfers and moderate radius sizes to facilitate insertion and nonfunctional external elements to help detect hidden features. For the assembly process, selection of a rigid-base part, tactile sensing capabilities, and vision systems are examples of compliance (Chang et al., 2005).

7.6 VET THE DESIGN VIA PRESENTATIONS

Once the design is completed, it is sometimes valuable to create a presentation describing the design and all of its features. In doing this, the design team can describe the design and obtain immediate input/feedback. In any presentation, reviewing of all of the requirements and how each requirement is addressed can become a good format. However, extra information should also be included, such as

- Who the product will help (population subgroups who will benefit from the treatment)
- How the product will help (scientific basis for the treatment)
- What kinds of results or observations should a practitioner expect
- What the device will look like
- How the device is administered
- Recommended skill levels of the individual who is operating the device
- Any cautionary issues for the practitioner or the patient

- SWAP (size, weight, and power)
- Sterility considerations
- Reuse of the system
- Storage of the device
- Battery management/replacement
- Shelf life of the device
- Overall device life
- Operator controls, device programming, and treatment options
- User parameter control limits (how the device is designed so that the user cannot create a hazardous condition)
- Safety controls built in
- Built in test (BIT) and/or calibration issues
- Emissions from, or susceptibility to, electromagnetic interference
- Restrictions on where or when the device can be used (if any)
- Export restrictions (if any)
- Regulatory conformance in the United States
- International regulatory compliance
- Any technical risk

This is an important step in the product development. There are medical device professionals who can steer the design clear of expensive development errors. These errors can be caused by incorrect assumptions, as well as incorrect solutions. It is recommended that before the development goes too far past this stage, input from manufacturers and other professionals should be accepted.

7.7 CREATE A PROTOTYPE

The designer has many options when it comes to development of medical devices. The newest capability in the marketplace is 3D printing. While 3D printed materials may not have all of the characteristics the final device may need, it is a quick and inexpensive way to *get a device prototype into your hands*. The electronics needed to power the device are a different matter. Conventional methodology is to

- Develop a schematic of the circuits needed
- Select the needed components
- Determine end of life for the components selected
- Estimate any firmware or software needs

Then, once the circuit design has been reviewed and approved, the printed circuit board can be laid out.

From here, the path taken by the designer may vary depending on the actual device. It may be prudent to initiate a software development team to begin developing code for the device simultaneously with the hardware development. Or it may be efficient to wait until the prototype printed circuit boards are populated and given to the software team. In any case, ultimately, there will be a functional prototype.

7.8 VERIFY CONFORMANCE TO REQUIREMENTS

Once a functional prototype has been made, it is always a good plan to verify the prototype capabilities against the requirements. This is conveniently done with the classical *conformance to requirements* matrix. An example of one is as follows (Figure 7.17).

Item	Requirement	Requirement Flowdown	Verification Method	Acceptance Criteria	Test Record Document	Pass/Fail Comments

FIGURE 7.17   Conformance to requirements matrix.

The definitions of the column headers are as follows:

- *Item*: This is an item number used to track the requirement in all associated documents.
- *Requirement*: This is the description of the requirement.
- *Requirement flow down*: This is the name/title of the requirement document from which this requirement is derived. For example, a regulation may be the source for a particular requirement.
- *Verification method*: This is a procedure that is used to measure/validate that the device conforms to the requirement.
- *Acceptance criteria*: This is the actual values/behavior the device must exhibit in order to pass the requirement.
- *Test record document*: This is the document that contains the data collected during the procedure.
- *Pass/fail comments*: This is a short description of the performance of the device when the procedure was executed.

The conformance to requirements matrix is also a good management reporting tool. It can become a summarized view of the status of the development. It is most useful when approaching the end of the development cycle. From the matrix, it is easy to pick out what features of the device are tested and verified as acceptable. Therefore, going forward, only the nontested or the *not yet accepted* requirements can be brought into focus.

There is no reason why the requirements cannot change during the development. Most developers consider this idea as *feature creep*, but there can be a healthy side to the changes. Some aspects of the medical device may not have been thought of early on in the program. Some existing parameters may be found to be too difficult to meet, and research indicates that tolerances on various parameters can be relaxed. In other situations, tolerances may need to be tightened. Care should be used to limit the number and kind of changes, since too many changes to the requirements indicate that the device is evolving as it is being developed. Generally, if the requirements need to evolve significantly during the development, then a new set of requirements will need to be released. The matrix is also a part of the next section, since it defines the metrics used to optimize the design.

7.9   ITERATE AND IMPROVE THE DEVICE BASED ON MEASURED PERFORMANCE

The conformance to requirements matrix is the method used to document which requirements are being met. The challenge is to use the matrix to make improvements to the device. Naturally, there is always a chance that any improvement made can adversely affect an already acceptable feature. Therefore, each improvement needs to be thought through with the idea in mind that adverse effects should not be produced.

In order to establish whether or not any change will cause any adverse effects, regression-testing plans may need to be developed. These regression tests are a subset of the *full requirements* testing, and they are designed to point out any changes in performance that are likely to be caused by the design changes. When any design change is made, it is good practice to understand what adverse effects the change may have and to develop a regression test to verify that the adverse effect did, or did not, occur.

The concept of regression testing is normally used in a software context. In this chapter, the concept is expanded to include hardware and software. Once all of the data on the device performance are collected, a design review will need to be performed and documented.

Assuming there are aspects of the device that do not conform to requirements, then the device will need to be changed. Exactly how the device is changed is up to the designers.

Each device change will need to be properly documented. The *traceability* of the device development process is a large part of FDA GMP protocols. Revisions of drawings and software/user manuals all need to be traceable. One needs to think of this design cycle as a completely transparent process in which all of the problems and solutions are well documented.

All of these data are usually kept in the *device master record* (Food and Drug Administration, 2011).

When the next revision of the device is ready for testing, it is up to the designer to validate all requirements over again. For more detail on this topic, see Part 820 Quality System Regulation (21CFR820) (Code of Federal Regulations).

At the end of the iteration cycle, it is assumed that there is now a medical device that conforms to requirements.

7.10   FINALIZING THE DOCUMENTATION

Every good design will have a complete set of documentation. In the case of a Class III medical device intended to go to the marketplace, the following documentation will usually be needed:

Design history file (DHF) is a compilation of records that describes the design history of a finished device.

Device master record (DMR) is a compilation of records containing the procedures and specifications for a finished device.

Design output means the results of a design effort at each design phase and at the end of the total design effort. The finished design output is the basis for the device master record. The total finished design output consists of the device, its packaging and labeling, and the device master record.

7.11   APPLY FOR REGULATORY APPROVALS

Depending on whether the device is to be marketed, or if the device is only to be used in a lab environment, it may or may not be necessary to apply for premarketing approval. If the device is a new design, it may be necessary to take the device through the premarket approval process. In the case of PEMF devices, unless otherwise defined, the device is a Class III device listed in the FDA Product Code MBQ:

Device	Peripheral electromagnetic field (PEMF) to aid wound healing
Review panel	Physical medicine
Product code	MBQ
Premarket review	Office of Device Evaluation (ODE)
	Division of Neurological and Physical Medicine Devices (DNPMD)
	Physical Medicine and Neurotherapeutic Devices Branch (PNDB)
Submission type	Contact ODE
Device class	3
Total product life cycle (TPLC)	TPLC product code report
GMP exempt?	No
Third-party review	Not third party eligible

Data extracted from Code of Federal Regulations.

ACKNOWLEDGMENT

The author would like to thank Muriel Kahler B.A. for her support and English grammar expertise.

APPENDIX 7.A

ENGINEERING DEFINITIONS

Note: In mathematics, the *root mean square* (abbreviated *RMS* or *rms*), also known as the *quadratic mean*, is a statistical measure of the magnitude of a varying quantity. It is especially useful when variants are positive and negative, for example, sinusoids. RMS is used in various fields, including electrical engineering (Wikipedia, 2014).

The RMS value of a set of values (or a continuous-time waveform) is the square root of the arithmetic mean (average) of the squares of the original values (or the square of the function that defines the continuous waveform).

In the case of a set of n values, the RMS value is given by this formula:

$$\{x_1, x_2, ..., x_n\}$$

$$x_{\text{RMS}} = \sqrt{\frac{1}{n}\left(x_1^2 + x_2^2 + \cdots + x_n^2\right)}.$$

The corresponding formula for a continuous function (or waveform) $f(t)$ defined over the interval $T_1 \le t \le T_2$ is

$$f_{\text{RMS}} = \sqrt{\frac{1}{T_2 - T_1}\int_{T_1}^{T_2}\left[f(t)\right]^2 dt},$$

and the RMS for a function over all time is

$$f_{\text{RMS}} = \lim_{x \to \infty}\sqrt{\frac{1}{T}\int_0^T\left[f(t)\right]^2 dt}.$$

The RMS over all time of a periodic function is equal to the RMS of one period of the function. The RMS value of a continuous function or signal can be approximated by taking the RMS of a series of equally spaced samples. Additionally, the RMS value of various waveforms can also be determined without calculus, as shown by Cartwright (2007).

REFERENCES

Cartwright, K.V. (Fall 2007). Determining the effective or RMS voltage of various waveforms without calculus. *Technology Interface* 8 (1): 20.

Chang, T.-C., Wysk, R.A., and Wang, H.P. (2005). *Computer-Aided Manufacturing*, 2nd edn. Prentice Hall, pp. 596–598.

Code of Federal Regulations. Title 21—Food and Drugs Chapter I—Food and Drug Administration, Department of Health and Human Services, Subchapter H—Medical Devices. Part 820 Quality System Regulation (21CFR820)   http://www.accessdata.fda.gov/scripts/cdrh/cfdocs/cfcfr/CFRSearch.cfm?CFRPart=820. Accessed February 15, 2014.

Crosby, P. (August 22, 2001). *Developer of the Zero-Defects Concept*. The New York Times, New York. Retrieved September 1, 2012.

FEMM 4.2. Finite Element Method Magnetics. http://www.femm.info/wiki/HomePage or http://ww1.foster-miller.net/. Accessed February 15, 2014.

Food and Drug Administration. Design Considerations for Pivotal Clinical Investigations for Medical Devices—Guidance for Industry, Clinical Investigators, Institutional Review Boards and Food and Drug Administration Staff. http://www.fda.gov/downloads/Training/CDRHLearn/UCM377490.pdf. Accessed February 20, 2014.

Food and Drug Administration. (2011). Quality System (QS) Regulation/Medical Device Good Manufacturing Practices. Last updated April 28, 2011. http://www.fda.gov/medicaldevices/deviceregulationandguidance/postmarketrequirements/qualitysystemsregulations/#introduction. Accessed February 20, 2014.

http://www.accessdata.fda.gov/scripts/cdrh/cfdocs/cfPCD/classification.cfm?ID=4769. Accessed February 20, 2014.

International Standards Organization (ISO 13485:2003). Medical Devices—Quality Management Systems—Requirements for Regulatory Purposes. http://www.iso.org/iso/catalogue_detail?csnumber=36786. Accessed February 20, 2014.

Markov, M.S. (2007). Pulsed electromagnetic field therapy history, state of the art and future. *Environmentalist* 27: 465–475. DOI: 10.1007/s10669-007-9128-2.

NASA-STD-3000 (July 1995). Man-Systems Integration Standards Revision B. http://www.nasa.gov/centers/johnson/slsd/about/divisions/hefd/standards/index.html. Accessed March 3, 2014.

Parker, R. and Markov, M. (2011). An analytical design techniques for magnetic field therapy devices. *Environmentalist* 31 (2): 155–160 (Special issue: Biological Effects of Electromagnetic Fields).

Pilla, A.A. et al. (1999). EMF signals and ion/ligand binding kinetics: Prediction of bioeffective waveform parameters. *Bioelectrochemistry and Bioenergetics* 48 (1): 27–34.

Roth, B.J. (2000). The electrical conductivity of tissues. In: J.D. Bronzino (ed.), *The Biomedical Engineering Handbook*, 2nd edn. CRC Press LLC, Boca Raton, FL, 2000.

Wheeler, H.A. (1928). Simple inductance formulas for radio coils. *Proceedings of the IRE*, October 1928, pp. 1398–1400.

Wikipedia. (2014). Root mean square. http://en.wikipedia.org/wiki/Root_mean_square. Last Updated April 6, 2014.

Zelnik, M. and Panero, J. (November 1, 1979). *Human Dimension and Interior Space: A Source Book of Design Reference Standards*. Crown Publishing Group.

8 Magnetic Field Influences on the Microcirculation

Chiyoji Ohkubo and Hideyuki Okano

CONTENTS

8.1 INTRODUCTION

Microcirculation consists of structurally and functionally differentiated small blood vessels: small muscular arteries, arterioles, metarterioles, capillaries, postcapillary venules, venules, and lymphatic capillaries (reviewed by Ohkubo and Okano, 2004). In the cutaneous microvascular beds, the connection between the arterioles and the venules is made by some thoroughfare channels including capillaries or arteriolar–venular shunts. In most vascular beds, the precapillary resistance vessels are responsible for the largest function of the resistance in a vascular bed and hence are the major components that influence regional hemodynamics and total peripheral resistance. Smooth muscle cells are found in all of these except the blood capillaries and lymphatic capillaries. The blood capillary wall is composed of a single layer of endothelial cells. The lymphatic capillaries are composed of endothelium-lined vessels similar to blood capillaries. Fluid and protein that have extravasated from the blood capillaries partially enter the lymphatic capillaries and are transported via the lymphatic system back to the blood vascular system. Postcapillary venules play an important role in fluid and cellular exchange and are the major site of leukocyte migration into tissue spaces.

Rhythmical and spontaneous changes in both the diameter of arterioles and the volume and velocity of blood flow due to constriction and dilation of the vascular smooth muscle are known as vasomotion (Asano and Brånemark, 1972). The quantitative description of spontaneous arteriolar vasomotion requires data on frequency, amplitude, diameter, and branching order of the vessels observed. Fluid absorbed into lymphatic capillaries is passively transported through dynamic changes due to arteriolar vasomotion in cutaneous tissue. Frequency and amplitude of spontaneous vasomotion could play an important role in disease. In microangiopathies, lymphedema, and essential hypertension, altered patterns of arteriolar vasomotion could constitute an additional pathogenetic factor (Funk and Intaglietta, 1983). Extracellular control of the smooth muscle cells is exerted through neurogenic, hormonal, local, and myogenic mechanisms (Mulvany, 1983).

The circulatory system is the transport system of the organism that supplies O_2, ions, hormones, and substances absorbed from the gastrointestinal tract to the tissues and returns CO_2 to the lungs and other products of metabolism to the liver and kidneys (reviewed by WHO, 2006). It also has a central role in the regulation of body temperature and the distribution of the hormones and other agents that regulate tissue and cell function. The principal function of microcirculation is exchanging physiological substances between blood and tissues, and the compensatory adjustments should contribute to the efficacy of the exchange process: lumen dimensions, length, tortuosity, diameter of branch ratios, vascular density, wall thickness, vessel diameter (vasomotion), blood flow velocity, blood viscosity, intramicrovascular hematocrit, leukocyte–endothelial cell interaction, and others. Furthermore, when taken with the presence of water channels, aquaporins in the continuous endothelium of microvascular beds, it is most likely that these molecules are responsible for the exclusive water pathways through the walls of all microvessels with continuous endothelium (Michel and Curry, 1999).

For evaluating the effects of magnetic fields including static magnetic fields (SMFs) and electromagnetic fields (EMFs) on microcirculatory system, various microcirculatory preparations, for example, rabbit ear chamber (REC) (Asano et al., 1965; Okano et al., 1999; Gmitrov and Ohkubo, 2002a,b; Gmitrov et al., 2002); dorsal skinfold chamber (DSC) in mice (Ushiyama and Ohkubo, 2004; Ushiyama et al., 2004a,b; Traikov et al., 2005, 2006; Morris and Skalak, 2007), rats (Morris and Skalak, 2005), and hamsters (Brix et al., 2008; Strieth et al., 2008; Strelczyk et al., 2009; Gellrich et al., 2014); and cranial window (CW) in rats (Masuda et al., 2007a–c, 2009, 2011), have been used to observe and analyze microcirculation. These preparations allow noninvasive, continuous measurement of hemodynamics, blood velocity, angiogenesis, and metabolites, for example, pH and pO_2, transport of molecules and particles, and cell-to-cell interactions in vivo (Jain, 1997).

Effects of magnetic fields on the circulatory system have been reviewed in experimental animals (Ohkubo and Okano, 2004, 2011; Saunders, 2005; Tenforde, 2005; WHO, 2006; McKay et al., 2007; Ohkubo et al., 2007; Robertson et al., 2007; Nittby et al., 2008; Ruggiero, 2008; Funk et al., 2009; McNamee et al., 2009; Ohkubo and Okano, 2011; Yu and Shang, 2014) and in humans (Jauchem, 1997; Chakeres and de Vocht, 2005; Crozier and Liu, 2005; van Rongen, 2005; McKay et al., 2007; Robertson et al., 2007; Funk et al., 2009; McNamee et al., 2009; Jokela and Saunders, 2011; Ohkubo and Okano, 2011), and possible interpretations of the effects have been discussed using theoretical models (Tenforde et al., 1983; Kinouchi et al., 1996; ICNIRP, 1998, 2009, 2010, 2014; Luo et al., 2005; Tenforde, 2005; Kainz et al., 2010). There are many therapeutic applications of locally increased blood flow. It has been suggested that magnetic fields could have the potential to modify microcirculatory perfusion. Regarding the effects of magnetic fields on microcirculation and microvasculature, McKay et al. (2007) reviewed that nearly half of the cited experiments (10/27) are related to either a vasodilatory effect, increased blood flow, or increased blood pressure (BP). Conversely, 3 of the 27 studies reported a decrease in blood perfusion/pressure. Four studies reported no effect. The remaining 10 studies found that magnetic fields could trigger either vasodilation or vasoconstriction depending on the initial tone of the vessel. In terms of cellular effects of magnetic fields related to perfusion, 4 of a total of 19 studies reported an increase in nitric oxide (NO) activity from magnetic field exposures (one of these studies used a model with an altered vessel state prior to exposure), 1 found a bidirectional effect, and 5 found no effect. Nine studies reported vascular development effects (seven reported increased angiogenesis; two reported decreased angiogenesis). Other cellular effects were reported in three studies.

8.2 STATIC MAGNETIC FIELDS

This is an area of research that would benefit from increased investigation because SMF therapy could be useful for circulatory diseases including ischemic pain and hypertension, primarily due to the modulation of blood flow or BP (reviewed by Ohkubo and Okano, 2004, 2011). There are many therapeutic applications for locally increased blood flow. It is suggested that magnetic fields have the potential to modify microcirculatory perfusion. However, there might be safety concerns on magnetic resonance imaging (MRI) and magnetic levitation for transportation using strong-intensity SMF with Tesla (T) levels (reviewed by Silva et al., 2006). Therefore, the knowledge of the SMF effects on microcirculation is extremely important in consideration of the human health and the relation of capillary formation and tumor growth.

Several attempts have been made to explore the parameters of microcirculation and microvasculature when tissue and/or blood vessels have been exposed to SMF. In particular, the REC offers the advantages of superior optical quality. Due to the longer duration of an individual measurement, we have exclusively utilized REC to investigate the effects of SMF on microcirculation using micro-photoelectric plethysmography (MPPG) monitoring system. REC is a round-table chamber made of acryl resin for disk with an observing table and three holding pillars, a sustaining ring, and a glass window. The methods for installation of REC and its availability to the bioelectromagnetic research have been published in detail (Ohkubo and Xu, 1997; Xu et al., 1998; Okano et al., 1999; Gmitrov and Ohkubo, 2002a,b; Gmitrov et al., 2002). BP in a central artery contralateral to that of an ear lobe having the REC, fixed on the microscope stage, was monitored by a BP monitoring system.

Using the REC methods, Ohkubo and Xu (1997) firstly demonstrated that cutaneous microcirculation was modulated by moderate-intensity SMF with millitesla (mT) levels: the bidirectional effects of a 1, 5, and 10 mT SMF on cutaneous microcirculation were found in conscious rabbits. A 10 min exposure to SMF induced changes in vasomotion in a non-dose-dependent manner. When the initial vessel diameter was less than a certain value, SMF exposure caused an increase in vessel diameter (vasodilation). In contrast, when the initial diameter was greater than a certain value, SMF exposure caused a decrease in vessel diameter (vasoconstriction). Based on these results, it would appear that the initial state of the vessel is of importance when considering SMF effects on microcirculation and microvasculature.

Likewise, this observation using REC is reflected in the following studies: the bidirectional effects (activation/inhibition) of 10 min exposure to 1.0 mT SMF on cutaneous microcirculation were found in conscious rabbits treated with vasoactive agents (Okano et al., 1999). When high vascular tone was induced by norepinephrine to cause vasoconstriction, the SMF exposure led to increased vasomotion and caused vasodilation. In contrast, low vascular tone was induced by acetylcholine (ACh) to cause vasodilation, and the SMF exposure led to decreased vasomotion and caused vasoconstriction. Other studies without using REC have also been well reviewed by McKay et al. (2007). Our studies described earlier were performed at Department of Environmental Health (the former Department of Physiological Hygiene), National Institute of Public Health, Japan.

Similar findings were reported by an independent laboratory using different techniques without using REC for the microvessels of rat skeletal muscle: an SMF exposure (70 mT for 15 min) had a restorative effect on microvascular tone (Morris and Skalak, 2005). When vessels had high tone (constricted), the SMF acted to reduce tone, and when vessels had low tone (dilated), the SMF acted to increase tone. This response was amplified when the vessels had an initial diameter of less than 30 μm.

The effects of higher SMF applying for MRI on blood–brain barrier (BBB) permeability were investigated in rats using a radioactive tracer, ^{153}Gd-DTPA (Prato et al., 1990, 1994). Prato et al. (1990) reported that exposure to low-field (0.15 T, for 23.2 min) MRI increases BBB permeability. In addition, Prato et al. (1994) found that exposures to SMF alone for 22.5 min, without radio frequency (RF)-specific absorption rate (SAR) and temporal gradient, increased BBB permeability at 1.5 and 1.89 T. In contrast, Yamaguchi-Sekino et al. (2012) obtained the following results: acute exposure (up to 3 h) to ultrahigh magnetic field (17.2 T generated by an MRI system) does not alter BBB permeability in rats. However, so far, there has been no explanation available for these inconsistent results of BBB alteration between different intensities of SMF.

In another aspect, previously, researches have shown that acute exposure (30–40 min) to SMF (30–223 mT) increased skin vasomotion amplitude (Li et al., 2007; Yan et al., 2011). In resting skin blood flow in healthy young men, Yan et al. (2011) found that SMF exposure at the maximum magnetic flux density B_{max} of 223 mT for 30 min induced a significant increase in vasomotion amplitude, mainly reflecting the intrinsic myogenic and endothelial-related metabolic activities, by placing a magnet to the center of the middle finger prominence, and, after removal of the SMF, the vasomotion amplitude vanished gradually. In an animal (rat) model, Li et al. (2007) reported significant enhancement of vasomotion amplitude, mainly reflecting the endothelial-related metabolic activity (0.01–0.05 Hz) in the skin stressed by pressure loading over the trochanter area upon exposure to an SMF at B_{max} of 30 mT for 40 min. In their study, prolonged surface loading caused significant reduction of the endothelial-related metabolic activity and increased the myogenic activity; that is, it induced a higher vascular tone in tissues that had been stressed as compared with the unstressed ones (Li et al., 2007). In contrast, SMF significantly increased the endothelial-dependent vasodilation and subsequently increased blood flow in the stressed skin (Li et al., 2007). The modulating effect of SMF on the vasomotion amplitude might be related to the vascular tone modified by prolonged compressive loading (Li et al., 2007).

In contrast to the aforementioned studies (Li et al., 2007; Yan et al., 2011), Xu et al. (2013) focused on examining the subchronic effects of moderate-intensity inhomogeneous SMF on peripheral hemodynamics. Their study indicated that SMF exposure of B_{max} 160 mT for 3 weeks seems to have tendency to modulate the vasomotion amplitude at 0.05 Hz in the range of endothelial-related metabolic activity for rats and contract the increased vasomotion amplitude in the ischemic area, but did not induce significant change in any one of these parameters during the SMF exposure period of 3–7 weeks investigated. These results suggest that SMF may have a regulatory effect on rhythmic vasomotion in the ischemic area by smoothing the vasomotion amplitude, mainly reflecting the endothelial-related metabolic activity, in the early stage of the wound healing process. The physiological implication is that smoothing or buffering the vasomotion amplitude may play a key role on inherent hemodynamic control mechanisms for rhythmic vasomotion and endothelial-dependent vasodilation.

In the context of angiogenesis, Ruggiero et al. (2004) reported that 3 h exposure to 0.2 T SMF did not apparently affect the basal pattern of vascularization or chick embryo viability using the chick embryo chorioallantoic membrane (CAM) assay. Prostaglandin E_1 and fetal calf serum elicited a strong angiogenic response in sham-exposed eggs. This angiogenic response was significantly inhibited by 3 h exposure to 0.2 T SMF. These findings point to possible use of SMF in inhibiting angiogenesis. Preclinical studies examining the inhibition of angiogenesis by SMF exposure have turned to the use of SMF in the treatment of cancer in terms of tumor angiogenesis. Brix et al. (2008) evaluated SMF effects on capillary flow of red blood cells (RBCs) in unanesthetized Syrian golden hamsters, using a DSC technique for intravital fluorescence microscopy. Capillary RBC velocities (v_{RBC}), capillary diameters (D), arteriolar diameters (D_{art}), and functional vessel densities (FVDs) were measured in striated skin muscle at different magnetic flux densities. Exposure above 500 mT for 1 h resulted in a significant reduction of v_{RBC} in capillaries compared with the baseline value. At B_{max} of 587 mT, v_{RBC} was reduced (40%). Flow reduction was reversible when the field strength was decreased below the threshold level. In contrast, mean values determined at different exposure levels for the parameters D, D_{art}, and FVD did not vary (5%). Blood flow through capillary networks is affected by SMF directed perpendicular to the vessels.

The same research group further analyzed the effects of SMF (≤587 mT) on tumor microcirculation (Strieth et al., 2008). In vivo fluorescence microscopy was performed in A-Mel-3 tumors growing in DSC preparations of hamsters. Short time exposure (≥150 mT) resulted in a significant reduction of v_{RBC} and segmental blood flow in tumor microvessels. At B_{max} of 587 mT, a reversible reduction of v_{RBC} (40%) and of FVD (15%) was observed. Prolongation of the exposure time (1 min to 3 h) resulted in reductions. Microvessel diameters and leukocyte–endothelial cell interactions remained unaffected by SMF exposures. However, in contrast to tumor-free striated muscle controls, exposure at B_{max} of 587 mT induced a significant increase in platelet–endothelial cell adherence in a time-dependent manner that was reversible after reducing the strength of the SMF.

The authors assumed that these reversible changes may have implications for functional measurements of tumor microcirculation by MRI and new therapeutic strategies using strong SMF.

The same research group further evaluated the effects of an SMF (586 mT, for 3 h) on tumor angiogenesis and growth (Strelczyk et al., 2009). The analysis of microcirculatory parameters revealed a significant reduction of parameters, FVD, D, and v_{RBC}, in tumors after SMF exposure compared with the control tumors. These changes reflect retarded vessel maturation by antiangiogenesis. The increased edema after SMF exposure indicated an increased tumor microvessel leakiness possibly enhancing drug uptake.

In addition, combining SMF exposure (587 mT, for 2–3 h) with paclitaxel chemotherapy, tumor growth was analyzed (Gellrich et al., 2014). SMF inhibited tumor angiogenesis and increased tumor microvessel permeability significantly. This was not mediated by inflammatory leukocyte–endothelial cell interactions. Further, SMF increased the effectiveness of paclitaxel chemotherapy significantly. These findings support that SMF possibly open the blood–tumor barrier to small molecular therapeutics (Brix et al., 2008; Strieth et al., 2008; Strelczyk et al., 2009; Gellrich et al., 2014).

Atef et al. (1995) showed that the kinetics of oxyhemoglobin auto-oxidation decreased in the auto-oxidation reaction rate of 2%–5.9% and 10%–17%, under the SMF exposure of 100–250 and 350–400 mT, respectively. Djordjevich et al. (2012) indicated that subchronic continuous exposure to 16 mT SMF for 28 days caused lymphocyte and granulocyte redistribution between the spleen and blood in mice. These results suggested that observed changes were not due to an unspecific stress response, but that they were rather caused by specific adaptation to subchronic SMF exposure.

In another aspect of SMF effects on erythrocytes, Lin et al. (2013a,b) found that 0.8 T SMF coupled with the slow cooling procedure increased survival rates of frozen–thawed human erythrocytes without any negative effects on cell morphology or function. They suggest that the SMF cryoprotective effect is due to enhanced biophysical stability of the cell membrane, which reduces dehydration damage to the erythrocyte membrane during the slow cooling procedure.

As for clinical trials, Kim et al. (2010) performed the following experiment using single photon emission computed tomography (SPECT): the permanent magnet (0.3 T, unipolar, disk shaped, 4 cm diameter, and 1 cm thick) was placed on the right frontotemporal region of the brain for 20 min for each of 14 healthy subjects. Technetium (Tc)-99m ethylcysteinate dimer (ECD) perfusion SPECT was taken to compare the cerebral blood flow (CBF) patterns in the subjects exposed to the SMF with those of the resting and sham conditions. They found that the regional CBF (rCBF) was significantly increased in the right frontal and parietal regions and also the right insula. In contrast, rCBF was rather decreased in the left frontal and left parietal regions. These results suggested that 0.3 T SMF induces an increase in rCBF in the targeted brain areas noninvasively, which may result from a decrease in rCBF in contralateral regions.

Wang et al. (2009a) investigated the effects of gradient SMF (0.2–0.4 T, 2.09 T/m, 1–11 days) on angiogenesis both in vitro and in vivo. An 3-(4,5-dimethyl-2-thiazolyl)-2,5-diphenyl-2H-tetrazolium bromide (MTT) assay was used as an in vitro method to detect the proliferation ability of human umbilical vein endothelial cells (HUVECs). Two kinds of in vivo models, a chick CAM and a matrigel plug, were used to detect the effects of gradient SMF on angiogenesis. The results showed that the proliferation ability of HUVEC was significantly inhibited 24 h after the onset of exposure. With regard to the CAM model, vascular numbers in the CAM that was continuously exposed to the gradient SMF were all less than those in normal condition. In accordance with the gross appearance, the contents of hemoglobin (Hb) in the models exposed to gradient SMF for 7–9 days were also less. In addition, similar to the CAM model, the results of vascular density and Hb contents in the mouse matrigel plug also demonstrated that the gradient SMF exposure for 7–11 days inhibited vascularization. These findings indicate that gradient SMF might inhibit or prevent new blood vessel formation and could be helpful for the treatment of some diseases relevant to pathological angiogenesis.

Other reports have shown that an SMF of strong intensity higher than a few Tesla has bioeffects (Tenforde et al., 1983; Higashi et al., 1993, 1995; Shiga et al., 1993; Denegre et al., 1998;

Ichioka et al., 2000, 2003; Snyder et al., 2000; Haik et al., 2001; Kotani et al., 2002; Zborowski et al., 2003; Yamamoto et al., 2004; Valiron et al., 2005; Hsieh et al., 2008; ICNIRP, 2009, 2014; Tao and Huang, 2011; Mian et al., 2013; Glover et al., 2014; Theysohn et al., 2014). For instance, an SMF of strong intensity with an extremely high magnetic gradient (B_{max} of 8 T, the maximum gradient G_{max} of 400 T^2/m) could induce some bioeffects on paramagnetic Hb by magnetic attraction in a high gradient or diamagnetic Hb by magnetic repulsion in a high gradient, retarding the mean blood velocity in peripheral circulation, partly due to the asymmetric distribution of RBCs with different magnetic susceptibilities, and magnetically induced movement of diamagnetic water vapor at the skin surface, which may lead to a skin temperature decrease (Ichioka et al., 2003). Moreover, it has been shown that RBCs rotate and orient so that the concave surface is aligned parallel to strong uniform SMF due to magnetic torque (Higashi et al., 1993). From a rheological point of view, concerning the SMF effects on blood viscosity, different results were obtained, depending on the exposure conditions (Haik et al., 2001; Yamamoto et al., 2004; Tao and Huang, 2011). Haik et al. (2001) reported an increase in viscosity of blood flowing parallel to inhomogeneous strong SMF at B_{max} of 3, 5, and 10 T. Yamamoto et al. (2004) also indicated an increased viscosity of both fully oxygenated and fully deoxygenated blood and greater increase in blood viscosity of deoxygenated blood relative to that of oxygenated blood in 1.5 T homogeneous SMF exposure for 36 min. In contrast, Tao and Huang (2011) demonstrated that acute exposure to 1.33 T homogeneous SMF reduced blood viscosity when it was applied parallel to the flow direction (for an exposure duration of 1 min, short chains of RBC were formed; for an exposure duration of 12 min, long cluster chains of RBC were formed) (Tao and Huang, 2011). This process could enable the RBC to pass through the blood vessels in a more streamlined fashion, thereby reducing the blood viscosity (Tao and Huang, 2011).

These observations suggested that both inhomogeneous and homogeneous high-intensity SMFs (Tesla level) may modulate in vivo hemodynamics. However, the underlying mechanisms and physiological consequences have not yet been fully understood. Because our applied SMF at B_{max} of 160 mT was much lower in the magnetic force compared with the SMF of several Tesla, different plausible mechanisms might exist between them, such as through a Ca^{2+}–calmodulin-regulated NO–cGMP signaling pathway. Takeshige and Sato (1996) suggested a mechanism of SMF action for the promotion of hemodynamic responses in which an SMF at B_{max} of 130 mT might inhibit acetylcholinesterase (AChE). The recovery of circulation is assumed to be partly due to the enhanced release of ACh by the SMF exposure, activating the cholinergic vasodilator nerve endings innervated to the muscle artery (Takeshige and Sato, 1996). The inhibitory effect of SMF on AChE was also observed in the magnetic flux density of 0.8 mT or more (Ravera et al., 2010). In addition, it is also suggested that an SMF at B_{max} of 5.5 mT should have a potential to counteract the action of a nitric oxide synthase (NOS) inhibitor L-NAME, presumably via increased endogenous ACh release (Okano and Ohkubo, 2005b, 2006). The increased (upregulated) effect of a 120 µT SMF on endothelial nitric oxide synthase (eNOS) expression was also confirmed in HUVEC (Martino et al., 2010).

Muehsam et al. (2013) found that exposure for 10–30 min to SMF of B_{max} 186 mT resulted in more rapid Hb deoxygenation, using a reducing agent dithiothreitol in an in vitro cell-free preparation. Thus, SMF significantly increased the rate of Hb deoxygenation occurring several minutes to several hours after the end of SMF exposure (Muehsam et al., 2013). The observation that SMF pretreatment of Hb alone, or of the deoxygenation solution itself, failed to yield a significant effect suggests that SMF exposure acted upon the interaction of Hb with the deoxygenation solution (Muehsam et al., 2013). With regard to the mechanism, they speculated that these SMF modalities modified protein/solvation structure in a manner that altered the energy required for deoxygenation (Muehsam et al., 2013). They further speculated that the mT-range SMF could induce the action of the Lorentz force on charges bound at the protein/water interface based on the Lorentz–Langevin model for weak magnetic field bioeffects (Muehsam and Pilla, 2009a,b). This model suggested that weak exogenous ac/dc magnetic fields can act on an ion/ligand bound in a molecular cleft, based upon the assumption that the receptor molecule is able to detect the Larmor trajectory of an ion or ligand within the binding site. To date, however, there is insufficient direct experimental evidence pertaining to this model. Further studies are required to better understand the mechanisms of SMF bioeffects on hemodynamic function. Spectral analysis would be useful to examine these effects in more detail with a view of hemodynamics.

8.2.1 DISCUSSION AND CONCLUSION

In summary, significant circulatory system responses to SMF have been recently reviewed in experimental animals and/or humans: SMF exposures between 1.0 and 8 T for anywhere between 10 min and 12 weeks can influence cutaneous microcirculation, hemodynamics, arterial BP, and/or angiogenesis in vivo (Ohkubo and Okano, 2004, 2011; Chakeres and de Vocht, 2005; Crozier and Liu, 2005; Saunders, 2005; Tenforde, 2005; van Rongen, 2005; WHO, 2006; McKay et al., 2007; Ohkubo et al., 2007; Robertson et al., 2007; Ruggiero, 2008; Funk et al., 2009; McNamee et al., 2009; Ohkubo and Okano, 2011; Yu and Shang, 2014).

Ten of a total of 38 studies in the last two decades report either an increase in blood flow, an elevation in BP, or increased angiogenesis. In contrast, 14 of the 38 studies indicate either a decrease in blood flow, a reduction in BP, or reduced angiogenesis. Eight studies report no effect. The remaining six studies found that SMF exposures could trigger either vasodilation or vasoconstriction depending on the initial tone of the vessel, which induce bidirectional effect. For a summary, please refer to Table 8.1.

In particular, a series of studies, as shown in Table 8.1, have demonstrated that cutaneous microcirculation, hemodynamics, and/or arterial BP were modulated by moderate-intensity SMF with mT levels in pharmacologically treated animals and genetically hypertensive animals, while no changes were observed in normal animals when the initial state of the vessel was not identified. Therefore, it was concluded that significant bioresponses to therapeutic signals occur when the state of the target is far from the homeostasis or the equilibrium of hemodynamics (Ohkubo and Okano, 2004).

In the context of neoplasia, Strelczyk group reported that SMF reduces RBC velocity and increases platelet adhesion without affecting the leukocytes or smooth muscle cells of the vessel walls, as demonstrated by the lack of changes in mean arterial BP and vessel diameters (Brix et al., 2008; Strieth et al., 2008; Strelczyk et al., 2009; Gellrich et al., 2014). These findings support that SMF possibly open the blood–tumor barrier to small molecular therapeutics (Brix et al., 2008; Strieth et al., 2008; Strelczyk et al., 2009; Gellrich et al., 2014).

The mechanisms of moderate-intensity SMF effects could be mediated by suppressing or enhancing the action of biochemical effectors, thereby inducing homeostatic effects bidirectionally. The potent mechanisms of SMF effects have often been linked to NO pathway, Ca^{2+}-dependent pathway, sympathetic nervous system (e.g., baroreflex sensitivity and the action of sympathetic agonists or antagonists), and neurohumoral regulatory system (e.g., production and secretion of angiotensin II and aldosterone), as reviewed by McKay et al. (2007). In addition, moderate-intensity SMF exposures have been reported to perturb distribution of membrane proteins and glycoproteins, receptors, cytoskeleton, and transmembrane fluxes of different ions, especially calcium $[Ca^{2+}]_i$, that, in turn, interfere with many different physiological activities, including phagocytosis (Dini, 2010). Wang et al. (2009b) reported that over time periods of many hours to several days, moderate-intensity SMF-initiated changes to Ca^{2+} can modulate signaling pathways, leading to significant changes in gene expression, cell behavior, and phenotype. Furthermore, Wang et al. (2010) revealed that moderate-intensity SMF reproduced several responses elicited by ZM241385, a selective $A_{2A}R$ antagonist and also a Parkinson's disease (PD) drug candidate, in PC12 cells including altered Ca^{2+} flux, increased ATP levels, reduced cAMP levels, reduced NO production, reduced p44/42 mitogen-activated protein kinase (MAPK) phosphorylation, inhibited proliferation, and reduced iron uptake. SMF also counteracted several PD-relevant endpoints exacerbated by $A_{2A}R$ agonist CGS21680 in a manner similar to ZM241385; these include reduction of increased expression of $A_{2A}R$, reversal of altered Ca^{2+} efflux, dampening of increased adenosine production, reduction of enhanced proliferation and associated p44/42 MAPK phosphorylation, and inhibition of neurite outgrowth (Wang et al., 2010).

TABLE 8.1
Summary of SMF Effects on Hemodynamics, BP, and/or Angiogenesis In Vivo

Increased Effect (n = 10)	Exposure Conditions: Intensity; Duration; Subjects
Prato et al. (1990)	0.15 T; 23.2 min; rats (BBB)
Prato et al. (1994)	1.5 and 1.89 T; 22.5 min; rats (BBB)
Xu et al. (1998)	180 mT; 1–3 weeks; rabbits (REC)*
Xu et al. (2001)	1.0 and 10.0 mT; 10 min; mice
Gmitrov (2002)	250 mT; 40 min; rabbits (REC)
Gmitrov and Ohkubo (2002a)	250 mT; 40 min; rabbits (REC)
Gmitrov and Ohkubo (2002b)	250 mT; 40 min; rabbits (REC)
Okano et al. (2005a)[a]	25 mT; 2–12 weeks; rats*
Li et al. (2007)	30 mT; 40 min; rats*
Yan et al. (2011)	223 mT; 30 min; humans*

Decreased Effect (n = 14)	Exposure Conditions: Intensity; Duration; Subjects
Ichioka et al. (2000)	8 T; 20 min; rats
Ichioka et al. (2003)	8 T; 20 min; rats
Okano and Ohkubo (2003a)[b]	5.5 mT; 30 min; rabbits*
Okano and Ohkubo (2003b)[c]	10 and 25 mT; 2–8 weeks; SHR*
Okano and Ohkubo (2005b)[c]	180 mT; 5–8 weeks; SHR*
Okano and Ohkubo (2006)[c]	180 mT; 6 weeks; SHR*
Okano et al. (2005b)[c]	5 mT; 2–8 weeks; SHR*
Okano and Ohkubo (2007)[b]	12 mT; 2–10 weeks; rats*
Brix et al. (2008)[d]	538 and 587 mT; 1 h; hamsters (DSC)*
Strieth et al. (2008)[d]	149–587 mT; 3 h; hamsters (DSC)*
Strelczyk et al. (2009)[d]	587 mT; 3 h; hamsters (DSC)*
Ruggiero et al. (2004)	200 mT; 3 h; chick embryo*
Wang et al. (2009a,b)	400 mT; 7–11 days; mouse* and chick embryo*
Mayrovitz and Groseclose (2005)	400 mT; 45 min; humans*

Bidirectional Effect (n = 6)	Exposure Conditions: Intensity; Duration; Subjects
Ohkubo and Xu (1997)	1, 5, and 10 mT; 10 min; rabbits (REC)*
Okano et al. (1999)	1 mT; 10 min; rabbits (REC)*
Okano and Ohkubo (2001)	5.5 mT; 30 min; rabbits*
Okano and Ohkubo (2005a)	5.5 mT; 30 min; rabbits*
Morris and Skalak (2005)	70 mT; 15 min; rats (DSC)
Kim et al. (2010)	300 mT; 20 min; humans*

No Effect (n = 8)	Exposure Conditions: Intensity; Duration; Subjects
Yamaguchi-Sekino et al. (2012)	17.2 T; 3 h; rats (BBB)
Kangarlu et al. (1999)	8 T; 1 h for humans*; 3 h for pigs*
Steyn et al. (2000)	27 mT; 46 h; horses*
Kuipers et al. (2007)	60 mT; 1 h; humans*
Mayrovitz et al. (2001)	100 mT; 36 min; humans*
Mayrovitz et al. (2005)	85 mT: 20 min; humans*
Hinman (2002)	50 mT; 15 min; humans*
Martel et al. (2002)	80 mT; 30 min; humans*

BBB, blood–brain barrier; REC, rabbit ear chamber; DSC, dorsal skinfold chamber.

[a] Initial state of subject: pharmacologically induced hypotension.
[b] Initial state of subject: pharmacologically induced hypertension.
[c] Initial state of subject: spontaneous hypertensive rats (SHR).
[d] Tumor microvessels.
* Conscious conditions (without anesthesia).

Csillag et al. (2014) investigated that effects of SMF exposure on a murine model of allergic inflammation and also on human provoked skin allergy. Inhomogeneous SMF was generated with an apparatus validated previously, providing a peak-to-peak magnetic induction of the dominant SMF component 389 mT by 39 T/m lateral gradient in the in vivo and in vitro experiments and 192 mT by 19 T/m in the human study at the 3 mm target distance. They found that even a single 30 min exposure of mice to SMF immediately following intranasal ragweed pollen extract (RWPE) challenge significantly lowered the increase in the total antioxidant capacity of the airways and decreased allergic inflammation. Repeated (on 3 consecutive days) or prolonged (60 min) exposure to SMF after RWPE challenge decreased the severity of allergic responses more efficiently than a single 30 min treatment. SMF exposure did not alter reactive oxygen species (ROS) production by RWPE under cell-free conditions, while diminished RWPE induced increase in the ROS levels in A549 epithelial cells. Results of the human skin prick tests indicated that SMF exposure had no significant direct effect on provoked mast cell degranulation. The observed beneficial effects of SMF are likely owing to the mobilization of cellular ROS-eliminating mechanisms rather than direct modulation of ROS production by pollen NAD(P)H oxidases.

Ruggiero (2008) proposed that some proteins might not be the sole candidates for this role of sensory molecules: large and complex polymers like glycosaminoglycans, which show a precise array of electric charges on their surface, as well as small second messengers with stereospecific positioning of charges such as inositol 1,4,5-trisphosphate, might play a direct or indirect role in the cell response to SMF. Ruggiero (2008) further speculated that some genes are sensitive to SMF and others to variable EMF. Ruggiero (2008) termed these genes as *magnetic-sensitivity-conferring genes* and proposed that an approach should first identify which genes are expressed in magnetic field–sensitive cells, but not in nonsensitive cells.

Further study concerning the magnetic-sensitivity-conferring genes is an intriguing areas of research and also shed light on the signal transduction and biophysical mechanisms of the magnetic sense in animals (Ritz et al., 2004; Rodgers et al., 2009; Yoshii et al., 2009; Gegear et al., 2010; Nishimura et al., 2010; Eder et al., 2012; Maeda et al., 2012; Stoneham et al., 2012; Wiltschko and Wiltschko, 2012; Dodson et al., 2013; Lee et al., 2014; Neil et al., 2014; Solov'yov et al., 2014) and humans (Foley et al., 2011). In particular, recent studies on the effects of magnetic field on the biological clock through the radical pair mechanism have been in progress in the relatively weak magnetic fields less than 1 mT (Yoshii et al., 2009). Furthermore, studies on the subtle effects of RF–EMFs have been recently focused on the mechanism of spin biochemistry (Usselman et al., 2014). Technological advances in spin biochemistry will enable to understand the influence of various kinds of magnetic fields on biological systems.

One of the other established physical mechanisms is *magnetic induction* (van Rongen et al., 2007) in relatively high intensity of SMF over 1 T generating from MRI devices. This mechanism originates from the following two types of interaction:

1. *Electrodynamic interactions with moving electrolytes*: An SMF exerts Lorentz forces on moving ionic charge carriers and thereby gives rise to induced electric fields and currents. This interaction is the basis of magnetically induced potentials associated with flowing blood. More recently, using a uniform SMF of 0.2 T, Kainz et al. (2010) successfully demonstrated the experimental and theoretical validation of a magnetohydrodynamic (MHD) solver for blood flow analysis. The measured voltage value probably induced by MHD signal was 245 µV. The computational MHD results can then be correlated with the actual measurements. The authors hope to develop an MHD-based biomarker to noninvasively estimate the blood flow for the evaluation of heart failure. This MHD theory is also supported by other studies (Oster et al., 2014; Srivastava, 2014).
2. *Induced electric fields and currents*: Time-varying magnetic fields induce electric currents in living tissues in accordance with Faraday's law of induction. Electric currents may also be induced by SMF (Gupta et al., 2008). Gupta et al. (2008) suggested that, in

the bore of 1.5 T MRI magnet, this induced voltage distorts the electrocardiogram (ECG) signal of the patient and appears as an elevation of the T-wave of the ECG signal. The flow of blood through the aortic arch is perpendicular to the SMF and coincides with the occurrence of the T-wave of the ECG. The induced electric fields could be enhanced by movement in an SMF (Glover and Bowtell, 2008). Glover and Bowtell (2008) measured in situ surface electric fields induced by typical human body movements such as walking or turning in the *fringe* magnetic fields of a whole-body 3 T MRI scanner. These values were 0.15, 0.077, and 0.015 V/m for the upper abdomen, head, and across the tongue, respectively. A peak electric field of 0.3 V/m was measured for the chest. The speed of movements was not specified in this study. In a body moving at a constant speed of 0.5 m/s into a 4 T magnet, Crozier and Liu (2005) estimated the maximum induced electric field strength to be 2 V/m, which is equal to the apparent threshold for peripheral nerve stimulation in the frequency range from 10 Hz to 1 kHz (ICNIRP, 1998). It should be noted, however, that frequencies associated with body movements are likely to be ≤10 Hz, the frequency below which more negative resting membrane potential decreases the electrical excitability of neurons due to the slow inactivation of voltage-gated sodium ion channels (Bezanilla, 2002). Head translational and rotational frequencies during walking, for example, vary from 0.4 to 4.0 Hz (Grossman et al., 1988; Pozzo et al., 1990; MacDougall and Moore, 2005).

Moreover, when considering a MHD Lorentz force acting on ions, Roberts et al. (2011) recently discovered that SMF from high-strength MRI machines (3 and 7 T, for 25 min) induces nystagmus in all normal humans and that a Lorentz force, derived from ionic currents in the endolymph and pushing on the cupula, best explains this effect. Furthermore, Ward et al. (2014) recorded eye movements in the SMF of a 7 T MRI machine for 25 min in nine individuals with unilateral labyrinthine hypofunction, as determined by head impulse testing and vestibular-evoked myogenic potentials (VEMPs). Eye movements were recorded using infrared video-oculography. Static head positions were varied in pitch with the body supine, and slow-phase eye velocity (SPV) was assessed. All subjects exhibited predominantly horizontal nystagmus after entering the magnet head first, lying supine. The SPV direction reversed when entering feet first. Pitching chin to chest caused subjects to reach a null point for horizontal SPV. Right unilateral vestibular hypofunction (UVH) subjects developed slow-phase-up nystagmus and left UVH subjects, slow-phase-down nystagmus. Vertical and torsional components were consistent with superior semicircular canal excitation or inhibition, respectively, of the intact ear. These findings provide compelling support for the hypothesis that magnetic vestibular stimulation (MVS) is a result of a Lorentz force and suggest that the function of individual structures within the labyrinth can be assessed with MVS. We speculate that a Lorentz force acting on ions in the endolymph could also be induced in the microvessels.

In another aspect of other medical applications, magnetic drug delivery system (MDDS), which is a technology to control the drug kinetics from outside the body by external magnetic force using spatially inhomogeneous SMF or dc/ac magnetic fields, has been investigated (Mishima et al., 2007; Nishijima et al., 2008, 2009; Hirota et al., 2009; Chorny et al., 2010; Morais, 2010; Nakagawa et al., 2012; Wang et al., 2014). In particular, under SMF exposure conditions with regard to MDDS using suspension of the ferromagnetic particles, the trajectory of the ferromagnetic particles in the blood vessel was calculated, and the possibility of the navigation of the drug has been discussed (Mishima et al., 2007; Nishijima et al., 2008, 2009; Hirota et al., 2009; Nakagawa et al., 2012). The drug navigation probability to the desired direction was confirmed to be higher than 80% using a high-temperature superconducting magnet of B_{max} 4.5 T (Nishijima et al., 2008). A rat experiment was also performed successfully using a permanent magnet of B_{max} 0.3 T (Nishijima et al., 2008). Moreover, the suspension of the magnetite was injected into the blood vessel of the pig, and the magnetite was successfully navigated and/or accumulated by the high-temperature superconducting magnet (Nishijima et al., 2009).

There are a number of important dosimetry issues that could exhibit significant effects on micro-circulation. It has been shown that it is more appropriate to consider biological responses to SMF through the hypothesis of intensity windows, instead of intensity-response dependence (Markov et al., 2004). Furthermore, it has been reported that the gradient component of SMF might be responsible for the physiological responses in vivo (Okano and Ohkubo, 2005a,b), because the in vitro effects of gradient fields on action potential generation (Cavopol et al., 1995; McLean et al., 1995) and myosin phosphorylation (Engström et al., 2002) have been found mostly in the absolute field gradient range of more than 1.0 mT/mm in the target tissues or cells. However, these hypotheses have not been tested well in vivo compared with in vitro, and the effects and the underlying mechanisms remain elusive. In particular, to reveal and clarify the effects and mechanisms of spatial magnetic gradient in vitro as well as in vivo, it is necessary to carry out the experiments comparing the spatially homogeneous and inhomogeneous SMF.

8.3 EXTREMELY LOW-FREQUENCY ELECTROMAGNETIC FIELDS

Many researches for exploring biological and health hazardous effects on extremely low-frequency EMFs (ELF–EMF) have been done. It has been reported that the suppression of natural killer (NK) cell activity in mice after subchronic and chronic exposures to 60 Hz EMF at 1.0 mT (House et al., 1996; House and McCormick, 2000), although the specific mechanisms related to these phenomena are unclear. Moreover, it is uncertain whether the same effects occur under normal physiological conditions.

In the fields of ELF–EMF exposure effects on the microcirculatory system in vivo, Ohkubo research group (Xu et al., 2001; Ushiyama and Ohkubo, 2004; Ushiyama et al., 2004a,b) has investigated the effects. Xu et al. (2001) investigated acute hemodynamic effects of not only SMF but also ELF–EMF at a threshold level on modulating the muscle capillary microcirculation in pentobarbital-anesthetized mice. The skin in a tibialis anterior was circularly removed with 1.5 mm diameter for intravital-microscopic recording of the capillary blood velocity in the tibialis anterior muscle. Fluorescein isothiocyanate-labeled dextran (FITC-Dx) was used for an in vivo fluorescent plasma marker of the muscle capillaries. Following a bolus injection of FITC–Dx solution into the caudal vein, the peak blood velocity in the muscle capillaries was measured prior to, during, and following exposure to SMF or 50 Hz EMF using a fluorescence epi-illumination system. The whole body of experimental animals, placed on the observing stage of a fluorescence microscope, was exposed to SMF of 0.3, 1.0, and 10 mT or 50 Hz EMF of 0.3 and 1.0 mT for 10 min using a specially devised electromagnet. For sham exposure, the electromagnet was not energized. During exposure and postexposure to SMF of 10.0 mT, the peak blood velocity significantly increased as compared to sham exposure. After the withdrawal of SMF and 50 Hz EMF of 1.0 mT, significant similar effects on the blood velocity were present or enhanced. These findings suggest that field intensity of 1.0 mT might be considered as a threshold level for enhancing muscle microcirculation under pentobarbital-induced hypnosis.

Furthermore, Ohkubo research group employed an in vivo microscopic approach using DSC method and quantified the effect on immune responses at microcircular level in subcutaneous tissue (Ushiyama and Ohkubo, 2004; Ushiyama et al., 2004a,b). They explored the effect of ELF–EMF (0.3, 1.0, 3.0, 10, or 30 mT) of 50 Hz on microcirculatory system, acute and subchronic effects on leukocyte–endothelium interactions in subcutaneous tissue of mice. They developed a nonmetallic DSC, which is made of polyacetal resin (Duracon™). This nonmetallic frame chamber can be applied to studies for ELF–EMF exposure without any resultant thermal effects. The adherent leukocyte count to the endothelium is one of the good indicators for estimating pathophysiological conditions; particularly, rolling counts always increased when the immune system is activated. In a series of experiments, they focused on free-flowing leukocytes in venules, to investigate the effects of short ELF–EMF exposure periods on the interaction between leukocytes and endothelium. Leukocyte–endothelium interactions mainly occur under conditions of inflammation, in regions where leukocytes secrete cytokines, which induce expression of the cell adhesion molecules on endothelial cell surfaces. Under inflammatory conditions, this process permits leukocytes to adhere to tissues (Waldman and Knight, 1996).

Ushiyama and Ohkubo (2004) reported that significant increase in endothelium-adhering and rolling leukocytes was detected in venules following acute EMF exposure (50 Hz, 30 mT, for 30 min). These vessels showed no abnormal pathophysiology or inflammation under the intravital microscopy. Since the velocities of free-flowing leukocytes were unchanged, as determined by mean blood velocity results, the increased leukocyte–endothelium interaction observed may be attributed to EMF exposure, subsequently indicating that EMFs trigger the modulation of endothelial cell adhesion. Additionally, this effect will not be directly related to the health effect because increased interaction was found only in the 30 mT exposure group.

Ushiyama et al. (2004a) reported similar result from the experiments of subchronic exposure. In this experiment, mice with DSC were maintained under the various intensity (0.3, 1.0, and 3.0 mT) of 50 Hz EMF for 17 consecutive days. They demonstrated that a significant increased number of rolling leukocytes compared to preexposure status was observed only in the highest exposure level group (3.0 mT). In other groups, no significant change in rolling count was observed in any measuring time point. They also measured TNF-α and IL-1β concentration in serum; however, there was no significant change among three groups, suggesting the sensitivity against these molecules is not good enough or other unknown mechanism is underlied.

In a recent in vivo study, Gutiérrez-Mercado et al. (2013) investigated the effects of ELF–EMF on the capillaries of some circumventricular organs (CVOs). They showed that an ELF–EMF (120 Hz, 0.66 mT, for 2 h/day for a 7-day period) induces a vasodilation as well as an increase in the permeability to nonliposoluble substances in rats. All animals were administered colloidal carbon (CC) intravenously to study, through optical and transmission electron microscopy, the capillary permeability in CVO and the BBB in brain areas. An increase in capillary permeability to CC was detected in the ELF–EMF-exposed group as well as a significant increase in vascular area (capillary vasodilation); none of these effects were observed in individuals of the control (no treatment) and sham ELF–EMF groups. They suggested that the ELF–EMF has over structural and permeability characteristics of CVO capillaries and to a lesser extent on brain regions with BBB.

Using a combination of in vivo and in vitro approaches, Delle Monache et al. (2013) investigated the effect and underlying mechanism of ELF–EMF (50 Hz, 2 mT) on endothelial cell models HUVEC and MS1 (mouse pancreatic endothelial cells) measuring cell status and proliferation, motility, and tubule formation ability. MS1 cells when injected in mice determined a rapid tumorlike growth that was significantly reduced in mice inoculated with ELF–EMF-exposed cells. In particular, histological analysis of tumors derived from mice inoculated with ELF–EMF-exposed MS1 cells indicated a reduction of hemangioma size and of blood-filled spaces and hemorrhage. In parallel, in vitro proliferation of MS1 treated with ELF–EMF was significantly inhibited. They also found that the ELF–EMF exposure downregulated the process of proliferation, migration, and formation of tubule-like structures in HUVEC. In particular, ELF–EMF exposure significantly reduced the expression and activation levels of vascular endothelial growth factor receptor 2 (VEGFR2), suggesting a direct or indirect influence of ELF–EMF on VEGF receptors placed on cellular membrane. In conclusion, ELF–EMF reduced, in vivo and in vitro, the ability of endothelial cells to form new vessels, most probably affecting VEGF signal transduction pathway that was less responsive to activation. They suggested that these findings could not only explain the mechanism of antiangiogenic action exerted by ELF–EMF but also promote the possible development of new therapeutic applications for treatment of those diseases where excessive angiogenesis is involved.

Robertson et al. (2007) reviewed several mechanisms of protection by various types of ELF–EMF, such as heat shock proteins (HSPs), opioids, collateral blood flow, and NO induction, and the evidence supporting the use of ELF–EMF as a means of providing protection in each of these mechanisms. The comments were that, although there are few studies demonstrating direct protection with ELF–EMF therapies, there are many published reports suggesting that ELF–EMF may be able to influence some of the biochemical systems with protective applications (Robertson et al., 2007).

As for clinical studies, McNamee et al. (2010, 2011) indicated that ELF–EMF (60 Hz, 0.2 and 1.8 mT) did not significantly affect perfusion, heart rate, or mean arterial pressure. The decrease in perfusion and heart rate trends over time appears to be associated with a combination of inactivity (resulting in decreased body temperatures) and reduced physiological arousal. In contrast, Nishimura et al. (2011) suggested that repeated exposure to an ELF–EMF (6 and 8 Hz, 1 μT, 10 V/m, for at least two 10–15 min sessions per week, over a period of 4 weeks) has a BP-lowering effect on humans with mild-to-moderate hypertension.

8.3.1 DISCUSSION AND CONCLUSION

An early in vivo study reported that during the postexposure period of 50 Hz EMF at 1.0 mT, the peak blood velocity significantly increased as compared to sham exposure (Xu et al., 2001). In contrast, the microcirculatory parameters of pia mater were not affected by similar condition (Ushiyama and Ohkubo, 2004; Ushiyama et al., 2004a). Therefore, it was concluded that no evidence about health hazardous effect by low intensity of ELF–EMF exposure is observed (Ushiyama and Ohkubo, 2004; Ushiyama et al., 2004a). However, Gutiérrez-Mercado et al. (2013) suggested that ELF–EMF (120 Hz, 0.66 mT, for 2 h/day for a 7-day period) has over structural and permeability characteristics of some CVO capillaries and to a lesser extent on brain regions with BBB. In vivo and in vitro studies, ELF–EMF (50 Hz, 2 mT) reduced the ability of endothelial cells to form new vessels, most probably affecting VEGF signal transduction pathway that was less responsive to activation (Delle Monache et al., 2013). Several mechanisms of protection by ELF–EMF have been reported, such as HSP, opioids, collateral blood flow, and NO induction (Robertson et al., 2007). The comments were that, although there are few studies demonstrating direct protection with ELF–EMF therapies, there are many published reports suggesting that ELF–EMF may be able to influence some of the biochemical systems with protective applications (Robertson et al., 2007). Few clinical reports have addressed the effects of ELF–EMF. Therefore, further exploration is required to comprehensively evaluate and understand the effects of ELF–EMF exposure on microcirculation along with the resulting physiological consequences.

8.4 PULSED ELECTROMAGNETIC FIELD

In the past few decades since Bassett's reports (Bassett et al., 1982a,b), therapy with pulsed electromagnetic fields (PEMFs) has led to clinical trials, commercial production, and availability of devices for promoting the healing of bone nonunions in the clinic (Chalidia et al., 2011; Griffin et al., 2011). Accordingly, it has been clarified that PEMF stimulation could promote neovascularization and improve the perfusion (Lin et al., 1993; Smith et al., 2004; Kavak et al., 2009; McKay et al., 2010; Nikolaeva et al., 2010; Pan et al., 2013).

Lin et al. (1993) reported that PEMF exposure (10 Hz, 5 mT, for 6 h/day) in rabbits induced the increase of blood flow and fibroblasts at the defects from 2 to 4 weeks after operation. These results suggested that PEMF enhanced the blood flow and increased the fibroblasts at the defect. At the same time, PEMFs directly stimulated the collagen production from the fibroblasts, thus accelerating the healing process of the ligament.

Smith et al. (2004) reported that local PEMF stimulation (18.8 T/s [positive amplitude] and 8 T/s [negative amplitude] for 2 min) produced significant vasodilation, compared to prestimulation values, in cremasteric arterioles in anesthetized rats. This dilation occurred after 2 min of stimulation (9% diameter increase) and after 1 h of stimulation (8.7% diameter increase). Rats receiving sham stimulation demonstrated no statistically significant change in arteriolar diameter following either sham stimulation period. PEMF stimulation of the cremaster did not affect systemic arterial pressure or heart rate, nor was it associated with a change in tissue environmental temperature.

These results suggested that local application of a specific PEMF waveform can elicit significant arteriolar vasodilation.

Kavak et al. (2009) examined the PEMF (50 Hz, 5 mT) effects of thoracic aorta rings obtained from streptozotocin-induced diabetic and healthy control rats to determine if PEMF could ameliorate problems associated with diabetes. Streptozotocin was given via tail vein to produce diabetes mellitus (DM). The PEMF stimulation occurred four times daily for 30 min at 15 min intervals repeated daily for 30 days. Thoracic aorta rings from both DM and non-DM rats exposed to PEMF were evaluated for contraction and relaxation responses and membrane potential changes in the presence or absence of chemical agents that were selected to test various modes of action. Relaxation response of thoracic aorta rings was significantly reduced in DM than non-DM group. PEMF significantly increased the relaxation response of the diabetic rings to ACh and reduced the concentration response to phenylephrine. Resting membrane potential was significantly higher in DM than in non-DM group. Inhibitors of NO, both nitro-l-arginine (l-NO-ARG) and l-NO-ARG + indometacin combination, produced a significant transient hyperpolarization in all groups. Inhibitors of potassium channel activity, charybdotoxin or apamin, produced a membrane depolarization. However, PEMF did not induce any significant effect on the membrane potential in DM group. It was concluded that treatment with PEMF ameliorated the diabetes-induced impairments in the relaxation response of these rings.

Pan et al. (2013) indicated that PEMF stimulation (15 Hz, 1.2 mT, 8 h/day for 28 days) enhanced acute hind limb ischemia-related perfusion and angiogenesis, associated with upregulating fibroblast growth factor (FGF)-2 expression and activating the extracellular signal-regulated kinase (ERK)1/2 pathway in diabetic rats. They suggested that PEMF stimulation may be valuable for the treatment of diabetic patients with ischemic injury.

McKay et al. (2010) investigated the acute effect of a PEMF (72 Hz, 225 µT, 6.7 mV/m, for 30 and 60 min) on blood flow in the skeletal microvasculature of a male Sprague–Dawley rat model. ACh (0.1, 1.0, and 10.0 mM) was used to perturb normal blood flow and to delineate the differential effects of PEMF, based on the degree of vessel dilation. The authors found that there were no significant effects of PEMF on peak blood flow, heart rate, and myogenic activity, but a small attenuation effect on anesthetic-induced respiratory depression was noted.

Borsody et al. (2013) investigated an effective means of PEMF stimulation of the facial nerve for the purpose of increasing CBF. In normal anesthetized dog and sheep, a focal PEMF was directed toward the facial nerve within the temporal bone by placing a 6.5 cm figure 8 stimulation coil over the ear. In an initial set of experiments, CBF was measured by laser Doppler flowmetry and the cerebral vasculature was visualized by angiography. The effect of facial nerve stimulation was found to be dependent on stimulation power, frequency, and the precise positioning of the stimulation coil. Furthermore, an increase in CBF was not observed after direct electrical stimulation in the middle ear space, indicating that nonspecific stimulation of the tympanic plexus, an intervening neural structure with vasoactive effects, was not responsible for the increase in CBF after PEMF stimulation. Subsequent experiments using perfusion MRI demonstrated reproducible increases in CBF throughout the forebrain that manifested bilaterally, albeit with an ipsilateral predominance. Moreover, Borsody et al. (2014) performed an additional experiment using an ischemic stroke dog model involving injection of autologous blood clot into the internal carotid artery that reliably embolizes to the middle cerebral artery. Facial nerve stimulation caused a significant improvement in perfusion in the hemisphere affected by ischemic stroke and a reduction in ischemic core volume in comparison to sham stimulation control. The ATP/total phosphate ratio showed a large decrease poststroke in the control group versus a normal level in the stimulation group. The same stimulation administered to dogs with brain hemorrhage did not cause hematoma enlargement. They concluded that these results support the development and evaluation of a noninvasive facial nerve stimulator device as a treatment of ischemic stroke (Borsody et al., 2013, 2014).

For clinical evaluation, Nikolaeva et al. (2010) reported that for children with diabetic polyneuropathy, after local PEMF (16 Hz, 45 mT, for 15 min), along with improvements in clinical measures, the microcirculatory bed showed a virtually twofold increase in blood influx with predominant changes in efflux with maintained high shunting parameter. Myogenic tone reacted most strongly to PEMF, with weak reactions from the endothelium and neurogenic tone. This state of the microcirculatory bed can be evaluated as an adaptive mechanism, ensuring effective circulation, its shunting, and activation of the myogenic component of regulation.

Mesquita et al. (2013) investigated the effects of ELF (1 Hz)–PEMF called repetitive transcranial magnetic stimulation (rTMS) on motor cortex CBF and tissue oxygenation in seven healthy adults, during/after 20 min stimulation. Noninvasive optical methods are employed: diffuse correlation spectroscopy (DCS) for blood flow and diffuse optical spectroscopy (DOS) for Hb concentrations. A significant increase in median CBF (33%) on the side ipsilateral to stimulation was observed during rTMS and persisted after discontinuation. The measured hemodynamic parameter variations enabled computation of relative changes in cerebral metabolic rate of oxygen consumption during rTMS, which increased significantly (28%) in the stimulated hemisphere. By contrast, hemodynamic changes from baseline were not observed contralateral to rTMS administration. They suggested that these findings provide new information about hemodynamic/metabolic responses to rTMS and demonstrated the feasibility of DCS/DOS for noninvasive monitoring of TMS-induced physiological effects.

8.4.1 DISCUSSION AND CONCLUSION

Though the available evidence suggests that specific PEMF stimulation may offer some benefit in the treatment of microcirculatory disorders, it is inconclusive and insufficient to inform current practice. More definitive conclusions on treatment effect await further well-conducted randomized controlled trials. Further studies are required to explore the effects and underlying mechanisms.

8.5 RADIO FREQUENCY–ELECTROMAGNETIC FIELDS

There is some concern that the exposure to RF–EMFs emitted by cellular phones could cause adverse health effects (Mann et al., 2000). The BBB function has been focused on as one of the important research topic related to the adverse health effects of RF–EMF on tissue or organ microcirculation (Ohkubo et al., 2007). The BBB function is important for maintenance of brain homeostasis (Mayhan, 2001). Many researchers have reported that RF–EMF exposures increase the blood flow in the microcirculation of musculature over the past two decades (Shrivastav et al., 1983; Sharma and Hoopes, 2003). Most of the results were, however, obtained under thermal conditions induced by high intensity of RF–EMF exposure. Thus, those responses were attributed to the increase in body or local region temperature.

There had been little information about the effects of RF–EMF exposures on microcirculation under nonthermal conditions until 1994. In this year, Salford et al. (1994) reported that albumin leakage sites were found in the rat brain after 915 MHz–EMF exposure for 2 h even under nonthermal intensity levels, which is less than 0.08 W/kg of whole-body averaged SAR. Because permeability change of BBB has been a matter of concern as it could result in health hazard on the brain, many research groups attempted to confirm their results. However, a few studies (Schirmacher et al., 2000; Aubineau and Tore, 2005) found in the low-level RF–EMF affect BBB permeability in vivo and in vitro, whereas others (Tsurita et al., 2000; Franke et al., 2005a,b; Kuribayashi et al., 2005; Finnie et al., 2006) failed to replicate Salford's findings (Salford et al., 1994, 2003).

The cerebral microcirculatory dynamics including BBB function have been a target to evaluate the biological effects of RF–EMF (Masuda et al., 2007a–c, 2009, 2011). As indicators of dynamic changes in microcirculation, there are several parameters, such as blood flow velocity, vessel diameter changes, and leukocyte behavior. Those changes are often observed in the tissue or organs under pathophysiological conditions. For example, the BBB disruption and the increase in leukocyte adhesiveness to endothelium in pial venules were found in inflammatory brain (Mayhan, 2000;

Gaber et al., 2004). Therefore, simultaneous investigation of several parameters is helpful to assess the effects of RF–EMF exposures on the microcirculation.

The closed cranial window (CCW) method is one of the useful techniques to evaluate the cerebral microcirculatory parameters in experimental animals in vivo (Yuan et al., 2003; Gaber et al., 2004; Masuda et al., 2007a–c, 2009, 2011). This method allows for direct observation of pial microvasculature and several blood cells via a transparent glass window implanted on the parietal region of the brain. To investigate the possible effects of the exposure to RF–EMF on cerebral microcirculation including several parameters mentioned earlier, Masuda et al. (2007a–c) introduced the CCW method into rats and observed the changes in the parameters in the brain after acute or subchronic exposure to RF–EMF using the exposure system consisted of a small anechoic chamber and a monopole antenna. The head of each rat was positioned toward the central antenna and was locally exposed to 1439 MHz electromagnetic near-field of the time division multiple access (TDMA) signal controlled by mean SAR of the brain. The values of brain-averaged SAR were 0.6, 2.4, and 4.8 W/kg for acute exposure experiment and 2.4 W/kg for subchronic exposure experiment, respectively. The exposure duration was 10 min for the acute exposure and was 60 min everyday, 5 days a week for 4 weeks for subchronic exposure. The pial microcirculation including vascular diameters, plasma velocities, leukocyte behavior, and BBB function within the CW was observed using intravital fluorescence microscopy. As results in acute exposure experiment, the values in the diameters and maximal plasma velocity of the pial venule of pre- and postexposures did not significantly differ from each other for any tested SAR. Corresponding to the increase in SAR, the number of rolling leukocytes on the venular endothelia tended to decrease; however, no significant differences were recognized between the values for pre- and postexposures. No extravasation of two kinds of fluorescence dyes, FITC–Dx (MW: 250 kDa) and sodium fluorescein (MW: 376 kDa), from the pial venule was noticed due to any SAR. Furthermore, in subchronic exposure experiment, no significant differences were recognized between the values for pre- and postexposure in plasma velocities or adherent leukocyte counts. No extravasation of the two kinds of fluorescence dyes from the pial venule was noticed.

McQuade et al. (2009) carried out a study designed to confirm whether exposure to 915 MHz radiation, using a similar transverse electromagnetic (TEM) transmission line exposure cell and similar exposure parameters to those used by Salford and colleagues, caused the extravasation of albumin in rat brain tissue. These authors exposed or sham-exposed the rats (28–46 per group) for 30 min to CW 915 MHz or 915 MHz radiation pulse modulated at 16 or 270 Hz at whole-body SARs ranging between 1.8 mW/kg and 20 W/kg and examined the brain tissue shortly after exposure. The authors examined coronal sections from three or more regions along the rostro-caudal axis, assigning scores for extracellular extravasation across the whole section. Separate brain regions in each section were distinguished but these results were not presented. Overall, McQuade et al. (2009) reported little or no extracellular extravasation of albumin in the brain tissue of any exposure group compared to sham-exposed animals, in contrast to the effects seen in the positive control groups.

de Gannes et al. (2009) also used improved staining techniques, as well as those originally used by the Salford research group, in order to identify albumin extravasation and the presence of dark neurons in rat brains 14 or 50 days after the head-only exposure or sham exposure of rats (8 rats per group) for 2 h to a Global System for Mobile Communications (GSM)-900 signal at brain-averaged SARs of 140 mW/kg and 2.0 W/kg. In addition, de Gannes et al. (2009) used a more specific marker for neuronal degeneration than the one used by the Salford research group and also looked for the presence of apoptotic neurons. Like McQuade et al. (2009) and Masuda et al. (2009), de Gannes et al. (2009) also used a cage-control group and a positive control group. The authors reported that they were unable to find any evidence of the increase in albumin extravasation or number of dark neurons in 12 different regions of the brain tissue of exposed animals, although clear increases in both were seen in the positive control group.

Huber et al. (2002, 2003, 2005) examined the effects of RF–EMF exposures on rCBF, electroencephalogram (EEG), and heart rate variability (HRV). They reported that (1) pulse-modulated

RF–EMF alters waking rCBF and (2) pulse modulation of RF–EMF is necessary to induce waking and sleep EEG changes and (3) affects HRV. They speculated that pulse-modulated RF–EMF exposure may provide a new, noninvasive method for modifying brain function for experimental, diagnostic, and therapeutic purposes.

The effects of RF–EMF exposures on skin temperature were measured under normal blood flow and without blood flow in the rabbit ear (Jia et al., 2007). The results showed the following: (1) physiological blood flow clearly modified RF–EMF-induced thermal elevation in the pinna as blood flow significantly suppressed temperature increases even at 34.3 W/kg and (2) under normal blood flow conditions, exposures at 2.3 and 10.0 W/kg, approximating existing safety limits for the general public (2 W/kg) and occupational exposure (10 W/kg), did not induce significant temperature rises in the rabbit ear. However, 2.3 W/kg induced local skin temperature elevation under no blood flow conditions. The results demonstrate that the physiological effects of blood flow should be considered when extrapolating modeling data to living animals and particular caution is needed when interpreting the results of modeling studies that do not include blood flow.

Other in vivo effects of RF–EMF (GSM signals) have been investigated in the skin of hairless rat (Masuda et al., 2006; Sanchez et al., 2006, 2008). The results of 2 h acute exposure (Masuda et al., 2006) and 12-week chronic exposure (Sanchez et al., 2006) did not demonstrate major histological variations, and there was no evidence that GSM signals alter HSP expression in rat skin (Sanchez et al., 2008).

8.5.1 Discussion and Conclusion

Masuda et al. (2007a–c, 2009, 2011) focused on the cerebral microcirculatory parameters and investigated the acute and subchronic effects on the exposure to RF–EMF on those parameters. According to the results of these studies, no significant changes were found at least in vascular diameters, plasma velocities, leukocyte behavior, or BBB function either after acute or subchronic exposure experiment. These findings lead to the following two suggestions.

The first is that the RF–EMF exposures lower than the local permissible level (2.0 W/kg) in the ICNIRP guidelines (ICNIRP, 1998) do not induce any changes in cerebral microcirculation, if a presumption for dose–response relationship between intensities of the RF–EMF exposures and biological responses is accepted. The values of brain-averaged SAR in the present exposures were 0.6, 2.4, and 4.8 W/kg. These exposure levels range from a low level comparable to the study by Salford et al. (1994) to a moderate level of 2.4 times higher than that of the safety guideline. Several studies that evaluated changes in BBB permeability after RF–EMF exposure found that the exposure under lower SAR level than the 2.0 W/kg did not modify the BBB permeability (Tsurita et al., 2000; Kuribayashi et al., 2005; Franke et al., 2005a,b; Finnie et al., 2006). These results not only support the previous findings but also provide new information for considering the effects of RF–EMF exposures on the microcirculation.

The second is that the multiparameter evaluation supports the lack of increase in the BBB permeability under RF–EMF exposures. Although many investigators have reported the effects of RF–EMF on BBB permeability, these studies mainly used histological evaluation (Salford et al., 1994, 2003; Fritze et al., 1997; Tsurita et al., 2000; Kuribayashi et al., 2005; Franke et al., 2005a,b; Finnie et al., 2006; Nittby et al., 2008). On the contrary, Masuda et al. (2007a–c, 2009, 2011), McQuade et al. (2009), and de Gannes et al. (2009) examined not only BBB permeability but also other microcirculatory parameters in vivo. Several reports have shown that the BBB disruption is accompanied with changes in leukocyte behaviors (Mayhan, 2000; Gaber et al., 2004) or hemodynamics (Mayhan, 1998) under inflammatory condition in the rat brain. Therefore, the findings by these investigators strengthen the negative results that the RF–EMF exposure does not induce BBB disruption. However, further studies are required under other exposure conditions to confirm these phenomena.

Thus, many confirmation studies in vivo and in vitro have been performed to confirm since Salford's findings (Salford et al., 1994). Most of them suggest no effects of RF–EMF exposure on the BBB. However, these investigators recognize that no one can replicate the Salford's studies (1994, 2003) with the same methodology they used, because their methodology involves much uncertainty (SSM, 2009; Stam, 2010). Therefore, there is still considerable disagreement about the Salford's reports (SSM, 2009; Stam, 2010).

8.6 OVERALL CONCLUSION

There is an importance of understanding the effects of various kinds of magnetic fields with different intensities and frequencies on microcirculatory system. Methodologically speaking, pharmacological treatments as well as pathological animal models are useful for experimental evaluation of the effects and mechanisms of magnetic fields. It may have direct and indirect role in the interaction of magnetic fields with different tissues and organs. The results could be useful in applying specific SMF and PEMF for microcirculatory disorders. In contrast, the results obtained from exposure to most of ELF–EMF and RF–EMF failed to show any changes in microcirculatory system except for leukocyte and endothelial cell interaction. The range of EMF investigated is very higher than that of the international exposure guidelines. These microcirculatory studies in the animal models, combined with in vivo real-time molecular imaging techniques and in vitro pathway analysis, can contribute to evaluate therapeutic application and possible health risks of magnetic fields.

REFERENCES

Asano, M. and P. I. Brånemark. 1972. Microphotoelectric plethysmography using a titanium chamber in man. In *Advances in Microcirculation*, Vol. 4. H. Harders (ed.). S. Karger, Basel, Switzerland, pp. 131–160.

Asano, M., K. Yoshida, and K. Tatai. 1965. Microphotoelectric plethysmography using a rabbit ear chamber. *J Appl Physiol* 20:1056–1062.

Atef, M. M., M. S. Abd el-Baset, A. el-Kareem, S. Aida, and M. A. Fadel. 1995. Effects of a static magnetic field on hemoglobin structure and function. *Int J Biol Macromol* 17:105–111.

Aubineau, P. and F. Tore. 2005. Head exposure to 900 MHz microwaves induces plasma protein extravasation in the rat brain and dura mater at non-thermal SAR levels. In *Monografie Workshops: Influences of RF- and Electromagnetic Field Interaction*. Berufsgenossenschaft der Feinmechanik und Elektrotechnik, Köln, Germany, pp. 82–83.

Bassett, C. A., S. N. Mitchell, and S. R. Gaston. 1982b. Pulsing electromagnetic field treatment in ununited fractures and failed arthrodeses. *JAMA* 247:623–638.

Bassett, C. A., M. G. Valdes, and E. Hernandez. 1982a. Modification of fracture repair with selected pulsing electromagnetic fields. *J Bone Joint Surg Am* 64:888–895.

Bezanilla, F. 2002. Voltage sensor movements. *J Gen Physiol* 120:465–473.

Borsody, M. K., C. Yamada, D. Bielawski, T. Heaton, F. Castro Prado, A. Garcia, J. Azpiroz, and E. Sacristan. 2014. Effects of noninvasive facial nerve stimulation in the dog middle cerebral artery occlusion model of ischemic stroke. *Stroke* 45:1102–1107.

Borsody, M. K., C. Yamada, D. Bielawski, T. Heaton, B. Lyeth, A. Garcia, F. Castro Prado, J. Azpiroz, and E. Sacristan. 2013. Effect of pulsed magnetic stimulation of the facial nerve on cerebral blood flow. *Brain Res* 1528:58–67.

Brix, G., S. Strieth, D. Strelczyk, M. Dellian, J. Griebel, M. E. Eichhorn, W. Andrä, and M. E. Bellemann. 2008. Static magnetic fields affect capillary flow of red blood cells in striated skin muscle. *Microcirculation* 15:15–26.

Cavopol, A. V., A. W. Wamil, R. R. Holcomb, and M. J. McLean. 1995. Measurement and analysis of static magnetic fields that block action potentials in cultured neurons. *Bioelectromagnetics* 16:197–206.

Chakeres, D. W. and F. de Vocht. 2005. Static magnetic field effects on human subjects related to magnetic resonance imaging systems. *Prog Biophys Mol Biol* 87:255–265.

Chalidia, B., N. Sachinis, A. Assiotis, and G. Maccauro. 2011. Stimulation of bone formation and fracture healing with pulsed electromagnetic fields: Biologic responses and clinical implications. *Int J Immunopathol Pharmacol* 24:17–20.

Chorny, M., I. Fishbein, B. B. Yellen, I. S. Alferiev, M. Bakay, S. Ganta, R. Adamo, M. Amiji, G. Friedman, and R. J. Levy. 2010. Targeting stents with local delivery of paclitaxel-loaded magnetic nanoparticles using uniform fields. *Proc Natl Acad Sci USA* 107:8346–8351.

Crozier, S. and F. Liu. 2005. Numerical evaluation of the fields induced by body motion in or near high-field MRI scanners. *Prog Biophys Mol Biol* 87:267–278.

Csillag, A., B. V. Kumar, K. Szabo, M. Szilasi, Z. Papp, M. E. Szilasi, K. Pazmandi et al. 2014. Exposure to inhomogeneous static magnetic field beneficially affects allergic inflammation in a murine model. *J R Soc Interface* 11:Article ID 20140097.

de Gannes, F. P., B. Billaudel, M. Taxile, E. Haro, G. Ruffié, P. Lévêque, B. Veyret, and I. Lagroye. 2009. Effects of head-only exposure of rats to GSM-900 on blood-brain barrier permeability and neuronal degeneration. *Radiat Res* 172:359–367.

Delle Monache, S., A. Angelucci, P. Sanità, R. Iorio, F. Bennato, F. Mancini, G. Gualtieri, and R. C. Colonna. 2013. Inhibition of angiogenesis mediated by extremely low-frequency magnetic fields (ELF-MFs). *PLoS ONE* 8:Article ID e79309.

Denegre, J. M., J. M. Valles Jr., K. Lin, W. B. Jordan, and K. L. Mowry. 1998. Cleavage planes in frog eggs are altered by strong magnetic fields. *Proc Natl Acad Sci USA* 95:14729–14732.

Dini, L. 2010. Phagocytosis of dying cells: Influence of smoking and static magnetic fields. *Apoptosis* 15:1147–1164.

Djordjevich, D. M., S. R. De Luka, I. D. Milovanovich, S. Janković, S. Stefanović, S. Vesković-Moračanin, S. Cirković, A. Ž. Ilić, J. L. Ristić-Djurović, and A. M. Trbovich. 2012. Hematological parameters' changes in mice subchronically exposed to static magnetic fields of different orientations. *Ecotoxicol Environ Safety* 81:98–105.

Dodson, C. A., P. J. Hore, and M. I. Wallace. 2013. A radical sense of direction: Signalling and mechanism in cryptochrome magnetoreception. *Trends Biochem Sci* 38:435–446.

Eder, S. H., H. Cadiou, A. Muhamad, P. A. McNaughton, J. L. Kirschvink, and M. Winklhofer. 2012. Magnetic characterization of isolated candidate vertebrate magnetoreceptor cells. *Proc Natl Acad Sci USA* 109:12022–12027.

Engström, S., M. S. Markov, M. J. McLean, R. R. Holcomb, and J. M. Markov. 2002. Effects of non-uniform static magnetic fields on the rate of myosin phosphorylation. *Bioelectromagnetics* 23:475–479.

Finnie, J. W., P. C. Blumbergs, Z. Cai, J. Manavis, and T. R. Kuchel. 2006. Effect of mobile telephony on blood-brain barrier permeability in the fetal mouse brain. *Pathology* 38:63–65.

Foley, L. E., R. J. Gegear, and S. M. Reppert. 2011. Human cryptochrome exhibits light-dependent magneto-sensitivity. *Nat Commun* 2:Article ID 356.

Franke, H., E. B. Ringelstein, and F. Stögbauer. 2005a. Electromagnetic fields (GSM 1800) do not alter blood-brain barrier permeability to sucrose in models in vitro with high barrier tightness. *Bioelectromagnetics* 26:529–535.

Franke, H., J. Streckert, A. Bitz, J. Goeke, V. Hansen, E. B. Ringelstein, H. Nattkamper, H. J. Galla, and F. Stögbauer. 2005b. Effects of universal mobile telecommunications system (UMTS) electromagnetic fields on the blood-brain barrier in vitro. *Radiat Res* 164:258–269.

Fritze, K., C. Sommer, B. Schmitz, G. Mies, K. A. Hossmann, M. Kiessling, and C. Wiessner. 1997. Effect of global system for mobile communication (GSM) microwave exposure on blood-brain barrier permeability in rat. *Acta Neuropathol (Berl)* 94:465–470.

Funk, R. H., T. Monsees, and N. Ozkucur. 2009. Electromagnetic effects—From cell biology to medicine. *Prog Histochem Cytochem* 43:177–264.

Funk, W. and M. Intaglietta. 1983. Spontaneous arteriolar vasomotion. In *Progress in Applied Microcirculation*, Vol. 3. K. Messemer and F. Hammersen (eds.). S. Karger, Basel, Switzerland, pp. 66–82.

Gaber, M. W., H. Yuan, J. T. Killmar, M. D. Naimark, M. F. Kiani, and T. E. Merchant. 2004. An intravital microscopy study of radiation-induced changes in permeability and leukocyte-endothelial cell interactions in the microvessels of the rat pia mater and cremaster muscle. *Brain Res Brain Res Protoc* 13:1–10.

Gegear, R. J., L. E. Foley, A. Casselman, and S. M. Reppert. 2010. Animal cryptochromes mediate magnetoreception by an unconventional photochemical mechanism. *Nature* 463:804–807.

Gellrich, D., S. Becker, and S. Strieth. 2014. Static magnetic fields increase tumor microvessel leakiness and improve antitumoral efficacy in combination with paclitaxel. *Cancer Lett* 343:107–114.

Glover, P. M. and R. Bowtell. 2008. Measurement of electric fields induced in a human subject due to natural movements in static magnetic fields or exposure to alternating magnetic field gradients. *Phys Med Biol* 53:361–373.

Glover, P. M., Y. Li, A. Antunes, O. S. Mian, and B. L. Day. 2014. A dynamic model of the eye nystagmus response to high magnetic fields. *Phys Med Biol* 59:631–645.

Gmitrov, J. and C. Ohkubo. 2002a. Artificial static and geomagnetic field interrelated impact on cardiovascular regulation. *Bioelectromagnetics* 23:329–338.

Gmitrov, J. and C. Ohkubo. 2002b. Effects of 12 mT static magnetic field on sympathetic verapamil protective effect on natural and artificial magnetic field cardiovascular impact. *Bioelectromagnetics* 23:531–541.

Gmitrov, J., C. Ohkubo, and H. Okano. 2002. Effect of 0.25 T static magnetic field on microcirculation in rabbits. *Bioelectromagnetics* 23:224–229.

Griffin, X. L., M. L. Costa, N. Parsons, and N. Smith. 2011. Electromagnetic field stimulation for treating delayed union or non-union of long bone fractures in adults. *Cochrane Database Syst Rev* 4:Article ID CD008471.

Grossman, G. E., R. J. Leigh, L. A. Abel, D. J. Lanska, and S. E. Thurston. 1988. Frequency and velocity of rotational head perturbations during locomotion. *Exp Brain Res* 70:470–476.

Gupta, A., A. R. Weeks, and S. M. Richie. 2008. Simulation of elevated T-waves of an ECG inside a static magnetic field (MRI). *IEEE Trans Biomed Eng* 55:1890–1896.

Gutiérrez-Mercado, Y. K., L. Cañedo-Dorantes, U. Gómez-Pinedo, G. Serrano-Luna, J. Bañuelos-Pineda, and A. Feria-Velasco. 2013. Increased vascular permeability in the circumventricular organs of adult rat brain due to stimulation by extremely low frequency magnetic fields. *Bioelectromagnetics* 34:145–155.

Haik, Y., V. Pai, and C. Chen. 2001. Apparent viscosity of human blood in a high static magnetic field. *J Magn Magn Mater* 225:180–186.

Higashi, T., A. Yamagishi, T. Takeuchi, and M. Date. 1995. Effects of static magnetic fields of erythrocyte rheology. *Bioelectrochem Bioenerg* 36:101–108.

Higashi, T., A. Yamagishi, T. Takeuchi, N. Kawaguchi, S. Sagawa, S. Onishi, and M. Date. 1993. Orientation of erythrocytes in a strong static magnetic field. *Blood* 82:1328–1334.

Hinman, M. R. 2002. Comparative effect of positive and negative static magnetic fields on heart rate and blood pressure in healthy adults. *Clin Rehabil* 16:669–674.

Hirota, Y., Y. Akiyamaa, Y. Izumi, and S. Nishijima. 2009. Fundamental study for development magnetic drug delivery system. *Phys C Superconduct* 469:1853–1856.

House, R. V. and D. L. McCormick. 2000. Modulation of natural killer cell function after exposure to 60 Hz magnetic fields: Confirmation of the effect in mature B6C3F1 mice. *Radiat Res* 153:722–724.

House, R. V., H. V. Ratajczak, J. R. Gauger, T. R. Johnson, P. T. Thomas, and D. L. McCormick. 1996. Immune function and host defense in rodents exposed to 60-Hz magnetic fields. *Fundam Appl Toxicol* 34:228–239.

Hsieh, C. H., M. C. Lee, J. J. Tsai-Wu, M. H. Chen, H. S. Lee, H. Chiang, C. H. Herbert Wu, and C. C. Jiang. 2008. Deleterious effects of MRI on chondrocytes. *Osteoarthr Cartil* 16:343–351.

Huber, R., J. Schuderer, T. Graf, K. Jütz, A. A. Borbély, N. Kuster, and P. Achermann. 2003. Radio frequency electromagnetic field exposure in humans: Estimation of SAR distribution in the brain, effects on sleep and heart rate. *Bioelectromagnetics* 24:262–276.

Huber, R., V. Treyer, A. A. Borbély, J. Schuderer, J. M. Gottselig, H. P. Landolt, E. Werth et al. 2002. Electromagnetic fields, such as those from mobile phones, alter regional cerebral blood flow and sleep and waking EEG. *J Sleep Res* 11:289–295.

Huber, R., V. Treyer, J. Schuderer, T. Berthold, A. Buck, N. Kuster, H. P. Landolt, P. Achermann. 2005. Exposure to pulse-modulated radio frequency electromagnetic fields affects regional cerebral blood flow. *Eur J Neurosci* 21:1000–1006.

Ichioka, S., M. Minegishi, M. Iwasaka, M. Shibata, T. Nakatsuka, J. Ando, and S. Ueno. 2003. Skin temperature changes induced by strong static magnetic field exposure. *Bioelectromagnetics* 24:380–386.

Ichioka, S., M. Minegishi, M. Iwasaka, M. Shibata, T. Nakatsuka, K. Harii, A. Kamiya, S. Ueno. 2000. High-intensity static magnetic fields modulate skin microcirculation and temperature in vivo. *Bioelectromagnetics* 21:183–188.

ICNIRP (International Commission on Non-Ionizing Radiation Protection). 1998. ICNIRP statement on the "Guidelines for limiting exposure to time-varying electric, magnetic, and electromagnetic fields (up to 300 GHz)." *Health Phys* 74:494–522.

ICNIRP (International Commission on Non-Ionizing Radiation Protection). 2009. Guidelines on limits of exposure to static magnetic fields. *Health Phys* 96:504–514.

ICNIRP (International Commission on Non-Ionizing Radiation Protection). 2010. Guidelines for limiting exposure to time-varying electric and magnetic fields (1 Hz to 100 kHz). *Health Phys* 99:818–836.

ICNIRP (International Commission on Non-Ionizing Radiation Protection). 2014. Guidelines for limiting exposure to electric fields induced by movement of the human body in a static magnetic field and to time-varying magnetic fields below 1 Hz. *Health Phys* 106:418–425.

Jain, R. K. 1997. The Eugene M. Landis Award Lecture 1996. Delivery of molecular and cellular medicine to solid tumors. *Microcirculation* 4:1–23.

Jauchem, J. R. 1997. Exposure to extremely-low-frequency electromagnetic fields and radiofrequency radiation: Cardiovascular effects in humans. *Int Arch Occup Environ Health* 70:9–21.

Jia, F., A. Ushiyama, H. Masuda, G. F. Lawlor, and C. Ohkubo. 2007. Role of blood flow on RF exposure induced skin temperature elevations in rabbit ears. *Bioelectromagnetics* 28:163–172.

Jokela, K., R. D. Saunders. 2011. Physiologic and dosimetric considerations for limiting electric fields induced in the body by movement in a static magnetic field. *Health Phys* 100:641–653.

Kainz, W., J. Guag, S. Benkler, D. Szczerba, E. Neufeld, V. Krauthamer, J. Myklebust et al. 2010. Development and validation of a magnetohydrodynamic solver for blood flow analysis. *Phys Med Biol* 55:7253–7261.

Kangarlu, A., R. E. Burgess, H. Zhu, T. Nakayama, R. L. Hamlin, A. M. Abduljalil, and P. M. Robitaille. 1999. Cognitive, cardiac, and physiological safety studies in ultra high field magnetic resonance imaging. *Magn Reson Imaging* 17:1407–1416.

Kavak, S., M. Emre, I. Meral, H. Unlugenc, A. Pelit, and A. Demirkazik. 2009. Repetitive 50 Hz pulsed electromagnetic field ameliorates the diabetes-induced impairments in the relaxation response of rat thoracic aorta rings. *Int J Radiat Biol* 85:672–679.

Kim, S., Y. A. Chung, C. U. Lee, J. H. Chae, R. Juh, and J. Jeong. 2010. Target-specific rCBF changes induced by 0.3-T static magnetic field exposure on the brain. *Brain Res* 1317:211–217.

Kinouchi, Y., H. Yamaguchi, and T. S. Tenforde. 1996. Theoretical analysis of magnetic field interactions with aortic blood flow. *Bioelectromagnetics* 17:21–32.

Kotani H., H. Kawaguchi, T. Shimoaka, M. Iwasaka, S. Ueno, H. Ozawa, K. Nakamura, and K. Hoshi. 2002. Strong static magnetic field stimulates bone formation to a definite orientation in vitro and in vivo. *J Bone Miner Res* 17:1814–1821.

Kuipers, N. T., C. L. Sauder, and C. A. Ray. 2007. Influence of static magnetic fields on pain perception and sympathetic nerve activity in humans. *J Appl Physiol* 102:1410–1415.

Kuribayashi, M., J. Wang, O. Fujiwara, Y. Doi, K. Nabae, S. Tamano, T. Ogiso, M. Asamoto, and T. Shirai. 2005. Lack of effects of 1439 MHz electromagnetic near field exposure on the blood-brain barrier in immature and young rats. *Bioelectromagnetics* 26:578–588.

Lee, A. A., J. C. Lau, H. J. Hogben, T. Biskup, D. R. Kattnig, and P. J. Hore. 2014. Alternative radical pairs for cryptochrome-based magnetoreception. *J R Soc Interface* 11:Article ID 20131063.

Li, Z., E. W. Tam, A. F. Mak, and R. Y. Lau. 2007. Wavelet analysis of the effects of static magnetic field on skin blood flowmotion: Investigation using an in vivo rat model. *In Vivo* 21:61–68.

Lin, C. Y., W. J. Chang, S. Y. Lee, S. W. Feng, C. T. Lin, K. S. Fan, and H. M. Huang. 2013a. Influence of a static magnetic field on the slow freezing of human erythrocytes. *Int J Radiat Biol* 89:51–56.

Lin, C. Y., P. L. Wei, W. J. Chang, Y. K. Huang, S. W. Feng, C. T. Lin, S. Y. Lee, and H. M. Huang. 2013b. Slow freezing coupled static magnetic field exposure enhances cryopreservative efficiency—A study on human erythrocytes. *PLoS ONE* 8:Article ID e58988.

Lin, Y., R. Nishimura, K. Nozaki, N. Sasaki, T. Kadosawa, N. Goto, M. Date, and A. Takeuchi. 1993. Collagen production and maturation at the experimental ligament defect stimulated by pulsing electromagnetic fields in rabbits. *J Vet Med Sci* 55:527–531.

Luo, R., Y. Zhang, and L. Xia. 2005. Electrophysiological modeling study of ECG T-wave alternation caused by ultrahigh static magnetic fields. *Conf Proc IEEE Eng Med Biol Soc* 3:3012–3015.

MacDougall, H. G. and S. T. Moore. 2005. Functional assessment of head-eye coordination during vehicle operation. *Optom Vis Sci* 82:706–715.

Maeda, K., A. J. Robinson, K. B. Henbest, H. J. Hogben, T. Biskup, M. Ahmad, E. Schleicher, S. Weber, C. R. Timmel, and P. J. Hore. 2012. Magnetically sensitive light-induced reactions in cryptochrome are consistent with its proposed role as a magnetoreceptor. *Proc Natl Acad Sci USA* 109:4774–4779.

Mann, S. M., T. G. Cooper, S. G. Allen, R. P. Blackwell, and A. J. Lowe. 2000. Exposure to radio waves near mobile phone base stations. NRPB-R321, National Radiological Protection Board.

Markov, M. S., C. D. Williams, I. L. Cameron, W. E. Hardman, and J. R. Salvatore. 2004. Can magnetic fields inhibit angiogenesis and tumor growth? In *Bioelectromagnetic Medicine*. P. J. Rosch and M. S. Markov (eds.). Marcel Dekker, New York, pp. 625–636.

Martel, G. F., S. C. Andrews, and C. G. Roseboom. 2002. Comparison of static and placebo magnets on resting forearm blood flow in young, healthy men. *J Orthop Sports Phys Ther* 32:518–524.

Martino, C. F., H. Perea, U. Hopfner, V. L. Ferguson, and E. Wintermantel. 2010. Effects of weak static magnetic fields on endothelial cells. *Bioelectromagnetics* 31:296–301.

Masuda, H., A. Hirata, H. Kawai, K. Wake, S. Watanabe, T. Arima, F. Poulletier de Gannes, I. Lagroye, and B. Veyret. 2011. Local exposure of the rat cortex to radiofrequency electromagnetic fields increases local cerebral blood flow along with temperature. *J Appl Physiol* (*1985*) 110:142–148.

Masuda, H., S. Sanchez, P. E. Dulou, E. Haro, R. Anane, B. Billaudel, P. Lévêque, and B. Veyret. 2006. Effect of GSM-900 and -1800 signals on the skin of hairless rats. I: 2-hour acute exposures. *Int J Radiat Biol* 82:669–674.

Masuda, H., A. Ushiyama, S. Hirota, G. F. Lawlor, and C. Ohkubo. 2007a. Long-term observation of pial microcirculatory parameters using an implanted cranial window method in the rat. *In Vivo* 21:471–479.

Masuda, H., A. Ushiyama, S. Hirota, K. Wake, S. Watanabe, Y. Yamanaka, M. Taki, and C. Ohkubo. 2007b. Effects of acute exposure to a 1439 MHz electromagnetic field on the microcirculatory parameters in rat brain. *In Vivo* 21:555–562.

Masuda, H., A. Ushiyama, S. Hirota, K. Wake, S. Watanabe, Y. Yamanaka, M. Taki, and C. Ohkubo. 2007c. Effects of subchronic exposure to a 1439 MHz electromagnetic field on the microcirculatory parameters in rat brain. *In Vivo* 21:563–570.

Masuda, H., A. Ushiyama, M. Takahashi, J. Wang, O. Fujiwara, T. Hikage, T. Nojima, K. Fujita, M. Kudo, and C. Ohkubo. 2009. Effects of 915 MHz electromagnetic-field radiation in TEM cell on the blood-brain barrier and neurons in the rat brain. *Radiat Res* 172:66–73.

Mayhan, W. G. 1998. Effect of lipopolysaccharide on the permeability and reactivity of the cerebral microcirculation: Role of inducible nitric oxide synthase. *Brain Res* 792:353–357.

Mayhan, W. G. 2000. Leukocyte adherence contributes to disruption of the blood-brain barrier during activation of mast cells. *Brain Res* 869:112–120.

Mayhan, W. G. 2001. Regulation of blood-brain barrier permeability. *Microcirculation* 8:89–104.

Mayrovitz, H. N. and E. E. Groseclose. 2005. Effects of a static magnetic field of either polarity on skin microcirculation. *Microvasc Res* 69:24–27.

Mayrovitz, H. N., E. E. Groseclose, and D. King. 2005. No effect of 85 mT permanent magnets on laser-Doppler measured blood flow response to inspiratory gasps. *Bioelectromagnetics* 26:331–335.

Mayrovitz, H. N., E. E. Groseclose, M. Markov, and A. A. Pilla. 2001. Effects of permanent magnets on resting skin blood perfusion in healthy persons assessed by laser Doppler flowmetry and imaging. *Bioelectromagnetics* 22:494–502.

McKay, J. C., M. Corbacio, K. Tyml, F. S. Prato, and A. W. Thomas. 2010. Extremely low frequency pulsed electromagnetic field designed for antinociception does not affect microvascular responsiveness to the vasodilator acetylcholine. *Bioelectromagnetics* 31:64–76.

McKay, J. C., F. S. Prato, and A. W. Thomas. 2007. A literature review: The effects of magnetic field exposure on blood flow and blood vessels in the microvasculature. *Bioelectromagnetics* 28:81–98.

McLean, M. J., R. R. Holcomb, A. W. Wamil, J. D. Pickett, and A. V. Cavopol. 1995. Blockade of sensory neuron action potentials by a static magnetic field in the 10 mT range. *Bioelectromagnetics* 16:20–32.

McNamee, D. A., M. Corbacio, J. K. Weller, S. Brown, F. S. Prato, A. W. Thomas, and A. G. Legros. 2010. The cardiovascular response to an acute 1800-µT, 60-Hz magnetic field exposure in humans. *Int Arch Occup Environ Health* 83:441–454.

McNamee, D. A., M. Corbacio, J. K. Weller, S. Brown, R. Z. Stodilka, F. S. Prato, Y. Bureau, A. W. Thomas, and A. G. Legros. 2011. The response of the human circulatory system to an acute 200-µT, 60-Hz magnetic field exposure. *Int Arch Occup Environ Health* 84:267–277.

McNamee, D. A., A. G. Legros, D. R. Krewski, G. Wisenberg, F. S. Prato, and A. W. Thomas. 2009. A literature review: The cardiovascular effects of exposure to extremely low frequency electromagnetic fields. *Int Arch Occup Environ Health* 82:919–933.

McQuade, J. M., J. H. Merritt, S. A. Miller, T. Scholin, M. C. Cook, A. Salazar, O. B. Rahimi, M. R. Murphy, and P. A. Mason. 2009. Radiofrequency-radiation exposure does not induce detectable leakage of albumin across the blood-brain barrier. *Radiat Res* 171:615–621.

Mesquita, R. C., O. K. Faseyitan, P. E. Turkeltaub, E. M. Buckley, A. Thomas, M. N. Kim, T. Durduran et al. 2013. Blood flow and oxygenation changes due to low-frequency repetitive transcranial magnetic stimulation of the cerebral cortex. *J Biomed Opt* 18:Article ID 067006.

Mian, O. S., Y. Li, A. Antunes, P. M. Glover, and B. L. Day. 2013. On the vertigo due to static magnetic fields. *PLoS ONE* 8:Article ID e78748.

Michel, C. C. and F. E. Curry. 1999. Microvascular permeability. *Physiol Rev* 79:703–761.

Mishima, F., S. Fujimoto, S. Takeda, Y. Izumi, and S. Nishijima. 2007. Development of magnetic field control for magnetically targeted drug delivery system using a superconducting magnet. *IEEE Trans Appl Supercond* 17:2303–2306.

Morais, P. C. 2010. Nanoparticulated magnetic drug delivery systems: Preparation and magnetic characterization. *J Phys Conf Ser* 217:Article ID 012091.

Morris, C. and T. Skalak. 2005. Static magnetic fields alter arteriolar tone in vivo. *Bioelectromagnetics* 26:1–9.

Morris, C. E. and T. C. Skalak. 2007. Chronic static magnetic field exposure alters microvessel enlargement resulting from surgical intervention. *J Appl Physiol (1985)* 103:629–636.

Muehsam, D. J. and A. A. Pilla. 2009a. A Lorentz model for weak magnetic field bioeffects: Part I—Thermal noise is an essential component of AC/DC effects on bound ion trajectory. *Bioelectromagnetics* 30:462–475.

Muehsam, D. J. and A. A. Pilla. 2009b. A Lorentz model for weak magnetic field bioeffects: Part II—Secondary transduction mechanisms and measures of reactivity. *Bioelectromagnetics* 30:476–488.

Muehsam, D., P. Lalezari, R. Lekhraj, P. Abruzzo, A. Bolotta, M. Marini, F. Bersani, G. Aicardi, A. Pilla, and D. Casper. 2013. Non-thermal radiofrequency and static magnetic fields increase rate of haemoglobin deoxygenation in a cell-free preparation. *PLoS ONE* 8:Article ID e61752.

Mulvany, M. J. 1983. Functional characteristics of vascular smooth muscle. In *Progress in Applied Microcirculation*, Vol. 3. K. Messemer and F. Hammersen (eds.). S. Karger, Basel, Switzerland, pp. 4–18.

Nakagawa, K., F. Mishima, Y. Akiyama, and S. Nishijima. 2012. Study on magnetic drug delivery system using HTS bulk magnet. *IEEE Trans Appl Supercond* 22:Article ID 4903804.

Neil, S. R., J. Li, D. M. Sheppard, J. Storey, K. Maeda, K. B. Henbest, P. J. Hore, C. R. Timmel, and S. R. Mackenzie. 2014. Broadband cavity-enhanced detection of magnetic field effects in chemical models of a cryptochrome magnetoreceptor. *J Phys Chem B* 118:4177–4184.

Nikolaeva, N. V., N. V. Bolotova, V. F. Luk'yanov, Y. M. Raigorodskii, and E. N. Tkacheva. 2010. Non-pharmacological correction of impaired microcirculation in children with diabetic polyneuropathy. *Neurosci Behav Physiol* 40:347–350.

Nishijima, S., F. Mishima, Y. Tabata, H. Iseki, Y. Muragaki, A. Sasaki, and N. Saho. 2009. Research and development of magnetic drug delivery system using bulk high temperature superconducting magnet. *IEEE Trans Appl Supercond* 19:2257–2260.

Nishijima, S., S. Takeda, F. Mishima, Y. Tabata, M. Yamamoto, J. Joh, H. Iseki et al. 2008. A study of magnetic drug delivery system using bulk high temperature superconducting magnet. *IEEE Trans Appl Supercond* 18:874–877.

Nishimura, T., H. Okano, H. Tada, E. Nishimura, K. Sugimoto, K. Mohri, and M. Fukushima. 2010. Lizards respond to an extremely low-frequency electromagnetic field. *J Exp Biol* 213:1985–1990.

Nishimura, T., H. Tada, X. Guo, T. Murayama, S. Teramukai, H. Okano, J. Yamada, K. Mohri, and M. Fukushima. 2011. A 1-µT extremely low-frequency electromagnetic field vs. sham control for mild-to-moderate hypertension: A double-blind, randomized study. *Hypertens Res* 34:372–377.

Nittby, H., G. Grafström, J. L. Eberhardt, L. Malmgren, A. Brun, B. R. Persson, L. G. Salford. 2008. Radiofrequency and extremely low-frequency electromagnetic field effects on the blood-brain barrier. *Electromagn Biol Med* 27:103–126.

Ohkubo, C. and H. Okano. 2004. Static magnetic field and microcirculation. In *Bioelectromagnetic Medicine*. P. J. Rosch and M. S. Markov (eds.). Marcel Dekker, New York, pp. 563–591.

Ohkubo, C. and H. Okano. 2011. Clinical aspects of static magnetic field effects on circulatory system. *The Environmentalist* 31:97–106.

Ohkubo, C., H. Okano, A. Ushiyama, and H. Masuda. 2007. EMF effects on microcirculatory system. *The Environmentalist* 27:395–402.

Ohkubo, C. and S. Xu. 1997. Acute effects of static magnetic fields on cutaneous microcirculation in rabbits. *In Vivo* 11:221–225.

Okano, H., J. Gmitrov, and C. Ohkubo. 1999. Biphasic effects of static magnetic fields on cutaneous microcirculation in rabbits. *Bioelectromagnetics* 20:161–171.

Okano, H., H. Masuda, and C. Ohkubo. 2005a. Effects of 25 mT static magnetic field on blood pressure in reserpine-induced hypotensive Wistar-Kyoto rats. *Bioelectromagnetics* 26:36–48.

Okano, H., H. Masuda, and C. Ohkubo. 2005b. Decreased plasma levels of nitric oxide metabolites, angiotensin II, and aldosterone in spontaneously hypertensive rats exposed to 5 mT static magnetic field. *Bioelectromagnetics* 26:161–172.

Okano, H. and C. Ohkubo. 2001. Modulatory effects of static magnetic fields on blood pressure in rabbits. *Bioelectromagnetics* 22:408–418.

Okano, H. and C. Ohkubo. 2003a. Anti-pressor effects of whole body exposure to static magnetic field on pharmacologically induced hypertension in conscious rabbits. *Bioelectromagnetics* 24:139–147.

Okano, H. and C. Ohkubo. 2003b. Effects of static magnetic fields on plasma levels of angiotensin II and aldosterone associated with arterial blood pressure in genetically hypertensive rats. *Bioelectromagnetics* 24:403–412.

Okano, H. and C. Ohkubo. 2005a. Effects of neck exposure to 5.5 mT static magnetic field on pharmacologically modulated blood pressure in conscious rabbits. *Bioelectromagnetics* 26:469–480.

Okano, H. and C. Ohkubo. 2005b. Exposure to a moderate intensity static magnetic field enhances the hypotensive effect of a calcium channel blocker in spontaneously hypertensive rats. *Bioelectromagnetics* 26:611–623.

Okano, H. and C. Ohkubo. 2006. Elevated plasma nitric oxide metabolites in hypertension: Synergistic vasodepressor effects of a static magnetic field and nicardipine in spontaneously hypertensive rats. *Clin Hemorheol Microcirc* 34:303–308.

Okano, H. and C. Ohkubo. 2007. Effects of 12 mT static magnetic field on sympathetic agonist-induced hypertension in Wistar rats. *Bioelectromagnetics* 28:369–378.

Oster, J., R. Llinares, S. Payne, Z. T. Tse, E. J. Schmidt, and G. D. Clifford. 2014. Comparison of three artificial models of the magnetohydrodynamic effect on the electrocardiogram. *Comput Methods Biomech Biomed Eng*, April 24: 1–18. [Epub ahead of print].

Pan, Y., Y. Dong, W. Hou, Z. Ji, K. Zhi, Z. Yin, H. Wen, and Y. Chen. 2013. Effects of PEMF on microcirculation and angiogenesis in a model of acute hindlimb ischemia in diabetic rats. *Bioelectromagnetics* 34:180–188.

Pozzo, T., A. Berthoz, and L. Lefort. 1990. Head stabilisation during various locomotor tasks in humans. *Exp Brain Res* 82:97–106.

Prato, F. S., J. R. Frappier, R. R. Shivers, M. Kavaliers, P. Zabel, D. Drost, and T. Y. Lee. 1990. Magnetic resonance imaging increases the blood-brain barrier permeability to 153-gadolinium diethylenetriaminepentaacetic acid in rats. *Brain Res* 523:301–304.

Prato, F. S., J. M. Wills, J. Roger, H. Frappier, D. J. Drost, T. Y. Lee, R. R. Shivers, and P. Zabel. 1994. Blood-brain barrier permeability in rats is altered by exposure to magnetic fields associated with magnetic resonance imaging at 1.5 T. *Microsc Res Tech* 27:528–534.

Ravera, S., B. Bianco, C. Cugnoli, I. Panfoli, D. Calzia, A. Morelli, and I. M. Pepe. 2010. Sinusoidal ELF magnetic fields affect acetylcholinesterase activity in cerebellum synaptosomal membranes. *Bioelectromagnetics* 31:270–276.

Ritz, T., P. Thalau, J. B. Phillips, R. Wiltschko, and W. Wiltschko. 2004. Resonance effects indicate a radical-pair mechanism for avian magnetic compass. *Nature* 429:177–180.

Roberts, D. C., V. Marcelli, J. S. Gillen, J. P. Carey, C. C. Della Santina, and D. S. Zee. 2011. MRI magnetic field stimulates rotational sensors of the brain. *Curr Biol* 21:1635–1640.

Robertson, J. A., A. W. Thomas, Y. Bureau, and F. S. Prato. 2007. The influence of extremely low frequency magnetic fields on cytoprotection and repair. *Bioelectromagnetics* 28:16.

Rodgers, C. T. and P. J. Hore. 2009. Chemical magnetoreception in birds: The radical pair mechanism. *Proc Natl Acad Sci USA* 106:353–360.

Ruggiero, M. 2008. Static magnetic fields, blood and genes: An intriguing relationship. *Cancer Biol Ther* 7:820–821.

Ruggiero, M., D. P. Bottaro, G. Liguri, M. Gulisano, B. Peruzzi, and S. Pacini. 2004. 0.2 T magnetic field inhibits angiogenesis in chick embryo chorioallantoic membrane. *Bioelectromagnetics* 25:390–396.

Salford, L. G., A. Brun, K. Sturesson, J. L. Eberhardt, and B. R. Persson. 1994. Permeability of the blood-brain barrier induced by 915 MHz electromagnetic radiation, continuous wave and modulated at 8, 16, 50, and 200 Hz. *Microsc Res Tech* 27:535–542.

Salford, L. G., A. E. Brun, J. L. Eberhardt, L. Malmgren, and B. R. Persson. 2003. Nerve cell damage in mammalian brain after exposure to microwaves from GSM mobile phones. *Environ Health Perspect* 111:881–883.

Sanchez, S., H. Masuda, B. Billaudel, E. Haro, R. Anane, P. Lévêque, G. Ruffie, I. Lagroye, and B. Veyret. 2006. Effect of GSM-900 and -1800 signals on the skin of hairless rats. II: 12-week chronic exposures. *Int J Radiat Biol* 82:675–680.

Sanchez, S., H. Masuda, G. Ruffié, F. P. de Gannes, B. Billaudel, E. Haro, P. Lévêque, I. Lagroye, and B. Veyret. 2008. Effect of GSM-900 and -1800 signals on the skin of hairless rats. III: Expression of heat shock proteins. *Radiat Biol* 84:61–68.

Saunders, R. 2005. Static magnetic fields: Animal studies. *Prog Biophys Mol Biol* 87:225–239.

Schirmacher, A., S. Winters, S. Fischer, J. Goeke, H. J. Galla, U. Kullnick, E. B. Ringelstein, and F. Stogbauer. 2000. Electromagnetic fields (1.8 GHz) increase the permeability to sucrose of the blood-brain barrier in vitro. *Bioelectromagnetics* 21:338–345.

Sharma, H. S. and P. J. Hoopes. 2003. Hyperthermia induced pathophysiology of the central nervous system. *Int J Hyperthermia* 19:325–354.

Shiga, T., M. Okazaki, A. Seiyama, and N. Maeda. 1993. Paramagnetic attraction of erythrocyte flow due to an inhomogeneous magnetic field. *Bioelectrochem Bioenerg* 30:181–188.

Shrivastav, S., W. G. Kaelin Jr., W. T. Joines, and R. L. Jirtle. 1983. Microwave hyperthermia and its effect on tumor blood flow in rats. *Cancer Res* 43:4665–4669.

Silva, A. K., E. L. Silva, E. S. Egito, and A. S. Carriço. 2006. Safety concerns related to magnetic field exposure. *Radiat Environ Biophys* 45:245–252.

Smith, T. L., D. Wong-Gibbons, and J. Maultsby. 2004. Microcirculatory effects of pulsed electromagnetic fields. *J Orthop Res* 22:80–84.

Snyder, D. J., J. W. Jahng, J. C. Smith, and T. A. Houpt. 2000. c-Fos induction in visceral and vestibular nuclei of the rat brain stem by a 9.4 T magnetic field. *NeuroReport* 11:2681–2685.

Solov'yov, I. A., T. Domratcheva, and K. Schulten. 2014. Separation of photo-induced radical pair in cryptochrome to a functionally critical distance. *Sci Rep* 4:Article ID 3845.

Srivastava, N. 2014. Analysis of flow characteristics of the blood flowing through an inclined tapered porous artery with mild stenosis under the influence of an inclined magnetic field. *J Biophys* 2014:Article ID 797142.

SSM (Swedish Radiation Safety Authority). 2009. Recent Research on EMF and Health Risks. Sixth annual report from SSMs: Independent Expert Group on Electromagnetic Fields. *SSM Report* 2009:36.

Stam, R. 2010. Electromagnetic fields and the blood-brain barrier. *Brain Res Rev* 65:80–97.

Steyn, P. F., D. W. Ramey, J. Kirschvink, and J. Uhrig. 2000. Effect of a static magnetic field on blood flow to the metacarpus in horses. *J Am Vet Med Assoc* 217:874–877.

Stoneham, A. M., E. M. Gauger, K. Porfyrakis, S. C. Benjamin, and B. W. Lovett. 2012. A new type of radical-pair-based model for magnetoreception. *Biophys J* 102:961–968.

Strelczyk, D., M. E. Eichhorn, S. Luedemann, G. Brix, M. Dellian, A. Berghaus, and S. Strieth. 2009. Static magnetic fields impair angiogenesis and growth of solid tumors in vivo. *Cancer Biol Ther* 8:1756–1762.

Strieth, S., D. Strelczyk, M. E. Eichhorn, M. Dellian, S. Luedemann, J. Griebel, M. Bellemann, A. Berghaus, and G. Brix. 2008. Static magnetic fields induce blood flow decrease and platelet adherence in tumor microvessels. *Cancer Biol Ther* 7:814–819.

Takeshige, C. and M. Sato. 1996. Comparisons of pain relief mechanisms between needling to the muscle, static magnetic field, external qigong and needling to the acupuncture point. *Acupunct Electrother Res* 21:119–131.

Tao, R. and K. Huang. 2011. Reducing blood viscosity with magnetic fields. *Phys Rev E* 84:Article ID 011905.

Tenforde, T. S. 2005. Magnetically induced electric fields and currents in the circulatory system. *Prog Biophys Mol Biol* 87:279–288.

Tenforde, T. S., C. T. Gaffey, B. R. Moyer, and T. F. Budinger. 1983. Cardiovascular alterations in Macaca monkeys exposed to stationary magnetic fields: Experimental observations and theoretical analysis. *Bioelectromagnetics* 4:1–9.

Theysohn, J. M., O. Kraff, K. Eilers, D. Andrade, M. Gerwig, D. Timmann, F. Schmitt, M. E. Ladd, S. C. Ladd, and A. K. Bitz. 2014. Vestibular effects of a 7 tesla MRI examination compared to 1.5 T and 0 T in healthy volunteers. *PLoS ONE* 9:Article ID e92104.

Traikov, L., A. Ushiyama, G. F. Lawlor, and C. Ohkubo. 2006. Changes of the magnitude of arteriolar vasomotion during and after ELF-EMF exposure in vivo. In *Bioelectromagnetics Current Concepts: The Mechanisms of the Biological Effect of Extremely High Power Pulses (NATO Science for Peace and Security Series B: Physics and Biophysics)*. S. N. Ayrapetyan and M. S. Markov (eds.). Springer, Heidelberg, Germany, pp. 377–389.

Traikov, L., A. Ushiyama, G. F. Lawlor, R. Sasaki, and C. Ohkubo. 2005. Subcutaneous arteriolar vasomotion changes during and after ELF-EMF exposure in mice in vivo. *Environmentalist* 25:93–101.

Tsurita, G., H. Nagawa, S. Ueno, S. Watanabe, and M. Taki. 2000. Biological and morphological effects on the brain after exposure of rats to a 1439 MHz TDMA field. *Bioelectromagnetics* 21:364–371.

Ushiyama, A. and C. Ohkubo. 2004. Acute effects of low-frequency electromagnetic fields on leukocyte-endothelial interactions in vivo. *In Vivo* 18:125–132.

Ushiyama, A., H. Masuda, S. Hirota, and C. Ohkubo. 2004a. Subchronic effects on leukocyte-endothelial interactions in mice by whole body exposure to extremely low frequency electromagnetic fields. *In Vivo* 18:425–432.

Ushiyama, A., S. Yamada, and C. Ohkubo. 2004b. Microcirculatory parameters measured in subcutaneous tissue of the mouse using a novel dorsal skinfold chamber. *Microvasc Res* 68:147–152.

Usselman, R. J., I. Hill, D. J. Singel, and C. F. Martino. 2014. Spin biochemistry modulates reactive oxygen species (ROS) production by radio frequency magnetic fields. *PLoS ONE* 9:Article ID e93065.

Valiron, O., L. Peris, G. Rikken, A. Schweitzer, Y. Saoudi, C. Remy, and D. Job. 2005. Cellular disorders induced by high magnetic fields. *J Magn Reson Imaging* 22:334–340.

van Rongen, E. 2005. International workshop "effects of static magnetic fields relevant to human health" Rapporteur's report: Dosimetry and volunteer studies. *Prog Biophys Mol Biol* 87:329–333.

van Rongen, E., R. D. Saunders, E. T. van Deventer, and M. H. Repacholi. 2007. Static fields: Biological effects and mechanisms relevant to exposure limits. *Health Phys* 92:584–590.

Waldman, W. J. and D. A. Knight. 1996. Cytokine-mediated induction of endothelial adhesion molecule and histocompatibility leukocyte antigen expression by cytomegalovirus-activated T cells. *Am J Pathol* 148:105–119.

Wang, S., Y. Zhou, J. Tan, J. Xu, J. Yang, and Y. Liu. 2014. Computational modeling of magnetic nanoparticle targeting to stent surface under high gradient field. *Comput Mech* 53:403–412.

Wang, Z., P. L. Che, J. Du, B. Ha, and K. J. Yarema. 2010. Static magnetic field exposure reproduces cellular effects of the Parkinson's disease drug candidate ZM241385. *PLoS ONE* 5:Article ID e13883.

Wang, Z., A. Sarje, P. L. Che, and K. J. Yarema. 2009b. Moderate strength (0.23–0.28 T) static magnetic fields (SMF) modulate signaling and differentiation in human embryonic cells. *BMC Genomics* 10:356.

Wang, Z., P. Yang, H. Xu, A. Qian, L. Hu, and P. Shang. 2009a. Inhibitory effects of a gradient static magnetic field on normal angiogenesis. *Bioelectromagnetics* 30:446–453.

Ward, B. K., D. C. Roberts, C. C. Della Santina, J. P. Carey, and D. S. Zee. 2014. Magnetic vestibular stimulation in subjects with unilateral labyrinthine disorders. *Front Neurol* 5:1–8.

WHO. 2006. Chapter 7: Cellular and animal studies. *Static Fields. Environmental Health Criteria 232*. World Health Organization, Geneva, Switzerland.

Wiltschko, R. and W. Wiltschko. 2012. Magnetoreception. *Adv Exp Med Biol* 739:126–141.

Xu, S., H. Okano, M. Nakajima, N. Hatano, N. Tomita, and Y. Ikada. 2013. Static magnetic field effects on impaired peripheral vasomotion in conscious rats. *Evid Based Complement Alternat Med* 2013:Article ID 746968.

Xu, S., H. Okano, and C. Ohkubo. 1998. Subchronic effects of static magnetic fields on cutaneous microcirculation in rabbits. *In Vivo* 12:383–389.

Xu, S., H. Okano, and C. Ohkubo. 2001. Acute effects of whole-body exposure to static magnetic fields and 50-Hz electromagnetic fields on muscle microcirculation in anesthetized mice. *Bioelectrochemistry* 53:127–135.

Yamaguchi-Sekino, S., L. Ciobanu, M. Sekino, B. Djemai, F. Geffroy, S. Meriaux, T. Okuno, and D. Le Bihan. 2012. Acute exposure to ultra-high magnetic field (17.2T) does not open the blood brain barrier. *Biol Biomed Rep* 2:295–300.

Yamamoto, T., Y. Nagayama, and M. Tamura. 2004. A blood oxygenation-dependent increase in blood viscosity due to a static magnetic field. *Phys Med Biol* 49:3267–3277.

Yan, Y., G. Shen, K. Xie, C. Tang, X. Wu, Q. Xu, J. Liu, J. Song, X. Jiang, and E. Luo. 2011. Wavelet analysis of acute effects of static magnetic field on resting skin blood flow at the nail wall in young men. *Microvasc Res* 82:277–283.

Yoshii, T., M. Ahmad, and C. Helfrich-Förster. 2009. Cryptochrome mediates light-dependent magnetosensitivity of Drosophila's circadian clock. *PLoS Biol* 7:Article ID e1000086.

Yu, S. and P. Shang. 2014. A review of bioeffects of static magnetic field on rodent models. *Prog Biophys Mol Biol* 114:14–24.

Yuan, H., M. W. Gaber, T. McColgan, M. D. Naimark, M. F. Kiani, and T. E. Merchant. 2003. Radiation-induced permeability and leukocyte adhesion in the rat blood-brain barrier: Modulation with anti-ICAM-1 antibodies. *Brain Res* 969:59–69.

Zborowski, M., G. R. Ostera, L. R. Moore, S. Milliron, J. J. Chalmers, and A. N. Schechter. 2003. Red blood cell magnetophoresis. *Biophys J* 84:2638–2645.

9 Extremely High-Frequency Electromagnetic Fields in Immune and Anti-Inflammatory Response

Andrew B. Gapeyev

CONTENTS

The chapter is devoted to research into the mechanisms of immunomodulating and anti-inflammatory effects of low-intensity extremely high-frequency electromagnetic radiation (EHF EMR). With the use of immunological tests, it was shown that pronounced immunotropic action of low-intensity EHF EMR is expressed as modification of reactions of cellular and non-specific immunity. It was revealed that the exposure of animals to low-intensity EHF EMR (42.2 GHz, 0.1 mW/cm², 20 min) causes changes in chromatin organization of lymphoid cells, reduces an intensity of the cellular immune response in delayed-type hypersensitivity reaction, reduces the phagocytic activity of peripheral blood neutrophils, and does not influence the humoral immune response to thymus-dependent antigen. It was found that local EHF EMR exposure induces the degranulation of skin mast cells that is the important amplifying mechanism in realization of EHF EMR action at the level of whole organism with participation of nervous, endocrine, and immune systems. A *histamine model* of biological effects of low-intensity EHF EMR was proposed, which joins cellular reactions and systemic reaction of an organism to EHF EMR. Using methods of comet assay, flow cytometry, gas chromatography, and reverse transcription polymerase chain reaction, responses of effector immune cells to the influence of low-intensity EHF EMR were studied in health and disease. A number of key links responsible for realization of additive and synergistic effects of EHF EMR on the cellular and whole organism levels were found. Specifically, the EHF EMR caused structural changes in chromatin of lymphoid cells, providing a genoprotective action during inflammatory process; activated T-cellular immunity manifested as an increase in the number of CD4+ and CD8+ T-cells in thymus of exposed animals; substantially increased the content of some n-3 and n-6 polyunsaturated fatty acids in the thymus of exposed mice; and created the specific profile of cytokines, providing anti-inflammatory effect. It is supposed that by means of initiation of certain signaling and effector systems, the directed answer of an organism to a specific combination of effective parameters of the electromagnetic radiation can be carried out. The material presented will be useful for a better understanding of the mechanisms of therapeutic action of EHF EMR and will stimulate further research into the mechanisms of biological action of weak electromagnetic fields.

9.1 INTRODUCTION

The development of telecommunication technologies, communications, radiolocation, and radio-navigation entails the mastering of new frequency bands of radio-frequency electromagnetic radiation (EMR), an increase in the radiated power, complexity of the structure of electromagnetic signals, and an increase in the area covered by radio transmission systems. All this leads to an obvious question about the safety of application and dissemination of man-made EMR of various frequency ranges. Strict hygienic standards for EMR based on the results of biomedical and epidemiological research carried out under controlled conditions are necessary. On the other hand, the therapeutic effects of EMR of certain frequency ranges are of great interest as an important direction in the practical application of EMR in medicine. The principal problem is the lack of fundamental knowledge about the mechanisms of biological effects of EMR at different levels of biological systems. Understanding of the laws of biomedical effects and development of unified concept of biophysical mechanisms would allow purposeful assessment of adverse effects of EMR and development of application technologies for the prevention, diagnosis, and treatment of a wide spectrum of diseases.

Due to the pioneering works of Russian academician N.D. Devyatkov in the 1960s of twentieth century in the former USSR, the development of millimeter wavelength range had begun, and after the establishment of radiating devices, a new line of research on interaction of electromagnetic waves of extremely high frequencies (EHF) with biological structures and organisms has been discovered in order to effectively treat a number of diseases. The foundations of EHF therapy method opened a new era in shaping the understanding of the role of electromagnetic fields in biology and medicine. Currently, EHF EMR is widely used in medical practice for prevention, diagnosis, and treatment of various diseases (Rojavin and Ziskin, 1998; Pakhomov and Murphy, 2000; Betskii and Lebedeva, 2007). High efficacy of EHF therapy was demonstrated in the treatment of cardiovascular, neurological, urological, gynecological, cutaneous, gastrointestinal, and dental diseases and cancers. In this connection, it should be noted that one of the problems of medical applications of EHF EMR is weak justification of the physical parameters of the radiation used for the treatment. The second problem is lack of knowledge about the specific mechanisms of therapeutic effects of EHF EMR. All this leads to an empirical approach on clinical application of EHF therapy and specific hardware. However, in spite of this practice, the efficacy of EHF therapy is not in doubt. The original biomedical studies made in recent years and a wide range of therapeutic equipment created confirm the reasonability and necessity for further investigations of the mechanisms of biological effects of EHF EMR at all levels of organization of living systems.

Features of biological effects of EHF EMR in vitro are usually associated with one or several of its parameters (Betskii and Lebedeva, 2001; Gapeyev, 2014). However, a wide range of objects under study in vitro, a large dispersion of EMR parameters (frequency, intensity, exposure duration, etc.) used in various studies, and phenomenological focus of most studies lead to enormous amount of data, which are often contradictory. The aforementioned discussion shows the need for a systematic approach to research into the mechanisms of biological effects of EHF EMR in a wide range of parameters, applying the correct dosimetry supply, using different experimental methods for analyzing changes in activity of biological object after the exposure to EMR. The results of such studies should be the revealing of ranges of physical parameters of EHF EMR that will be most effective in respect to certain functions of biological systems under controlled conditions. When conducting preclinical testing, this approach will determine the most effective parameters of EHF EMR for EHF therapy aimed at specific nosologies (Betskii et al., 2005). We believe that the efficacy of EHF therapy can be substantially increased if certain parameters of EHF EMR will be related to certain types of pathological conditions. This requires targeted studies of the mechanisms of action of EHF EMR in vivo at the level of the regulatory systems of the organism.

In terms of application of EHF EMR in therapeutic practice, the question on sensitivity to EHF EMR of various biological processes and systems at the level of whole organism is of particular interest. Analysis of the published data shows that the action of low-intensity EHF EMR both in norm and at pathology may induce significant changes in the functioning of the immune system (Devyatkov et al., 1991; Lushnikov et al., 2002). It is possible that EHF EMR affects the rate of synthesis and the balance of various regulatory molecules in the organism that defines the functional status of the immune system in general.

The aim of this paper is to systematize the results of our studies of the effects of low-intensity EHF EMR with effective parameters on the basic reactions of the immune system of laboratory animals. We believe that these data not only will be useful for physicians and researchers for a better understanding of the mechanisms of therapeutic effects of EHF EMR but will stimulate further studies of the immunomodulatory effects of weak electromagnetic fields and low-intensity radiation.

9.2 GENERAL STATEMENTS ON BIOLOGICAL OBJECTS AND EXPOSURE

9.2.1 BIOLOGICAL OBJECTS

Adult male NMRI mice (2 months of age, 25–30 g in body weight) and Wistar rats (300–350 g in body weight) were purchased from the Laboratory Animal Breeding Facility (Branch of Shemyakin–Ovchinnikov Institute of Bioorganic Chemistry, Pushchino, Moscow Region, Russia). The animals were housed in an air-conditioned room with a controlled 12 h light–dark cycle and free access to standard chow and tap water. All manipulations with the animals were conducted in accordance with experimental protocols approved by the Local Animal Care and Use Committee (Branch of Shemyakin–Ovchinnikov Institute of Bioorganic Chemistry, Pushchino, Moscow Region, Russia).

All experiments were conducted utilizing the *blind* experimental protocol, when an investigator making the measurements did not know which treatments were made. All data are given as the mean ± standard error of the mean (SEM). The normality of data was analyzed using the Kolmogorov–Smirnov test. If the data matched the normal distribution, the statistical analysis was performed using the parametric Student's *t*-test. Otherwise, nonparametric statistical methods were used: Mann–Whitney *u*-test for pairwise comparison of two groups of data or Kruskal–Wallis test for pairwise comparison of multiple groups of data. Differences were considered significant at the significance level $p < 0.05$.

9.2.2 Generating Devices

Dosimetric tests and irradiation of biological objects were conducted on experimental setups, including (1) a high-frequency generator G4-141 (Istok, Fryazino, Russia); (2) a panoramic meter R2-68 bundled with a standing wave ratio indicator YA2R-67 (Istok, Fryazino, Russia) to measure the standing wave ratio and attenuation; (3) a power meter (thermistor head M5-49 with the absorbed power wattmeter M3-22A); (4) a wave meter CH2-25; and (5) emitters of standard and special shapes (waveguide radiator—open end of the waveguide section of 5.2×2.6 mm^2 without flange, pyramidal horn antenna with an aperture of 32×32 mm^2, trough radiator based on the trough waveguide with an aperture of 17.5×12.5 mm^2 [NPP "Pirs," Moscow, Russia]). All experiments were carried out under controlled all parameters of EMR, as well as under controlled static magnetic field measured by ferromagnetic probe (Istok, Fryazino, Russia).

9.2.3 EHF EMR Exposure

The whole-body exposure of mice to EHF EMR was conducted in the far-field zone of a pyramidal horn antenna with an aperture of 32×32 mm^2 at a distance of 300 mm from the radiating end of the antenna. The mice were exposed from the top in plastic containers (LaboratorSnab, Moscow, Russia) with a size of $100 \times 100 \times 130$ mm^3 where animals moved freely. The breadth of the directional diagram of the electric field vector for the pyramidal horn antenna was $2\theta_{0.1}^E \sim 25°$. Accordingly, the major lobe width (0.1 level) was about 130 mm at a distance of 300 mm from the antenna (Gapeyev and Chemeris, 2010). The bottom square of the animal container corresponded to the square of the exposed zone created by the major lobe of the antenna. To eliminate the interference in the plane of an exposed object, an effective multilayer absorbent was placed between the animal container and the floor; therefore, the conditions of exposure were close to free-field conditions. The frequency of the output signal was controlled by a CH2-25 wavemeter. The frequency stability of the generator in the continuous wave mode was ±15 MHz. The output power of the generator was set at 8 mW and controlled with a M5-49 thermistor head and a M3-22A wattmeter so that the incident power density (IPD) in the plane of the exposed object was 0.1 mW/cm^2. The IPD distribution in the plane of the exposed object was approximately normal with a maximal value in the center of the exposure zone and a twofold smaller value near the walls of the animal container. The IPD in the plane of exposed objects and the surface specific absorption rate (SAR) in the mouse skin were calculated according to previous dosimetric studies, which were carried out using microthermometric, infrared thermographic, and calculation methods (Gapeyev et al., 2002). The surface SAR was about 1.5 W/kg at an IPD of 0.1 mW/cm^2. Control animals were sham-exposed by placing the mice into the exposure zone when the generator was turned on, but the output power was maximally attenuated (to <1 µW). The background induction of the geomagnetic field was 45 ± 3 µT.

Cell suspensions were exposed in special flat-bottomed cylindrical plastic cuvettes ($\varnothing = 10$ mm, wall and bottom thickness of 0.2–0.3 mm) in the far-field zone of trough radiator. When placing the cell suspension in the cuvette, the cells settled on the bottom are a layer of thickness less than 50 µm, and the overall height of the solution in the cuvette was 2 mm. The exposure of the cells to EHF EMR was conducted from the cuvette bottom. Three to six cuvettes were fixed in a special foam plastic holder and irradiated simultaneously at room temperature of 20°C–22°C. The breadth of the directional diagram of the electric field vector for the trough radiator was $2\theta_{0.1}^E \sim 28°$. Accordingly, the major lobe width (0.1 level) was about 200 mm at a distance of 400 mm from the radiator (Gapeyev et al., 1996; Gapeyev and Chemeris, 2010). The square of the cuvette holder corresponded to the square of the exposed zone created by the major lobe of the radiator. As the sham controls, we used 3–6 cuvettes in similar conditions except for radiation.

9.3 IMMUNOMODULATING EFFECTS OF EHF EMR

A number of studies have demonstrated that low-intensity EHF EMR can change indices of the immune and neuroendocrine systems (Devyatkov et al., 1991; Rojavin and Ziskin, 1998). The immune system plays a special role in maintaining the normal functioning of the organism; it controls the integrity and genetic homogeneity of the organism and is involved in all pathological and reparative processes (Roitt et al., 1998; Abbas and Lichtman, 2004). Quantitative relationships and interactions between immune cells, the formation of a specific cytokine profile, and functional activity of various immune cells are supported by numerous positive and negative feedbacks (Yarilin, 1999). Low-intensity EHF EMR, providing regulatory action, is able to cause immunomodulating effects, which is one of the basic mechanisms of correction of the organism's state during EHF therapy (Lushnikov et al., 2002). Our objective was to determine the changes in the activity of the basic reactions of humoral, cell-mediated, and nonspecific immunity under the action of low-intensity EHF EMR with parameters whose efficacy in relation to immune cells in vitro has been previously demonstrated (Gapeyev et al., 1996).

9.3.1 EHF EMR EFFECTS ON THE HUMORAL IMMUNITY

The effects of low-intensity EHF EMR (42.2 GHz, 0.1 mW/cm^2) on specific defense mechanisms (adaptive immunity) were examined in a model of development of the humoral immune response to thymus-dependent antigen sheep red blood cells (SRBCs). The humoral immune response was evaluated on day 5 after immunization of mice with SRBCs by a number of antibody-producing cells (APCs) in the spleen using a method of local hemolysis in agarose gel and by hemagglutination test (Klaus, 1987). The number of nucleated cells (NCs) in the spleen, thymus, and bone marrow was also determined (Friemel, 1984). The animals were exposed to EHF EMR before immunization (for 20 min once, for 20 min daily during 5 consecutive days, and for 20 min daily during 20 consecutive days) and after immunization (for 20 min daily during 5 consecutive days). After a single exposure and a series of five daily exposures before and after immunization, indices of humoral immune response (hemagglutinating antibody titers and a number of APCs) and the number of NCs in the lymphoid organs were not significantly changed. These results indicate that the short-term exposure of mice to EHF EMR does not introduce significant changes in the mechanisms of antigen perception by the organism and does not affect the formation of immune response by humoral way (Lushnikov et al., 2001). After prolonged fractionated EHF EMR exposure before immunization (for 20 min daily during 20 consecutive days), we found a decrease in the number of NCs in the thymus and spleen by 18%±5% and 14%±4%, respectively ($p < 0.05$ compared to the sham control), which can be explained by adaptogenic activity of EHF EMR (Devyatkov et al., 1991). For example, the efficacy of long-term courses of irradiation on the survival of animals after exposure to γ-radiation and influenzal infection was shown in several studies (Ryzhkova et al., 1991; Sazonov and Ryzhkova, 1995).

Taking into account that the number of APCs in the spleen did not change, we can assume that the decrease in the number of NCs in the spleen did not occur at the expense of B-cells. Decrease in the number of NCs in the thymus shows the sensitivity of T-cells to low-intensity EHF EMR. This corresponds to a notion about an effective influence of EHF EMR on cell-mediated immunity at various pathologies (Rojavin and Ziskin, 1998). Our results indicate a cumulative character of EHF EMR effect at the level of whole organism, that is, the ability of an organism to accumulate and store information about the exposure. The process of reception of EHF EMR by an organism apparently is systemic and can include reactions from the neuroendocrine system with changes in the content and/or the synthesis of biologically active substances (hormones, cytokines, neurotransmitters, etc.) (Lushnikov et al., 2002). Decrease in the number of NCs in the thymus and spleen of animals with chronic exposure to low-intensity EHF EMR may be related to the migration of cells outside the lymphoid organs, as well as changes at the level of the cellular genome (abnormal cell cycle, apoptotic processes). Therefore, in the next series of experimental studies, we assessed the effect of EHF EMR on the chromatin state of lymphoid cells.

9.3.2 EHF EMR EFFECTS ON THE CHROMATIN STATE OF LYMPHOID CELLS

An activity of the immune system cells is largely determined by the rate of their mobilization, the expression level, and repertoire of cellular receptors. It is believed that EHF EMR can influence the genetic apparatus of cells and can change the structure and functions of chromosomes, cellular resistance to standard mutagens, and damaging agents (Shckorbatov et al., 1998; Fedorov et al., 2001; Belyaev, 2005, 2010). To assess the effect of low-intensity EHF EMR on cellular DNA, we used an express method of molecular genotoxicology comet assay. The method allows to detect DNA damage and changes in the structure of nucleoids (Ostling and Johanson, 1984; Singh et al., 1988; Sirota et al., 1991; Tice et al., 2000). Modification of the alkaline version of the comet assay allowed us to significantly improve the sensitivity of the method, which permits to reliably detection of even slight changes in chromatin structure (Sirota et al., 1996; Gapeyev et al., 2003; Chemeris et al., 2006).

We have found that the whole-body exposure of intact animals to EHF EMR (42.2 GHz, 0.1 mW/cm^2, 20 min exposure) induced multidirectional effects on chromatin structure of lymphoid cells. There was chromatin condensation in thymocytes, which was manifested as an increase in the fluorescence intensity of nucleoids by 16% ($p < 0.03$), and chromatin decondensation in splenocytes, which was manifested as a decrease in the fluorescence intensity of nucleoids by 16% ($p < 0.001$) as compared to the cells of sham-exposed animals. However, EHF EMR had no effect on the chromatin structure of peripheral blood leukocytes of exposed animals.

To assess the direct effects of EHF EMR (42.2 GHz, 0.1 mW/cm^2, 20 min exposure) on the chromatin structure, we exposed different cells in vitro. Exposure of cultured human lymphoid cells *Raji* and mouse whole blood leukocytes resulted in a decrease in fluorescence intensity of nucleoids by 22.6% ± 2.3% ($p < 0.005$) and 17.6% ± 2.7% ($p < 0.001$), respectively. Similar direction of the EHF EMR effects on different cells in vitro indicates the presence of the same mechanism for realization of the effect, which can be connected with the changes in the level of intracellular signaling (Alovskaya et al., 1998). The lack of EHF EMR effect on the chromatin structure of peripheral blood leukocytes in vivo and the existence of the effect in vitro indicate different mechanisms involved in the effect realization in vitro and in vivo. Differently directed reaction of thymocytes and splenocytes in the absence of the effect of EHF EMR on peripheral blood leukocytes in vivo suggests that the effect of EHF EMR on the lymphoid organs is not implemented directly, but with the participation of the regulatory systems of the organism, and can be explained by the different innervation of lymphoid organs, different population composition of white blood cells, and specific receptor repertoire of thymocytes and splenocytes (Shurlygina et al., 1999). We believe that the underlying mechanisms of the observed effects of EHF EMR on DNA of lymphoid cells in vitro and in vivo are the physiological responses of the cells to biologically active substances that can be induced by the exposure (Govallo et al., 1991). In this regard, the cell-mediated immune responses as the most reactive chain of the immunity might be potentially more sensitive to low-intensity EHF EMR.

9.3.3 EHF EMR EFFECTS ON CELL-MEDIATED IMMUNE RESPONSES

An influence of EHF EMR on the cell-mediated immunity was studied using delayed-type hypersensitivity (DTH) reaction. Animals were sensitized by SRBCs in the left paw and then whole body exposed to EHF EMR (42.2 GHz, 0.1 mW/cm^2, for 20 min daily during 5 consecutive days) up to challenging injection. Control animals were also injected and sham-exposed. Changes in paw edema were determined calculating a relative increase in the thickness of inflamed paw compared to contralateral and expressed in percentage (Friemel, 1984). Sensitizing injection caused

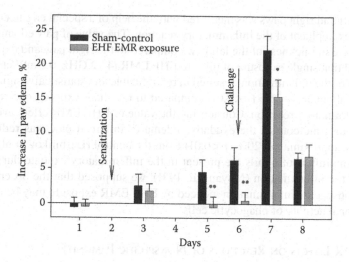

FIGURE 9.1 Relative increase in paw edema in sham-exposed and EHF EMR–exposed mice in DTH reaction. The mice were sensitized by SRBCs in the left paw on day 2 of the experiment; challenging injection was made on day 6. During 2–6 days, animals were sham-exposed or exposed to EHF EMR (42.2 GHz, 0.1 mW/cm^2, 20 min daily). *$p < 0.05$, **$p < 0.01$ by the Student's t-test, $n = 14$.

local nonspecific inflammatory response in the paw. On days 3–4 after the sensitizing injection, the local inflammatory response was more significantly expressed in sham-exposed animals ($p < 0.01$ compared to exposed animals). At 24 h after challenging injection, the paw edema in animals exposed to EHF EMR increased by 15% ± 3%, which was significantly lower ($p < 0.05$) compared to sham-exposed animals in which the paw edema increased by about 22% ± 2% (Figure 9.1). At 48 h after the challenging injection, the edema dissipated and was approximately the same in the exposed and sham-exposed animals. Thus, we have shown that EHF EMR exposure almost entirely (by 92% ± 19% compared to the control) suppresses the nonspecific inflammation and significantly reduced specific inflammation in the DTH reaction by 30% ± 12% compared to the control, which may be due to a more complex cascade of reactions that provide specific inflammation. Importantly, the decrease in the intensity of inflammation under the influence of EHF EMR observed at the sensitization persists after the introduction of the challenging antigen dose (Figure 9.1).

Suppression of nonspecific inflammation and decrease in the intensity of specific inflammation under the EHF EMR exposure may be due to reduction in functional activity of phagocytes, because the main cells involved in the development of DTH reaction along with T-effectors are macrophages (Abbas and Lichtman, 2004). Overall, we showed that EHF EMR is capable to reduce the intensity of cell-mediated immune responses (Lushnikov et al., 2003).

9.3.4 ANTI-INFLAMMATORY EFFECTS OF EHF EMR

The inflammatory reaction is the first level of immune response reactions. It intended for attraction to the point of introducing foreign agents of immune cells (neutrophils, monocytes and macrophages, and then lymphocytes) and its activation. The significant suppression of nonspecific inflammatory reaction found by us under the action of EHF EMR in DTH reaction prompted us to further investigate the effect of EHF EMR on the development of nonspecific inflammation.

For this purpose, we used a model of acute inflammation induced by intraplantar injection of zymosan suspension (25 μL at 5 mg/mL) into the left hind paw of mice (Ibrahim et al., 2002). The animals were exposed to EHF EMR for 20 min at 1 h after zymosan injection. An equal volume of sterile physiological saline was injected into the contralateral paw. To assess a value of paw edema, thickness of the left and right paws was measured with the help of a special engineering micrometer during 3–8 h of development of the inflammatory reaction. The value of paw edema was calculated as a relative increase in thickness of the left paw compared to the right paw and expressed as a percentage. We found that single exposure of mice to EHF EMR (42.2 GHz, 0.1 mW/cm^2, for 20 min at 1 h after initiation of the inflammation) resulted in reproducible and statistically significant decrease in paw edema by about 18%–22% ($p < 0.01$) compared to the sham-exposed control at 3–7 h after initiating the inflammatory reaction. Considering the value of EHF EMR effect averaged by 3–7 h of the inflammation time course, the exudative edema of inflamed paw was reduced after EHF EMR exposure by approximately 20% ($p < 0.01$) from the control (Lushnikov et al., 2005). Taking into account that mainly neutrophils are present in the inflammatory exudate during the first day after initiation of the inflammation (Mayanskii, 1991), we supposed that the observed decrease in the intensity of nonspecific inflammation induced by EHF EMR exposure may be due to modification of the functional activity of phagocytic cells.

9.3.5 EHF EMR Effects on Reactions of Nonspecific Immunity

Considering that the water-containing media at a depth of 1 mm attenuates EHF EMR by approximately 100 times, skin structures, subcutaneous tissue, fat, and muscle tissues entering the zone of direct action of the radiation can take part in the reception at the level of whole organism. In the skin, numerous structures that transmit information to regulatory systems of the organism by different ways including the nervous transmission, humoral signals, and presentation of antigens to the immune system cells are present. These are receptors of nervous system (mechanoreceptors, nociceptors, free nerve endings, etc.), cells of the diffuse neuroendocrine system (Merkel cells), fibroblasts, Langerhans cells (antigen presenting cells of the monocyte–macrophage type), mast cells, and blood cells of microvascular bed. We studied the phagocytic activity of peripheral blood neutrophils and neutrophils from inflammatory exudate after exposure of mice to EHF EMR using bacterial cells (*Escherichia coli*, strain E15) as a phagocytosis object. Percentage content of phagocyting neutrophils from a number of potential phagocytes was determined by means of light microscopy (Lushnikov et al., 2003). To determine a background phagocytic activity of neutrophils within 2 days prior to exposure procedures, sham-exposure procedures were carried out (for 20 min daily), and blood samples (30 μL) from the tail vein of mice were collected. On day 3, the mice from exposure group were exposed to EHF EMR (42.2 GHz, 0.1 mW/cm^2, 20 min exposure), and blood was collected after 30 min, 1 h, and 3 h. It was found that phagocytic activity of peripheral blood neutrophils has not changed for 2 days prior to exposure, that is, procedures for blood sampling, as well as the sham exposure, had no significant effect on neutrophils' phagocytic activity. At 30 min after exposure of animals to EHF EMR, we registered almost twofold decrease in neutrophils' phagocytic activity ($p < 0.05$ compared to the sham-exposure control) (Table 9.1). The effect lasted

---

TABLE 9.1

Changes in Phagocytic Activity of Neutrophils (in % of the Control) under the Exposure to EHF EMR in Norm and on the Background against Inflammatory Process

Intact peripheral blood neutrophils	In vivo ($n = 12$)	57 ± 9[a]
	In vitro ($n = 9$)	105 ± 6
Peripheral blood neutrophils at inflammation	In vivo ($n = 14$)	98 ± 12
Peritoneal neutrophils at inflammation	In peritoneal exudate ($n = 14$)	109 ± 6
	After washing ($n = 14$)	86 ± 5[a]

[a] $p < 0.05$ compared to corresponding sham control by the Student's t-test.

for 3 h after irradiation and was not significantly changed in a day after exposure (Kolomytseva et al., 2002; Lushnikov et al., 2003). We supposed that the changes in phagocytic activity are caused by rapid initiation of functional rearrangements at a level of intracellular signaling systems of neutrophils that leads to the inhibition of phagocytosis (Kolomytseva et al., 2002).

We have shown that the direct exposure of whole blood samples to EHF EMR in vitro did not change the phagocytic activity of neutrophils (Table 9.1). Comparison of the results obtained by the irradiation in vivo and in vitro suggests that the perception of low-intensity EHF EMR by the organism is systemic. We assume that the effect of EHF EMR in vivo may be mediated by changes in the content of biologically active substances in the blood plasma (Lushnikov et al., 2003).

Studies of phagocytic activity of peripheral blood neutrophils under the action of EHF EMR (42.2 GHz, 0.1 mW/cm^2, 20 min exposure) on the background against inflammation induced by intraperitoneal zymosan injection did not reveal any significant changes (Table 9.1). Under the same conditions, phagocytic activity of peritoneal neutrophils was also not significantly different from the control when the phagocytosis reaction was set in peritoneal exudate (Table 9.1). However, after washing the cells from the peritoneal exudate, the phagocytic activity of peritoneal neutrophils of exposed animals was lower by about 13.7% ± 4.5% ($p < 0.05$) compared to the control group (Table 9.1).

Analysis of the kinetics of production of reactive oxygen species (ROS) by neutrophils showed that cells of mice exposed to EHF EMR after induction of inflammatory response demonstrated suppression of the first phase of the respiratory burst associated with the activation of NADPH-oxidase involving activation of protein kinase C and rapid mobilization of $[Ca^{2+}]_i$, and a reduction of the second phase related to extracellular calcium entry (Lushnikov et al., 2004). Thus, we have showed that EHF EMR can significantly reduce the activity of phagocytic cells (phagocytosis and ROS production by neutrophils), which is due to changes at the level of intracellular signaling systems. It is logical to assume that these changes at the level of intracellular signaling systems of phagocytic cells are probably also reflected on the other functions of these cells such as degranulation, synthesis, and secretion of inflammatory cytokines. In general, anti-inflammatory effect of EHF EMR can be determined particularly by changing the functional activity of phagocytic cells (Lushnikov et al., 2003).

9.4 DEGRANULATION OF SKIN MAST CELLS UNDER THE INFLUENCE OF EHF EMR

It is suggested that the structural elements of the skin may play a role of primary receptors of low-intensity EHF EMR (Lushnikov et al., 2002). The skin contains a large number of nerve endings. Mast cells of the skin can mediate a complex cascade of biochemical reactions that provides pain syndrome; mast cells contain histamine and several other biologically active substances (Chernukh and Frolov, 1982). Previously it was shown that the exposure of rat skin to EHF EMR (42, 52, and 60 GHz; 50 mW/cm^2, 5 min exposure, skin surface overheating of 3°C–4°C) causes dermal mast cell degranulation, vasodilation of the capillaries, and changes in the structure of the myelin of encapsulated nerve endings (Zavgorodny et al., 2000). Our task was to carry out qualitative and quantitative analyses of morphological changes in skin mast cells under the influence of low-intensity EHF EMR that does not induce heating of the irradiated region.

The inner surface of the rat paw was exposed to EHF EMR (42.2 GHz, 0.05 mW/cm^2, 20 min exposure) in the far-field zone of the waveguide radiator. The rest of the body of anesthetized rats was screened from EHF EMR by an effective multilayer absorber. For the sham-control group of rats, sham-exposure procedures were carried out. Biopsy of the exposed skin was taken at 15 min after exposure or sham-exposure procedures. Serial semithin or ultrathin (70–90 nm) sections were prepared from these samples on an ultramicrotome Reichert (Austria). Sections were stained and analyzed by light microscopy or using an electron microscope JEOL 100B (Japan). Quantitative analysis was made by counting the number of granules on the profile of the mast cell

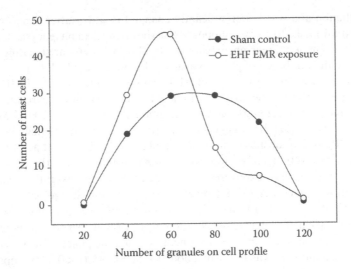

FIGURE 9.2 Distribution of skin mast cells by the number of granules on the cell profile in the sham-control samples and samples exposed to EHF EMR (42.2 GHz, 0.05 mW/cm^2, 20 min exposure).

using 2–3 projections of cells randomly selected on each of the sections. After identification of the samples, the summary distribution histograms of mast cells by the number of granules per cell profile were plotted for control and exposed samples.

It was found that after exposure to low-intensity EHF EMR, skin mast cell degranulation is observed. The average number of granules per mast cell profile was about 62 ± 2 in the sham control ($n = 137$) and 52 ± 2 ($n = 146$) in the exposed samples ($p < 0.001$). Distribution of mast cells by the number of granules per cell profile in the sham-control samples resembles a normal Gaussian distribution (Figure 9.2). For the exposed samples, the distribution maximum is shifted toward smaller number of granules on the cell profile. This indicates degranulation of mast cells containing the greatest number of granules or more mature mast cells (Popov et al., 2001). It should be noted that in vitro exposure of isolated peritoneal rat mast cells to visible radiation (630 nm) or EHF EMR (Graevskaya et al., 2000) did not initiate appreciable degranulation. However, mast cell degranulation was observed in vivo when the skin was exposed to low-intensity EHF EMR and laser light (632 or 820 nm) (Sayed and Dyson, 1996). Thus, it is possible that mast cell degranulation by the action of EHF EMR in vivo is the result of a chain of events induced by the exposure in the skin (or in the body) but is not the result of a direct action of the radiation on mast cells, which could play a role of primary cellular targets. Trigger mechanism may be the formation of ROS and free radicals in the skin by the action of EHF EMR, the possibility of which has been demonstrated in model systems (Potselueva et al., 1998; Gugkova et al., 2005).

9.5 INFLUENCE OF EHF EMR ON REGENERATION OF FULL-THICKNESS SKIN WOUNDS

Histamine release from skin mast cells under the action of low-intensity EHF EMR may play an important role in the course of various inflammatory processes. It is known that histamine entering the blood microvascular bed, along with proinflammatory vascular effects, causes inflammatory cellular effects (Gushchin, 1998). As a result, in the inflammation focus, microcirculation, blood flow, and processes of energy and plastic metabolism are enhanced, and cytotoxic activity of white blood cells is reduced, limiting the damage to own tissues by produced ROS. All this may lead to positive changes in the local inflammatory response to injury during recovery phase of wound healing.

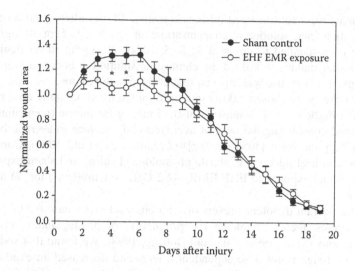

FIGURE 9.3 Dynamics of contraction of wound surface in sham-exposed mice and mice exposed to EHF EMR (42.2 GHz, 0.1 mW/cm², exposure for 20 min daily during 10 consecutive days). $n = 10$, *$p < 0.05$ by the Student's t-test.

We studied the effects of EHF EMR on the dynamics of contraction of wound surface during posttraumatic regeneration of the skin on mouse back. Wound process was simulated in the interscapular region of mice by excising skin flap area of about 70 mm² with a damaged underlying fascia and muscle layer under ether anesthesia. The dynamics of the healing process was determined by the change in the area of the wound defect, which is daily photographed with a digital camera Nikon Coolpix 990 (Japan). After applying the full-thickness skin wounds, mice were exposed to EHF EMR (42.2 GHz, 0.1 mW/cm², for 20 min daily during 10 consecutive days). Control mice were sham-exposed. During development of the inflammatory process on days 3–6, the wound surface area of the control animals had increased on average by 30% compared to the initial area (Figure 9.3). Inflammatory process in the exposed animals was less expressed; the wounds were dry and clean, with smooth edges. Wound surface area was not practically increased, which is significantly different from the control ($p < 0.05$) on days 4–5 (Figure 9.3). Subsequently, the area of wound defect and the final terms of wound healing in control and exposed mice did not differ on the average, but earlier maturation of granulation tissue, ordering of its structure, and increase in the number of mature collagen fibers with a gradual full recovery of the epithelium occurred in exposed animals (Bessonov et al., 2003). General condition of the exposed mice was significantly better than the control, as reflected in their more active behavior and intensive consumption of food and water.

Thus, we have shown that EHF EMR reduces the intensity of the inflammatory process that occurs in the injured area. Based on these data, we can assume that it is starting chain of reactions to EHF EMR exposure, including the release of biologically active substances from mast cells, leading to the subsequent decrease in acuity of inflammation, and alleviation of the pathological process, thus creating favorable conditions for the normalization of disturbed functions.

9.6 PHARMACOLOGICAL ANALYSIS OF ANTI-INFLAMMATORY EFFECTS OF EHF EMR

To investigate the mechanisms of realization of anti-inflammatory effects of EHF EMR, we conducted a comparative analysis of the effects of radiation with known nonsteroidal anti-inflammatory drug (NSAID) sodium diclofenac and antihistamine clemastine in a model of acute zymosan-induced inflammation in mice described earlier. Sodium diclofenac and clemastine in different series of experiments were administered intraperitoneally, 30 min after zymosan injection. Sixty microliters of the diclofenac solution in concentrations of 1, 1.5, 2.5, 5, or 10 mg/mL per mouse was injected that corresponded to doses of 2, 3, 5, 10, or 20 mg/kg in recalculation for mice (25–30 g). In a similar manner, 60 μL of the clemastine solution in concentrations of 0.01, 0.05, 0.1, 0.2, or 0.3 mg/mL per mouse was injected that corresponded to doses of 0.02, 0.1, 0.2, 0.4, or 0.6 mg/kg. These doses were chosen, taking into account the maximal daily dose of diclofenac of 25 mg/kg and the middle dose of clemastine of 0.34 mg/kg for mouse (recalculation coefficient of 11.9 was obtained considering the ratio of averaged body surface to average body weight for humans [257 cm^2/kg] and mice [3050 cm^2/kg]) (Arzamastsev et al., 2005). Control mice were subjected to intraperitoneal injection of sterile physiological saline and sham-exposed. The mice were sham-exposed and exposed to EHF EMR (42.2 GHz, 0.1 mW/cm^2) for 20 min at 1 h after zymosan injection.

It is known that sodium diclofenac reversibly inhibits cyclooxygenase (COX), one of the key inflammatory enzymes, thereby reducing the production of prostaglandins involved in most inflammatory reactions (Vane, 1998; Vane and Botting, 1998). We found that sodium diclofenac (2–20 mg/kg) caused significant dose-dependent effects and decreased an exudation of inflammatory focus (Figure 9.4). Starting with a dose of 5 mg/kg, the effect of sodium diclofenac on exudative edema reaches a plateau with an average value of about 26% ($p < 0.01$ compared to the control) (Figure 9.4).

Comparative analysis showed that the kinetics of the effects of EHF EMR and sodium diclofenac at doses of 3–5 mg/kg on exudative edema is very close. Exposure of mice to EHF EMR caused a decrease in paw edema by about 20% that is comparable with that induced by sodium diclofenac at doses of 3–5 mg/kg (Figure 9.4), which are close to single therapeutic doses of the drug. Based on these findings, we can hypothesize that anti-inflammatory effects of EHF EMR, similar to effects of diclofenac, were realized through inhibition of COX. To test this hypothesis, we studied the combined effects of diclofenac and EHF EMR.

The combined action of sodium diclofenac at a dose of 3 mg/kg and EHF EMR decreased the paw edema up to a value that was the same as the effect of diclofenac at doses of 5–20 mg/kg (Figure 9.4). However, the combined action of diclofenac at doses of 5–20 mg/kg and EHF EMR

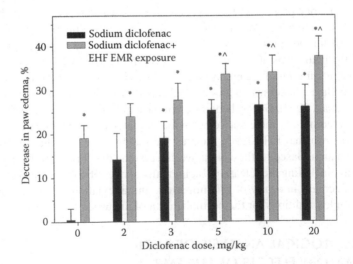

FIGURE 9.4 Decrease in paw edema (averaged by 3–7 h of the inflammation time course, as % of the control) depending on diclofenac doses in sham-exposed and EHF EMR–exposed (42.2 GHz, 0.1 mW/cm^2, 20 min) mice. $n = 15$, *$p < 0.02$ from the control (at diclofenac dose of 0 mg/kg), ^$p < 0.03$ from the value of EHF EMR effect at diclofenac dose of 0 mg/kg by the Student's t-test. (Adapted from Gapeyev, A.B. et al., *Bioelectromagnetics*, 29(3), 197, 2008.)

exposure generated a partial additive effect (Gapeyev et al., 2008). This partial additive effect decreased paw edema by about 35%, exceeded the effect of diclofenac alone on the average by 10% ($p < 0.05$), and was independent of diclofenac dose within the range of 5–20 mg/kg (Figure 9.4).

These results suggest that the anti-inflammatory effect of EHF EMR, at least in part, may be due to decreased activity of COX. However, the partial additive effect found under the combined action of diclofenac and EHF EMR exposure points to involvement of other mechanisms of EHF EMR influence on the inflammation, which are not connected with inhibition of COX. Degranulation of skin mast cells under the influence of low-intensity EHF EMR (Popov et al., 2001) suggests that histamine and serotonin, as base components of mast cell granules, may take part in the realization of biological effects of EHF EMR. To assess the role of histamine in the realization of the anti-inflammatory effects of EHF EMR, we studied the effects of antihistaminic drug clemastine and combined effects of clemastine and EHF EMR exposure with the use of our model of acute inflammation.

We found that clemastine at doses of 0.02–0.6 mg/kg caused dose-dependent decrease in exudative edema (Figure 9.5) (Gapeyev et al., 2006). Studying the combined action of clemastine and EHF EMR exposure, we revealed that clemastine dose-dependently abolished the anti-inflammatory effect of EHF EMR (Figure 9.5). Clemastine at doses of 0.2–0.6 mg/kg practically fully abolished the effect of EHF EMR (Gapeyev et al., 2008). These results indicate the involvement of histamine in the realization of anti-inflammatory effects of low-intensity EHF EMR.

The effects revealed could be explained by the following way. EHF EMR exposure of animals causes the histamine release from skin mast cells in sites of direct action of EMR (Popov et al., 2001). Histamine is released to blood flow and reduces the functional activity of phagocytes and T-lymphocytes (Bury et al., 1992; Hirasawa et al., 2002; Hori et al., 2002). This release of histamine to blood flow causes the anti-inflammatory effects of EHF EMR manifested as inhibition of migration of leukocytes to the inflammatory focus and reduction of their functional activity, which were registered as decrease in exudative edema of inflamed region. Low doses of clemastine, producing comparatively low concentrations of the drug in blood, dose-dependently inhibit the mast cell degranulation, presumably by the receptor-independent mechanism (Hagermark et al., 1992; Graziano et al., 2000; Assanasen and Naclerio, 2002). These low doses of clemastine weakly

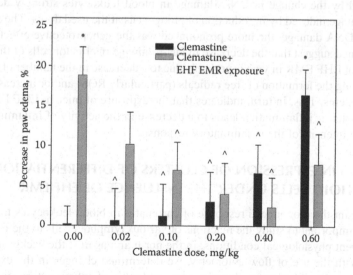

FIGURE 9.5 Decrease in paw edema (averaged by 3–7 h of the inflammation, as % of the control) depending on clemastine doses in sham-exposed and EHF EMR–exposed (42.2 GHz, 0.1 mW/cm², 20 min) mice. $n = 15$, *$p < 0.01$ from the control (at clemastine dose of 0 mg/kg), ^$p < 0.05$ from the value of EHF EMR effect at clemastine dose of 0 mg/kg by the Student's t-test. (Adapted from Gapeyev, A.B. et al., *Bioelectromagnetics*, 29(3), 197, 2008.)

influence the edema development at zymosan-induced inflammation (Erdo et al., 1993). High doses of clemastine, producing comparatively higher concentrations of the drug in blood, cause an anti-inflammatory effect itself by direct suppression of neutrophil activity. At combined action of clemastine and EHF EMR, clemastine dose-dependently abolishes the anti-inflammatory effect of EHF EMR, apparently due to the ability to inhibit the degranulation of mast cells.

Thus, our studies indicate the involvement of a great number of signaling and effector systems in the realization of biological and therapeutic effects of EHF EMR. Further, we consider results of the study of the mechanisms of anti-inflammatory effects of the EMR in terms of identifying key systems involved in the realization of EHF EMR effects at the cellular and whole organism levels.

9.7 STRUCTURAL CHANGES IN CHROMATIN OF LYMPHOID CELLS UNDER THE INFLUENCE OF EHF EMR AT THE INFLAMMATION

Our experimental data indicate that EHF EMR can influence the chromatin state of thymocytes and splenocytes under whole-body exposure of intact mice (Gapeyev et al., 2003). The EMR caused a multidirectional effect, chromatin condensation in thymocytes, and chromatin decondensation in splenocytes and did not affect the chromatin structure of peripheral blood leukocytes. Taking into account the possibility of the influence of low-intensity EHF EMR on chromatin structure of lymphoid cells with subsequent modifications of DNA–protein interaction and gene expression of several cytokines (Gapeyev et al., 2010), there is a reason to believe that the mechanisms of anti-inflammatory effects of EHF EMR may be associated with structural changes in chromatin of effector cells of inflammation.

Using the alkaline comet assay and systemic inflammatory model induced by intraperitoneal injection of lipopolysaccharide at a dose of 1 mg/kg or zymosan suspension in doses of 100 and 250 mg/kg of body weight in mice, it was shown that exposure of animals to EHF EMR (42.2 GHz, 0.1 mW/cm^2, 20 min exposure) at 1 h after induction of the inflammatory response resulted in a significant decrease in DNA damage in peripheral blood leukocytes and peritoneal neutrophils (Gapeyev et al., 2011b). It is important to note that the value of the anti-inflammatory effect of EHF EMR determined by the change in DNA damage in blood leukocytes strongly depended on the level of DNA damage induced by introduction of phlogogens at the used doses. The smaller was the induced level of DNA damage, the more pronounced was the genoprotective effect of EHF EMR. The results obtained suggest that the decrease in DNA damage in effector cells of the inflammation under the action of EHF EMR in vivo is probably due to a decrease in the number of oxidative DNA damage by reducing the formation of free radicals (particularly ROS) and/or increasing the activity of antioxidant systems. This, in turn, indicates that the exposure of mice to EHF EMR against the background of systemic inflammation leads to a decrease in the activity of inflammatory cells and a reduction in the intensity of the inflammatory response.

9.8 CHANGES IN EXPRESSION OF CLUSTERS OF DIFFERENTIATION IN LYMPHOID CELLS UNDER THE INFLUENCE OF EHF EMR

Taking into account the pronounced response of chromatin in blood leukocytes to the EHF EMR exposure, it was important to study the reactions of various lymphoid cells to the EHF EMR exposure under different physiological conditions, in the norm and against the background of systemic inflammation. With the use of flow cytometry, we determined changes in the expression of CD markers in thymocytes, splenocytes, lymphocytes, and cells of inflammatory exudate (Gapeyev et al., 2010). Phenotypic analysis of thymocytes and splenocytes was conducted with the use of commercial monoclonal antibodies on a laser flow cytometer (PAS-III, Germany). For staining surface antigens, we used rat monoclonal antibodies specific to mouse CD4, CD8, or CD25 labeled with phycoerythrin (R-PE) (Invitrogen Co., United States). Staining of freshly isolated cells was conducted in accordance with the generally accepted methods (Givan, 2001). A total of not less than 50,000 events per sample were collected on the flow cytometer. Dead cells and tissue debris were excluded according to forward and side-scatter properties. Systemic inflammation was induced in mice by intraperitoneal injection of zymosan suspension at a dose of 50 mg/kg of body weight. Control mice were injected by equal volume of physiological saline and sham-exposed.

It was found that the exposure of normal and inflamed mice to EHF EMR (42.2 GHz, 0.1 mW/cm², 20 min exposure) leads to an increase in the number of CD4+ and CD8+ T-lymphocytes on the average by 9%–13% ($p < 0.02$ compared to the control) and to a decrease in the number of CD25+ T-lymphocytes in the thymus (Figure 9.6). The ratio of CD4+/CD8+ T-lymphocytes in the thymus of animals was not significantly changed. Thus, when the mice were exposed to EHF EMR in the norm or against the background of the inflammation, the value and direction of the changes in the expression of CD4, CD8, and CD25 markers in thymus cells did not differ substantially. In the spleen of exposed animals, the number of CD4+ and CD8+ T-lymphocytes was significantly decreased that was especially pronounced after exposure against the background of the inflammation, by twofold and fourfold ($p < 0.05$ and $p < 0.01$ compared to the control), respectively (Figure 9.6). The ratio of CD4+/CD8+ T-lymphocytes in the spleen of exposed animals increased more than twofold ($p < 0.01$ compared to the control), which may indicate a stimulating effect of EHF EMR. These data are in good agreement with the literature, demonstrating the growth of helper/suppressor ratio and activity of T-lymphocytes in the spleen of mice exposed to EHF EMR (42.2 GHz, exposure for 30 min daily during 3 days) at thermal intensity of about 30 mW/cm² (Makar et al., 2003). At the same time, it should be noted that the EHF EMR exposure had no effect on the expression of CD25 marker of lymphocytic activation in the spleen of exposed animals (Figure 9.6).

Change in the expression of CD markers in blood lymphocytes under the influence of EHF EMR had a character similar to the changes occurring in the spleens of the exposed animals. The number of CD4+ and CD8+ T-lymphocytes was decreased approximately by 25% ($p < 0.02$ compared to the control) in blood of exposed mice, similar to that was done at the inflammatory process (Figure 9.7). When the inflamed mice were exposed, these changes were more pronounced and reached 50% or more (Figure 9.7). The ratio of CD4+/CD8+ T-lymphocytes in the blood of animals was

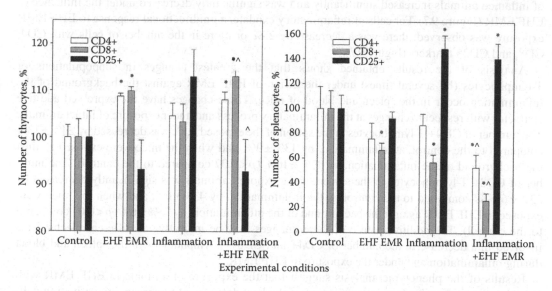

FIGURE 9.6 Number of thymocytes and splenocytes (in % of the control) expressing CD4, CD8, or CD25 markers in mice: in the norm (control), exposed to EHF EMR, with systemic zymosan-induced inflammation, and exposed to EHF EMR (42.2 GHz, 0.1 mW/cm²) for 20 min at 1 h after induction of the inflammation. *$p < 0.02$ compared to the control; ^$p < 0.02$ compared to the level at the inflammation by the Kruskal–Wallis test, $n = 7$ for each point.

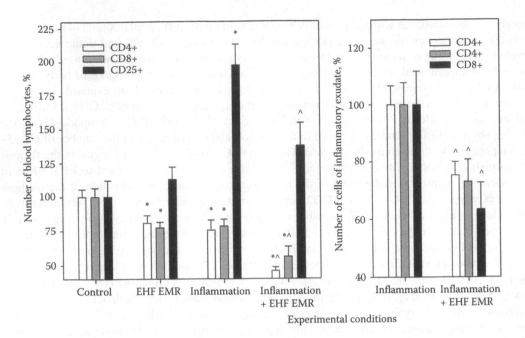

FIGURE 9.7 Number of blood lymphocytes and cells of inflammatory exudate (in % of the control) express-ing CD4, CD8, or CD25 markers in mice: in the norm (control), exposed to EHF EMR, with systemic zymosan-induced inflammation, and exposed to EHF EMR (42.2 GHz, 0.1 mW/cm²) for 20 min at 1 h after induction of the inflammation. $*p < 0.02$ compared to the control; $^p < 0.02$ compared to the level at the inflammation by the Kruskal–Wallis test, $n = 7$ for each point.

not significantly changed. An expression of CD25 marker of lymphocytic activation in the blood of inflamed animals increased significantly and was significantly decreased under the influence of EHF EMR (Figure 9.7). For cells of inflammatory exudate, a unidirectional response to EHF EMR exposure was observed: there was a decrease by 25% or more in the number of cells with CD4, CD8, and CD25 markers (Figure 9.7).

Analysis of the results obtained shows that the greatest changes in subpopulations of T-lymphocytes (by several times) under the action of EHF EMR against the background of the inflammation occur in the spleen and blood of mice. These changes have the expressed additive character with respect to changes at the inflammatory process and under exposure of intact animals. The number of CD4+ T-lymphocytes in the spleen of the exposed mice was decreased by 24% ± 13% compared to the control, at inflammation, by 13% ± 9%, and when the mice were exposed against the background of the inflammation, by 47% ± 10% ($p < 0.02$ compared to the control). The num-ber of CD8+ T-lymphocytes in the spleen of the exposed animals was significantly decreased by 33% ± 6% as compared to the control, at the inflammation, by 43% ± 6%, and when the mice were exposed to EHF EMR against the background of the inflammation, by 73% ± 6% ($p < 0.01$ compared to the control). The additive nature of these changes may be indicative of different mechanisms involved in a decrease in the number of CD4+ and CD8+ T-lymphocytes in the spleen and blood during inflammation and under the exposure to EHF EMR.

Results of the phenotypic analysis suggest that the exposure of animals to EHF EMR with effective parameters leads to intensification of the host defenses of the organism, activating cell-mediated immunity. This is evidenced by increasing the number of CD4+ and CD8+ T-cells in the thymus and the growth of the ratio of CD4+/CD8+ T-lymphocytes in the spleen of exposed animals. The significant decrease in the number of CD4+ and CD8+ T-lymphocytes in the spleen and blood may be due to either reduction in expression of these differentiation markers or as a result of migration of inflammatory cells into the inflammatory region to implement their functions in inflammation. Taking into account the fact that the exposure to EHF EMR against the background of the inflammation leads to decrease in the number of CD4+ and CD8+ cells of the inflammatory exudate, we assume that the radiation influences the expression of these markers. The characteristic effect of EHF EMR against the background of the inflammation in almost all tissues examined except the spleen is a decrease in the expression of CD25 marker of lymphocytic activation, which in turn may support the reduced activity of effector cells of the inflammation and, as a consequence, the lowering of an intensity of inflammatory reactions induced by the radiation.

9.9 CHANGES IN THE BALANCE OF MONO- AND POLYUNSATURATED FATTY ACIDS UNDER THE INFLUENCE OF EHF EMR

Recent pharmacological analysis of anti-inflammatory effects of EHF EMR using a model of acute zymosan-induced inflammation in mice and NSAID sodium diclofenac allows us to suggest that the anti-inflammatory effect of EHF EMR may be connected with an inhibition of the activity and/or expression of the inducible form of cyclooxygenase (Gapeyev et al., 2006). It was also suggested that besides the influence of EHF EMR on the cyclooxygenase activity, the anti-inflammatory effect may also be caused by changes in the levels of some fatty acids (FAs), substrata for cyclooxygenase, due to the influence of radiation on the composition and phase state of the membrane lipids and/or activity of phospholipases, possibly with the participation of calcium-dependent membrane-associated processes or glucocorticoids (Gapeyev et al., 2006, 2008).

With regard to the inflammatory response, its initiation, development, duration, and resolution are strongly dependent on broad spectrum of different mediators that are produced locally in the tissues by leukocytes and endothelial cells (Li et al., 1996). The detection of such mediators and studying their role in inflammation is the basis for modern approaches to development of new drugs for immunopharmacological correction of inflammation in various diseases. Two different types of these mediators, FAs and cytokines, are of greatest interest in this respect. It is known that FAs are important components in immune reactions both in normal and pathological processes, as they have diverse activity in living cells and influence membrane composition and function, eicosanoid synthesis, cellular signaling, and regulation of gene expression (Galli and Calder, 2009). Analysis of published data indicates that biomedical effects of some polyunsaturated FAs (PUFAs), in particular as a part of various diets (Galli and Calder, 2009), have pleiotropic nature and are comparable with the effects of low-intensity EHF EMR (Gapeyev and Chemeris, 2007). This similarity has led us to assume that PUFAs may be involved in the realization of biological effects of EHF EMR at the level of the whole organism. To test the hypothesis with the use of a method of gas chromatography, we studied changes in the content of FAs in thymic cells of mice exposed to EHF EMR in the norm and during systemic inflammation.

Analysis of FA composition in thymocytes was conducted according to the usual procedures (Knapp, 1979; Gapeyev et al., 2011a). Systemic inflammation was induced in mice by intraperitoneal injection of zymosan suspension at a dose of 50 mg/kg of body weight. Control mice were injected by equal volume of sterile physiological saline and sham-exposed.

It was found that a single exposure of intact mice to EHF EMR with effective parameters (42.2 GHz, 0.1 mW/cm^2, 20 min exposure) led to an increase in the content of linoleic, dihomo-γ-linolenic, and arachidonic $n-6$ PUFAs and in the content of eicosapentaenoic and docosapentaenoic $n-3$ PUFAs in the thymus of animals. In mice with systemic inflammation, the similar changes in FA composition were observed. In mice exposed against the background of the inflammation, the decrease in the content of palmitoleic and oleic monounsaturated FAs (MUFAs) and even more significant increase in PUFA content were found. The results suggest a modifying effect of EHF EMR on FA composition in thymus cells of normal and inflamed mice. The data obtained support the notion that MUFAs might be replaced by PUFAs that can enter into the thymic cells from the external media. Taking into account the fact that the metabolites of PUFAs are lipid messengers actively involved in inflammatory and immune reactions, we assume that the increase in the content of $n-3$ and $n-6$ PUFAs in phospholipids of cellular membranes facilitates the realization of anti-inflammatory effects of EHF EMR. Recent studies have identified a novel group of protective lipid mediators, lipoxins, formed from arachidonic acid, and E- and D-series resolvins, formed from eicosapentaenoic and docosahexaenoic acids, respectively, which dampen acute leukocyte responses, switch the synthesis of proinflammatory mediators to anti-inflammatory ones, and facilitate the resolution of inflammation (Serhan and Chiang, 2008; Ratnayake and Galli, 2009).

The phenomenon of the significant increase in the content of PUFAs in thymic cells of mice exposed to EHF EMR, which was first discovered in the study (Gapeyev et al., 2011a), is of special interest in terms of the new mechanisms of biological effects of the radiation and the effective application of EHF EMR in medical practice. It was shown that the increase in the content of PUFAs induced by EHF EMR exposure in mice with inflammation was additive, corresponding to the sum of changes in the content of PUFAs in normal mice exposed to EHF EMR and in mice with inflammation separately. This can indicate that different mechanisms of initiation processes cause increases in the content of PUFAs in inflammation and by the action of EHF radiation. The possibility to change the content and proportion of various PUFAs, the precursors of many signaling molecules, will make it possible to influence the composition of bioactive mediators synthesized in an organism, switch between different signaling systems, and regulate gene expression. It is noted that with increasing content of $n-3$ PUFAs, platelet aggregation is decreased and blood clotting is suppressed, which can provide prevention of thrombosis in diseases of the heart and kidneys (Serhan and Chiang, 2008). The positive effect of $n-3$ PUFAs on profile of synthesized eicosanoids and cytokines, as well as on the stability of endogenous proteins to denaturation, can be the basis for their application in the treatment of immune-dependent inflammatory diseases (Grimm et al., 2002). The positive effect of PUFAs is often achieved in the treatment of allergic dermatitis, chronic arthritis, autoimmune hepatitis, ulcerative colitis, pancreatitis, nephropathy, and other diseases for which EHF therapy is also effective (Rojavin and Ziskin, 1998; Pakhomov and Murphy, 2000; Betskii and Lebedeva, 2007).

The results obtained confirm the assumptions about the importance of the FAs in realization of a number of biological effects of low-intensity EHF EMR, in particular anti-inflammatory effects of the radiation. It was interesting to check the role of FAs in the implementation of the therapeutic effects of EHF EMR in other pathological processes. The influence of low-intensity EHF EMR on FA composition in cells of thymus and solid Ehrlich carcinoma in mice was studied (Gapeyev et al., 2013). It was shown for the first time that exposure of tumor-bearing mice to EHF EMR with effective parameters (42.2 GHz, 0.1 mW/cm^2, exposure for 20 min daily during 5 consecutive days after tumor inoculation) resulted in the recovery of modified by tumor growth FA composition in thymocytes to the state that is typical for normal animals. In tumor tissue that is characterized by elevated level of MUFAs, the exposure to EHF EMR significantly decreased the summary content of MUFAs and increased the summary content of PUFAs. These results suggest that the recovery of the FA composition in thymocytes and the modification of the FA composition in the tumor under the influence of EHF EMR on tumor-bearing animals are key elements in the mechanisms of realization of antitumor effects of the radiation.

9.10 INFLUENCE OF EHF EMR ON CYTOKINE PROFILE

Considering the changes in the subpopulation composition and activity of lymphocytes in the thymus and spleen of exposed animals, it was important to check how the effector functions of these cells are changed, in particular the production of several cytokines, which play a key role in the development and resolution of the inflammation. With the use of the reverse transcription reaction and real-time polymerase chain reaction (RT-PCR), changes in gene expression of inflammatory mediators (IL-1β, IL-6, IL-10, TNFα, and IFNγ) were studied in thymocytes and splenocytes of normal mice and in mice with systemic inflammation exposed to EHF EMR. RT-PCR was performed on specialized equipment Real-Time PCR System 7300 (Applied Biosystems, United States) with specialized primers (Overbergh et al., 2003) using a reaction Power SYBR Green PCR Master Mix (Applied Biosystems, United States). RT-PCR was performed in accordance with the requirements of the manufacturer. Each sample was amplified in triplicate using GAPDH as a control housekeeping gene (Mamo et al., 2007) for the purpose of standardization (Gapeyev et al., 2010).

In the norm, we did not observe marked constitutive expression of genes of studied cytokines. Development of the inflammatory process led to a marked increase in the expression of genes of IFNγ both in thymocytes and splenocytes of mice. In mice exposed to EHF EMR (42.2 GHz, 0.1 mW/cm^2, 20 min exposure), we found an increased expression of genes of IL-1β and IFNγ in thymocytes and significantly enhanced expression of genes of IL-1β, IL-10, and TNFα in splenocytes, indicating the involvement of the transcription factor NF-κB in the realization of the effects (Vallabhapurapu and Karin, 2009). These data are consistent with the previously observed fact of chromatin decondensation in splenocytes under the influence of EHF EMR (Gapeyev et al., 2003), which indirectly testified an activation of intracellular synthesis. Increased expression of genes of IFNγ and TNFα in lymphocytes of exposed animals indicates that EHF EMR can modulate immune responses, shifting the balance of immune response in the direction to the pathway mediated by Th1. This confirms earlier findings of high sensitivity of T-mediated and nonspecific immunity and low sensitivity of humoral immunity to the action of low-intensity EHF EMR (Lushnikov et al., 2001, 2003; Kolomytseva et al., 2002; Gapeyev and Chemeris, 2007). Significant enhancement of the expression of IL-10 in splenocytes of animals exposed against the background of the inflammation indicates on the initiation of anti-inflammatory immune reactions. Thus, by means of activating certain subpopulations of lymphocytes due to the action of the radiation, the specific cytokine profile can be created that produces the anti-inflammatory effect of EHF EMR. High anti-inflammatory potential of low-intensity EHF EMR with certain physical parameters in relation to the local inflammatory process has been demonstrated in previous studies (Gapeyev et al., 2006, 2008, 2009).

9.11 CONCLUSION

Thus, it was shown that the EHF EMR with effective parameters causes expressed immunotropic action in vivo. It was found that EHF EMR does not appreciably alter the formation of humoral immune response to thymus-dependent antigen (Lushnikov et al., 2001) but reduces the intensity of cell-mediated immune response in DTH reaction (Lushnikov et al., 2003). It was demonstrated that the exposure of mice to EHF EMR reduces the phagocytic activity of peripheral blood neutrophils in intact animals and alters the functional state of neutrophils in the inflammation focus (Kolomytseva et al., 2002; Lushnikov et al., 2003). The results obtained suggest that the most sensitive to the EHF EMR exposure are reactions of phagocytosis and cell-mediated immunity.

Comparing the results obtained in vivo and in vitro, we can conclude that the perception of low-intensity EHF EMR by the organism is systemic (Lushnikov et al., 2002). We believe that the effects of EHF EMR at the level of individual cells are due to changes in the intracellular signaling systems but, at the level of the whole organism, are due to the changes in the content of biologically active substances in the blood plasma and microenvironment of cells in the lymphoid organs. Primary processes leading to changes in the synthesis/secretion of biologically active substances by the action of EHF EMR occur at nerve endings and secretory cells of the skin (Radzievsky et al., 2001, 2004, 2008). We have demonstrated that the degranulation of skin mast cell may be an important amplification mechanism in the chain of events leading to a systemic response of the organism to the exposure (Popov et al., 2001; Gapeyev et al., 2006).

Using models of acute inflammation and posttraumatic regeneration of the skin, we showed for the first time that EHF EMR reduces the intensity of nonspecific inflammation (Bessonov et al., 2003; Lushnikov et al., 2003; 2004). Reducing acuity of the inflammation, EHF EMR can facilitate the pathologic process and thus create favorable conditions for the normalization of disturbed functions in different pathologies. From these positions, a universality and pleiotropy of EHF-therapy can be explained, which is used effectively in the treatment of diseases with expressed inflammation of wide localization in the pathogenesis (Betskii and Lebedeva, 2007). Pharmacological analysis showed that the kinetics and magnitude of anti-inflammatory effects of EHF EMR are close to those induced by a single therapeutic dose of NSAID sodium diclofenac (Gapeyev et al., 2006, 2008). We found that the cellular mechanisms of realization of anti-inflammatory effects of EHF EMR are associated with changes in the functional activity of phagocytic cells in the inflammation focus (decrease in phagocytic activity and ROS production), probably under the influence of biologically active substances, which enter into microvascular system of the skin, and blood flow is induced by EHF EMR exposure. This is confirmed by a dose-dependent abolishment of anti-inflammatory effects of the radiation when antihistamine clemastine is administered. Based on these data, we conclude that the release of biologically active substances, including histamine, from skin mast cell under the influence of low-intensity EHF EMR is a key point in the implementation of the biological action of EHF radiation and, in particular, its anti-inflammatory effects (Gapeyev et al., 2006).

Analysis of published data (Lushnikov et al., 2002) and our own results allows us to formulate *histaminic model* of biological action of EHF EMR (Gapeyev et al., 2006) linking the implementation mechanisms of EHF EMR effects at the cellular and whole organism levels. According to this model, the EHF EMR exposure of the skin causes a degranulation of skin mast cell in the exposed area (Popov et al., 2001) and the entry of released histamine and other physiologically active compounds into the blood flow. These substances, including histamine, can initiate a series of reactions leading to a decrease in the functional activity of phagocytes and T-lymphocytes, inducing the anti-inflammatory effect.

Taken together, these results demonstrate that the basic mechanisms for the implementation of low-intensity EHF EMR effects are associated with the modification of the immune status of the organism as a result of systemic reaction to the exposure. Among the key elements responsible for the implementation of EHF EMR effects at the cellular and whole organism levels, it should be highlighted: (1) the pronounced response of chromatin of lymphoid cells providing a genoprotective action of the radiation during inflammatory process; (2) the activation of T-cellular immunity manifested as an increase in the number of CD4+ and CD8+ T-cells in thymus of exposed animals; (3) changes in a balance of MUFAs and PUFAs in the thymus of exposed mice; and (4) the creation of the specific profile of cytokines providing anti-inflammatory effect of EHF EMR. It is supposed that by means of initiation of certain signaling and effector systems, the directed answer of an organism to a specific combination of effective parameters of the EMR can be carried out.

The results obtained have significantly deepened the fundamental understanding of the mechanisms of biological action of EHF EMR at the cellular and organismal levels and may serve as a basis for the development of therapeutic recommendations on the combined use of EHF EMR and drugs. It should be noted that the observed mechanisms of action of low-intensity EHF EMR are not the only possibility, however, as we have shown, to make a significant contribution to the realization of biological effects of the radiation at the whole organism level. The observed mechanisms of effects of low-intensity EHF EMR in vivo will be of unquestionable value to further studies on the role of regulatory systems (nervous, endocrine, and immune) in the realization of EHF EMR effects.

ACKNOWLEDGMENTS

The author is very grateful to Aripovsky A.V., Kolomytseva M.P., Kudryavtsev A.A., Kulagina T.P., Lushnikov K.V., Popov V.I., Rogachevskii V.V., Romanova N.A., Sadovnikov V.B., Safronova V.G., Sirota N.P., Sokolov P.A., Fesenko E.E., Khramov R.N., Chemeris N.K., and Shumilina Yu.V. for their participation in certain studies.

REFERENCES

Abbas A.K., Lichtman A.H. *Basic Immunology: Functions and Disorders of the Immune System.* Saunders, an Imprint of Elsevier, Philadelphia, 2004.

Alovskaya A.A., Gabdoulkhakova A.G., Gapeyev A.B. et al. Biological effect of EHF EMR is determined by cell functional status. *Herald New Med. Technol.* 5(2) (1998): 11–15 (in Russian).

Arzamastsev E.V., Gus'kova T.A., Berezovskaya I.V., Lubimov B.I., Liberman S.S., Verstakova O.L. Methodical instructions for studying general toxic effects of pharmacological substances. In *Manual on Experimental (Pre-Clinical) Study of the New Pharmacological Substances*, Khabriev R.U. ed. Moscow, Russia: Medicine Ltd., 2005, pp. 41–54 (in Russian).

Assanasen P., Naclerio R.M. Antiallergic anti-inflammatory effects of H_1-antihistamines in humans. *Clin. Allergy Immunol.* 17 (2002): 101–139.

Belyaev I.Y. Dependence of non-thermal biological effects of microwaves on physical and biological variables: Implications for reproducibility and safety standards. *Eur. J. Oncol.* 5 (2010): 187–217.

Belyaev I.Ya. Non-thermal biological effects of microwaves. *Microw. Rev.* 11 (2005): 13–29.

Bessonov A.E., Gapeyev A.B., Lushnikov K.V. et al. The influence of wide-band modulated radiation generated by a minitag therapeutic apparatus and continuous extremely high-frequency electromagnetic radiation on regeneration of full-thickness skin wounds in laboratory mice. *Herald New Med. Technol.* 10(1–2) (2003): 14–15 (in Russian).

Betskii O.V., Lebedeva N.N. Modern views on the mechanisms of influence of low-intensity millimeter waves on biological objects. *Millimeter Waves Biol. Med.* 3(23) (2001): 5–18 (in Russian).

Betskii O.V., Lebedeva N.N. Application of low-intensity millimeter waves in biology and medicine. *Biomed. Radioelectron.* 8–9 (2007): 6–25 (in Russian).

Betskii O.V., Lebedeva N.N., Kotrovskaya T.I. Application of low-intensity millimeter waves in medicine (retrospective review). *Millimeter Waves Biol. Med.* 2(38) (2005): 23–39 (in Russian).

Bury T.B., Corhay J.L., Radermecker M.F. Histamine-induced inhibition of neutrophil chemotaxis and T-lymphocyte proliferation in man. *Allergy* 47 (1992): 624–629.

Chemeris N.K., Gapeyev A.B., Sirota N.P. et al. Lack of direct DNA damage in human blood leukocytes and lymphocytes after in vitro exposure to high power microwave pulses. *Bioelectromagnetics* 27(3) (2006): 197–203.

Chernukh A.M., Frolov E.P., eds. *The Skin (Structure, Function, Common Pathology and Therapy).* Moscow, Russia: Medicine, 1982 (in Russian).

Devyatkov N.D., Golant M.B., Betskii O.V. *Millimeter Waves and Their Role in the Life.* Moscow, Russia: Radio i Svyas', 1991.

Erdo F., Torok K., Aranyi P., Szekely J.I. A new assay for antiphlogistic activity: Zymosan-induced mouse ear inflammation. *Agents Actions* 39 (1993): 137–142.

Fedorov V.I., Pogodin A.S., Dubatolova T.D., Verlamov A.V., Leontiev K.V., Khamoyan A.G. Comparative study of the effects of electromagnetic radiation of infra-red, sub millimeter and millimeter ranges on γ-irradiation induced somatic mutations of wing cells of *Drosophila melanogaster. Biofizika* 46(2) (2001): 298–302 (in Russian).

Friemel H., ed. *Immunological Methods.* Jena, Germany: Veb Gustav Fischer Verlag, 1984.

Galli C., Calder P.C. Effects of fat and fatty acid intake on inflammatory and immune responses: A critical review. *Ann. Nutr. Metab.* 55(1–3) (2009): 123–139.

Gapeyev A.B. Study of the mechanisms of biological effects of low-intensity extremely high-frequency electromagnetic radiation: Progress, problems and prospects. *Biomed. Technol. Radioelectron.* 6 (2014): 20–30 (in Russian).

Gapeyev A.B., Chemeris N.K. Mechanisms of biological effects of extremely high-frequency electromagnetic radiation at the whole organism level. *Biomed. Radioelectron.* 8–9 (2007): 30–46 (in Russian).

Gapeyev A.B., Chemeris N.K. Dosimetry questions at studying biological effects of extremely high-frequency electromagnetic radiation. *Biomed. Radioelectron.* 1 (2010): 13–36 (in Russian).

Gapeyev A.B., Kulagina T.P., Aripovsky A.V. Exposure of tumor-bearing mice to extremely high-frequency electromagnetic radiation modifies the composition of fatty acids in thymocytes and tumor tissue. *Int. J. Radiat. Biol.* 89(8) (2013): 602–610.

Gapeyev A.B., Kulagina T.P., Aripovsky A.V., Chemeris N.K. The role of fatty acids in anti-inflammatory effects of low-intensity extremely high-frequency electromagnetic radiation. *Bioelectromagnetics* 32(5) (2011a): 388–395.

Gapeyev A.B., Lushnikov K.V., Shumilina Iu.V., Chemeris N.K. Pharmacological analysis of anti-inflammatory effects of low-intensity extremely high-frequency electromagnetic radiation. *Biophysics* 51(6) (2006): 927–939.

Gapeyev A.B., Lushnikov K.V., Shumilina Yu.V., Sirota N.P., Sadovnikov V.B., Chemeris N.K. The influence of low-intensity extremely high-frequency electromagnetic radiation on chromatin structure in lymphoid cells in vivo and in vitro. *Radiat. Biol. Radioecol.* 43(1) (2003): 87–92 (in Russian).

Gapeyev A.B., Mikhailik E.N., Chemeris N.K. Anti-inflammatory effects of low-intensity extremely high-frequency electromagnetic radiation: Frequency and power dependence. *Bioelectromagnetics* 29(3) (2008): 197–206.

Gapeyev A.B., Mikhailik E.N., Chemeris N.K. Features of anti-inflammatory effects of modulated extremely high-frequency electromagnetic radiation. *Bioelectromagnetics* 30(6) (2009): 454–461.

Gapeyev A.B., Romanova N.A., Chemeris N.K. Changes in the chromatin structure of lymphoid cells under the influence of low-intensity extremely high-frequency electromagnetic radiation against the background of inflammatory process. *Biophysics* 56(4) (2011b): 672–678.

Gapeyev A.B., Safronova V.G., Chemeris N.K., Fesenko E.E. Modification of the activity of mouse peritoneal neutrophils on exposure to millimetre waves in the near and far zones of the emitter. *Biophysics* 41(1) (1996): 219–234.

Gapeyev A.B., Sirota N.P., Kudryavtsev A.A., Chemeris N.K. Responses of thymocytes and splenocytes to low-intensity extremely high-frequency electromagnetic radiation in normal mice and in mice with systemic inflammation. *Biophysics* 55(4) (2010): 577–582.

Gapeyev A.B., Sokolov P.A., Chemeris N.K. A study of absorption of energy of the extremely high frequency electromagnetic radiation in the rat skin by various dosimetric methods and approaches. *Biophysics* 47(4) (2002): 706–715.

Givan A.L. Principles of flow cytometry: An overview. *Methods Cell Biol.* 63 (2001): 19–50.

Govallo V.I., Barer F.C., Volchek I.A., Baranovskaya V.T., Malyavko T.P. Production of proliferation activating factor by EMR-exposed human lymphocytes and fibroblasts. In *Reports of International Symposium on Millimeter Waves of Non-Thermal Intensity in Medicine, Part 2*, Moscow, Russia, 1991, pp. 340–344 (in Russian).

Graevskaya E.E., Akhalaya M.Y., Ensu C., Parkhomenko I.M., Strakhovskaya M.G., Goncharenko E.N. Effect of middle-wave ultraviolet irradiation and red light on degranulation of peritoneal mast cell in rats. *Bull. Exp. Biol. Med.* 129(4) (2000): 357–358.

Graziano F.M., Cook E.B., Stahl J.L. Antihistamines and epithelial cells. *Allergy Asthma Proc.* 21 (2000): 129–133.

Grimm H., Mayer K., Mayser P., Eigenbrodt E. Regulatory potential of N-3 fatty acids in immunological and inflammatory processes. *Br. J. Nutr.* 87(suppl. 1) (2002): S59–S67.

Gugkova O.Yu., Gudkov S.V., Gapeyev A.B., Bruskov V.I., Rubanik A.V., Chemeris N.K. Study of the mechanisms of formation of reactive oxygen species in aqueous solutions exposed to high peak-power pulsed electromagnetic radiation of extremely high frequencies. *Biophysics* 50(5) (2005): 679–684.

Gushchin I.S. *Allergic Inflammation and their Pharmacological Control.* Moscow, Russia: Pharmus Print, 1998.

Hagermark O., Wahlgren C.F., Gios I. Inhibitory effect of loratadine and clemastine on histamine release in human skin. *Skin Pharmacol.* 5 (1992): 93–98.

Hirasawa N., Ohtsu H., Watanabe T., Ohuchi K. Enhancement of neutrophil infiltration in histidine decarboxylase-deficient mice. *Immunology* 107 (2002): 217–221.

Hori Y., Nihei Y., Kurokawa Y. et al. Accelerated clearance of *Escherichia coli* in experimental peritonitis of histamine-deficient mice. *J. Immunol.* 169 (2002): 1978–1983.

Ibrahim T., Cunha J.M., Madi K., da Fonseca L.M., Costa S.S., Goncalves Koatz V.L. Immunomodulatory and anti-inflammatory effects of *Kalanchoe brasiliensis*. *Int. Immunopharmacol.* 2(7) (2002): 875–883.

Klaus G.G.B., ed. *Lymphocytes: A Practical Approach.* Oxford, U.K.: IRL Press Ltd., 1987.

Knapp D.R. *Handbook of Analytical Derivatization Reactions*, Vol. 154. New York: John Wiley & Sons Inc., 1979, pp. 164–167.

Kolomytseva M.P., Gapeyev A.B., Sadovnikov V.B., Chemeris N.K. Suppression of nonspecific resistance in vivo by weak extremely high frequency electromagnetic radiation. *Biophysics* 47(1) (2002): 64–69.

Li Y., Ferrante A., Poulos A., Harvey D.P. Neutrophil oxygen radical generation. Synergistic responses to tumor necrosis factor and mono/polyunsaturated fatty acids. *J. Clin. Invest.* 97(7) (1996): 1605–1609.

Lushnikov K.V., Gapeyev A.B., Chemeris N.K. Effects of extremely high-frequency electromagnetic radiation on the immune system and systemic regulation of the homeostasis. *Radiat. Biol. Radioecol.* 42(5) (2002): 533–545 (in Russian).

Lushnikov K.V., Gapeyev A.B., Sadovnikov V.B., Chemeris N.K. Effect of the weak extremely-high frequency electromagnetic radiation on the indices of the humoral immunity in healthy mice. *Biophysics* 46(4) (2001): 711–719.

Lushnikov K.V., Gapeyev A.B., Shumilina Yu.V., Shibaev N.V., Sadovnikov V.B., Chemeris N.K. Suppression of cell-mediated immune response and nonspecific inflammation on exposure to extremely high frequency electromagnetic radiation. *Biophysics* 48(5) (2003): 856–863.

Lushnikov K.V., Shumilina J.V., Yakushev E.Yu., Gapeyev A.B., Sadovnikov V.B., Chemeris N.K. Comparative study of anti-inflammatory effects of low-intensity extremely high-frequency electromagnetic radiation and diclofenac on footpad edema in mice. *Electromagnet. Biol. Med.* 24(2) (2005): 143–157.

Lushnikov K.V., Shumilina Y.V., Yakushina V.S., Gapeyev A.B., Sadovnikov V.B., Chemeris N.K. Effects of low-intensity ultrahigh frequency electromagnetic radiation on inflammatory processes. *Bull. Exp. Biol. Med.* 137(4) (2004): 364–366.

Makar V., Logani M., Szabo I., Ziskin M. Effect of millimeter waves on cyclophosphamide induced suppression of T cell functions. *Bioelectromagnetics* 24 (2003): 356–365.

Mamo S., Gal A.B., Bodo S., Dinnyes A. Quantitative evaluation and selection of reference genes in mouse oocytes and embryos cultured in vivo and in vitro. *BMC Dev. Biol.* 7 (2007): 14.

Mayanskii D.N. *Chronic Inflammation.* Moscow, Russia: Medicine 1991 (in Russian).

Ostling O., Johanson K.J. Microelectrophoretic study of radiation-induced DNA damages in individual mammalian cells. *BBRC* 123(1) (1984): 291–298.

Overbergh L., Giulietti A., Valckx D., Decallonne B., Bouillon R., Mathieu C. The use of real-time reverse transcriptase PCR for the quantification of cytokine gene expression. *J. Biomol. Tech.* 14 (2003): 33–43.

Pakhomov A.G., Murphy M.R. Low-intensity millimeter waves as a novel therapeutic modality. *IEEE Trans. Plasma Sci.* 28(1) (2000): 34–40.

Popov V.I., Rogachevskii V.V., Gapeyev A.B., Khramov R.N., Chemeris N.K., Fesenko E.E. Degranulation of dermal mast cells caused by the low-intensity electromagnetic radiation of extremely high frequency. *Biophysics* 46(6) (2001): 1041–1046.

Potselueva M.M., Pustovidko A.V., Evtodienko Iu.V., Khramov R.N., Chaĭlakhian L.M. Formation of reactive oxygen species in aqueous solutions after exposure to extremely-high frequency electromagnetic fields. *Dokl. Akad. Nauk.* 359 (1998): 415–418 (in Russian).

Radzievsky A.A., Gordiienko O.V., Cowan A., Alekseev S.I., Ziskin M.C. Millimeter-wave-induced hypoalgesia in mice: Dependence on type of experimental pain. *IEEE Trans. Plasma Sci.* 32(4) (2004): 1634–1643.

Radzievsky A.A., Gordiienko O.V., Alekseev S.I., Szabo I., Cowan A., Ziskin M.C. Electro-magnetic millimeter wave induced hypoalgesia: Frequency dependence and involvement of endogenous opioids. *Bioelectromagnetics* 29 (2008): 284–295.

Radzievsky A.A., Rojavin M.A., Cowan A., Alekseev S.I., Radzievsky A.A. Jr., Ziskin M.C. Peripheral neural system involvement in hypoalgesic effect of electromagnetic millimeter waves. *Life Sci.* 68(10) (2001): 1143–1153.

Ratnayake W.M., Galli C. Fat and fatty acid terminology, methods of analysis and fat digestion and metabolism: A background review paper. *Ann. Nutr. Metab.* 55(1–3) (2009): 8–43.

Roitt I., Brostoff J., Mail D. *Immunology.* London, U.K.: Mosby International Ltd., 1998.

Rojavin M.A., Ziskin M.C. Medical application of millimetre waves. *Quart. J. Med.* 91 (1998): 57–66.

Ryzhkova L.V., Starik A.M., Volgarev A.P., Galchenko S.V., Sazonov Yu.A. Protective effect of EHF-therapy at lethal influenza infection. In *Reports of International Symposium on Millimeter Waves of Non-Thermal Intensity in Medicine, Part 2*, Moscow, Russia, 1991, pp. 373–378 (in Russian).

Sayed S.O., Dyson M. Effect of laser pulse repetition rate and pulse duration on mast cell number and degranulation. *Lasers Surg. Med.* 19(4) (1996): 433–437.

Sazonov A.Yu., Ryzhkova L.V. Influence of EMR of MM-range on biological objects of various complexity. In *Reports of 10 Russian Symposium with International Participation Millimeter Waves in Medicine and Biology*, Moscow, Russia, 1995, pp. 112–114.

Serhan C.N., Chiang N. Endogenous pro-resolving and anti-inflammatory lipid mediators: A new pharmacologic genus. *Br. J. Pharmacol.* 153(suppl. 1) (2008): S200–S215.

Shckorbatov Y.G., Grigoryeva N.N., Shakhbazov V.G., Grabina V.A., Bogoslavsky A.M. Microwave irradiation influences on the state of human cell nuclei. *Bioelectromagnetics* 19 (1998): 414–419.

Shurlygina A.V., Truphakin V.A., Guschin G.V., Korneva E.A. Daily variations of adrenaline, noradrenalin and β-adrenoreceptor contents in blood and lymphoid organs of healthy rats. *Bull. Exp. Biol. Med.* 128(9) (1999): 344–346 (in Russian).

Singh N.P., McCoy M.T., Tice R.R., Schneider E.L. A simple technique for quantification of levels of DNA damage in individual cells. *Exp. Cell Res.* 175 (1988): 184–191.

Sirota N.P., Bezlepkin V.G., Kuznetsova E.A. et al. Modifying effect in vivo of interferon alpha on induction and repair of lesions of DNA of lymphoid cells of gamma-irradiated mice. *Radiat. Res.* 146 (1996): 100–105.

Sirota N.P., Podlutskii A.Ya., Gasiev A.I. DNA damage in single mammalian cells. *Radiobiology* 31(5) (1991): 722–727 (in Russian).

Tice R.R., Agurell E., Anderson D. et al. Single cell gel/comet assay: Guidelines for in vitro and in vivo genetic toxicology testing. *Environ. Mol. Mutagen.* 35 (2000): 206–221.

Vallabhapurapu S., Karin M. Regulation and function of NF-kB transcription factors in the immune system. *Annu. Rev. Immunol.* 27 (2009): 693–733.

Vane J.R. Cyclooxygenase 1 and 2. *Annu. Rev. Pharmacol. Toxicol.* 38 (1998): 97–120.

Vane J.R., Botting R.M. Anti-inflammatory drugs and their mechanism of action. *Inflamm. Res.* 47(suppl. 2) (1998): S78–S87.

Yarilin A.A. *Fundamental Immunology.* Moscow, Russia: Medicine, 1999.

Zavgorodny S.V., Khizhnyak Y.P., Voronkov V.N., Sadovnikov V.B. Morphological changes in skin nerves caused by electromagnetic radiation of the millimeter range. *Crit. Rev. Biomed. Eng.* 28(3–4) (2000): 641–658.

10 Effects of Therapeutic and Low-Frequency Electromagnetic Fields on Immune and Anti-Inflammatory Systems

Walter X. Balcavage, Gabi N. Waite,
and Stéphane J.-P. Egot-Lemaire

CONTENTS

10.1 INTRODUCTION

This chapter aims (1) to discuss the effect of electromagnetic fields (EMFs) on the biological formation and regulation of reactive oxygen species (ROS) generated by cells of the immune system, (2) to review recent EMF literature related to this topic, and (3) to report on previously unpublished EMF/immune system studies performed in our laboratory at Indiana University. To facilitate this discussion, we will first outline key features of the immune system to provide a background for our consideration of EMF effects on ROS, such as superoxide radical, hydrogen peroxide, and hydroxyl radical. These ROS, which are intrinsically highly cytotoxic, give rise to additional cytotoxins such as perchlorates but can also act as autocrine or cytokine signaling molecules to enhance an immune response. Next, we will review the recent literature dealing with EMF effects on cells of the immune system, including our unpublished experiments with the transformed THP-1 immune cell line and normal human immune cells. Finally, we will consider EMF-based physicochemical mechanisms that are likely to underpin the reported observations and conclude with our view of the future role of EMFs in the therapy of inflammatory diseases.

10.2 IMMUNE SYSTEM AND ROS

There are two main functions of an individual's immune system. The first is to discriminate normal healthy cells belonging to that individual (self cells) from pathogens (nonself cells), damaged self cells, malignant self cells, and a variety of toxins. The second function is to eliminate pathogenic, damaged, and malignant cells and their debris, while leaving normal healthy cells undisturbed. To accomplish these tasks, the immune system has evolved a number of cellular and chemical recognition and defense mechanisms that are conventionally categorized into the immediate responding *innate* and the delayed *adaptive* (or *acquired*) immune subsystems. The innate system is highly associated with ROS generation. It is characterized as being constitutive, rapidly triggered, and acting immunologically in a nonspecific fashion against a broad spectrum of immunogens. The adaptive system is characterized by traits and processes that are developmentally acquired during the life of the individual, and, in sharp contrast to the innate system, the adaptive system specifically targets each unique immunogen encountered during life. Cell-based processes and humoral mechanisms, such as antibody binding reactions, comprise the tools, or defense mechanisms, of both the innate and adaptive immune systems, and they act in concert to recognize and eliminate pathogenic substances.

While EMFs might play a role in the expression and/or activity of humoral (e.g., antibody) components of the immune system, here, we are mainly concerned with ROS generation, an important and robust intra- and extracellular process of many cells of the innate and adaptive immune systems. EMF modulation of ROS has been documented for cells of both the innate and adaptive immune systems, as discussed in more detail later. And so a brief enumeration of these key cell types and their key biological properties follows. We begin by outlining the origin, development, and principal function of T cells, B cells, and antibodies as the main cellular and humoral components of the adaptive immune system. We then outline development maturation and the function of cellular components of the innate immune system, including macrophages, dendritic cells, and natural killer (NK) cells. The humoral components, while important, will not be further presented, as they are only peripherally related to our discussion on the role of EMFs for immune system ROS production.

All immune system cells derive from bone marrow stem cells and differentiate into their final active state via a series of processes, driven by the need to maintain the organism's homeostatic state. These processes are regulated mainly by extracellular ligands that initiate intracellular signal transduction cascades. These cascades are often initiated as maturing immune system cells (known as leukocytes) move from their site of origin in the bone marrow through the lymph and blood circulatory systems to inflamed and/or diseased tissues. These cascades generate new DNA transcription factors and gene rearrangement processes whose activity culminates in genetically determined changes in cell identity (differentiation) and elaboration of new extracellular antibodies, cytokines, and autocrines, including ROS. In concert, these events enhance the immune response.

T lymphocytes and B lymphocytes are the principal cells of the acquired immune system and are derived from circulatory system leukocytes. In response to pathogenic insult, T cells in lymph nodes or tissue-resident T cells sense pathogen-related immunogens (or antigens) through their high-affinity binding to specific receptors on T cells. After this cell-specific, immunogen/antigen binding, T cell receptors (TCRs) transmit the TCR–antigen binding event to the cell interior through molecular changes in the transmembrane TCR. These high-affinity TCR binding events activate intracellular signal cascades that are modulated by other ligands (autocrines or cytokines) in the extracellular environment, leading to further differentiation of the previously *naïve* T cells into one of three *effector* cell types, cytotoxic T cells (CTLs) and two types of T helper cells (Th1 and Th2 cells). CTLs have the capability of recognizing self cells infected by pathogens and of killing such pathogen-infected cells by inducing cell lysis or apoptotic death. Th1 cells secrete pro-inflammatory cytokines including interferon gamma (IFNγ), which potentiates the response of innate immune system cells against intracellular pathogens (bacteria and viruses localized in self cells [Nathan et al. 1983]). Th2 cells also produce a number of cytokines, including interleukin 4 (IL-4), which acts as a Th2 autocrine and has a key role in helping stimulate development of antibody-producing B cells (Paul 1987). The antibodies produced by these cells are a key humoral component of the adaptive immune system. Secreted antibodies act mainly against extracellular immunogens and immunogens on the cell surface of pathogens.

The main role of B cells is to circulate throughout the body and to differentiate into antibody-producing cells in response to immunogenic challenge. To accomplish this task, bone marrow stem cells continually differentiate into new B cells, each with a cell surface B cell receptor (BCR) that is molecularly unique and binds a specific immunogen. These unique B cells and their direct progeny, which express the identical BCR, are released into the lymphatic circulation and ultimately into the blood circulatory system. Each population of B cells having the same BCR is known as a B cell clone. When circulating members of a B cell clone encounter their BCR-specific antigen (normally nonself in origin), they undergo massive clonal expansion. With the aid of cytokines from Th2 cells, the expanded clonal cells further differentiate into short-lived plasma B cells that secrete antibodies and into memory B cells, some of which survive for the life of the individual as a part of their acquired immune system repertoire.

Like the acquired immune system, the cells of the innate immune system are also derived from bone marrow stem cells and appear in the circulation either as granulocytic leukocytes, characterized by prominent intracellular secretory granules and a multilobed nucleus, or as agranular, monocytic leukocytes with a smooth ellipsoidal nucleus, or as lymphocytic leukocytes known as NK cells. The humoral components of the innate system include the complement system, the coagulation cascade, lysozyme, and interferons. While the latter humoral processes and substances are important components of the innate immune system, they will not be considered further as they are only peripherally related to our current consideration of EMF effects on immune system ROS production. For a discussion of innate system humoral components, see Janeway and Medzhitov (2002).

Neutrophils are the most prominent granulocytes. They produce substantial quantities of ROS and are the first cell responders to infection. Neutrophils at a site of tissue damage phagocytize pathogens and toxins, and in addition to producing ROS, they secrete a variety of other cytotoxic agents. ROS generation requires an immune triggering event which is followed by rapid intracellular assembly of preexisting proteins into a functional multienzyme system that consumes molecular oxygen and produces ROS. This process is known as the respiratory burst.

Classically, monocytic leukocytes have been differentiated into discrete cell types known as macrophages and dendritic cells. After neutrophils, macrophages are viewed as the innate system's second line of phagocyte defense, while dendritic cells are mainly viewed as providing an interface, linking with and modulating the slower-acting acquired immune system. Like neutrophils, immunogen-activated macrophages phagocytize pathogens and other toxins and exhibit a respiratory burst culminating in ROS production. There are two categories of macrophages with opposing activities. Macrophages of the M1 type have the killer phenotype that is typically associated with this cell type, in that they scavenge debris and microbes. They foster the progression of the immune reaction and are hence characterized as pro-inflammatory cells. Macrophages of the M2 type have a function in tissue repair and healing. Since they contribute to the termination of the immune response, they are characterized as anti-inflammatory cells. While macrophages and dendritic cells are still generally considered to be separate cell types, they increasingly appear to be very similar and may in fact be alternatively activated states of the same cell (Segerer et al. 2008).

Morphologically, NK cells are granulocytic, but they mature via a lineage distinct from that of classical granulocytes, such as neutrophils. NK cells sense and bind to foreign peptides displayed on the surface of tumorigenic and virus-infected self cells. When activated, these innate system NK cells release the cytokine content of granules, including porins and granzymes, which respectively cause lysis and apoptosis of infected or tumorigenic self cells.

10.3 ROS BIOCHEMISTRY

While molecular oxygen (O_2) is an exception, most stable ground-state atoms and molecules are generally characterized as having outer orbital valence electrons paired with opposing quantum spin (spin up and spin down, or $+\frac{1}{2}$ and $-\frac{1}{2}$). Such substances are said to be in the singlet state. With regard to EMFs, ground-state singlets (i.e., with no unpaired electrons) they are mildly repelled by external magnetic fields, are termed diamagnetic, and are not viewed to be appreciably effected by magnetic fields. When the electron configuration of a singlet, ground-state atom or molecule is rearranged (i.e., by gain or loss of one electron), so that it has one or more unpaired electrons, it becomes less stable, more energetic, and more reactive and is known as a radical or free radical. Such substances are termed paramagnetic, meaning that they interact with and are attracted by EMFs. Thus, paramagnetic biomolecules have the potential for having their chemical reactivity altered by external magnetic fields and are of interest in bioelectromagnetic studies.

In sharp contrast with the latter generalities, the most stable ground state of molecular oxygen (O_2) is unusual because its valence electron configuration has two unpaired electrons. Thus, in its ground state, with its two unpaired electrons, oxygen is a paramagnetic diradical. Diradicals have three isoenergetic, or degenerate, electron orbital configurations and so are said to be in a triplet state ($^3O_2$). Thus, ground-state oxygen is a paramagnetic diradical triplet. As shown in Figure 10.1, in biological systems, ground-state oxygen with its two unpaired electrons interacts with one-electron donors such as flavin nucleotides or coenzyme Q to form superoxide ($O_2^{-\bullet}$), which is the progenitor of most biological ROS. In the reaction, one of the oxygen's two unpaired electrons pairs with the added electron, converting it to $O_2^{-\bullet}$, which contains one unpaired electron and remains paramagnetic. Thus, ground-state oxygen and its one electron adduct, superoxide, are both paramagnetic and are both subject to EMF-induced changes in their chemical reactivity.

One-electron reduction of superoxide leads to the formation of peroxide ion (O_2^{-2}), which in its protonated (neutralized) form is called hydrogen peroxide (H_2O_2). H_2O_2 has a relatively long lifetime (seconds to minutes) compared to other ROS, whose lifetimes are only fractions of seconds. Consequently, H_2O_2 is less reactive than other ROS, and it can easily be transformed into water by catalases and peroxidases, making it an important messenger in ROS signaling cascades. On the other hand, H_2O_2 can be dangerous to the cell, since it can readily lose an electron and a proton to generate the highly reactive hydroxyl radical ($\bullet OH$) (Mulligan et al. 1991), which can directly initiate an H_2O_2-generated chain reaction. If the chain reactions remain unquenched by radical scavengers (such as melatonin or vitamin E), these damaging chain reactions will terminate with bond scission of nucleic acids, proteins, and other biomolecules. ROS-initiated bond scission reactions produce a plethora of products, the most representative of which are aldehydes like malondialdehyde. Thus, the relatively long-lived H_2O_2, produced from $O_2^{-\bullet}$, acts as a free-radical carrier that can

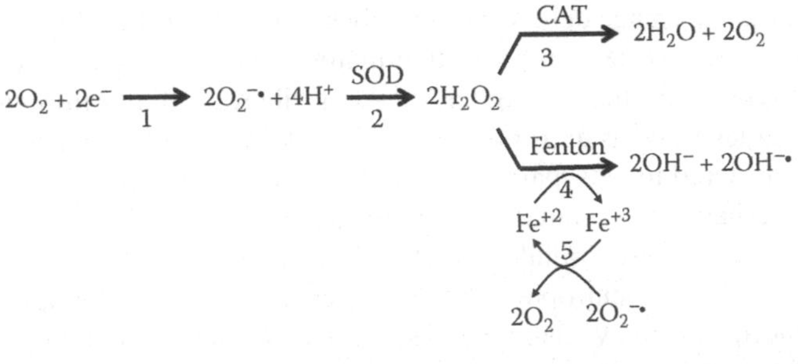

$$2O_2 + 2e^- \xrightarrow{\ 1\ } 2O_2^{-\bullet} + 4H^+ \xrightarrow[\ 2\ ]{SOD} 2H_2O_2 \nearrow^{\ \xrightarrow[\ 3\ ]{CAT} 2H_2O + 2O_2}_{\ \searrow \xrightarrow[\ 4\ ]{Fenton} 2OH^- + 2OH^{-\bullet}}$$

$$\begin{array}{c} Fe^{+2} \quad Fe^{+3} \\ \diagdown\ 5\ \diagup \\ 2O_2 \quad 2O_2^{-\bullet} \end{array}$$

FIGURE 10.1 Common ROS-related reactions. Type 1 reactions produce superoxide ($O_2^{-\bullet}$) and are generated via diverse reactions such as NADPH-dependent oxidases, xanthine oxidase, and e^- leaks from the mitochondria. Reaction 2, catalyzed by the ubiquitous enzyme superoxide dismutase, converts short-lived $O_2^{-\bullet}$ to H_2O_2, a more long-lived molecule with better second messenger characteristics than $O_2^{-\bullet}$. H_2O_2 is degraded by at least two types of reactions: Reaction 3 catalyzed by catalase converts H_2O_2 to the innocuous products H_2O and O_2. Reaction 4, the classical Fe^{++}-dependent Fenton reaction, converts H_2O_2 to hydroxide (OH^-) and the extremely reactive hydroxyl radical ($OH^{-\bullet}$ that can react with cellular molecules to produce an innumerable number and variety of biologically aberrant products). Reaction 5 is a second dismutase-like reaction that consumes $O_2^{-\bullet}$ and regenerates Fe^{++} required for further cycles of H_2O_2 processing via the Fenton reaction.

migrate long distances, such as from the mitochondrion to the nucleus, where it can be reconverted to hydroxyl radical and initiate numerous DNA-based reactions including nucleotide mutations leading to cancer (Kryston et al. 2011).

Most biological molecules are diamagnetic and not significantly influenced by EMFs. However, living organisms do contain O_2 and small numbers and quantities of paramagnetic molecules, like most ROS mentioned earlier. The effect of magnetic fields on these substances is to alter their electron spin properties, changing their quantum spin state, which often lengthens their lifetime and extends their sphere of reactivity. In early biological studies, it was speculated that the lifetime-lengthening effect of EMFs could provide greater opportunity for novel radical reactions to proceed and thus exaggerate the molecular damaging effect of radicals such as $O_2^{-\bullet}$ and its downstream reactive products (Lacy-Hulbert et al. 1998). However, over the last several decades, our understanding of how ROS impact cells and organisms has evolved from a perception that ROS generation is mainly cytotoxic and injurious, to our current view in which ROS, especially H_2O_2, are collectively viewed as indispensable regulators of diverse cellular activities via their impact on signal transduction cascades in many mammalian cells (Finkel and Holbrook 2000), including immune system cells (Tatla et al. 1999). Thus, the potential for bioeffects of static and extremely low frequency (ELF) EMFs has correspondingly expanded to include much more significant, long-lasting effects, mediated by their impact on radical lifetimes, which are in turn coupled to signal transduction cascades. In addition, early (Boxer et al. 1983) and more recent studies (Woodward and Vink 2007) have greatly improved our understanding of how low field effects (LFEs) of EMFs, nominally less than 1 milliTesla (mT), can have a substantial impact on the reactivity of radicals and how LFEs and high field effects (HFE) can generate opposing effects on the chemical reactivity of a specific radical. For example, Woodward and Vink have shown that increasing EMF strength in the LFE region can lead to decreased radical reactivity, which switches to increased reactivity in the HFE region. With the burgeoning interest in the biochemistry of radical signaling, new and more sensitive techniques are being exploited to help better understand the implications of biological radicals. For example, Maeda et al. have recently used cavity ring-down spectroscopy to directly assay the effect of a 10 Hz EMF on the lifetime of triplet-state flavin mononucleotide in aqueous solution (Maeda et al. 2011). In their experiments, significant LFE lifetime lengthening was observed, and the lifetime lengthening continued to increase with increasing field intensity and was not saturated at 52 mT, the highest intensity studied.

As illustrated by the reactions shown in Figure 10.1, we currently recognize two ubiquitous, well-described sources of biologically derived ROS: The first is intracellular $O_2^{-\bullet}$, generated by one-electron transfer reactions from mitochondrial electron transport chain components to molecular oxygen (O_2) (Brand 2010). The second is $O_2^{-\bullet}$ generated by one-electron transfer reactions from NADPH to O_2, catalyzed by a family of cell membrane–localized NADPH oxidases (NOX) that are widely distributed in mammalian tissues (Bedard and Krause 2007; Krause et al. 2012). In 2014, both of these $O_2^{-\bullet}$-generating processes continue to be subjects of vigorous investigation. Mitochondrial ROS, although still widely viewed as deleterious, has recently been shown to regulate postsynaptic neuronal signaling (Accardi et al. 2014) and to initiate apoptosis in neural progenitor cells exposed to amyloid beta protein (Hou et al. 2014). ROS produced by NOX is most extensively studied in immune system cells (T cells, neutrophils, and macrophages) (Nathan and Cunningham-Bussel 2013). When released into the extracellular space, $O_2^{-\bullet}$-derived radicals, such as peroxides, perchlorates, and peroxynitrites (Davies 2011; Mulligan et al. 1991), are responsible for the innate immune system's primary microbicide activity, as well as differentiation of monocytes into M2 macrophages (Zhang et al. 2013). More recently, it has been observed that mitochondria also produce $O_2^{-\bullet}$ via a process different from $O_2^{-\bullet}$ generated by *electron leaks* from the mitochondrial electron transport chain. Cooperatively with Ca^{++}, this recently discovered source of $O_2^{-\bullet}$ appears to regulate the activity of the mitochondrial permeability pore (Hou et al. 2013) which, when activated, leads to mitochondrial swelling and cell death. This is another example of the broadening role of ROS in regulating cell activity.

10.4   EMF EFFECTS ON IMMUNE CELLS

While it is now recognized that macrophages, dendritic cells, and granulocytes are essential for defense against invading pathogens, as well as protection from developing cancer cells (Gabrilovich et al. 2012), our ability to translate this knowledge into therapeutic benefits remains limited. We are only now beginning to appreciate the tight integration that exists between tissue metabolism and immune function (Osborn and Olefsky 2012), how cytokines like IL-10 might be used against infection (Ouyang et al. 2011), or how immunotherapies might be used to treat cancer (Pardoll 2012). Most efforts at regulating the immune system for therapeutic benefit have exploited the potential benefits of cytokines (Chen et al. 2012), and more recently, our ability to engineer T cells has put us on the threshold of major breakthroughs in cancer immunotherapy (Kloss et al. 2013). Less well appreciated is the role of ROS in regulating the immune system and the associated potential of regulating ROS signaling activity through the use of EMFs.

While extracellular H_2O_2 derived from NOX has long been recognized as one of the earliest acting innate immune system defenses against invasion by pathogens (Lambeth 2004), intracellular ROS derived from NOX and mitochondrial activity have assumed increasingly more attention as the breadth of ROS signaling is uncovered.

West and coworkers (West et al. 2011a, 2011b) have shown that the bactericidal activity of macrophages is initiated by the interaction of bacteria with macrophage toll-like receptors (TLR), which recruits mitochondria to the phagosome, where mitochondrial ROS augments that formed via NOX. Since ELF EMFs are known to alter the reactivity of radicals like ROS (Canseven et al. 2008; Coskun et al. 2009; Poniedzialek et al. 2013a,b), it is possible that EMFs may ultimately provide a noninvasive route to regulate the TLR-linked intracellular ROS, generated by activated macrophages, and so modulate the innate immune response to infection.

Additionally, it has recently been demonstrated that ELF EMFs can modulate mitochondrial energy transduction and so regulate the swimming activity of sperm (Iorio et al. 2011). In associated studies, sperm motility after cryopreservation has been shown to be improved by the addition of the ROS scavenger butylated hydroxytoluene (BHT) during cryopreservation (Merino et al. 2014). Although the link between EMF exposure, reduced ROS levels, and improved sperm motility has not been made at this writing, it can be anticipated that studies bearing on this linkage will soon be forthcoming.

Even more impressive than the ephemeral ROS-regulated phenotypic effects outlined in the preceding paragraph are the more permanent genotypic/cell differentiating effects initiated by intracellular or autocrine ROS. For instance, Zhang et al. have recently shown that ROS generation is a key step in monocyte differentiation into M2 macrophages (Zhang et al. 2013). Transcription of new genes leading to inflammatory cell differentiation or cancer transformation is induced by O_2-dependent changes in the intracellular redox state (Brigelius-Flohe and Flohe 2011). Myant and coworkers report that ROS-regulated NF-κB is critical in triggering colorectal cancers (Myant et al. 2013). Activation of the ERK/NF-κB pathway by arsenite-induced ROS generation results in transformation of human embryo lung fibroblasts (Ling et al. 2012). From studies such as the latter, it is clear that regulating production of ROS has the potential of altering NF-κB activity and its associated pathologies. Not surprisingly, a 5 Hz EMF that can have profound effects on $O_2^{-\bullet}$ and its downstream products has recently been shown to downregulate NF-κB, the dominant transcription factor, in an inflammatory disease cell model (Ross and Harrison 2013a). The latter results suggest a wide-ranging application of EMF therapy to human disease states associated with ROS signaling (Ross and Harrison 2013b).

10.5   OUR CURRENT WORK

In our laboratory at Indiana University, we have a long-standing interest in how EMFs might alter biological processes and how EMF therapy might be applied to human pathologies (Markov et al. 2006; Nindl et al. 2003; Waite et al. 2011). Recently, we tested a commercially available 12.5 µT

EMF-generating device (Immunent BV, Veldhoven, the Netherlands) that has been used to improve animal growth and survival in the husbandry industry. A detailed description of the device has recently been provided by Bouwens et al. (2012). The ability of the device to generate a 40% reduction in parasitic, inflammatory intestinal lesions in chickens (Elmusharaf et al. 2007), a 60% reduction in infection-related mortality in a goldfish hatchery, and a 42% increase in the respiratory burst of isolated fish phagocytes (Cuppen et al. 2007) led to the hypothesis that the Immunent EMF might target the immune system. We tested that hypothesis by studying the influence of the Immunent field on H_2O_2 generated by human donor macrophages and the human lymphocytic THP-1 cell line subjected to inflammatory conditions.

In earlier work, we showed that hypoxia and extracellular H_2O_2, both characteristics of inflamed tissues, increased the respiratory burst in human acute monocytic leukemia THP-1 cells (Owegi et al. 2010), and we employed those culture conditions to test the effect of the Immunent field on the in vitro immune cell response.

THP-1 monocytes in 96-well plates were differentiated into macrophages by exposure to 0.1 μM phorbol myristate acetate (PMA) for 48 h (Dreskin et al. 2001), washed and subjected to inflammatory conditions (hypoxia with and without added H_2O_2) as follows. Hypoxic cells and normoxic (21% O_2) controls, $\pm H_2O_2$, as shown in the accompanying figure and table, were Immunent field treated or mock exposed for 30 min near the middle of a 4 h inflammatory condition exposure. At the end of the 4 h, the cells were washed with phosphate-buffered saline (PBS), resuspended in PBS-containing 50 μM luminol (the H_2O_2 reporter) and 100 μM NaN_3 (a catalase inhibitor). Subsequently, a PMA-dependent H_2O_2 burst was initiated by adding 1 μM PMA, and after an additional 45 min, the H_2O_2 in the culture well was assayed (using a LUMIstar Omega plate reader) by adding 10 μM NaOCl, which initiates the H_2O_2-dependent luminol chemiluminescence (Mueller 2000). Chemiluminescence differences due to different cell numbers or other variations between replicate culture wells were corrected as described earlier (Owegi et al. 2010).

Human monocytes were obtained from 12 human donors (IRBnet 197455–3) using a Ficoll–Hypaque gradient (1.077 g/mL) followed by a 48% Percoll/NaCl gradient, essentially as described by de Almeida et al. (2000). Isolated and washed monocytes were resuspended in growth medium (Owegi et al. 2010) and incubated overnight (5% CO_2, 37°C) in 96-well culture plates (5×10^4 cells/well). The wells were washed with medium, and the adherent naïve macrophages remained untreated or were differentiated into M1 and M2 macrophages as follows. Naïve macrophages were differentiated into mature macrophages by the addition of macrophage-colony-stimulating-factor (M-CSF, 20 ng/mL) for 5 days, during which cells elongated and developed pseudopodia. For polarization toward classically activated M1 macrophages, 100 ng/mL LPS and 20 ng/mL IFN-γ were added for the last day; for polarization toward alternatively activated M2 macrophages, 20 ng/mL IL-4 was added (Mantovani et al. 2004). Naïve M1 and M2 macrophages were subjected to inflammatory conditions with mock or Immunent field exposure and subsequently assayed for their ability to release H_2O_2 as described earlier for THP-1 cells.

Figure 10.2 illustrates the effect of the Immunent field on ROS (H_2O_2) generated by PMA-differentiated THP-1 macrophages, with the EMF applied for 30 min near the middle of a 4 h inflammatory hypoxic or a noninflammatory normoxic incubation at 37°C. In these experiments, the cells were equilibrated at the indicated O_2 levels for 2 h; during which period, it is generally anticipated that increasingly stringent inflammatory conditions, for example, increasing hypoxia, will lead to parallel activation of PKCs (Sumagin et al. 2013). Such activated PKCs, including the isoforms β1, δ, and ζ (Dang et al. 2001), will catalyze phosphorylation and membrane localization of the cytosolic NOX subunit p47[phox], followed by the phosphorylated p47[phox]-dependent membrane assembly of NOX and the production of H_2O_2 from NOX-generated $O_2^{-\bullet}$ (Fontayne et al. 2002). Subsequent to the 4 h experimental exposure, the culture wells were washed with PBS and the wells supplemented with PBS-containing luminol and NaN_3. Using luminol to quantitate H_2O_2 (Mueller 2000), the adherent cells were tested for their capacity to produce PMA-induced H_2O_2 during a 45 min incubation. Figure 10.2, Panel a, illustrates the results of a representative experiment comparing

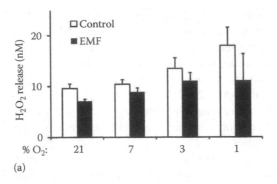

(a)

% O$_2$:	21	7	3	1
% Difference of 9 exp.	−14.4 (0.002)	−10.8 (0.03)	−7.3 (0.08)	−14.9 (0.01)

(b)

FIGURE 10.2 Effect of Immunent EMF on ROS (H$_2$O$_2$), generated by PMA-differentiated THP-1 macrophages. EMF-exposed cells exhibit about a 10% decrease in H$_2$O$_2$ release, at normoxic (21% O$_2$) or increasingly hypoxic (7%, 3%, and 1% O$_2$) conditions. (a) Representative experiment, n = 10 for each treatment condition. (b) Average percent differences of nine independent experiments. P values (t-test) are shown in parentheses.

mock EMF exposure with Immunent EMF exposure. Each treatment condition had an n = 10. Figure 10.2, Panel b, presents the tabulated results of nine such replicate experiments.

A number of observations flow from the data shown in Figure 10.2: First, compared to mock exposures, the Immunent-exposed cells exhibit a highly significant and repeatable (ca. 10%) decrease in ability to express PMA-stimulated H$_2$O$_2$, regardless of the O$_2$ content of the prior 4 h incubation. This result clearly demonstrates a significant EMF effect on the maturation of biochemical processes that lead to PMA-associated, NOX-generated ROS. Second, increasingly stringent hypoxia is expected to lead to increased ROS (Owegi et al. 2010), and as evidenced by the control experiments (i.e., the mock exposures), increasingly stringent hypoxia led to the anticipated parallel increase in THP-1 macrophage ability to generate H$_2$O$_2$, with a doubling of H$_2$O$_2$ produced under the most stringent hypoxic condition (1% O$_2$) as compared with normoxic or minimally hypoxic (7% O$_2$) cells.

What was unanticipated in these experiments is that appreciable PMA-induced H$_2$O$_2$ was generated by normoxia-treated, control cells (21% O$_2$). Since the dominant role of PMA during the 45 min test period is activating a variety of PKC isoforms and since we have little evidence that PMA activates NOX or other H$_2$O$_2$-generating processes, we conclude that the hypoxia-associated, PMA-induced H$_2$O$_2$ generated by mock-exposed cells during the 45 min test period is a consequence of PKC activation during the preceding 4 h incubation period. This view is supported by work of Goldberg et al. (1997) showing that hypoxia induced the translocation of diacylglycerol (DAG) and PMA-dependent PKCs α and ε to the particulate fraction of cells in a phospholipase C (PLC)-dependent manner. This strongly supports the notion that during our 4 h hypoxic incubations, one or more NOX-activating PKCs were activated, likely as a result of PLC-dependent DAG production. Considering the importance of the timing involved in our incubation and test periods, it is important to note that Goldberg et al. also showed that translocation of PLC to the membrane takes place within 1 h and its activity remains stable for up to 24 h.

In additional experiments, THP-1 macrophages were incubated for 4 h and tested using the luminol assay exactly as described for Figure 10.2 experiments, except that various concentrations of H_2O_2 were included during the 4 h incubation. In these experiments, which are not shown, the addition of H_2O_2 completely eliminated the EMF effect reported in Figure 10.2. This result indicates that the EMF-regulated events in Figure 10.2 are upstream of H_2O_2 signaling. The principal reactions upstream from H_2O_2 generation are those producing $O_2^{-\bullet}$. Thus, the fact that added H_2O_2 abrogates the EMF effect again indicates that the EMF effect shown in Figure 10.2 is related to an EMF effect on $O_2^{-\bullet}$, or $O_2^{-\bullet}$ production. A likely explanation of these events is that the EMF field induced conversion of appreciable $O_2^{-\bullet}$ to a triplet state that is not amenable to dismutation by superoxide dismutase and that allows for radical oxygen to return to the ground state without production of the messenger H_2O_2.

With increased interest in intracellular ROS signaling, a plethora of new, parallel routes has been uncovered for stressor-related (e.g., hypoxia) activation of PKCs leading to NOX activation and ROS signaling. For example, stressor-induced $O_2^{-\bullet}$ production has been widely linked to normal mitochondrial electron transport (Brand 2010), and xanthine oxidase catalyzes O_2-related free-radical reactions that result in H_2O_2 production (Kelley et al. 2010). The common feature in all of these NOX regulatory scenarios is the involvement of free-radical-based reactions that culminate in translocation of many PKCs to cell membranes with subsequent assembly of active NOXs. Since EMFs have the capacity to modulate radical-based reactions, it is not surprising to find that the Immunent EMF causes a long-lasting reproducible effect on PMA-induced H_2O_2 production during our test period.

To extend our studies with THP-1 cells to normal human macrophages, we obtained whole blood from human donors and prepared naïve M1 and M2 macrophages. Each of the latter cell types was subjected to the same 4 h incubation (normoxia and hypoxia, ±EMF) described earlier for THP-1 macrophages. However, since cells from human donors are limited in quantity, the number of EMF plus stressor (hypoxia) treatments we tested was limited to 21% O_2 (normoxic controls), 13% O_2, and 1% O_2. As with THP-1 macrophages, all macrophages from human donors were sensitive to hypoxia and exhibited a hypoxia-related increase in PMA-stimulated H_2O_2 generation. Conversely, in all trials with donor-derived macrophages, including naïve M1 and M2 cells from 12 different donors, we found no difference in luminol-reported H_2O_2 production between mock-exposed and Immunent-exposed cells that were subjected to PMA activation after the 4 h experimental incubations. However, in one series of experiments, we subjected naïve M1 and M2 cells to 1% O_2 plus 14 μM H_2O_2—an extreme stressor condition—during the 4 h stressor incubation. With M1 cells, but not with naïve or M2 cells, we found a highly significant inhibition of H_2O_2 production in Immunent-EMF-exposed cells (Table 10.1), much like that found with normoxic and hypoxic THP-1 macrophages.

Although the human leukemia THP-1 cell line provides a highly convenient model to study macrophage differentiation and polarization, there are substantial morphological and biochemical differences between PMA-differentiated THP-1 cells and macrophages differentiated from monocytes of human donors (Daigneault et al. 2010). Additionally, Daigneault et al. have shown that the time course of PMA-induced differentiation is critical in determining how closely PMA-differentiated THP-1 cells mimic normal human macrophages, with a 5-day differentiation period required for maximal macrophage-like characteristics to develop. Consequently, it is not surprising to find differences in response to stressor challenge between PMA-differentiated THP-1 macrophages and those derived from normal human donor peripheral blood lymphocytes, like those reported here. Conversely, it is important to emphasize that under conditions of maximal stressor stimulation, donor-derived M1 macrophages respond to the Immunent EMF with about a 10% downregulation of PMA-induced H_2O_2 production much like THP-1 macrophages. Clearly, a great deal remains to be learned about how EMFs impact cells and organisms, but the potential therapeutic benefits that will be gained when we finally understand how to apply the appropriate EMF to a particular therapeutic or other biological need makes this difficult journey worth the doing.

TABLE 10.1

H_2O_2 Production (in nM) of Immunent-EMF-Exposed Human Donor-Derived M1 Macrophages, Subjected to 1% O_2 Plus 14 µM H_2O_2, Compared with Mock-Exposed Cells

Mock Exposure	EMF Exposure	Difference (%)	Significance (p value)
162.4	127.1	−21.7	1.5×10^{-7}
130.1	108.9	−16.3	0.0003
135.7	102	−24.8	0.004
180.3	164.3	−8.9	0.002
144.2	113.9	−21.0	0.01
155	131.8	−15.0	0.03
224.1	201.2	−10.2	0.13
246.6	232.9	−5.6	0.35
181.6	176.7	−2.7	0.73
134.3	133.8	−0.4	0.96
	Average	−12.7	0.001

Notes: The results from 10 donors show a significant average decrease in H_2O_2 production associated with EMF exposure. Similar experiments were performed with normoxic (21% O_2) and hypoxic (13% and 1% O_2) cells and with naïve and M2 macrophages. Under the latter conditions, we found no EMF effect.

10.6 SUMMARY

Human health and well-being is critically dependent on timely expression of innate and acquired immune system processes that have evolved over millennia. Unfortunately, evolution has also provided challenges such as the evolution of methicillin-resistant *Staphylococcus aureus* and processes that allow tumor tissue to evade immune system killing and phagocytosis. Antibiotic-resistant strains of *S. aureus* evade human immune surveillance by secreting proteins that inhibit immune system cells and are lysozyme resistant, enabling survival in phagosomes (Foster 2005). Tumor cells downregulate phagocytic M1 macrophages and upregulate tumor-associated macrophages (TAMs) that activate angiogenesis and help provide a favorable environment for tumor proliferation (Gabrilovich et al. 2012). The development of specifically tailored EMFs that could assist the immune system in overcoming *S. aureus* and tumor pathology would be a therapeutic boon. The results we have observed with the Immunent EMF provides a glimpse into the future of EMF therapy, but that future will only be achieved by researchers with a thorough understanding of how biological processes are regulated and how EMFs modulate individual chemical reactions. In this chapter, we have attempted to provide a basis for continued development of bioelectromagnetics, a basal level of understanding of how the immune system works, and an overview of one way in which EMFs can modulate chemical reactions. We trust this effort will spur new and improved developments in EMF therapeutics.

REFERENCES

Accardi, M. V., B. A. Daniels et al. (2014). Mitochondrial reactive oxygen species regulate the strength of inhibitory GABA-mediated synaptic transmission. *Nat Commun* 5: 3168.

Bedard, K. and K. H. Krause (2007). The NOX family of ROS-generating NADPH oxidases: Physiology and pathophysiology. *Physiol Rev* 87(1): 245–313.

Bouwens, M., S. de Kleijn et al. (2012). Low-frequency electromagnetic fields do not alter responses of inflammatory genes and proteins in human monocytes and immune cell lines. *Bioelectromagnetics* 33(3): 226–237.

Boxer, S. G., E. Chidsey, D. et al. (1983). Magnetic field effects on reaction yields in the solid state.pdf. *Ann Rev Phys Chem* 34: 389–417.

Brand, M. D. (2010). The sites and topology of mitochondrial superoxide production. *Exp Gerontol* 45(7–8): 466–472.

Brigelius-Flohe, R. and L. Flohe (2011). Basic principles and emerging concepts in the redox control of transcription factors. *Antioxid Redox Signal* 15(8): 2335–2381.

Canseven, A. G., S. Coskun et al. (2008). Effects of various extremely low frequency magnetic fields on the free radical processes, natural antioxidant system and respiratory burst system activities in the heart and liver tissues. *Indian J Biochem Biophys* 45(5): 326–331.

Chen, L. L., X. Chen et al. (2012). Exploiting antitumor immunity to overcome relapse and improve remission duration. *Cancer Immunol Immunother* 61(7): 1113–1124.

Coskun, S., B. Balabanli et al. (2009). Effects of continuous and intermittent magnetic fields on oxidative parameters in vivo. *Neurochem Res* 34(2): 238–243.

Cuppen, J. J. M., G. F. Wiegertjes et al. (2007). Immune stimulation in fish and chicken through weak low frequency electromagnetic fields. *The Environmentalist* 27(4): 577–583.

Daigneault, M., J. A. Preston et al. (2010). The identification of markers of macrophage differentiation in PMA-stimulated THP-1 cells and monocyte-derived macrophages. *PLoS One* 5(1): e8668.

Dang, P. M., A. Fontayne et al. (2001). Protein kinase C zeta phosphorylates a subset of selective sites of the NADPH oxidase component p47phox and participates in formyl peptide-mediated neutrophil respiratory burst. *J Immunol* 166(2): 1206–1213.

Davies, M. J. (2011). Myeloperoxidase-derived oxidation: Mechanisms of biological damage and its prevention. *J Clin Biochem Nutr* 48(1): 8–19.

de Almeida, M. C., A. C. Silva et al. (2000). A simple method for human peripheral blood monocyte isolation. *Mem Inst Oswaldo Cruz* 95(2): 221–223.

Dreskin, S. C., G. W. Thomas et al. (2001). Isoforms of Jun kinase are differentially expressed and activated in human monocyte/macrophage (THP-1) cells. *J Immunol* 166(9): 5646–5653.

Elmusharaf, M. A., J. J. Cuppen et al. (2007). Antagonistic effect of electromagnetic field exposure on coccidiosis infection in broiler chickens. *Poult Sci* 86(10): 2139–2143.

Finkel, T. and N. J. Holbrook (2000). Oxidants, oxidative stress and the biology of ageing. *Nature* 408(6809): 239–247.

Fontayne, A., P. M. Dang et al. (2002). Phosphorylation of p47phox sites by PKC alpha, beta II, delta, and zeta: Effect on binding to p22phox and on NADPH oxidase activation. *Biochemistry* 41(24): 7743–7750.

Foster, T. J. (2005). Immune evasion by staphylococci. *Nat Rev Microbiol* 3(12): 948–958.

Gabrilovich, D. I., S. Ostrand-Rosenberg et al. (2012). Coordinated regulation of myeloid cells by tumours. *Nat Rev Immunol* 12(4): 253–268.

Goldberg, M., H. L. Zhang et al. (1997). Hypoxia alters the subcellular distribution of protein kinase C isoforms in neonatal rat ventricular myocytes. *J Clin Invest* 99(1): 55–61.

Hou, T., X. Zhang et al. (2013). Synergistic triggering of superoxide flashes by mitochondrial Ca2+ uniport and basal reactive oxygen species elevation. *J Biol Chem* 288(7): 4602–4612.

Hou, Y., P. Ghosh et al. (2014). Permeability transition pore-mediated mitochondrial superoxide flashes mediate an early inhibitory effect of amyloid beta1–42 on neural progenitor cell proliferation. *Neurobiol Aging* 35(5): 975–989.

Iorio, R., S. Delle Monache et al. (2011). Involvement of mitochondrial activity in mediating ELF-EMF stimulatory effect on human sperm motility. *Bioelectromagnetics* 32(1): 15–27.

Janeway, C. A., Jr. and R. Medzhitov (2002). Innate immune recognition. *Annu Rev Immunol* 20: 197–216.

Kelley, E. E., N. K. Khoo et al. (2010). Hydrogen peroxide is the major oxidant product of xanthine oxidase. *Free Radic Biol Med* 48(4): 493–498.

Kloss, C. C., M. Condomines et al. (2013). Combinatorial antigen recognition with balanced signaling promotes selective tumor eradication by engineered T cells. *Nat Biotechnol* 31(1): 71–75.

Krause, K. H., D. Lambeth et al. (2012). NOX enzymes as drug targets. *Cell Mol Life Sci* 69(14): 2279–2282.

Kryston, T. B., A. B. Georgiev et al. (2011). Role of oxidative stress and DNA damage in human carcinogenesis. *Mutat Res* 711(1–2): 193–201.

Lacy-Hulbert, A., J. C. Metcalfe et al. (1998). Biological responses to electromagnetic fields. *FASEB J* 12(6): 395–420.

Lambeth, J. D. (2004). NOX enzymes and the biology of reactive oxygen. *Nat Rev Immunol* 4(3): 181–189.

Ling, M., Y. Li et al. (2012). Regulation of miRNA-21 by reactive oxygen species-activated ERK/NF-kappaB in arsenite-induced cell transformation. *Free Radic Biol Med* 52(9): 1508–1518.

Maeda, K., S. R. Neil et al. (2011). Following radical pair reactions in solution: A step change in sensitivity using cavity ring-down detection. *J Am Chem Soc* 133(44): 17807–17815.

Mantovani, A., A. Sica et al. (2004). The chemokine system in diverse forms of macrophage activation and polarization. *Trends Immunol* 25(12): 677–686.

Markov, M., G. Nindl et al. (2006). Interactions between electromagnetic fields and the immune system: Possible mechanism for pain control. In: *Bioelectromagnetics Current Concepts of the Mechanisms of the Biological Effect of Extremely High Power Pulses in Bioelectromagnetic Medicine. NATO Advanced Research Workshop Publication Series*. Markov, M. and Ayrapetyan, S. (eds.). Springer Publishing, pp. 213–226.

Merino, O., W. E. Aguaguina et al. (2014). Protective effect of butylated hydroxytoluene on sperm function in human spermatozoa cryopreserved by vitrification technique. *Andrologia* 2014, Wiley Online Library.

Mueller, S. (2000). Sensitive and nonenzymatic measurement of hydrogen peroxide in biological systems. *Free Radic Biol Med* 29(5): 410–415.

Mulligan, M. S., J. M. Hevel et al. (1991). Tissue injury caused by deposition of immune complexes is L-arginine dependent. *Proc Natl Acad Sci USA* 88(14): 6338–6342.

Myant, K. B., P. Cammareri et al. (2013). ROS production and NF-kappaB activation triggered by RAC1 facilitate WNT-driven intestinal stem cell proliferation and colorectal cancer initiation. *Cell Stem Cell* 12(6): 761–773.

Nathan, C. and A. Cunningham-Bussel (2013). Beyond oxidative stress: An immunologist's guide to reactive oxygen species. *Nat Rev Immunol* 13(5): 349–361.

Nathan, C. F., H. W. Murray et al. (1983). Identification of interferon-gamma as the lymphokine that activates human macrophage oxidative metabolism and antimicrobial activity. *J Exp Med* 158(3): 670–689.

Nindl, G., M. T. Johnson et al. (2003). Low-frequency electromagnetic field effects on lymphocytes: Potential for treatment of inflammatory diseases. In: *Clinical Applications of Bioelectromagnetic Medicine*. Rosch, P. and Markov, M. (eds.). Marcel Dekker Inc., New York, pp. 369–389.

Osborn, O. and J. M. Olefsky (2012). The cellular and signaling networks linking the immune system and metabolism in disease. *Nat Med* 18(3): 363–374.

Ouyang, W., S. Rutz et al. (2011). Regulation and functions of the IL-10 family of cytokines in inflammation and disease. *Ann Rev Immunol* 29: 71–109.

Owegi, H. O., S. Egot-Lemaire et al. (2010). Macrophage activity in response to steady-state oxygen and hydrogen peroxide concentration. *Biomed Sci Instrum* 46: 57–63.

Pardoll, D. M. (2012). The blockade of immune checkpoints in cancer immunotherapy. *Nat Rev Cancer* 12(4): 252–264.

Paul, W. E. (1987). Interleukin 4/B cell stimulatory factor 1: One lymphokine, many functions. *FASEB J* 1(6): 456–461.

Poniedzialek, B., P. Rzymski et al. (2013a). Reactive oxygen species (ROS) production in human peripheral blood neutrophils exposed in vitro to static magnetic field. *Electromagn Biol Med* 32(4): 560–568.

Poniedzialek, B., P. Rzymski et al. (2013b). The effect of electromagnetic field on reactive oxygen species production in human neutrophils in vitro. *Electromagn Biol Med* 32(3): 333–341.

Ross, C. L. and B. S. Harrison (2013a). Effect of pulsed electromagnetic field on inflammatory pathway markers in RAW 264.7 murine macrophages. *J Inflamm Res* 6: 45–51.

Ross, C. L. and B. S. Harrison (2013b). The use of magnetic field for the reduction of inflammation: A review of the history and therapeutic results. *Altern Ther Health Med* 19(2): 47–54.

Segerer, S., F. Heller et al. (2008). Compartment specific expression of dendritic cell markers in human glomerulonephritis. *Kidney Int* 74(1): 37–46.

Sumagin, R., A. Z. Robin et al. (2013). Activation of PKC beta II by PMA facilitates enhanced epithelial wound repair through increased cell spreading and migration. *PLoS One* 8(2): e55775.

Tatla, S., V. Woodhead et al. (1999). The role of reactive oxygen species in triggering proliferation and IL-2 secretion in T cells. *Free Radic Biol Med* 26(1–2): 14–24.

Waite, G. N., S. J. P. Egot-Lemaire et al. (2011). A novel view of biologically active electromagnetic fields. *The Environmentalist* 31(2): 107. DOI: 10.1007/s10669-011-9319-8.

West, A. P., I. E. Brodsky et al. (2011a). TLR signalling augments macrophage bactericidal activity through mitochondrial ROS. *Nature* 472(7344): 476–480.

West, A. P., G. S. Shadel et al. (2011b). Mitochondria in innate immune responses. *Nat Rev Immunol* 11(6): 389–402.

Woodward, J. R. and C. B. Vink (2007). Hyperfine coupling dependence of the effects of weak magnetic fields on the recombination reactions of radicals generated from polymerisation photoinitiators. *Phys Chem Chem Phys* 9(47): 6272–6278.

Zhang, Y., S. Choksi et al. (2013). ROS play a critical role in the differentiation of alternatively activated macrophages and the occurrence of tumor-associated macrophages. *Cell Res* 23(7): 898–914.

11 Effect of PEMF on LPS-Induced Chronic Inflammation in Mice

Juan Carlos Pena-Philippides, Sean Hagberg, Edwin Nemoto, and Tamara Roitbak

CONTENTS

11.1 INTRODUCTION

Low-intensity pulsed electromagnetic fields (PEMFs) have been found to produce a variety of beneficial biological effects and have been successfully employed as adjunctive therapy for a variety of clinical conditions (Shupak, 2003; Guo et al., 2012). In the United States, PEMF is cleared by FDA for use in treating postoperative pain and edema and approved by the Centers for Medicare and Medicaid Services (CMS) for treating chronic wounds. PEMF signals induce electrical fields in tissue, generated by relatively simple devices, allowing noninvasive therapeutic application (Pilla, 2013). Unlike other electrotherapeutics, PEMF is not impeded by differences in types of tissue, so it appears within the tissue target virtually instantaneously. Properly configured PEMF signals have been demonstrated to regulate major cellular functions, including cell proliferation, differentiation, apoptosis, cell cycle, DNA replication, and cytokine/chemokine expression (Aaron et al., 2004; Pesce et al., 2013; Pena-Philippides et al., 2014). Among the possible mechanisms of PEMF-induced influence on the biological systems are modulation of calmodulin (CaM)-dependent nitric oxide (Pilla, 2012), increase in expression of protective stress protein hsp70 gene (George et al., 2008), and downregulation of proinflammatory NF-kB signaling pathway (Vianale et al., 2008). Exposure to PEMFs has been shown to attenuate tissue damage and inflammation following stroke (Grant et al., 1994; Pena-Philippides et al., 2014). Accumulated scientific data demonstrate the effect of electromagnetic fields in inflammatory diseases and conditions (Mizushima et al., 1975; Guo, 2011; Pilla, 2012). The anti-inflammatory effects of the PEMF treatment (using signals similar to those used in this study parameters) have demonstrated a decrease in proinflammatory cytokines and increase in anti-inflammatory cytokines after traumatic brain injury in rats (Rasouli et al., 2012) and stroke (Pena-Philippides et al., 2014), decreased pain in osteoarthritis (Nelson et al., 2013), wound healing (Pesce et al., 2013), and postsurgical recovery (Heden and Pilla, 2008; Rohde et al., 2010).

Several studies have reported that PEMF significantly affects inflammatory cytokine production during acute and chronic systemic inflammation, induced by lipopolysaccharides (LPSs) (Roman et al., 2002; Kaszuba-Zwoinska et al., 2008). LPSs, characteristic components of the cell wall of Gram-negative bacteria, are involved in host–parasite interactions and protect bacteria from phagocytosis and lysis (Rasanen et al., 1997). LPS administration is a well-established model for studying the acute (Copeland et al., 2005) and chronic (Qin et al., 2007; Belarbi et al., 2012) inflammatory responses in mice and human. LPS–host cell interaction is mediated by the increased production of cytokines and other mediators released by activated immune cells such as monocytes/macrophages. Recent studies suggest that the administration of LPS initiates the production and release of various cytokines, such as tumor necrosis factor alpha (TNF-α), interleukin 1 (IL-1), and IL-6. LPS-induced changes in cytokine profile are characterized by the so-called biphasic character: the initially strong acute inflammation is followed by decreased inflammatory response, when monocytes/macrophages reduce production of inflammatory cytokines in an attempt to return to homeostasis (Salomao et al., 2012). Recently, it was reported that a single peripheral LPS exposure in adult mice results in long-lasting elevation of TNF-1α and progression to chronic inflammation in mouse brain (Qin et al., 2007).

In the present study, we investigated the possible beneficial influence of PEMF treatment on chronic changes in inflammatory cytokine profiles in mouse peripheral blood and brain tissue, following single LPS injection.

11.2 MATERIALS AND METHODS

11.2.1 LPS Administration

The experimental procedures were approved by the University of New Mexico (UNM) Office of Animal Care Compliance. All institutional and national guidelines for the care and use of laboratory animals were followed. LPS (strain O111:B4) was purchased from Sigma. Two-month-old male C57BL/6 mice (N = 20) were intraperitoneally injected with a single dose of LPS (5 mg/kg). Control mice were injected with vehicle (0.9% saline). The analyses were performed at 1, 2, and 3 weeks after injections.

11.2.2 PEMF Stimulation

PEMF and sham treatments started immediately after injection. The PEMF signal was a 27.12 MHz carrier modulated by a 2 ms burst repeating at 2 bursts/s (2 Hz). The signal amplitude was adjusted to provide 3 ± 0.6 V/m within the mouse brain while providing full-body exposure. The PEMF exposure chamber was constructed such that free roaming mice were restricted to this field amplitude (Pena-Philippides et al., 2014). PEMF field characteristics were verified with a calibrated shielded loop probe 1 cm in diameter (model 100A, Beehive Electronics, Sebastopol, CA) connected to a calibrated 100 MHz oscilloscope (model 2012B, Tektronix, Beaverton, OR). In the previous studies, measurement of the PEMF signal distribution in a tissue phantom and in air showed that PEMF amplitude dose was uniform to within ±20% (Pilla, 2013; Pena-Philippides et al., 2014). Animals from the treatment (PEMF-treated, N = 10) group were subjected to 15 min PEMF, 4 times daily, with 2 h interval between the treatment sessions. Sham treatment group (N = 10) was subjected to 15 min of control treatment (*sham* mode of the applicator), with the same treatment intervals. Three to four mice from each group were sacrificed at each time point (7, 14, and 21 days after injections), for further analysis.

11.2.3 SAMPLE COLLECTION

Total brain tissue was homogenized in tissue extraction buffer (Life Tech/Invitrogen cat # FNN0071, 5 mL per 1 g of brain tissue) with the addition of protease inhibitor cocktail (Sigma). The samples were centrifuged at 10,000 rpm for 5 min, and the supernatant was collected and kept on ice. Protein concentration was determined for each sample, using DC protein assay kit from BioRad. *Blood* was collected via cardiac puncture, into anticoagulant-treated tubes (EDTA-treated), and centrifuged for 15 min at 2000 × g, using a refrigerated centrifuge, to remove red blood cells and platelets. Both brain tissue and plasma samples were aliquoted and frozen at −80°C.

11.2.3.1 Cytokine Expression Analysis

Cytokine expression analysis was performed using Mouse Cytokine Magnetic 20-Plex Panel Kit (Life Tech/Invitrogen), according to the manufacturer's recommendations. Brain tissue samples were normalized for total protein content and diluted at 1:10 in assay buffer. Blood plasma samples were measured undiluted. The measurements were done using Luminex xMAP 100 system, at the UNM Center for Molecular Discovery. Cytokine concentrations were calculated automatically, by the specialized Luminex system software. For statistical analysis, only cytokines with consistent expression throughout the samples were retained. Finally, t-test was conducted using Excel software, and statistical significance was detected.

11.3 RESULTS

In order to investigate an effect of PEMF on chronic inflammation, we used single systemic LPS administration, which was shown to induce a strong acute response, followed by long-lasting inflammatory response (Qin et al., 2007). Two-month-old male C57BL/6 mice (N = 20) were intraperitoneally injected with a single dose of LPS (5 mg/kg). Control mice were injected with vehicle (0.9% saline). PEMF or sham (*null* impulse) treatment was initiated immediately after the LPS injections. PEMF/sham impulse was administered for 15 min, 4 times daily (methods). Brain tissue samples and blood plasma samples were collected from PEMF-treated (N = 10) and sham-treated groups (N = 10 per group), at 7, 14, and 21 days after injections. Mouse Cytokine 20-Plex Panel (Invitrogen) was used to analyze cytokine profile in the collected samples. The panel is designed for the quantitative determination of FGF-basic, GM-CSF, IFN-γ, IL-1α, IL-1β, IL-2, IL-4, IL-5, IL-6, IL-10, IL-12 (p40/p70), IL-13, IL-17, IP-10, KC, MCP-1, MIG, MIP-1α, TNF-α, and VEGF expressions. According to the manufacturer, the utilized novel technology combines the efficiencies of multiplexing 20 different proteins for simultaneous analysis, with reproducibility similar to ELISA.

11.3.1 PLASMA SAMPLES

Figure 11.1 demonstrates the overall profile of the detected 12 major cytokines in the blood plasma samples collected from control (saline-injected/sham-treated, gray bars), sham (LPS-injected/sham-treated), and PEMF (LPS-injected/PEMF-treated) animal groups, at 7, 14, and 21 days after the LPS injection. No statistically significant differences in cytokine expression were found between saline-injected animals subjected to either sham or PEMF treatment (not shown). Cytokine 20-Plex Panel analysis of the LPS-injected animal samples demonstrated that expression (protein) levels of the major anti-inflammatory cytokine IL-10 were somewhat elevated in sham (134.5 pg/mL) and significantly increased in PEMF (174 pg/mL) animal groups, at 14 days after LPS injection, as compared to IL-10 expression at 7 days (114 and 92 pg/mL, respectively). At 21 days, IL-10 expression in sham group returned back to control values (102 pg/mL), while in PEMF-treated animals, IL-10 remained significantly higher (152 pg/mL), as compared to the control and sham animal groups (Figure 11.2a).

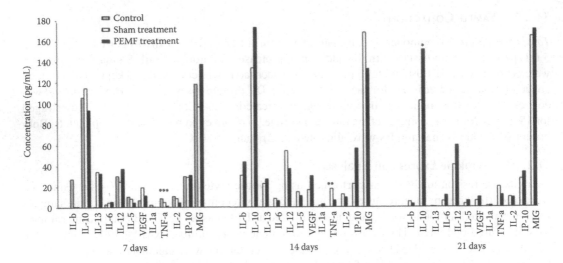

FIGURE 11.1 Protein expression profiles of 12 detected cytokines in the peripheral blood plasma, at 7, 14, and 21 days after LPS injections. Mouse Cytokine Magnetic 20-Plex Panel Kit (Life Tech/Invitrogen) and specialized Luminex system software were used for measurement, calculation, and quantification of the protein concentrations in the samples. Only cytokines with consistent expression throughout the samples were retained for quantification analysis. Gray bars, controls (saline injection/sham treatment); white bars, LPS-injected/sham-treated animals; black bars, LPS-injected/PEMF-treated animals. Student's t-test was performed to evaluate statistical significance of differences in PEMF versus sham animal groups. $p < 0.05$; **$p = 0.002$; ***$p < 0.0001$.

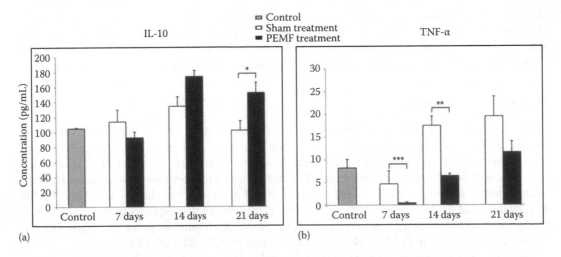

FIGURE 11.2 Detailed analysis of the expression of IL-10 (a) and TNF-1α (b) in control (saline injection/sham treatment, gray bar), sham (LPS injection/*null* signal treatment, white bars), and PEMF (LPS injection/PEMF treatment, black bars) animal groups, at 7, 14, and 21 days after LPS injections/treatment initiation. Student's t-test was performed for statistical analysis of differences between PEMF and sham animal groups. $p < 0.05$; **$p = 0.002$; ***$p < 0.0001$.

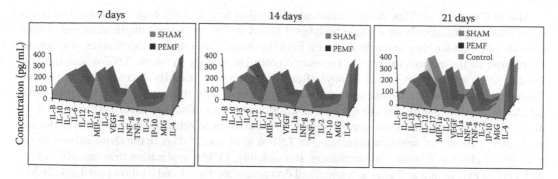

FIGURE 11.3 Protein expression profiles of 16 detected cytokines in mouse brain tissue, at 7, 14, and 21 days after LPS injections. Mouse Cytokine Magnetic 20-Plex Panel Kit (Life Tech/Invitrogen) and specialized Luminex system software were used for measurement, calculation, and quantification of the protein concentrations in the samples. Light gray chart, controls (saline injection/sham treatment); gray chart, sham (LPS-injected/sham-treated animals); dark gray chart, PEMF (LPS-injected/PEMF-treated animals). No significant differences between the three groups were detected.

Expression levels of proapoptotic and proinflammatory TNF-α significantly increased in the plasma of sham animal group (LPS-injected/sham-treated) at 14 days and remained elevated at 21 days after injection (17.5 and 19.6 pg/mL, respectively), as compared to expression at 7 days after injection (4.7 pg/mL). In contrast, in PEMF group (LPS-injected/PEMF-treated), TNF-α expression was almost undetectable (<1 pg/mL) at 7 days postinjection; at 14 days, it remained significantly lower (6.4 pg/mL), in comparison with sham-treated animal group (17.5 pg/mL). At 21 days postinjection, the concentration of TNF-α in the plasma of PEMF animals was 11.65 pg/mL versus 19.55 pg/mL in shams. It was remarkable that TNF-α levels at 7 days after LPS injection and PEMF treatment initiation were considerably lower than in sham-treated and even in control animals (Figure 11.2b).

Increase of anti-inflammatory IL-10 and downregulation of proinflammatory and proapoptotic TNF-α in PEMF-treated animals subjected to LPS injections demonstrate possible beneficial effect of electromagnetic field stimulation on chronic systemic inflammatory response.

11.3.2 BRAIN TISSUE SAMPLES

Figure 11.3 shows an overall expression profile of 16 detected cytokines in the mouse brain tissue at 7, 14, and 21 days after LPS injection. The profiles did not differ significantly between the sham, PEMF, and control groups. This indicates that there was no prolonged inflammatory reaction in the brain after a single injection of LPS. Even though no chronic changes were detected in the brain samples, these results show that PEMF exposure does not affect the cytokine expression, when no inflammation is present.

11.4 DISCUSSION

In the present study, we investigated possible influence of PEMF treatment on chronic inflammation following single LPS injection. Most studies utilize repetitive injections of LPS to induce chronic systemic inflammation (Belarbi et al., 2012). On the other hand, recent study showed that a single intraperitoneal injection of LPS is sufficient for the development of chronic changes in TNF-α expression in C57BL/6 mice (Qin et al., 2007). We decided to use this model of a single exposure to pathogen, as more relevant to clinical cases of the host–pathogen interaction. PEMF exposure regimen and treatment parameters were based on the requirements from the UNM Office of Animal Care Compliance as well as FDA-approved regulations.

Mouse Cytokine 20-Plex Panel technology allowed us to detect the expression of 20 different cytokines simultaneously in a single peripheral blood and brain tissue sample. Analysis results demonstrate that single intraperitoneal LPS injection results in long-lasting changes of cytokine expression in the peripheral blood. The major proinflammatory cytokine TNF-α was upregulated in sham animals, at 14 and 21 days after LPS injections. Initially decreased expression at 7 days (as compared to control expression) is in agreement with the knowledge about the biphasic character of LPS-induced changes, when after completion of acute phase, TNF-α release by macrophages is reduced below normal values, in an attempt to return to homeostasis (Salomao et al., 2012). However, subsequent increase in TNF-α at 14 and 21 days in the sham animal group is a sign of chronic systemic inflammation. Remarkably, PEMF application dramatically inhibited TNF-α expression at 7 days and remained downregulated at 14 and 21 days post-LPS. PEMF treatment also resulted in significant upregulation of major anti-inflammatory IL-10 at 14 and 21 days. Thus, the effect associated with PEMF treatment resulting in decreased TNF-1α and increased IL-10 expression could have a very beneficial influence on the suppression of chronic inflammation. Interestingly, this PEMF-induced influence on TNF-α and IL-10 expression was also seen in our previous studies on the effect of PEMF on postischemic brain inflammation (Pena-Philippides et al., 2014), as well as has been reported by other authors (Jonai et al., 1996; Pesce et al., 2013). Such effect on cytokine expression could be explained by proposed influence of PEMFs on the cells of the monocytic lineage and possible modulation of monocyte/macrophage transition (Di Luzio et al., 2001), a generalized acceleration in CaM-dependent anti-inflammatory activity (Pilla, 2013), or some combination. The PEMF-associated modulation of inflammatory cytokines and inflammation has been seen in multiple studies at different institutions with several *types* of PEMF devices ((Rohde et al., 2010; Pilla, 2012, 2013; Rasouli et al., 2012; Pena-Philippides et al., 2014; Taylor et al., 2014); see Pilla (2006) and Guo (2011) for a review), so our results are consistent with findings for this set of PEMF signal parameters and regimen.

In our study, PEMF treatment started immediately after the LPS injections. This decision was made based on our previous experiments demonstrating that PEMF provides a moderate reduction of inflammation during the acute stage of inflammatory response and a stronger suppression of the inflammation process at later stages (Pena-Philippides et al., 2014). We think that *fine-tuning* of the immediate inflammatory response will not suppress the initial positive effects of inflammation (and studies with the same technology demonstrate improved healing in wound (Strauch et al., 2007) and tendon models (Strauch et al., 2006) among others), but rather reduce negative effects of the increased production of inflammatory cytokines during the acute phase. On the other hand, stronger suppression (or resolution) of inflammation in later stages could prevent the development of chronic inflammation.

Our results are different from the reported effect of single LPS injection on chronic proinflammatory changes in Qin et al. (2007). In contrast with this research group, we did not find the traces of chronic inflammation in the mouse brain tissue in any of the groups, but instead detected them in peripheral blood. The differences in our findings in blood could be explained by the fact that we used peripheral blood plasma, in contrast with blood serum used in Qin et al. (2007). Even though no chronic inflammatory changes were detected in the brain samples, we found that PEMF exposure does not affect the cytokine expression in the brain, when no inflammation is present. This confirms the safety of PEMF administration when used in the intact, noninjured biological systems.

In summary, we propose that PEMF treatment may be potentially utilized as a noninvasive and long-lasting adjunctive treatment during recovery from systemic inflammation. The current clinical availability of the technology, the safe and effective use in acute conditions, and the ease of use and lack of known side effects make clinical evaluation in such conditions relatively simple and straightforward.

ACKNOWLEDGMENTS

This work was supported by Ivivi Health Sciences, LLC. We would like to thank Dr. Arthur Pilla for his valuable input and assistance. Analysis was performed in close collaboration with Life Tech/Invitrogen Company and UNM Center of Molecular Discovery.

REFERENCES

Aaron, R.K., Boyan, B.D., Ciombor, D.M., Schwartz, Z., and Simon, B.J. (2004). Stimulation of growth factor synthesis by electric and electromagnetic fields. *Clin Orthop Relat Res 419*, 30–37.

Belarbi, K., Jopson, T., Tweedie, D., Arellano, C., Luo, W.M., Greig, N.H., and Rosi, S. (2012). TNF-alpha protein synthesis inhibitor restores neuronal function and reverses cognitive deficits induced by chronic neuroinflammation. *J Neuroinflammation 9*, 1742–2094.

Copeland, S., Warren, H.S., Lowry, S.F., Calvano, S.E., and Remick, D. (2005). Acute inflammatory response to endotoxin in mice and humans. *Clin Diagn Lab Immunol 12*, 60–67.

Di Luzio, S., Felaco, M., Barbacane, R.C., Frydas, S., Grilli, A., Castellani, M.L., Macri, M.A. et al. (2001). Effects of 50 Hz sinusoidal electromagnetic fields on MCP-1 and RANTES generated from activated human macrophages. *Int J Immunopathol Pharmacol 14*, 169–172.

George, I., Geddis, M.S., Lill, Z., Lin, H., Gomez, T., Blank, M., Oz, M.C., and Goodman, R. (2008). Myocardial function improved by electromagnetic field induction of stress protein hsp70. *J Cell Physiol 216*, 816–823.

Grant, G., Cadossi, R., and Steinberg, G. (1994). Protection against focal cerebral ischemia following exposure to a pulsed electromagnetic field. *Bioelectromagnetics 15*, 205–216.

Guo, L., Kubat, N., and Isenberg, R. (2011). Pulsed radio frequency energy (PRFE) use in human medical applications. *Electromagn Biol Med 30*, 21–45.

Guo, L., Kubat, N.J., Nelson, T.R., and Isenberg, R.A. (2012). Meta-analysis of clinical efficacy of pulsed radio frequency energy treatment. *Ann Surg 255*, 457–467.

Heden, P. and Pilla, A.A. (2008). Effects of pulsed electromagnetic fields on postoperative pain: A double-blind randomized pilot study in breast augmentation patients. *Aesthetic Plast Surg 32*, 660–666.

Jonai, H., Villanueva, M.B., and Yasuda, A. (1996). Cytokine profile of human peripheral blood mononuclear cells exposed to 50 Hz EMF. *Ind Health 34*, 359–368.

Kaszuba-Zwoinska, J., Ciecko-Michalska, I., Madroszkiewicz, D., Mach, T., Slodowska-Hajduk, Z., Rokita, E., Zaraska, W., and Thor, P. (2008). Magnetic field anti-inflammatory effects in Crohn's disease depends upon viability and cytokine profile of the immune competent cells. *J Physiol Pharmacol 59*, 177–187.

Mizushima, Y., Akaoka, I., and Nishida, Y. (1975). Effects of magnetic field on inflammation. *Experientia 31*, 1411–1412.

Nelson, E.R., Zvirbulis, R., and Pilla, A.A. (2013). Non-invasive electromagnetic field therapy produces rapid and substantial pain reduction in early knee osteoarthritis: A randomized double-blind pilot study. *Rheumatol Int 33*, 2169–2173.

Pena-Philippides, J.C., Yang, Y., Bragina, O., Hagberg, S., Nemoto, E., and Roitbak, T. (2014). Effect of pulsed electromagnetic field (PEMF) on infarct size and inflammation after cerebral ischemia in mice. *Transl Stroke Res 5*, 491–500.

Pesce, M., Patruno, A., Speranza, L., and Reale, M. (2013). Extremely low frequency electromagnetic field and wound healing: Implication of cytokines as biological mediators. *Eur Cytokine Netw 24*, 1–10.

Pilla, A.A. (2006). *Mechanisms and Therapeutic Applications of Time Varying and Static Magnetic Fields*, CRC Press, Boca Raton, FL.

Pilla, A.A. (2012). Electromagnetic fields instantaneously modulate nitric oxide signaling in challenged biological systems. *Biochem Biophys Res Commun 426*, 330–333.

Pilla, A.A. (2013). Nonthermal electromagnetic fields: From first messenger to therapeutic applications. *Electromagn Biol Med 32*, 123–136.

Qin, L.Y., Wu, X.F., Block, M.L., Liu, Y.X., Bresse, G.R., Hong, J.S., Knapp, D.J., and Crews, F.T. (2007). Systemic LPS causes chronic neuroinflammation and progressive neurodegeneration. *Glia 55*, 453–462.

Rasanen, L.A., Russa, R., Urbanik, T., Choma, A., Mayer, H., and Lindstrom, K. (1997). Characterization of two lipopolysaccharide types isolated from Rhizobium galegae. *Acta Biochim Pol 44*, 819–825.

Rasouli, J., Lekhraj, R., White, N.M., Flamm, E.S., Pilla, A.A., Strauch, B., and Casper, D. (2012). Attenuation of interleukin-1beta by pulsed electromagnetic fields after traumatic brain injury. *Neurosci Lett 519*, 4–8.

Rohde, C., Chiang, A., Adipoju, O., Casper, D., and Pilla, A.A. (2010). Effects of pulsed electromagnetic fields on interleukin-1 beta and postoperative pain: A double-blind, placebo-controlled, pilot study in breast reduction patients. *Plast Reconstr Surg 125*, 1620–1629.

Roman, A., Vetulani, J., and Nalepa, I. (2002). Effect of combined treatment with paroxetine and transcranial magnetic stimulation (TMS) on the mitogen-induced proliferative response of rat lymphocytes. *Pol J Pharmacol 54*, 633–639.

Salomao, R., Brunialti, M.K., Rapozo, M.M., Baggio-Zappia, G.L., Galanos, C., and Freudenberg, M. (2012). Bacterial sensing, cell signaling, and modulation of the immune response during sepsis. *Shock 38*, 227–242.

Shupak, N. (2003). Therapeutic uses of pulsed magnetic field exposure: A review. *Radio Sci Bull 307*, 9–32.

Strauch, B., Patel, M.K., Navarro, J.A., Berdichevsky, M., Yu, H.L., and Pilla, A.A. (2007). Pulsed magnetic fields accelerate cutaneous wound healing in rats. *Plastic Reconstr Surg 120*, 425–430.

Strauch, B., Patel, M.K., Rosen, D.J., Mahadevia, S., Brindzei, N., and Pilla, A.A. (2006). Pulsed magnetic field therapy increases tensile strength in a rat Achilles' tendon repair model. *J Hand Surg Am 31A*, 1131–1135.

Taylor, E., Hardy, K., Alonso, A., Pilla, A., and Rohde, C. (2014). Abstract 15: Pulsed electromagnetic field (PEMF) dosing regimen impacts pain control in breast reduction patients. *Plast Reconstr Surg 133*, 983–984.

Vianale, G., Reale, M., Amerio, P., Stefanachi, M., Di Luzio, S., and Muraro, R. (2008). Extremely low frequency electromagnetic field enhances human keratinocyte cell growth and decreases proinflammatory chemokine production. *Br J Dermatol 158*, 1189–1196.

12 Computational Fluid Dynamics for Studying the Effects of EMF on Model Systems

L. Traikov, I. Antonov, E. Dzhambazova,
A. Ushiama, and Chiyoji Ohkubo

CONTENTS

12.1 INTRODUCTION

Electromagnetic fields (EMFs) of extremely low-frequency range are widely used in therapeutic medical applications. Proof of effectiveness has been demonstrated in numerous clinical applications where each treatment employs specific characteristics of frequency, modulation, and intensity to achieve its efficacy. Most profound clinical applications of extremely low-frequency electromagnetic field (ELF EMF) are related to bone repair, wound repair, inflammation, and pain management.

According to Pilla et al. (2011), there are a lot of scientific and clinical evidences that time-varying EMFs can modulate molecular, cellular, and tissue functions in a physiologically significant manner. Pilla and Markov (1994; Pilla, 2002) also reported that pulsed EMF has been successful in promoting bone repair and healing of spine fusions for the treatment of chronic back pain from worn and/or damaged spinal discs.

Liboff (2004, 2007) describes signal shapes in electromagnetic therapies that contribute greatly to our understanding of the various forms of EMF signal delivery that are fundamental to eliciting specific bioeffects. He simply and elegantly describes electric and magnetic signal characteristics, their signature shapes, and methods of delivery (time varying, oscillatory, or modulated) that create special interactions with human tissues and organs for healing.

Ten years ago, we started investigations of in vivo model systems in order to observe mutual interaction mechanisms of regulation foundations of the base of homeostasis; received signal has very complex and nonlinear nature, similar investigations on skin microcirculation are provided by (Braverman, 1997, 2000; Coca et al., 1998). The effects of ELF EMF also have a complex nature and basically impact single periodical chemical reactions (pacemaker reactions). The main aim of our investigations is evaluation of possible indirect mechanism of action of ELF EMF in living systems in vivo (Traikov et al., 2004a,b, 2005).

Investigations of ELF EMF biological effects (Lednev, 1991, 1996; Liboff and Parkinson, 1991; Liboff, 1997, 2009) claimed that specific combinations of low-level dc and ac magnetic fields could cause biologically significant effects that fulfill the theoretical conditions for classical cyclotron resonance. Calcium ions play an important role in the regulation of physiological processes and mediation of interactions of biological systems with external physical and chemical factors. It is well known that calcium ions play important roles in the aforementioned mechanisms, which are possible ways for the regulation of blood vessel diameter and vasomotion on cellular level (Messina and Gardner, 1992; Brekke et al., 2006; Kapela et al., 2008). Smooth muscle cell (SMC) calcium dynamics and diameter were measured in intact pressurized rat mesenteric artery segments during vasoconstriction and vasomotion by Schuster et al. (2004).

If one considers Ca^{2+}/CaM activation as a switching process, then increasing the magnetic field at Ca^{2+} levels in excess of optimal acts to bias this switch toward lower calcium concentrations (Liboff et al., 2003). Some relationship between the change in diameter of blood vessels and dilation of SMCs can be assumed by the conclusions of research of Markov and Pilla (1993, 1994).

Arteries from many vascular beds display vasomotion, that is, rhythmic oscillations superimposed on a tonic contraction. Vasomotion has been studied for more than 100 years, but the underlying mechanisms are still not fully understood; they may even differ between vascular beds (Aalkjaer and Nilsson, 2005).

12.1.1 Origin of Vasomotion

According to Peng et al. (2001), during agonist stimulation, Ca^{2+} is released intermittently from the sarcoplasmic reticulum (SR). Released Ca^{2+} can activate a membrane conductance carrying inward (depolarizing) current. When a sufficient number of cells become active at the same moment, the current will overcome the current sink in the preparation and depolarize all cells (coupled via gap junctions). Depolarization causes Ca^{2+} influx that activates Ca^{2+} release in all parts of all cells (also in inactive cells). This converts Ca^{2+} waves into global Ca^{2+} oscillations, synchronizes the cells, and causes contraction. All cells now have released Ca^{2+} and start refilling their SR simultaneously, which leads to a new synchronous release, thus reiterating the cycle (Figure 12.1).

This hypothesis was found to be consistent with the experimental data obtained. Vasomotion was correlated with the appearance of synchronous Ca^{2+} changes in the SMCs, which required a functional SR (Kapela et al., 2008, 2011). Vasomotion does not appear if the membrane potential is clamped and is only present when cGMP levels are above basal. Ca^{2+} release was shown to cause depolarization, but this occurred only when cGMP levels were elevated. Similarly, Ca^{2+} release evoked an inward current, but this occurred only in the presence of cGMP.

Schuster et al. analyzed the intercellular calcium communication between SMCs and endothelial cells (ECs) by simultaneously monitoring artery diameter and intracellular calcium concentration in a rat mesenteric arterial segment in vitro under physiological pressure (50 mmHg) and flow (50 μL/min) in a specially developed system. Intracellular calcium was expressed as the fura 2 ratio. The diameter was measured using a digital image acquisition system. Vasomotion originates in the SMCs; calcium oscillates in both SMCs and ECs during vasomotion, suggesting again a calcium flux from the SMCs to the ECs (Altura and Altura, 1982; Ursino et al., 1998; Schuster et al., 2001, 2003; Nilsson and Aalkjaer, 2003; Arciero and Secomb, 2012).

Analysis of the dynamics within the measured blood flow signal and blood vessel diameter has been introduced as an approach for the evaluation of microvascular control mechanisms

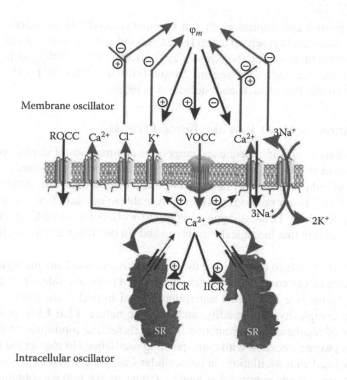

FIGURE 12.1 Main target of ELF EMF action, membrane calcium oscillator, membrane potential, and ion fluxes that contribute to the two oscillatory subsystems of model of vasomotion based on experimental pharmacological findings. Internal stores (i.e., SR) possessing a ryanodine-sensitive Ca^{2+}-induced Ca^{2+} release (CICR) mechanism are essential for genesis of intracellular Ca^{2+} oscillations. Agonists stimulate Ca^{2+} influx through voltage- and receptor-operated Ca^{2+} channels (VOCC and ROCC, respectively) and also cause Ca^{2+} release (IICR) from stores into cytosol. Cyclic changes in membrane potential (φ_m) and Ca^{2+} influx via VOCC underpin activity of membrane oscillator, with negative feedback on V_m provided by transport systems that promote hyperpolarization. These include Ca^{2+}-activated K^+ channels, Na^+-K^+-ATPase, and Cl^- channels. Ca^{2+}-ATPase pumps mediate uptake by stores and contribute to Ca^{2+} extrusion from cell. Na^+/Ca^{2+} exchange can promote Ca^{2+} extrusion or influx, depending on instantaneous value of V_m. Coupling between the two oscillatory subsystems is effected through cytosolic Ca^{2+} concentration (on the basis of theoretical concepts of Peng II. et al., 2001).

(Parthimos, et al., 1996; Pradhan and Chakravarthy, 2007, 2011; Pradhan et al., 2007). This approach is based on spectral analysis of the periodic oscillations of microvascular blood flow (Morlet, 1983; Meyer and Intaglietta, 1986; Hoffmann et al., 1990; Stefanovska, 1992; Bollinger et al., 1993; Muck-Weymann et al., 1994, 1996; Stefanovska and Kroselj, 1997; Landsverk et al., 2007; Stefanovska et al., 2011). The spectral analysis breaks down the steady fluctuating time series into its frequency elements, by computing amplitudes of signal components at predetermined frequency values (Oppenheim and Schaefer, 1975). The analysis of these periodic oscillations increases the knowledge of the dynamics of vascular control mechanisms (Akselrod, 1988; Colantuoni et al., 1990).

During the years, periodic oscillations in the microvasculature are detected both by the *non-invasive* technique of laser Doppler flowmeter (LDF) (Kaufman and Intaglietta, 1985; Meyer and Intaglietta, 1986; Kastrup et al., 1989) and invasive techniques of intravital microscopy (IVM) (Asano et al., 1987, 1980; Ushiyama et al., 2004). The spectral analysis of the laser doppler flowmeter (LDF) and IVM signal from human forearm skin and mice DSC has revealed five characteristic frequency bands (Xiu and Intaglietta, 1985, 1986; Stefanovska, 1992; Stefanovska and Kroselj, 1997; Bracic and Stefanovska, 1998; Traikov et al., 2005; Benedicic et al., 2007), called α-, β-, γ-, δ-, and θ.

In addition to the cardiac and respiratory rhythms around 0.3 and 1 Hz, in small animals, cardiac and respiratory rhythms could reach up to 2.6 Hz (Stefanovska, 1992; Bollinger et al., 1993; Muck-Weymann et al., 1996; Stefanovska and Kroselj, 1997; Traikov et al., 2005). In human skin, three frequencies have been detected in the regions around 0.1, 0.04, and 0.01 Hz (Stefanovska, 1992; Stefanovska and Kroselj, 1997; Bracic and Stefanovska, 1998).

12.1.2 OSCILLATIONS IN HEART RATE AND BLOOD PRESSURE

Rhythmic oscillations in specific frequency ranges are common to all cardiovascular and autonomic nervous system signals (Funk et al., 1983; Deriu et al., 1996). For instance, oscillations with 10 s period are noticeable in the heart rate and arterial pressure, related to increased sympathetic nervous system activity. In our previous works, we were able to measure frequencies for blood pressure between 40 and 80 Hz. Leading hypothesis of our work is based on Ca cyclotron resonance theory and understanding that biological membrane and Ca oscillator are the main targets of ELF EMF action.

The *aim of this study was* to characterize the compounds and mechanisms laying in the base of process of regulation of vasomotion in microcirculatory bed from one side. From another side how external physical factor like ELF EMF superimposed and impact these physiological processes manifested in the complexity, nonlinearity, and dynamic nature. ELF EMF possibly rearranges existing pathways of regulation of vasomotion in order increasing amplitude of this process. The tension oscillations were associated with corresponding oscillations in membrane potential and are known to be associated with oscillations in intracellular Ca^{2+} concentration. Upon the reduction of extracellular Ca^{2+}, vasomotion continued as long as a tonic contraction was obtained.

To fulfill our tasks, we summarize the results obtained by IVM in terms of ELF EMF effects. All studies have been conducted in the past 10 years in the laboratories of the National Institute of Public Health (NIPH), Tokyo, Japan, and the laboratories of the Medical University of Sofia, Bulgaria.

In this review, we present several stages:

Stage one: First, we should note that a system for real-time determination of blood vessel diameter was developed, on the basis of IVM. The main idea was the system should meet or follow requirements: to detect blood vessels with mean diameter 30–90 μm of microcirculatory bed for long enough periods of time (this measurement is necessary for subsequent fast Fourier transform [FFT] analysis and determination of the amplitude of low frequencies in the FFT frequency domain). System for real time video measurement of blood vessel diameter has been developed. This system allow 33 min video recording with acquisition time one video frame per 280 ms.

Stage two: Mathematical algorithm was developed based on system OriginPro for bandpass filtration of the raw signal and subsequent FFT analysis. Due to the highly nonlinear nature of the sampled signal, wavelet analysis was implemented, based on the system OriginPro again. Studied signals and blood vessel diameter changes are presented in this work as a normalized amplitude relative related to the mean diameter. By these data, further analysis was made of the change in amplitude and distribution of vasomotion frequencies.

Stage three: Image-matrix analysis is based on the transformation of individual video frames (one frame per 280 ms) in the digital matrix and subsequent correlation analysis of the values of the different matrices, as the end point result is formation of an image only by the portions (pixels of the matrix) in which there were statistically significant changes. By means of these data, we can calculate a coefficient of elasticity of the vessel wall. After acquiring image-matrix analysis results, we can discuss about possible changes in some parameters in fundamental Lame's equation, before, during, and after 10 min ELF EMF 20 mT exposure at different frequencies (10, 16, and 50 Hz).

Stage four: Computational flow dynamics (CFD) is a method based on the image-matrix analysis data; the basic idea of this type of analysis is to provide insight investigation into the complex changes in blood vessel diameter vascular resistance, vasomotion frequencies (probably related to interference with Ca oscillator) related to the regulation of vascular tone, and wall shear stress before, during, and after ELF EMF 20 mT action.

A lot of scientific evidence exists that lead to the conclusion that an improvement of the condition of the damaged tissue pathological tissue (healing effect) is realized by improving the blood, oxygen, and metabolite supply. By our study, we address the possible link between ELF EMF 20 mT action vascular tone changes and vascular wall resistance.

12.2 MATERIAL AND METHODS

Male BALB/c mice (8–12 weeks old, 22–25 g; Tokyo Zikken Doubutsu, Japan) were used, and a dorsal skinfold chamber (DSC), made of polyacetal resin, was surgically implanted 4 days before the experiment.

Mice were divided into four groups (control, 10, 16, and 50 Hz ELF EMF 20 mT exposed). (Figure 12.2 shows the protocol of the experiment.) Fluorescein isothiocyanate–labeled dextran 250 kDa (FITC-dextran-250, 2.5% [w/v] in PBS, 50 μL/25 g body weight) was injected into the caudal vein to visualize vasculature. Mice were fixed on a plastic stage, under conscious conditions, for IVM; IVM was provided by means of Olympus Ltd. fluorescent microscope.

One small arteriole (30–90 μm in diameter) was chosen under fluorescent microscopy, and vasomotion was recorded on digital videotape throughout the whole experiment for off-line analysis. One recording process period (total of 33 min) included the following: for off-line analysis, noise in recorded video images was reduced with a signal amplifier (DVS-3000, Hamamatsu Photonics Inc., Hamamatsu, Japan), and diameters were measured automatically and recorded with a high-speed digital machine vision system CV-2100 (KEYENCE Inc., Osaka, Japan); using an edge gap detection algorithm, gathering data is in single ASCII code.

For whole process of off-line video image analysis, digital image-matrix analysis, signal processing (FFT), wavelet analysis, and statistical correlation analysis, on the base of OriginPro (OriginLab Corporation), software was used.

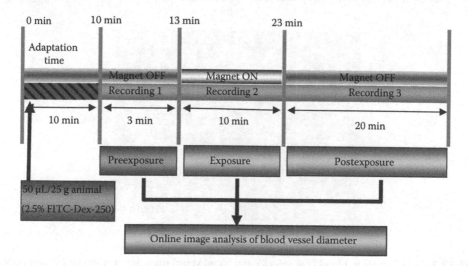

FIGURE 12.2 Experimental protocol of 33 min nonstop measurement of vasomotion. All experiments are provided at constant air temperature conditions (25°C).

12.3 RESULTS AND DISCUSSION

Obtained results are presented in chronological order in accordance with the development of CFD model.

System for real-time determination of blood vessel diameter on the base of Intravital microscopy-IVM in conscious animals was develop. The main idea was the system should meet or follow requirements: to detect blood vessels with mean diameter 30–90 μm of microcirculatory bed for long enough periods of time.

12.3.1 SYSTEM FOR ACQUIRING OF VIDEO DATA IN REAL TIME AND FOR IMAGE CAPTURING

At the system for real-time measurements, we used algorithm for pattern recognition: To find an X value for a given Y value, we enter the Y value in the bottom text box next to the Find Y button on GUI for NLSF analysis and then press Find X. When this action is performed, an iterative procedure is used to find the X value corresponding to the given Y value (Figure 12.3).

The X value of the midpoint of this interval is computed, $X_m = (X_1 + X_2)/2$, and the corresponding Y value, Y_m, is computed using the exact fit equation.

The interval to the right or left of this midpoint is chosen such that the user-specified Y value now falls within the new interval.

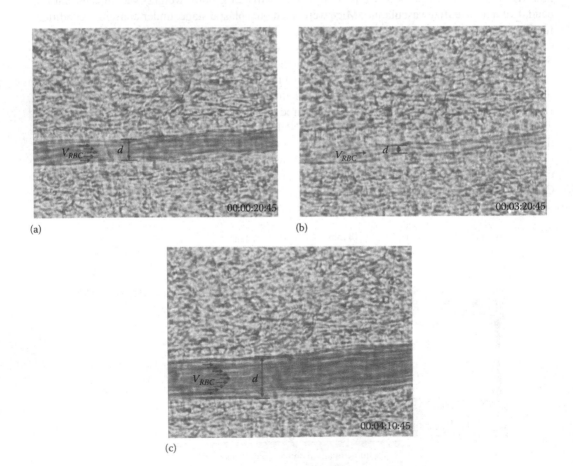

(a)

(b)

(c)

FIGURE 12.3 (a) Changes in blood flow velocity and blood vessel diameter at stationary vasculatory state. (b) Changes in blood flow velocity and blood vessel diameter at constricted vasculatory state. (c) Changes in blood flow velocity and blood vessel diameter at dilated vasculatory state.

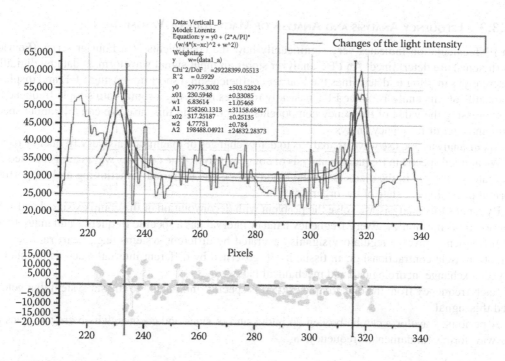

FIGURE 12.4 Densitometric profile of the image matrix and NLSF function for maximum finding and position calculation in pixels. NLSF algorithm found out two distinguished maximums with high probability and after that calculates distance between central points of these maximums in pixels, second phase converting distance from pixels to micrometers.

First, a check is performed to see if the Y value that we specified is inside the range of Y values described by the fit line. If the Y value passes this check, then the following steps are performed.

The fit line data set is scanned to find two points (X_1, Y_1) and (X_2, Y_2) such that the user-specified Y value lies in the interval $[Y_1, Y_2]$. In the case of a fit line that is not monotonic (i.e., multiple X value exist for the same Y value), the first interval that satisfies this criterion, starting from the lower end of the X axis, is selected.

Algorithm for Lorentz NLSF functions for curve fitting and pike finding (Figure 12.4).

By this tool, we can analyze amplitude of changes by processes with different time intervals or different frequencies, on the images with low magnification, and create activity maps of whole regions from microcirculatory bed, similar approach develop (Mayhew et al., 1998).

12.3.2 IMAGE-MATRIX ANALYSIS

Matrices have numbered columns that are mapped to linearly spaced X values and numbered rows that are mapped to linearly spaced Y values. Each cell value in a matrix represents a Z value that is located in the XY plane by the cell's X and Y values.

As a result of the work created by our algorithm for real-time signal processing of the change in diameter of blood vessels, we were able to get a raw signal of 33 min for each of the studied samples. The collection of data is according to the protocol of the experiment shown in Figure 12.2. Collection of recordings of 33 min measurements is important when we want to analyze raw signal and put it to a band-pass filtering and subsequent FFT analysis with the aim to distinguish embedded components of the raw complex signal.

Whereas these data have nonstationary nature, similar results are obtained by (Brauer and Hahn, 1999), we applied wavelet-based analysis, which is not the subject of this current work.

12.3.3 FREQUENCY ANALYSIS AND ANALYSIS OF MAGNITUDE OF VASOMOTION

An FFT calculation indicates very significantly how the coefficients of a Fourier series (Fourier coefficients) are determined. An FFT analyzer stores an input signal waveform as data by digitally (discretely) sampling it, determines the Fourier coefficients in a short time using FFT, and displays the results of this analysis. Since FFT basically involves resolving a signal into simple frequencies by expressing the value of frequency components (spectrum); in this case, we can talk about spectrum analyzer or frequency analysis.

In our analysis, we use wavelet analysis to detect changes of magnitude at every frequency band.

We applied spectrum frequency analysis of complex waves. The complex waves are generated by oscillatory changes in the diameter of the blood vessel, due to normal functioning and supporting normal physiological state of the organism.

By means this analysis, we solve the problem with deconvolution of the complex signal and subtraction of its harmonics. Every frequency band or interval has a specific frequency and magnitude; each frequency involve regulatory signals generated by different systems (e.g., heart rate beat or smooth muscle contractions) or, in tissue itself, governed by different internal, such as metabolic, oxygen exchange, neurological, and mechanical factors.

Each frequency from the one complex signal is specific and related to the subsystem that generated this signal.

A periodic waveform can be broken up into a sum of sines and cosines, which are multiples of the waveform's fundamental frequency, ω_o.

12.3.4 THEORETICAL BASIS

A periodic function, $F(t, \omega_o)$, can be written as

$$F\left(t,\omega_o\right) = \frac{a_o}{2} + \sum_{n=1}^{\infty} a_n \cos(n\omega_o t) + \sum_{n=1}^{\infty} b_n \sin(n\omega_o t) \tag{12.1}$$

where

$$a_n = \frac{1}{\pi} \int_{-\pi}^{+\pi} F(t,\omega_o) \cos(n\omega_o t) d(\omega_o t) \tag{12.2}$$

and

$$b_n = \frac{1}{\pi} \int_{-\pi}^{+\pi} F(t,\omega_o) \sin(n\omega_o t) d(\omega_o t) \tag{12.3}$$

ω_o is the fundamental frequency of the function $F(t, \omega_o)$, and a_o is the average magnitude of $F(t, \omega_o)$ and can be found by setting $n = 0$ and taking ½ of Equation 12.2. a_n's are symmetric coefficients because if $F(t, \omega_o)$ is entirely antisymmetric about the origin, then all the a_n's are zero. b_n's are antisymmetric coefficients because if $F(t, \omega_o)$ is entirely symmetric about the origin, then all the b_n's are zero. Often in carrying out the integrals of Equations 12.2 and 12.3, the range of integration can be cleverly reduced, and some integrals set equal to zero depending on the magnitude and symmetries of the function $F(t, \omega_o)$.

An alternate and sometimes more useful way to express a Fourier series is to consider phase and amplitude of each harmonic instead of cosine and sine. Equation 12.1 can be rewritten as

$$F(t,\omega_o) = \frac{a_o}{2} + \sum_{n=1}^{\infty} a_n \frac{e^{in\omega_o t} + e^{-in\omega_o t}}{2} + \sum_{n=1}^{\infty} b_n \frac{e^{in\omega_o t} - e^{-in\omega_o t}}{2i}$$

Then recollecting terms with the same exponential, one obtains

$$F(t,\omega_o) = \frac{a_o}{2} + \sum_{n=1}^{\infty} \frac{1}{2}(a_n - ib_n)e^{in\omega_o t} + \sum_{n=1}^{\infty} \frac{1}{2}(a_n + ib_n)e^{-in\omega_o t} \qquad (12.4)$$

All this can be written as a single sum if we define complex coefficients, c_n, such that

$$c_0 = \frac{a_0}{2}; \quad c_n = \frac{1}{2}(a_n - ib_n); \quad c_{-n} = \frac{1}{2}(a_n + ib_n) \qquad (12.4a)$$

and then

$$F(t,\omega_0) = \sum_{n=-\infty}^{\infty} c_n e^{in\omega_o t} \qquad (12.4b)$$

Using Equation 12.4, we can calculate a single new Fourier integral for all the c_n's:

$$c_n = \frac{1}{2\pi} \int_{-\pi}^{+\pi} F(t,\omega_o)e^{-in\omega_o t} d(\omega_o t) \qquad (12.5)$$

Frequency analysis (observing the waveform in the frequency domain or period) makes detection of even small changes in physiological significant parameters responsible for normal spectral distribution of vasomotion.

A waveform can be represented with three different parameters: amplitude, frequency (or period), and phase (or time difference). Amplitude is a function of magnitude of the waveform. Analysis of living systems is related to the strength of the process that we analyze or balance between the contracting powers in a system.

With assumption, we can state that for investigated system data obtained by intravital video fluorescence microscopy, amplitude of obtained signal is related to blood vessel diameter.

Before any calculations, our aim was to normalize raw signal of vasomotion and to transform it into oscillations around 0 or central axis; amplitude modulations are presented in μm.

In previous papers, we report that, after FFT analysis of the raw data, five frequency domains related to some specific physiological conditions were distinguished:

1. *Frequency domain* 0.0095–0.02 Hz -α. We hypothesize that this domain corresponds to metabolic activity—the rhythmic regulation of vessel resistance to the blood flow initiated by concentrations of metabolic substances in the blood.
2. *Frequency domain* 0.02–0.06 Hz -β. According to preliminary experiments, it has been observed to be diminishing after vegetative nervous system inhibition; we can conclude that this rhythm has neurogenic activity origin.
3. *Frequency domain* 0.06–0.15 Hz -γ. This frequency domain corresponds to intrinsic rhythmic activity of the vessels, caused by the pacemaker cells in the smooth muscles of their walls, which is called myogenic activity.
4. Frequency domain 0.9–1.2 Hz -δ. Results from respiratory activity.
5. Frequency band around 20–40 Hz band -θ. Results from the heart rate.

From a thorough analysis, we only take into account frequency domains in which we find statistically significant differences. Therefore, in this work, we omit and didn't discuss data with lack of statistically significant results obtained by exposure ELF EMF (20 mT, 10 and 50 Hz).

In this study, we examined 33 min nonstop measurements of arteriole diameters. No significant differences in vessel diameters throughout the whole measurement period were detected in the sham-exposure group, except for normal vasomotion, and amplitude of this process was stabile in time (Figure 12.5).

Amplitude of vasomotion 0.0095–0.02 Hz (alpha-dispersion) is 1.02 (rel.u) (Figure 12.6), and amplitude of vasomotion 0.06–0.15 Hz (gamma-dispersion) is 0.39 (a.u) (Figure 12.7).

At 16 Hz ELF EMF exposure group, significant increase of amplitude of vasomotion was observed in the postexposure period compared with the preexposure and exposure periods; all these changes were related to decreasing of the magnitude of rhythmic oscillations of the blood vessel diameter (Figure 12.8a). No significant effects on vascular diameter in the 10 and 50 Hz exposure groups were determined according to chi-square (χ^2) analysis. That is the reason to not discuss these frequencies at present work and new calculations.

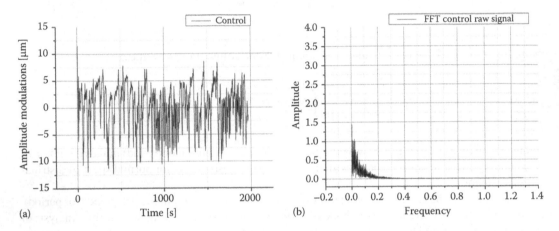

FIGURE 12.5 (a) Observation of 33 min nonstop changes in vasomotion. Normalized raw signal obtained by conscious mice in control/sham-exposed conditions (ambient temperature 25°C). Amplitude of vasomotion is presented in μm and is calculated by subtraction of any current value by mean blood vessel diameter value. (b) Frequency spectrum of the raw signal.

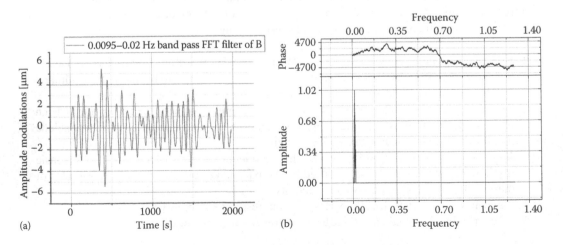

FIGURE 12.6 (a) Frequency domain 0.0095–0.02 Hz (alpha domain) obtained by raw signal of control/sham-exposed animals and (b) frequency spectrum of the frequency domain 0.0095–0.02 Hz (alpha domain) obtained by raw signal of control/sham-exposed animals.

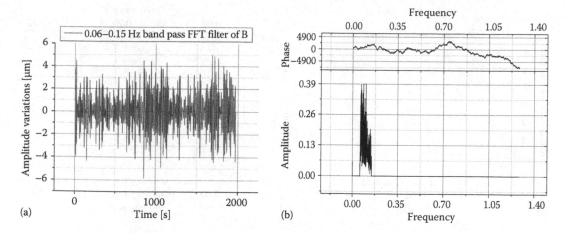

(a)

(b)

FIGURE 12.7 (a) Frequency domain 0.06–0.15 Hz (gamma domain) obtained by raw signal of control/sham-exposed animals and (b) frequency spectrum of the frequency domain 0.06–0.15 Hz (gamma domain) obtained by raw signal of control/sham-exposed animals.

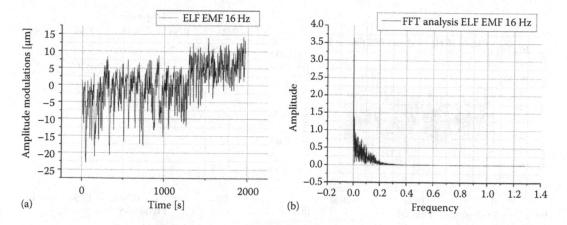

(a)

(b)

FIGURE 12.8 (a) Observation of 33 min nonstop changes in vasomotion. Normalized raw signal obtained by conscious mice at exposed conditions ELF EMF (20 mT; 16 Hz) (ambient temperature 25°C). Amplitude of vasomotion is presented in μm and is calculated by subtraction of any current value by mean blood vessel diameter value. (b) Frequency spectrum of the raw signal.

Amplitude of vasomotion 0.0095–0.02 Hz (alpha-dispersion) is 0.75 (rel.u) (Figure 12.9b), and amplitude of vasomotion 0.06–0.15 Hz (gamma-dispersion) is 1.35 (rel.u) (Figure 12.10b).

If we do a comparative analysis of the signal amplitude changes in the control and 16 Hz exposure group with respect to the amplitudes of frequency domain alpha and gamma (as shown in Figures 12.6, 12.7, 12.9b, and 12.10b), frequency domain alpha in control group is with greater amplitude than in the exposed group. While at exposed group, frequency domain gamma's amplitude is twice as large.

In Figure 12.11, it can be seen that in the area of the alpha-dispersion, amplitude increases under the control conditions, while in the range of gamma-dispersion, amplitude decreases in comparison to exposed group. Similar effect is observed during Hb fluctuations measured by IR spectroscopy under static magnetic field exposure by Xu et al. (2013). Opposite effect concerning amplitude of the frequency domain was observed in the ELF EMF (16 Hz; 20 mT)–exposed samples. The increase of the amplitude in the range of the gamma-dispersion under the action of the low-frequency EMF is

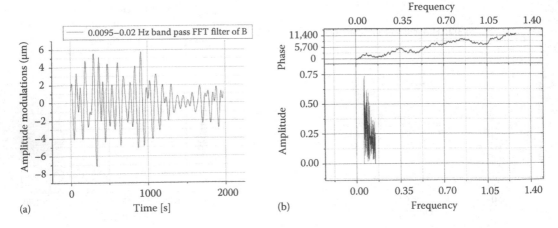

FIGURE 12.9 (a) Frequency domain 0.0095–0.02 Hz (alpha-dispersion) obtained by raw signal at exposed conditions ELF EMF (20 mT; 16 Hz). (b) Frequency spectrum of the frequency domain 0.0095–0.02 Hz (alpha-dispersion) obtained by raw signal at exposed conditions ELF EMF (20 mT; 16 Hz).

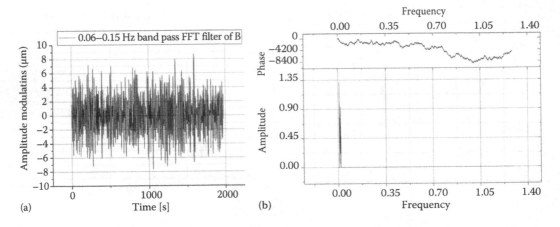

FIGURE 12.10 (a) Frequency domain 0.06–0.15 Hz (gamma-dispersion) obtained by raw signal at exposed conditions ELF EMF (20 mT; 16 Hz). (b) Frequency spectrum of the frequency domain 0.06–0.15 Hz (gamma-dispersion) obtained by raw signal at exposed conditions ELF EMF (20 mT; 16 Hz).

probably due to an indirect impact on membrane calcium oscillator (see Figure 12.1). These specific amplitude differences are probable explanation of existence of a biphasic effect on microcirculation after ELF EMF action accordingly (McKay et al., 2007; Figure 12.12).

Generation of image-matrix and subsequent image analysis plays a key role of grid generation by the ASCII raw data. Grids are vital for CFD modeling. Our benefits are that we are working with raw data obtained directly by the image of single blood vessel from microcirculatory bed, with diameter that we can calculate with 280 ms time definition, thickness of the blood microvessel wall. By means of image-matrix analysis, we calculated according to Lame's law coefficient of elasticity of blood vessel wall in control/sham-exposed conditions and during and after ELF EMF exposure (Figures 12.13 and 12.14).

We use finite-element method (FEM) for estimation of micro-biomechanical properties of the blood vessels at control/sham-exposed conditions and during and after ELF EMF exposure. Data gathered by FEM serve as a base for modeling of CFD.

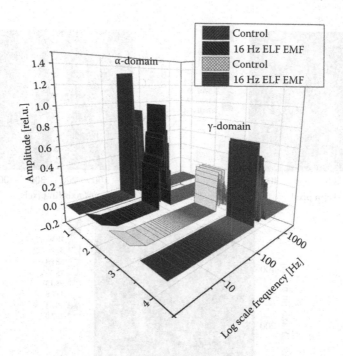

FIGURE 12.11 Comparative analysis of the amplitudes of control (sham-exposed) alpha and gamma frequency domains and exposed group with ELF EMF 20 mT 16 Hz for 10 min alpha and gamma frequency domains.

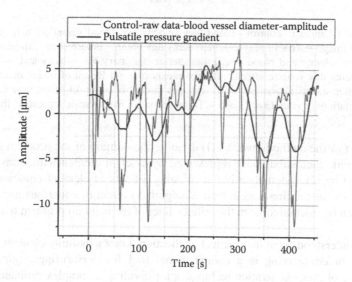

FIGURE 12.12 Changes of blood vessel diameter amplitude of vasomotion versus pulsatile pressure gradient. Pulsatile pressure gradient consists of a constant art and purely oscillatory part.

General principles of FEM have several advantages:

- Accurate representation of complex geometry
- Inclusion of dissimilar material properties
- Easy representation of the total solution
- Capture of local effects

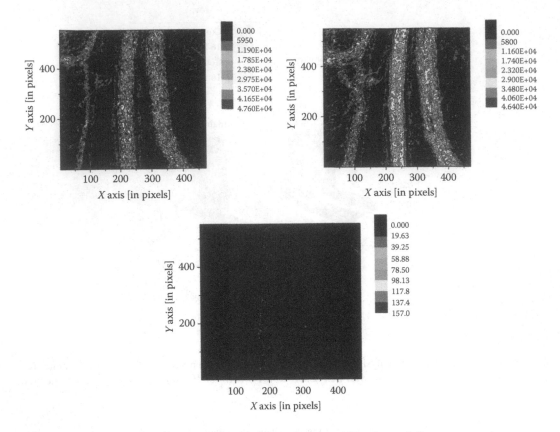

FIGURE 12.13 (a) Changes in control—blood flow velocity and blood vessel diameter at stationary vasculatory state—2D image matrix in grayscale represents flux volume before sham exposure. (b) Changes in control—blood flow velocity and blood vessel diameter at stationary vasculatory state—2D image matrix in grayscale represents flux volume after control/sham exposure. (c) Result of image matrix calculations—residual changes after statistical analysis of the changes in control group of blood flow velocity and blood vessel diameter at stationary vasculatory state—2D image matrix in grayscale represents flux volume.

A typical workout of the method involves (1) dividing the domain of the problem into a collection of subdomains, with each subdomain represented by a set of element equations to the original problem, followed by (2) systematically recombining all sets of element equations into a global system of equations for the final calculation. The global system of equations has known solution techniques and can be calculated from the initial values of the original problem to obtain a numerical answer.

FEM is best understood from its practical application, known as finite element analysis (FEA). FEA as applied in engineering is a computational tool for performing engineering analysis. It includes the use of mesh generation techniques for dividing a complex problem into small elements, as well as the use of software program coded with FEM algorithm (Chaplain et al., 2006).

FEA is a good choice for analyzing problems over complicated domains (like blood vessels), when the domain changes (as during a solid-state reaction with a moving boundary), when the desired precision varies over the entire domain, or when the solution lacks smoothness. For instance, in a frontal crash simulation, it is possible to increase prediction accuracy in *important* areas like the front of the car and reduce it in its rear (thus reducing cost of the simulation). Another example would be in numerical weather prediction, where it is more important to have accurate predictions over developing highly nonlinear phenomena (such as tropical cyclones in the atmosphere or eddies in the ocean) rather than relatively calm areas.

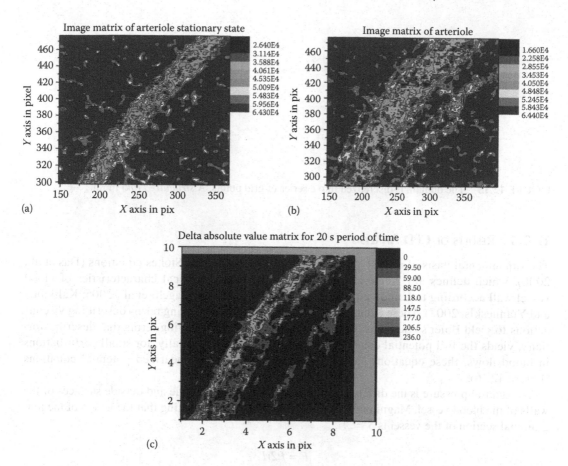

FIGURE 12.14 (a) Changes in exposed group in blood flow velocity and blood vessel diameter at stationary vasculatory state—2D image matrix in grayscale represents flux volume before ELF EMF exposure. (b) Changes in blood flow velocity and blood vessel diameter at dilated state—2D image matrix in grayscale represents flux volume 10 min after 10 min ELF EMF exposure (16 Hz; 20 mT). (c) Result of image matrix calculations—residual changes after statistical analysis of the changes in exposed group of blood flow velocity and blood vessel diameter at stationary vasculatory state—2D image matrix in grayscale represents flux volume 10 min after 10 min ELF EMF exposure (16 Hz; 20 mT).

While it is difficult to quote a date of the invention of the FEM, the method originated from the need to solve complex elasticity and structural analysis problems (Figure 12.15).

Using FEM, we find out approximate solutions to boundary value problems for differential equations. FEM uses variational methods (the calculus of variations) to minimize an error function and produce a stable solution. In our case, in the determination of the elasticity of the blood vessel wall, analogous to the idea that connecting many tiny straight lines can approximate a larger circle, FEM encompasses all the methods for connecting many simple element equations over many small subdomains, named finite elements, to approximate a more complex equation over a larger domain. Governing equations (in differential form) are discretized (converted into algebraic form).

The resulting set of linear algebraic equations is solved iteratively or simultaneously. The tendency that ELF-EMF action leads to increasing the natural variation of amplitude of vasomotion can be proven by ksi-square criterion of non-parametric statistical analysis, criteria for assessing differences between control and exposure group.

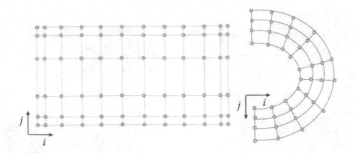

FIGURE 12.15   The domain is discretized into a series of grid points. A *structured* (*ijk*) mesh.

12.3.5   RESULTS OF CFD ANALYSIS

The fundamental basis of almost all CFD problems is the Navier–Stokes equations (Das et al., 2000), which defines any single-phase fluid flow and biomechanical characteristics of blood vessel wall according to Lame's equations and shear flow stress (Bagchi et al., 2005; Kaliviotis and Yianneskis, 2007). These equations can be simplified by removing terms describing viscous actions to yield Euler's equations. Further simplification, by removing terms that describe vorticity, yields the full potential equations (Debbaut et al., 2012). Finally, for small perturbations in blood flows, these equations can be linearized to yield the linearized potential equations (Figure 12.16).

Transmural pressure is the difference in pressure between the inside and outside surfaces of the walls of the blood vessel. Magnitude of F is easily determined, assuming that the action of the longitudinal section of the vessel is $S = 2rl$:

$$F = P2rl$$

For the elastic f forces opposing the force F, determined by mechanical stress σ and directed perpendicular to the longitudinal section, we get

$$f = \sigma2hl$$

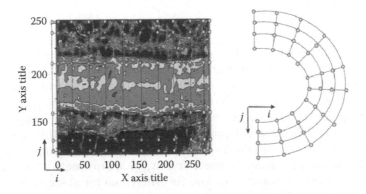

FIGURE 12.16   Transformation by the image-matrix analysis to FEM grid and to calculations of flow dynamics. FEM images obtained by IVM of the 50 µm arteriole.

Condition for equilibrium of forces is

$$F = f$$

Equation stated above gives distribution of transverse and longitudinal forces acting on the blood vessel walls at one same time. The condition of the balance of power is the dependence of mechanical stress in vascular pressure in them, the radius, and thickness of their walls (Lame's equation):

$$\sigma = \frac{pr}{h}$$

We did not obtain any significant changes for blood vessel wall elasticity σ when we compared control versus exposed group (at different exposure conditions, 10, 16, and 50 Hz).

12.4 CONCLUSION

Residual effects that we observed after low-frequency EMF action are probably not related to a change in blood vessel elasticity but with probable influence of the regulatory mechanisms governing vasomotion.

EMF acts as a vasodilator and blood vessel oscillation frequency modulator on subcutaneous microcirculation in mice. Similar effects but obtained for static magnetic field and by measurement of blood pressure was discovered by (Ohkubo and Xu, 1997; Ohkubo et al., 1997; Okano and Ohkubo, 1998; Xu et al., 2000, 2013).

A significant increase of the amplitude of low-frequency domain 0.06–0.15 Hz (gamma-dispersion) of blood vessel oscillations in postexposure period at ELF EMF 16 Hz exposed group, especially, compared with preexposure and exposure periods was observed.

Stated above results leads to the conclusion about the possible specific *window* effect at 16 Hz, both for the mean blood vessel diameter and related gamma frequency domain. This frequency band corresponds to intrinsic rhythmic activity of the vessels, caused by the pacemaker cells in the smooth muscles of their walls, which is called myogenic activity.

Increase of the blood vessel diameter and vascular resistance after 10 min ELF EMF (16 Hz; 20 mT) exposure is probably related to impairment of specific regulations of SMC Ca^{2+} channel.

Vasodilation refers to the widening of blood vessels. It results from relaxation of SMCs within the vessel walls. When blood vessels dilate, the flow of blood is increased due to a decrease in vascular resistance. Therefore, dilation of arterial blood vessels (mainly the arterioles) decreases blood pressure. Open Ca^{2+}-activated and voltage-gated K^+ channels → hyperpolarization → close voltage-dependent calcium channels (VDCCs) → ↓intracellular Ca^{2+}. VDCCs are one of the main channels of the SMCs; they are responsible for permeability of Ca^{2+}.

These channels are slightly permeable to sodium ions, so they are also called Ca^{2+}–Na^+ channels, but their permeability to calcium is about 1000-fold greater than sodium under normal physiological conditions. At resting membrane potential, VDCCs are normally closed. They are activated (opened) at depolarized membrane potentials (see Figure 12.1).

The concentration of calcium (Ca^{2+} ions) is normally several thousand times higher outside of the cell than inside. Activation of particular VDCCs allows Ca^{2+} to rush into the cell, which, depending on the cell type, results in activation of calcium-sensitive potassium channels, muscular contraction.

All those effects could be preceded by a series of metabolic and vascular tone changes in microvascular bed as we can see in changes (decreasing 20%) of the alpha domain after ELF EMF exposure. All those can lead to already known beneficial effects of ELF EMF at chronic exposures in vivo, related to improving blood supply and enervation of injured tissues, leading to better oxygen and metabolite supply.

REFERENCES

Aalkjaer C and Nilsson H (2005) Vasomotion: Cellular background for the oscillator and for the synchronization of smooth muscle cells. *Br J Pharmacol* 144: 605–616.

Akselrod S (1988) Spectral analysis of fluctuations in cardiovascular parameters: A quantitative tool for the investigation of autonomic control. *Trends Pharmacol Sci* 9: 6–9.

Altura BM and Altura BT (1982) Microvascular and vascular smooth muscle actions of ethanol, acetaldehyde, and acetate. *Fed Proc* 41: 2447–2451.

Arciero JC and Secomb TW (2012) Spontaneous oscillations in a model for active control of microvessel diameters. *Math Med Biol* 2: 163–180 (electronically published 2011).

Asano M, Ohkubo C, Sasaki A, Sawanobori K, and Nagano H (1987) Vasodilator effects of cepharanthine, a biscoclaurine alkaloid, on cutaneous microcirculation in the rabbit. *J Ethnopharmacol* 20: 107–120.

Asano M, Ohkubo C, Sawanobori K, and Yonekawa K (1980) Cutaneous microcirculatory effects of various vasodilator agents on the conscious rabbit, with special regard to changes in the rhythmic property of vasomotion. *Biochem Exp Biol* 16: 341–348.

Bagchi P, Johnson PC, and Popel AS (2005) Computational fluid dynamic simulation of aggregation of deformable cells in a shear flow. *J Biomech Eng* 127: 1070–1080.

Benedicic M, Bernjak A, Stefanovska A, and Bosnjak R (2007) Continuous wavelet transform of laser-Doppler signals from facial microcirculation reveals vasomotion asymmetry. *Microvasc Res* 74: 45–50.

Bollinger A, Yanar A, Hoffmann U, and Franzeck UK (1993) Is high-frequency fluxmotion due to respiration or to vasomotion activity? In *Progress in Applied Microcirculation*, Messmer, K. (ed.). Karger, Basel, Switzerland, pp. 52–58.

Bracic M and Stefanovska A (1998) Wavelet based analysis of human blood flow dynamics. *Bull Math Biol* 60: 417–433.

Brauer K and Hahn M (1999) Nonlinear analysis of blood flux in human vessels. *Phys Med Biol* 44: 1719–1733.

Braverman IM (1997) The cutaneous microcirculation: Ultrastructure and microanatomical organization. *Microcirculation* 4: 329–340.

Braverman IM (2000) The cutaneous microcirculation. *J Invest Dermatol Symp Proc* 5: 3–9.

Brekke JF, Jackson WF, and Segal SS (2006) Arteriolar smooth muscle Ca^{2+} dynamics during blood flow control in hamster cheek pouch. *J Appl Physiol* 101: 307–315.

Chaplain MA, McDougall SR, and Anderson AR (2006) Mathematical modeling of tumor-induced angiogenesis. *Annu Rev Biomed Eng* 8: 233–257.

Coca D, Zheng Y, Mayhew JE, and Billings SA (1998) Non-linear analysis of vasomotion oscillations in reflected light measurements. *Adv Exp Med Biol* 454: 571–582.

Colantuoni A, Bertuglia S, Coppini G, and Donato L (1990) Superposition of arteriolar vasomotion waves and regulation of blood flow in skeletal muscle microcirculation. *Adv Exp Med Biol* 277: 549–558.

Das B, Johnson PC, and Popel AS (2000) Computational fluid dynamic studies of leukocyte adhesion effects on non-Newtonian blood flow through microvessels. *Biorheology* 37: 239–258.

Debbaut C, Vierendeels J, Casteleyn C, Cornillie P, Van Loo D, Simoens P, Van Hoorebeke L, Monbaliu D, and Segers P (2012) Perfusion characteristics of the human hepatic microcirculation based on three-dimensional reconstructions and computational fluid dynamic analysis. *J Biomech Eng* 134: 011003.

Deriu F, Roatta S, Grassi C, Urciuoli R, Micieli G, and Passatore M (1996) Sympathetically-induced changes in microvascular cerebral blood flow and in the morphology of its low-frequency waves. *J Auton Nerv Syst* 59: 66–74.

Funk W, Endrich B, Messmer K, and Intaglietta M (1983) Spontaneous arteriolar vasomotion as a determinant of peripheral vascular resistance. *Int J Microcirc Clin Exp* 2: 11–25.

Hoffmann U, Franzeck UK, Geiger M, and Bollinger A (1990) Variability of different patterns of skin oscillatory flux in healthy controls and patients with peripheral arterial occlusive disease. *Int J Microcirc Clin Exp* 12: 255–273.

Kaliviotis E and Yianneskis M (2007) On the effect of dynamic flow conditions on blood microstructure investigated with optical shearing microscopy and rheometry. *Proc Inst Mech Eng H* 221: 887–897.

Kapela A, Bezerianos A, and Tsoukias NM (2008) A mathematical model of Ca^{2+} dynamics in rat mesenteric smooth muscle cell: Agonist and NO stimulation. *J Theor Biol* 253: 238–260.

Kapela A, Nagaraja S, Parikh J, and Tsoukias NM (2011) Modeling Ca^{2+} signaling in the microcirculation: Intercellular communication and vasoreactivity. *Crit Rev Biomed Eng* 39: 435–460.

Kastrup J, Buhlow J, and Lassen NA (1989) Vasomotion in human skin before and after local heating recorded with laser Doppler flowmetry. A method for induction of vasomotion. *Int J Microcirc Clin Exp* 8: 205–215.

Kaufman AG and Intaglietta M (1985) Automated diameter measurement of vasomotion by cross-correlation. *Int J Microcirc Clin Exp* 4: 45–53.

Landsverk SA, Kvandal P, Bernjak A, Stefanovska A, and Kirkeboen KA (2007) The effects of general anesthesia on human skin microcirculation evaluated by wavelet transform. *J Anesth Analg* 105(4): 1012–1019.

Lednev VV (1991) Possible mechanism for the influence of weak magnetic fields on biological systems. *Bioelectromagnetics* 12: 71–75.

Lednev VV (1996) Bioeffects of weak static and alternating magnetic fields. *Biofizika* 41: 224–232.

Liboff AR (1997) Electric-field ion cyclotron resonance. *Bioelectromagnetics* 18: 85–87.

Liboff AR (2004) Signal shapes in electromagnetic therapies: A primer. In *Bioelectromagnetic Medicine*, Rosch, P and Markov, M. (eds.). Marcel Dekker, New York, pp. 17–37.

Liboff AR (2007) Local and holistic electromagnetic therapies. *Electromagn Biol Med* 26: 315–325.

Liboff AR (2009) Electric polarization and the viability of living systems: Ion cyclotron resonance-like interactions. *Electromagn Biol Med* 28: 124–134.

Liboff AR, Cherng S, Jenrow KA, and Bull A (2003) Calmodulin-dependent cyclic nucleotide phosphodiesterase activity is altered by 20 microT magnetostatic fields. *Bioelectromagnetics* 24: 32–38.

Liboff AR and Parkinson WC (1991) Search for ion-cyclotron resonance in an Na(+)-transport system. *Bioelectromagnetics* 12: 77–83.

Markov MS and Pilla AA (1993) Ambient range sinusoidal and DC magnetic fields affect myosin phosphorylation in a cell-free preparation. In *Electricity and Magnetism in Biology and Medicine*, Blank, M. (ed.). San Francisco Press, San Francisco, CA, pp. 323–327.

Markov MS and Pilla AA (1994) Modulation of cell-free myosin light chain phosphorylation with weak low frequency and static magnetic fields. In *On the Nature of Electromagnetic Field Interactions with Biological Systems*, Frey, A. (ed.). R.G. Landes Co., Austin, TX, pp. 127–141.

Mayhew J, Hu D, Zheng Y, Askew S, Hou Y, Berwick J, Coffey PJ, and Brown N (1998) An evaluation of linear model analysis techniques for processing images of microcirculation activity. *Neuroimage* 7: 49–71.

McKay J, Prato F, and Thomas A (2007) A literature review: The effects of magnetic field exposure on blood flow and blood vessels in the microvasculature. *Bioelectromagnetics* 28: 81–98.

Messina LM and Gardner A (1992) Effect of calcium on the vasoactive response of arterioles to light during fluorescent intravital microscopy of the tibialis anterior muscle of the hamster. *Microvasc Res* 44: 274–285.

Meyer JU and Intaglietta M (1986) Measurement of the dynamics of arteriolar diameter. *Ann Biomed Eng* 14: 109–117.

Morlet J (1983) Sampling theory and wave propagation. In *Issues in Acoustic Signal/Image Processing and Recognition*, Chen, C.H. (ed.). Springer Verlag, Berlin, Germany, NATO ASI Series, Vol. I, pp. 233–253.

Muck-Weymann ME, Albrecht H-P, Hager D, Hiller D, Hornstein OP, and Bauer RD (1996) Respiratory-dependent laser-Doppler flux motion in different skin areas and its meaning to autonomic nervous control of the vessels of the skin. *Microvasc Res* 52: 69–78.

Muck-Weymann ME, Albrecht H-P, Hiller D, Hornstein OP, and Bauer RD (1994) Breath-dependent laser-Doppler-fluxmotion in skin. *VASA* 4: 299–304.

Nilsson H and Aalkjaer C (2003) Vasomotion: Mechanisms and physiological importance. *Mol. Interven* 3(2): 79–89.

Ohkubo C, Gmitrov J, Xu S, and Nakayama E (1997) *Vasodilator Effects of Static Magnetic Fields on Cutaneous Microcirculation Under Increased Vascular Tone in the Rabbits*. Nihon-Igakukan, Tokyo, Japan, pp. 233–253.

Ohkubo C and Xu S (1997) Acute effects of static magnetic fields on cutaneous microcirculation in rabbits. *In Vivo* 11(3): 221–225.

Okano H and Ohkubo C (1998) *Vasoconstrcting Effects of Static Magnetic Fields on Cutaneous Microcirculation under Decreased Vascular Tone in the Rabbit*. Nihon-Igakukan, Tokyo, Japan, pp. 133–148.

Oppenheim AV and Schaefer RW (1975) *Digital Signal Processing*. Prentice Hall, Upper Saddle River, NJ.

Parthimos D, Edwards DH, and Griffith TM (1996) Comparison of chaotic and sinusoidal vasomotion in the regulation of microvascular flow. *Cardiovasc Res* 31: 388–399.

Pilla A, Fitzsimmons R, Muehsam D, Wu J, Rohde C, and Casper D (2011) Electromagnetic fields as first messenger in biological signaling: Application to calmodulin-dependent signaling in tissue repair. *Biochim Biophys Acta* 1810: 1236–1245.

Pilla AA (2002) Low-intensity electromagnetic and mechanical modulation of bone growth and repair: Are they equivalent? *J Orthop Sci* 7: 420–428.

Pilla AA and Markov MS (1994) Bioeffects of weak electromagnetic fields. *Rev Environ Health* 10: 155–169.

Peng H, Matchkov V, Ivarsen A, Aalkjaer C, and Nilsson H (2001) Hypothesis for the initiation of vasomotion. *Circ Res* 88: 810–815.

Pradhan RK and Chakravarthy VS (2007) A computational model that links non-periodic vasomotion to enhanced oxygenation in skeletal muscle. *Math Biosci* 209: 486–499.

Pradhan RK and Chakravarthy VS (2011) Informational dynamics of vasomotion in microvascular networks: A review. *Acta Physiol* (Oxf) 201: 193–218.

Pradhan RK, Chakravarthy VS, and Prabhakar A (2007) Effect of chaotic vasomotion in skeletal muscle on tissue oxygenation. *Microvasc Res* 74: 51–64.

Schuster A, Beny JL, and Meister JJ (2003) Modelling the electrophysiological endothelial cell response to bradykinin. *Eur Biophys J* 32: 370–380.

Schuster A, Lamboley M, Grange C, Oishi H, Beny JL, Stergiopulos N, and Meister JJ (2004) Calcium dynamics and vasomotion in rat mesenteric arteries. *J Cardiovasc Pharmacol* 43: 539–548.

Schuster A, Oishi H, Beny JL, Stergiopulos N, Meister JJ (2001) Simultaneous arterial calcium dynamics and diameter measurements: Application to myoendothelial communication. *Am J Physiol Heart Circ Physiol* 280: H1088–H1096.

Stefanovska A (1992) Self-organization of biological systems influenced by electrical current. Thesis, Faculty of Electrical Engineering, University of Ljubljana, Slovenia, Vol. 4, pp. 457–478.

Stefanovska A. and Kroselj P (1997) Correlation integral and frequency analysis of cardiovascular functions. *Open Sys Inform Dyn* 4: 457–478.

Stefanovska A, Sheppard LW, Stankovski T, and McClintock PV (2011) Reproducibility of LDF blood flow measurements: Dynamical characterization versus averaging. *Microvasc Res* 82: 274–276.

Traikov L, Ushiyama A, Lawlor G, Sasaki R, and Ohkubo C (2004a) Frequency dependence of ELF-EMF action on blood vessel diameter oscillations in the cutaneous microcirculation in mice in vivo. In *Proceedings of Ninth National Conference on Biomedical Physics and Engineering*, Sofia, Bulgaria, Vol. 1, pp. 109–116.

Traikov L, Ushiyama A, Lawlor G, Sasaki R, and Ohkubo C (2004b) Acute effect of ELF-EMF Exposure on subcutaneous microvascular diameter in mice. In *Biological Effects of EMFS-Proceedings of International Workshop*, Kos, Greece, Vol. 2, pp. 741–749.

Traikov L, Ushiyama A, Lawlor G, Sasaki R, and Ohkubo C (2005) Subcutaneous arteriolar Vasomotion changes during and after ELF-EMF exposure in mice in vivo. *Environmentalist* 25: 93–101.

Ursino M, Colantuoni A, and Bertuglia S (1998) Vasomotion and blood flow regulation in hamster skeletal muscle microcirculation: A theoretical and experimental study. *Microvasc Res* 56: 233–252.

Ushiyama A, Yamada S, and Ohkubo C (2004) Microcirculatory parameters measured in subcutaneous tissue of the mouse using a novel dorsal skinfold chamber. *Microvasc Res* 68: 147–152.

Xiu RJ and Intaglietta M (1985) Microvascular vasomotion: I. Long-term observation and computer analysis of the vasomotion. *Zhonghua Yi Xue Za Zhi* 65: 129–135.

Xiu RJ and Intaglietta M (1986) Computer analysis of the microvascular vasomotion. *Chin Med J* 99: 351–360.

Xu S, Okano H, Nakajima M, Hatano N, Tomita N, and Ikada Y (2013) Static magnetic field effects on impaired peripheral vasomotion in conscious rats. *Eviden Complemen Altern Med* 2013(1): 1–6, Article ID 746968.

Xu S, Okano H, and Ohkubo C (2000) Acute effects of whole-body exposure to static magnetic fields and 50-Hz electromagnetic fields on muscle microcirculation in anesthetized mice. *Bioelectrochemistry* 53(1): 127–135.

13 Cell Hydration as a Marker for Nonionizing Radiation

Sinerik Ayrapetyan, Naira Baghdasaryan, Yerazik Mikayelyan,
Sedrak Barseghyan, Varsik Martirosyan, Armenuhi Heqimyan,
Lilia Narinyan, and Anna Nikoghosyan

CONTENTS

13.1 INTRODUCTION

As the biological effects of nonionizing radiation (NIR) can also be recorded at intensities much less than their thermal threshold (Devyatkov, 1973; Adey, 1981; Markov, 2004; Kaczmarek, 2006; Rehman et al., 2007; Gapeyev et al., 2009; Grigoryev, 2012), these effects cannot be explained by hypothesis from the point of the classical thermodynamic characteristics of NIR (Foster, 2006; Binhi and Rubin, 2007). Because of this, the elucidation of the mechanism of biological effects of NIR on cells and organisms still remains one of the core problems of modern magnetobiology. It is clear that target(s) of such weak signals could have quantum mechanical nature, since the theories based on classical thermodynamics fail to explain this phenomenon (see reviews by Belyaev, 2005, 2012; Gapeyev et al., 2009; Binhi, 2012; Blank and Goodman, 2012; Gapeyev et al., 2013). Thus, the issue of the primary target(s) for biological effects of NIR is still open to further studies. Some researchers try to explain these effects in terms of quantum mechanical approach, considering different biochemical reactions with participation of uncoupled electrons, such as the electron transferring from cytochrome C to cytochrome oxidize, oxidation of malonic acid, Na^+/K^+-ATPase, and others, as key targets (Belyaev, 2012; Binhi, 2012; Blank and Goodman, 2012). However, high NIR sensitivity of forming and breaking process of hydrogen bonds occurring in cell bathing medium and during collective dynamics of intracellular water molecules makes the essential role of individual biochemical reaction of NIR sensitivity in determination of cell metabolic activity less reliable.

Therefore, the so-called water hypothesis for explaining the biological effect of NIR seems more reliable in these aspects (Szent-Gyorgyi, 1968; Markov et al., 1975; Klassen, 1982; Lednev, 1991; Ayrapetyan et al., 1994; Markov, 2009; Chaplin, 2014). According to this hypothesis, the valence angle in water molecules between O–H bonds, which determines its dissociation, is highly sensitive to different environmental factors. This serves as a key target for NIR impacts on extracellular and intracellular aqua medium, which serves as a main medium for metabolic process.

Thus, high NIR sensitivity of water physicochemical properties determined by its molecule dissociation on the one hand and the cell membrane hyperpermeability for water molecules on the other hand makes cell hydration extremely sensitive to any factors inflicting on both intra- and extracellular water structure changes. Since the cell hydration is determining the cell metabolic activity via *folding–unfolding* mechanisms of intracellular macromolecules (Parsegian et al., 2000) and cell surface-dependent changes in a number of functionally active protein molecules in plasma membrane (Ayrapetyan, 1980; Ayrapetyan et al., 1984; Parton and Simon, 2007), the NIR-induced changes of cell hydration are suggested as a gate for metabolic cascade through which its biological effects on cells and organisms are realized.

Although in literature, there is a great number of data on NIR effects on physicochemical properties of water (Chaplin, 2006), the cell bathing medium is not widely recognized as a primary target for biological effects of NIR (see WHO-Bioinitiative, 2012). Therefore, in our previous experiments, the impacts of different types of NIR on water physicochemical properties in dependency of different environmental factors serve as the subject for study (Ayrapetyan et al., 1994; Ayrapetyan, 2006; Ayrapetyan et al., 2009; Baghdasaryan et al., 2012a,b, 2013).

13.2 NIR EFFECTS ON PHYSICOCHEMICAL PROPERTIES OF WATER

From the point of present knowledge on water structure, electromagnetic field (EMF) can modify the water structure by two pathways: (1) by changing the valence angle in water molecules and (2) by mechanical vibration (MV) of dipole molecules of water. Therefore, it was suggested that static magnetic field (SMF) effect on physicochemical properties of water could imitate the valence angle changes, while the effect of MV on it—the mechanical vibration of dipole molecules of water. The degree of water molecule dissociation (specific electrical conductivity [SEC]), heat fusion, pH, gas solubility, and hydrogen peroxide (H_2O_2) formation in it were used as markers for water structure changes (Ayrapetyan, 2006). Obtained data indicate that as in the case of NIR effects on cells and organisms (Adey, 1981; Liboff, 1985), their effects on physicochemical properties of water also have the intensity- and frequency *window*-dependent characters. The study of SMF, MV, and extremely low-frequency (ELF) EMF at frequencies less than 100 Hz on the decrease of conductivity of water solution, which is accompanied by the increase of pH as a result of decreasing of CO_2 solubility, has been carried out (Stepanyan and Ayrapetyan, 1999; Akopyan and Ayrapetyan, 2005; Ayrapetyan et al., 2006). By more detailed investigation of frequency-dependent effect of pulsing magnetic fields in the range of 0–10 Hz on physicochemical properties of water, 4 and 8 Hz *windows* were observed at which MV and ELF EMF have more pronounced modulation effects on it (Baghdasaryan et al., 2012a,b, 2013). It is known that NIR-induced changes in water molecule ionization can modulate the process of the reactive oxygen species (ROS) formation and change its level in water (Klassen, 1982; Domrachev et al., 1992, 1993; Bruskov et al., 2003; Gudkova et al., 2005). Therefore, the H_2O_2, one of the ROS to have the longest lifetime, was used as one of the sensitive markers for NIR effects on physicochemical properties of water (Ayrapetyan et al., 2009). As the effects of NIR-treated physiological solution (PS) on isolated neurons and heart muscle of snail were studied in our experiments, the study of snail's PS serves as a basic solution in the experiments of studying NIR effect on physicochemical properties of water solution.

Figure 13.1 shows that 10 min treated PS by 1–10 Hz, 30 dB MV has frequency-dependent decreasing effect on H_2O_2 contents in PS, which has more pronounced effect at 4 and 8 Hz frequencies. The same frequency *windows* were observed in the study of ELF EMF effect on PS (Baghdasaryan, 2012b). The comparative study of 4 and 8 Hz ELF EMF, MV, and microwave (MW) effect on heat fusion (marker for water density of PS) and H_2O_2 contents in it at 4°C and 18°C in different background radiations and illuminations has shown that NIR effects on PS depend not only on the characteristics of environmental medium (temperature, gas composition, background

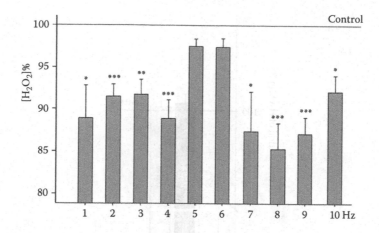

FIGURE 13.1 The frequency-dependent effects of 30 dB MV 10 min treatment of PS on its H_2O_2 contents. (From Baghdasaryan, N.S. et al., *Electromagnet. Biol. Med.*, 31(4), 310, 2012b, Figure 3.)

radiation, and illumination) in which experiments were performed but also on the characteristics of environmental medium at which PS samples were held before the experiments.

Figure 13.2 presents the data on the effect of 10 min incubation in normal background radiation (*NBGR*) and low background radiation (*LBGR*) in light and dark conditions on H_2O_2 content in non-treated (a) and 8 Hz MV-treated (b) PS samples at 18°C (Baghdasaryan et al., 2012b, 2013). As H_2O_2 content in PS depends on both background radiation and illumination (Figure 13.2a), it has been predicted that 8 Hz MV-induced changes of H_2O_2 content also depend on the background radiation and illumination of experimental medium (Figure 13.2b). Earlier by our work, the *aging*-dependent decrease of water SEC that leads to the decrease of its NIR sensitivity has been shown (Stepanyan

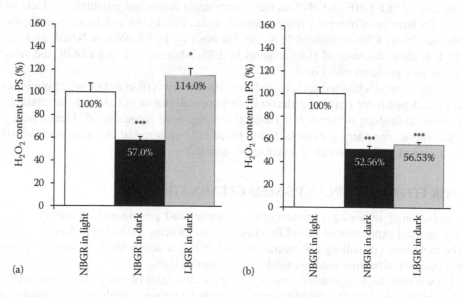

FIGURE 13.2 The effect of 10 min incubation in *NBGR in light* (white), *NBGR in dark* (black), and *LBGR in dark* (gray) conditions on H_2O_2 content in nontreated (a) and 8 Hz MV-treated (b) PS samples at 18°C. Bars indicate the mean of 10 independent measurements; vertical bars represent the standard error of mean (SEM), *$p < 0.05$, **$p < 0.01$ as compared with control (*NBGR in light*). (From Baghdasaryan, N.S. et al., *Electromagnet. Biol. Med.*, 31(4), 310, 2012b, Figure 5.)

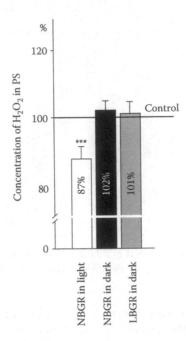

FIGURE 13.3   Eight hertz–modulated MW effect on H_2O_2 content in PS at *NBGR in light*, *NBGR in dark*, and *LBGR in dark* media at 18°C temperature condition. Columns indicate the mean of 10 independent measurements; vertical bars represent the standard error of mean, ***$p < 0.005$ as compared to control (*NBGR in light*).

and Ayrapetyan, 1999). These data allow us to suggest that the motion of water molecules, which is accompanied by the formation and breaking of hydrogen bonds, makes the water structure sensitive to NIR. In case of ELF EMF and MW, similar environment-dependent modulation effects on H_2O_2 contents in PS have been observed (Baghdasaryan et al., 2012b). As can be seen in Figure 13.3, 10 min-treated PS by 8 Hz-modulated MW with the intensity of 3.8 mW/g at NBGR and illumination at 18°C leads to decrease of H_2O_2 contents by 13%, while at dark and LBGR, the same MW radiation has no significant effect on it.

   Thus, on the basis of obtained data, it can be concluded that NIR effect on water physicochemical properties depends not only on different environmental factors but also on the characteristics of environmental medium in which experimental samples were preincubated. Therefore, it is suggested that without considering these facts and based only on physical characteristics of NIR, the characterization of NIR sensitivity of water seems unreliable.

13.3   NIR EFFECTS ON PLANT'S SEED GERMINATION

Plant's seeds, being in dormant (metabolically inactive) and germination (metabolically active) states, are an ideal experimental model for checking our aforementioned hypothesis, according to which the metabolic controlling cell hydration can serve as a sensor through which NIR-induced structural changes of bathing medium modulate cell metabolism.

   It is known that the temperature sensitivity (Q_{10} coefficient) of the process rate serves as a marker for detecting whether it has diffusion or metabolic nature, which is determined by Fick's law ($Q_{10} < 2$) or Arrhenius's equation ($Q_{10} > 2$), respectively. The time-dependent temperature sensitivity of seed hydration incubated in distilled water (DW) in cold (4°C) and at room (20°C) temperature is shown in Figure 13.4. The consequential changes in wet weights during seed incubation in control and experimental medium were expressed as a percentage of their initial

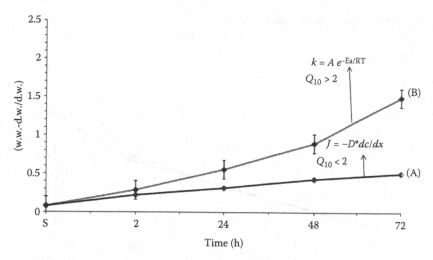

Seed hydration in nontreated (C = sham) cold (4°C) and warm (20°C) DW

FIGURE 13.4 Time-dependent seed hydration in nontreated DW in cold (A) and at room (B) temperature. On abscissa, the time (in hours) of seed incubation and, on ordinates, the value of seed hydration (mg of H_2O for 1 mg of dry weight) are presented. In curve A, experimental data were obtained in cold medium and in curve B at room temperature. (Amyan, A.M. and Ayrapetyan, S.N., *Physiol. Chem. Phys. Med. NMR*, 36, 69, 2004, Figure 3.)

value before incubation. It can be seen that during incubation at both temperatures (4°C and 20°C), there were increases in water absorption by seeds (2, 24, 48, and 72 h incubation). But in the cold medium, the rate of water absorption was much less than in the warm one, and these differences had time-dependent increasing character. At the end of the first 2 h of incubation, when seeds were in dormant state, Q_{10} was 1.25 for water absorption, while at the end of 72 h of incubation at room temperature, when seeds were in germination state, Q_{10} was 2.13. From these data, it can be concluded that in germination state, only water absorption by seeds has a metabolic controlling character.

In order to find out whether NIR has impacts on metabolic controlling of seed hydration (72 h incubation), the effect of different NIR (MV, SMF, ELF EMF, and MW)-treated aqua media on barley seed hydrations in its metabolically active and dormant states (in cold and at room temperature) was studied (Amyan and Ayrapetyan, 2004, 2006).

The data represented in Figure 13.5 show that the sensitivity of seed hydration to ELF EMF treatment of seed bathing water medium appeared only after seed wreaking and germination periods, at the ends of 48 and 72 h of incubation in aqua medium at 20°C, while in dormant state (in cold of 4°C and at room temperature in the first 2 h of incubation), this sensitivity was absent. Similar picture was observed by the study of mechanical MV- and MW-treated solutions (Amyan and Ayrapetyan, 2004; Ayrapetyan, 2006).

The fact that seed hydration directly correlates with its germination potential can be seen in Figure 13.6, where 15 Hz EMF-treated aqua medium has pronounced activation effect on seed hydration as well as a strong activation effect on seed germination.

Thus, obtained data allow us to suggest that the metabolic driving of intracellular water dynamics, which takes place during anabolic and catabolic processes, is a more sensitive target for NIR than simple thermodynamic process such as osmotic gradient–driven water absorption in seeds in dormant state.

Thus, summarizing these data with the literature ones on NIR-induced changes of water physicochemical properties brings us to the following suggestions:

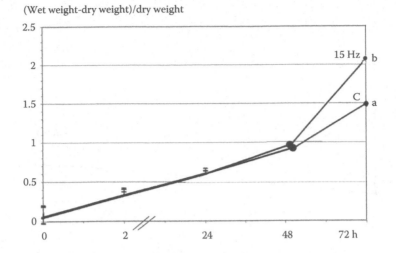

(Wet weight-dry weight)/dry weight

FIGURE 13.5 The effect of preliminary ELF EMF–treated DW on seed hydration during 72 h of incubation. (a) Seed hydration in cold (4°C) DW and (b) seed hydration in DW at room temperature (20°C). "C" means *control* sham treated.

(a)   (b)

FIGURE 13.6 Barley seed germination at the end of 72 h, in which seeds were 10 min preincubated in sham- (control) (a) and 15 Hz EMF-treated (b) DW.

1. Seed bathing aqua medium serves as a primary target for NIR effects on plant germination.
2. The metabolic driving intracellular collective water, continuously leading to the formation and breaking of hydrogen bonds, makes the cell hydration extra sensitive to NIR-induced water structural changes in bathing medium.
3. The metabolic controlling cell hydration can serve as an extra sensitive cellular marker for the biological effect of NIR on plants.

Similar conclusion brings us to the comparative study of the effects of direct NIR exposure on the culture medium of microbes and NIR-treated culture medium on microbe growth and developments.

13.4 NIR-TREATED CULTURE MEDIUM ON BACTERIAL GROWTH

It is known that microbes are extra sensitive to the changes of environmental medium, including the variations of NIR in medium (Trushin, 2003; Ilyin et al., 2010a,b; Segatore et al., 2012; Ahmed et al., 2013; Aslanimehr and Pahlevan, 2013; Nawrotek et al., 2014). Because of this, they are widely used as one of the sensitive markers for environmental control (Gromozova et al., 2011). Therefore, microorganisms are convenient objects for studying the effects of various stress factors including EMF. In the earlier study, it was also shown that ELF EMF and MV have frequency-dependent modulation effects on the growth and developments of *Escherichia coli* (Stepanyan and Ayrapetyan, 1999). However, the nature of both primary target and metabolic mechanism(s) that makes the bacteria extra sensitive to NIR was unclear. In order to elucidate the role of NIR-induced changes of cell bathing medium and direct exposure of NIR on microbes, proliferation and developments were studied (Martirosyan et al., 2013a,b).

The comparative study of ELF EMF and MV-treated medium and direct exposure of microbes has shown that these factors have frequency-dependent effect on bacterial growth. In Figures 13.7 and 13.8, the data on frequency-dependent effects of (2, 4, 6, 8, 10 Hz frequencies for 30 min) ELF EMF and MV-treated culture medium and culture on bacteria proliferation, respectively, are presented. It can be seen that 2 Hz ELF EMF (Figure 13.7) and MV stimulate bacterial growth (Figure 13.8). In the case of MV, the pretreatment of the culture media had about 7% stimulating effect, while in the case of EMF, the pretreatment had about 25.8% activation effect on the cell proliferation. It means that the nature of the stimulation at 2 Hz is different: probably the stimulation by MV was realized directly and nondirectly by the changes of the physicochemical properties of culture media, while the stimulation by EMF was realized only directly on bacteria itself (Figure 13.8). The treatment at 4 Hz EMF showed stimulation, while MV treatment has inhibition effects on cell proliferation. The pretreatment of the bacterial media by EMF had about 77.2% and by MV about 12.6% inhibiting effect on the cell proliferation compared to direct effect. It means that probably in both cases the effect of 4 Hz is realized directly on bacterial culture. MV at 6 Hz showed stimulating effect on cell proliferation, while ELF EMF had no significant effect on it. The treatment of EMF at 8 Hz (Figure 13.7) showed inhibiting

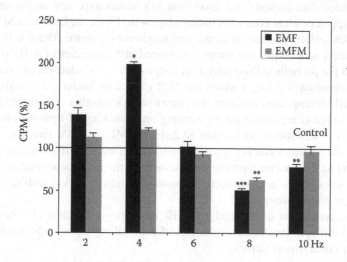

FIGURE 13.7 Frequency-dependent effect of ELF EMF on the bacterial growth calculated count per minute (CPM [%]). In the graph, the bacteria culture treated with culture media is marked as EMF, and beforehand treated culture media are marked as EMF-treated medium (EMFM). The incubation time after treatments was 18 h. Values in the graph are the means of these experiments standard deviation (±SD); at least three independent experiments performed in duplicate show similar results (From Martirosyan, V. et al., *Electromagnet. Biol. Med.* 32(3), 291, 2013b, Figure 7).

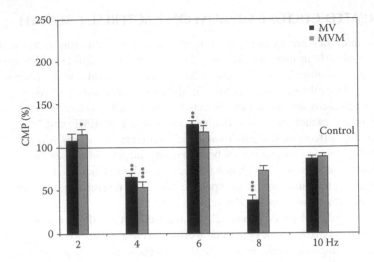

FIGURE 13.8 Frequency-dependent effect of MV treatment on the bacterial growth (CPM [%]). In the graph, the bacteria culture treated with culture media is marked as MV, and beforehand treated culture media are marked as MV-treated medium (MVM). The incubation time after treatments was 18 h. Values in the graph are the means of these experiments ±SD error. (From Martirosyan, V. et al., *Electromagnet. Biol. Med.* 32(1), 40, 2013a, Figure 7.)

effect on cell proliferation. The pretreatment of the bacterial media by EMF had about 13.1% stimulating effect on the cell proliferation compared to direct effect, while MV had about 34.3% (Figure 13.8). It means that the nature of the inhibition at 8 Hz is the same in both cases. Finally, EMF (Figure 13.7) and MV (Figure 13.8) at 10 Hz showed inhibiting effects on cell proliferation. The pretreatment of the bacterial media by EMF had about 18.6% and MV about 3.4% stimulating effect on the cell proliferation compared to direct effect.

Thus, the obtained data indicate that EMF and MV direct exposure on bacteria and bacteria media has frequency-dependent effects on bacterial growth. However, EMF and MV have different frequency-dependent effects in cases of direct and nondirect exposure. These differences between EMF and MV effects were expressed more pronounced at frequencies of 4 Hz (in case of direct exposure) and of 8 Hz (in both indirect and direct exposure). These data clearly indicate that there are minimum two pathways through which the NIR effects on bacteria are realized. One of the pathways is the cell bathing aqua medium structure changes; another is probably the intracellular water dynamics, which is accompanied by forming and breaking of hydrogen bonds (Orecchini et al., 2012a,b). As it was shown in Section 13.2, ELF EMF and MV have the same frequency *window* effect on SEC of water and H_2O_2 content in it. Therefore, the different frequency-dependent effects of EMF and MV on bacteria proliferation consist of the aforementioned suggestion on existence of two pathways (extra- and intracellular water medium changes) with different sensitivity through which their effects are realized.

Thus, the main conclusion of the study of NIR frequency-dependent effects on the bacterial growth is realized through the structural changes of cell bathing aqua medium and modulation of intracellular water collective dynamics.

13.5   NIR-TREATED PS EFFECT ON FUNCTIONAL ACTIVITY OF SNAIL ISOLATED NEURONS AND HEART MUSCLE CONTRACTILITY

It is known that giant neurons of mollusks serve as a classic model for the study of electrochemical properties of cell membrane because it can function in in vitro condition for hours. In order to estimate the possible biological effects of NIR-treated cell bathing medium on snail cells, its effects on isolated neuronal volume, potential-dependent and agonist-activated ionic membrane currents in isolated neurons, as well as on isolated heart muscle contractility were studied (Ayra-petyan et al., 1994, 2004, 2006; Hunanyan and Ayrapetyan, 2007).

As can be seen in Figure 13.9a, the replacement of isolated neuron bathing PS by PS 5 min pretreated 9.3 GHz MW (SAR = 50 kW/kg) at same temperature, the time-dependent changes of neuronal volume were observed: in the first 2–3 min, there was a shrinkage, and then in the following 5–7 min, it was more swollen than the initial one. It is worth to note that the equivalent PS preheated by infrared light radiation had no significant effect on cell volume. The similar picture was observed in case of incubation in PS 30 min pretreated 4 Hz–modulated 60 GHz MW with the intensity of SAR = 1.49 W/kg (Figure 13.9b). Microscopic picture in Fura-2 solution containing PS indicates that the cell shrinkage and swelling were accompanied by corresponding shrinking and swelling of neuronal nucleus. These data correspond with the literature on MW-induced increase of chromosomal density in cells (Belyaev, 2012). The studies of a Russian group from Institute of Cell Biophysics

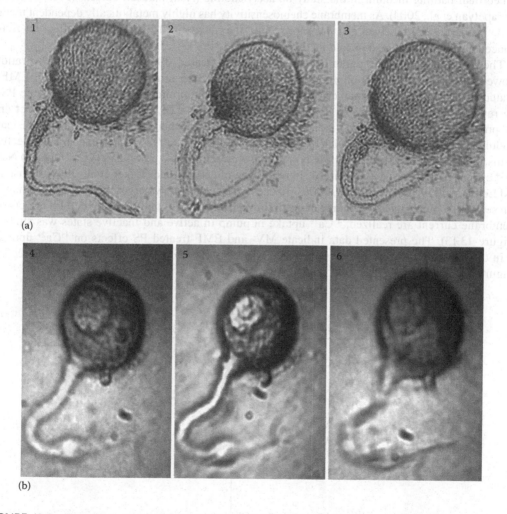

FIGURE 13.9 The effects of 9.3 GHz MW (SAR = 50 kW/kg) 5 min (a) and 4 Hz-modulated 60 GHz MW (SAR = 1.49 W/kg) 30 min (b) pretreated PS effects on snail neuronal volume. (a) In normal (sham-exposed) PS (1), after 2–3 min (2) and 5–7 min (3) of incubation in preliminary MW-treated PS (light microscopic picture). (b) In normal (sham-exposed) PS (4), after 2–3 min (5) and 5–7 min (6) of incubation in MW-treated PS (microscopic picture in Fura-2-containing PS).

at Pushchino have shown that high intensity of MW exposure leads to the generation of H_2O_2 in water, which plays an essential role in realization of the biological effects of higher intensity of MW (Gudkova et al., 2005; Gapeyev et al., 2009). In our laboratory, it was shown that the increase of H_2O_2 contents in PS in nanomolar concentrations took place even upon the effect of nonthermal intensity of MW (SAR = 1.49 W/kg; Ayrapetyan et al., 2009) that had a shrinkage effect on cells (Hunanyan and Ayrapetyan, 2007).

Earlier, it was shown that cell volume changes accompanied by the changes of number of functionally active proteins in the membranes have receptors, channels, and enzyme properties (Ayrapetyan, 1980; Ayrapetyan et al., 1984). It was previously shown that SMF exposure had dehydration effects on different cells of rats (Danielyan and Ayrapetyan, 1999). Therefore, NIR-treated PS effect on potential-activated ionic current and agonist-activated ionic current through the membrane in isolated neurons of snails was studied (Ayrapetyan et al., 2004; Hunanyan and Ayrapetyan, 2007). The *magnetized* PS depressing effect on inward (I_{Na}) and outward (I_K) currents can be seen in Figure 13.10. The same effect was observed when normal PS was replaced by 4 Hz-treated PS in neuronal bathing medium in the study on acetylcholine (Ach)-induced current (Figure 13.11a) (Ayrapetyan et al., 2004). As membrane chemosensitivity has highly metabolically dependent membrane properties (Ayrapetyan and Arvanov, 1979), the magnetized PS depression effect on Ach-induced current disappeared in cold medium (Figure 13.11b).

The study of the Na^+/K^+ pump dependency of NIR-treated PS effect on neuronal hydration showed that at normal functioning of Na^+/K^+ pump, both 15 min 4 Hz MV- and ELF EMF–treated PS have also dehydration effect on cells, while in pump in inactive state (in K^+-free PS), the reversal effect was observed (Figure 13.12). The data on ELF EMF–treated PS effect are not presented. These data clearly indicate that MV- and EMF-induced PS structure changes can modulate cell hydration by both Na^+/K^+ pump–dependent and Na^+/K^+ pump–independent mechanisms. It is known that there is a close correlation between Na^+/K^+ pump and ($^{45}Ca^{2+}$ uptake) Na^+/Ca^{2+} exchange that has an essential role in regulation of cell volume (Baker et al., 1969; Blaustein and Lederer, 1999; Takeuchi et al., 2006). To check the suggestion whether the Na^+/Ca^{2+} exchange can serve as a pump-independent mechanism through which MV- and EMF-treated PS effects on membrane current are realized, $^{45}Ca^{2+}$ uptake in pump in active and inactive states was studied (Figure 13.13). The presented data indicate MV- and EMF-treated PS effects on $^{45}Ca^{2+}$ uptake, as in the case of cell hydration, were reversed when Na^+/K^+ pump was inactivated by K^+-free PS (Figure 13.13b).

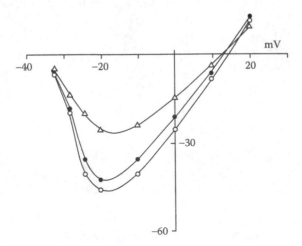

FIGURE 13.10 Current–voltage relations of the neuronal membrane in (O) control; (•) magnetized solution; (Δ) 30 mM sucrose-containing medium in voltage-clump experiments, where the clumping potential (E_c) is equal to the resting membrane potential ($E_r = -50$ mV).

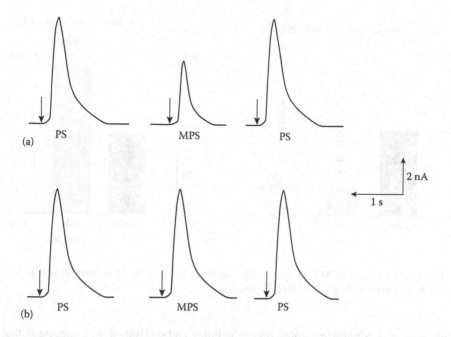

FIGURE 13.11 The effect of MPS on ACh-induced current in D-type neuron in voltage-clump experiments, where the clumping potential (E_c) is equal to the resting membrane potential ($E_r = -50$ mV) (a) at room temperature (23°C) and (b) in cold medium (12°C). The rows show the transit (30 ms) application of 10^{-4} M ACh-containing PS. The time of preincubation of neurons in tested physiological solution (PS and MPS), before the application of ACh was 3 min. The intervals between ACh applications were 5 min. (From Ayrapetyan, S.N. et al., *Bioelectromagnetics*, 25(5), 397, 2004, Figure 2.)

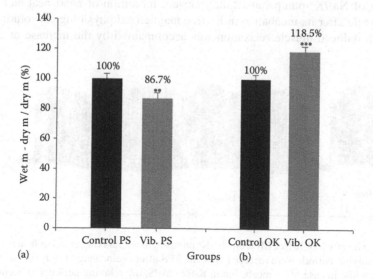

FIGURE 13.12 The effects of MV (15 min treatment) on ganglia's hydration in normal (a) and K-free PS (b). (Ayrapetyan, S.N., *Environ. J.*, 32, 210, 2012, Figure 5.)

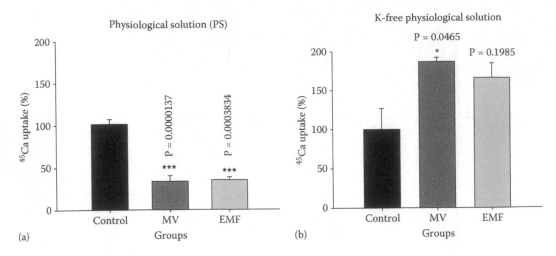

(a)

(b)

FIGURE 13.13 The effects of MV on ^{45}Ca uptake by neuronal ganglia in normal (a) and K-free PS (b). (Ayrapetyan, S.N., *Environ. J.*, 32, 210, 2012, Figure 6.)

As Ca^{2+} ions play a multifunctional role in cellular metabolism, it was suggested that SMF-induced effects on intracellular Ca^{2+} homeostasis can serve as one of the main metabolic pathways through which the biological effect of SMF is realized (Adey, 1981). Therefore, it is predicted that the membrane sensor, which is able to transfer SMF-induced water structural changes to cell metabolic cascades, can serve as a gate for the modulation of intracellular Ca^{2+} homeostasis. Previously, the cyclic nucleotide-dependent Na^+/Ca^{2+} exchanger was suggested as a candidate for this membrane sensor (Ayrapetyan et al., 1994; Ayrapetyan, 2001). As the heart muscle contractility strongly depends on intracellular Ca^{2+} concentration, it was chosen as a convenient model to check this hypothesis.

It is known that K^+ ion removal from extracellular medium leads to the inactivation of Na^+/K^+ pump, and the restoration of K^+ ions in the medium causes the reactivation of the pump, leading to hyperpolarization of the cell membrane (Thomas, 1972). As it can be seen in Figure 13.14, the reactivation of Na^+/K^+ pump caused the transient inhibition of heart beat and relaxation of heart muscle, while after its incubation in K^+-free magnetized physiological solution (MPS), the pump activation-induced muscle relaxation was accompanied by the increase of amplitudes of

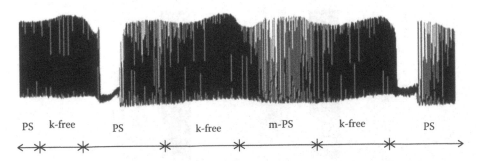

PS   k-free   PS   k-free   m-PS   k-free   PS

FIGURE 13.14 The effect of K-free MPS on Na^+/K^+ pump-induced relaxing effect of heart muscle. The transient silent and relaxing periods were recorded in normal PS after preincubation in K-free and sham-exposed K-free solutions, while in case of preincubation in K-free MPS, the relaxing period was accompanied by the contraction of the muscle with higher amplitudes than the initial one (a typical record of 1 of 10 studied heart muscles). (From Ayrapetyan, G.S. et al., *Bioelectromagnetics*, 26, 624, 2005, Figure 5.)

TABLE 13.1

MPS on Intracellular 45Ca^{2+} and Cyclic nucleotide in Heart Muscle and Neurons of Snail

Tissue Type	Exposure Medium	Intracellular $^{45}Ca^{2+}$ Content (cpm/mg Wet Weight) and Percentage of Their Change Compared to the Control	Cyclic Nucleotide Contents (pM/mg Wet Weight) and Percentage of Their Change Compared to the Control	
			cAMP	cGMP
Muscle	Sham exposed	1210±110 (100%)	0.53±0.04 (100%)	0.43±0.01 (100%)
		$T=6.399; P=0.03$	$T=3.532; P=0.24$	$T=3.554; P=0.24$
	SMF exposed	698±85 (~57%)	0.43±0.03 (82%)	0.61±0.08 (157.9%)
		$T=6.399; P=0.03$	$T=3.532; P=0.24$	$T=3.554; P=0.24$
Neuron	Sham exposed	100±10.07	100±9.6	100±11.6
		$P<0.05$ comp. to control		
	SMF exposed 4.6; 38 mT	74.33±9.22	79.1±9.5	164.9±41.6
		62.67±9.00	$P<0.02$	$P<0.05$
		$P<0.05$ comp. to control		

contractility without stopping the heart beat, and this effect was fully reversible by replacing the MPS by sham-exposed PS.

To estimate the role of Na$^+$/Ca^{2+} exchange in MPS-induced relaxing effect on muscles, the effect of MPS on $^{45}Ca^{2+}$ uptake by muscles on 10 dissected hearts was studied. In order to estimate the effect of Na$^+$/K$^+$ pump activity on intracellular Ca^{2+} contents, these series of experiments were performed in K$^+$-free medium, that is, when the pump was in inactive state. Data presented in Table 13.1 clearly demonstrate that $^{45}Ca^{2+}$ uptake by muscles in MPS was ~50% less than in nontreated PS (100%). These results are very similar to the early data obtained on neurons when MPS depressed $^{45}Ca^{2+}$ uptake by ~62% compared to sham exposed (Ayrapetyan et al., 1994).

To find out whether the MPS-induced relaxing effect on heart is also due to activation of cGMP-dependent Na$^+$/Ca^{2+} exchange, the effect of MPS on intracellular content of cyclic adenosine monophosphate (cAMP) and cyclic guanosine monophosphate (cGMP) was studied. As it can be seen from the data presented in Table 13.1, MPS had an elevating effect on intracellular cGMP content, which was accompanied by the decrease of intracellular level of cAMP. Thus, the comparative study of the impact of NIR-treated PS on neuronal and heart muscle activity of Helix snails and direct NIR exposure on their bathing in PS allows us to make the following conclusions:

1. NIR-treated PS and direct NIR exposure have synergic effect on neurons and heart muscles. They have dehydration effects on cells that are accompanied by decreasing nucleus volume.
2. NIR-treated PS and direct NIR exposure have depressive effect on both potential- and agonist-activated ionic channels in the membrane.
3. NIR-treated PS and direct NIR exposure activate cGMP-dependent Na$^+$/Ca^{2+} exchange in forward mode leading to the relaxing effect on heart muscle.

13.6 COMPARATIVE STUDY OF NIR-TREATED PS AND DIRECT EXPOSURE OF NIR ON MAMMALIAN TISSUES

In the present series of experiments, the comparative study of the effects of direct NIR exposure on animals and magnetized PS on cell proliferation and hydration, ^{3}H-ouabain binding with cell membrane, $^{45}Ca^{2+}$ uptake, and intracellular cyclic nucleotide contents in different tissues of rats was performed (Gharibova et al., 1996; Ayrapetyan et al., 2012; Heqimyan et al., 2012; Narinyan et al., 2012, 2013).

It is known that the deoxyribonucleic acid (DNA) hydration has a key role in regulation of its activity (Chaplin, 2006). In order to find out whether NIR is able to change DNA hydration, the frequency-dependent effect of MV on optical density of DNA-containing water solution was studied.

The DNA from calf thymus (Serva, United States) was used (2.5×10^{-5} M). The concentration was determined by absorption method (extinction coefficient at 260 nm–6400 M^{-1} sm^{-1}). The aquatic solution of DNA in 3 mL was treated by MV with the intensity of 90 dB for 30 min. Then the optical density of solutions was measured. The control solution was prepared in the same way and left without treatment for the same time until measuring. Ten experiments were carried out for each probe. As can be seen in Figure 13.15, the MV at frequencies of 4 and 10 Hz had depressing effect on optical density (at 260 nm) of DNA-containing solution by 4.2% ± 1.1% and 4.8% ± 1.2%, respectively, while nonsignificant changes in optical density were observed with 20 and 50 Hz frequencies of MV, which can be explained probably by the increase of hydrogenous bonds (between bases) formation. Therefore, from these data, it can be suggested that NIR can change DNA hydration and modulate its functional activity.

It has been shown in literature that the exposure to NIR in vitro and in vivo can lead to conformational and functional changes in the cell chromatin, depending on the intensity, doses, frequency bands, types, and parameters of modulation of NIR (Gapeyev et al., 2003, Gapeyev, 2011; Chemeris et al., 2004). Traditionally, the NIR effect on DNA reparation is explained by NIR-induced modulation of ROS in aqua medium (Gudkova et al., 2005). However, taking together the obtained data on NIR effect on DNA hydration with well-known dependency of DNA conformation and reparation on its hydration allows us to consider the DNA hydration as one of the essential pathways through which NIR effect on cell proliferation could be realized.

In order to find out whether NIR-induced changes of cell bathing structure could modulate the DNA functional activity, the ^{3}H-thymidine involvements in spleen DNA in vitro state, the effects of both direct SMF exposure spleen containing PS and magnetized PS were studied (Gharibova et al., 1996). As can be seen in Figure 13.16, the magnetized PS stimulates ^{3}H-thymidine involvements in DNA in spleen as in the case of direct SMF exposure of spleen bathing in PS. These data can be a strong evidence that cell bathing medium serves as a primary target for the biological effect of NIR in mammals as well. However, the nature of membrane sensors through which the signal

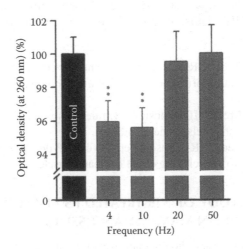

FIGURE 13.15 The frequency-dependent effects of MV (intensity 90 dB) on optical density of DNA-containing solution. (From Stepanyan, P.S. et al., *Radiat. Biol. Radioecol.*, 40, 435, 2000, Figure 3).

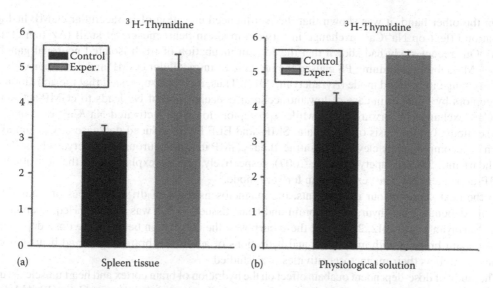

FIGURE 13.16 The effects of SMF (a) and MPS (b) on ^{3}H-thymidine involvement in DNA of spleen cells. (Ayrapetyan, S.N., *Environ. J.*, 32, 210, 2012, Figure 1.)

transduction of NIR-induced structural changes of cell bathing medium to cell hydration is realized stays unclear. Therefore, this question serves as a subject for our next series of study.

As was noted earlier (Section 13.4), the NIR-treated PS could modulate activities of Na$^+$/K$^+$ pump and Na$^+$/Ca^{2+} exchange that have crucial roles in regulation of cell volume in neuronal and muscle tissues. At present, it is well established that Na$^+$/K$^+$-adenosine triphosphatase (ATPase) (working molecules of Na$^+$/K$^+$ pump) in neurons and muscles of mammals has three catalytic isoforms ($\alpha1$, $\alpha2$, $\alpha3$) with different affinities to ouabain (specific inhibitor for Na$^+$/K$^+$-ATPase) and different roles in regulation of intracellular ionic homoeostasis (Juhaszova and Blaustein, 1982; Lucchesi and Sweadner, 1991; Blanco, 2005; Wymore et al., 2007; Blaustein et al., 2009). However, the individual roles of these isoforms in regulation of cell hydration as well as their NIR sensitivity are not studied in the literature available to us. It is known that $\alpha1$ and $\alpha2$ (low and middle affinity to ouabain, respectively) receptors are responsible for Na$^+$/K$^+$ pump activity; however, the real function of $\alpha3$ high-affinity receptors remains controversial (Juhaszova and Blaustein, 1982; Schwartz, 1989; Blaustein et al., 2009). But there is a great number of literature data that $\alpha1$, $\alpha2$, and $\alpha3$ isoforms also have a crucial role in the regulation of Na$^+$/Ca^{2+} exchange in cell (Adams et al., 1982; Blaustein and Lederer, 1999; Dipolo and Beaugé, 2006). Therefore, these problems were studied in our works (Heqimyan et al., 2012; Narinyan et al., 2012, 2013, 2014).

Traditionally, the pump-induced changes of Na$^+$ gradient on membrane are considered the main mechanism, through which the correlation between $\alpha1$ and $\alpha2$ isoforms and Na$^+$/Ca^{2+} exchange is explained (Baker et al., 1969; Blaustein et al., 2009). The nature of pathway through which $\alpha3$ isoforms could regulate the Na$^+$/Ca^{2+} exchange activity was recently elucidated by our work (Narinyan et al., 2014). It was shown that nM ouabain-induced formation of cAMP having activation effect on Ca^{2+} pump in membrane of endoplasmic reticulum (ER) (pushing these ions from cytoplasm into ER) brings to the activation of Na$^+$/Ca^{2+} exchange in reverse mode in cell membrane (Narinyan et al., 2014). The existence of three populations of ouabain receptors with different affinities in neuronal membranes was also demonstrated earlier in invertebrate neurons (Ayrapetyan et al., 1984). It was shown that ouabain at concentrations of $>10^{-7}$ M had an inhibitory effect, while at low concentrations, it activated ^{22}Na$^+$ efflux from the cells in exchange of Ca ions, which was accompanied by elevation of intracellular cAMP (Ayrapetyan et al., 1984; Saghyan et al., 1996).

On the other hand, it was shown that the NO-induced elevation of intracellular cGMP had an activation effect on Na^+/Ca^{2+} exchange in forward mode in heart muscle of snail (Azatian et al., 1998). Our recent study had shown that the 15 min incubation of fresh isolated neuronal ganglia in 10^{-11} M ouabain-containing PS caused the increase of intracellular cGMP and activation of Na^+/Ca^{2+} exchange in forward mode (Ayrapetyan, 2012). Thus, it can be suggested that the activation of α_3 receptors by ouabain in case of low intracellular concentration of Na leads to cGMP-activated Na^+/Ca^{2+} exchange in forward mode while α_2 receptors to cAMP-activated Na^+/Ca^{2+} exchange in reverse mode. On the basis of these data, SMF- and ELF EMF–induced decrease of Ca^{2+} uptake, which is accompanied by elevation of intracellular cGMP in snail neurons (Ayrapetyan et al., 1994) and heart muscles (Ayrapetyan et al., 2007), respectively, can be explained by the activation of cGMP-dependent Na^+/Ca^{2+} exchange in forward mode.

In the next series of our experiments, the magnetosensitivity of different types of catalytic α isoform–dependent cell hydration in brain and heart tissues of rats was studied (Heqimyan et al., 2012; Narinyan et al., 2012, 2013). For these purposes, the correlation between the dose-dependent [^{3}H]-ouabain binding with membrane and hydrations of adult rat's brain cortex and heart muscle tissues, as well as their magnetosensitivities, was studied.

The study of dose-dependent ouabain effect on the hydration of brain cortex and heart muscle tissues showed that even the intraperitoneal injection of extremely low concentrations of ouabain (10^{-11} M) had a strong modulation effects on cell hydration: in the cortex, it had dehydration (Figure 13.17a) while

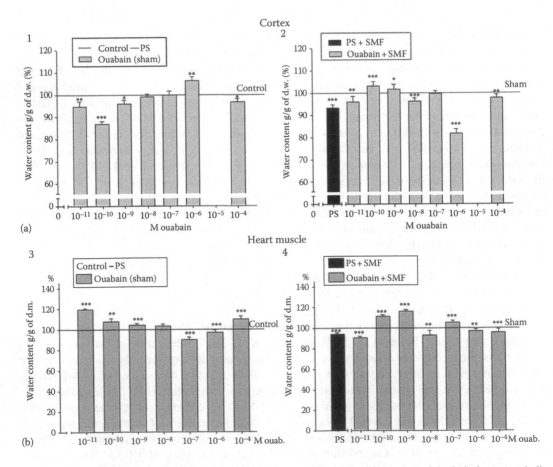

FIGURE 13.17 The effects of intraperitoneal injection of different concentration of ouabain in cortex (a-1) and heart muscle tissue (b-3) and the effects of SMF exposure on cell hydration in these tissues in intact and different concentrations of preliminary ouabain-injected rats (a-2, b-4).

in the heart, it had overhydration effects (Figure 13.17b). The facts that ouabain-induced activation of all the three receptors on cortex and heart tissues leads to different modulation effects on cell hydration indicate that each of these three types of receptors has its own specific metabolic pathways through which the modulation of cortex and muscle hydrations is realized. It is worth to note that although the ouabain binding with high-affinity receptors of cortex and heart tissues has the same character, their activations have different effects on the cell hydration in these tissues, which indicates that they have different nature of metabolic pathways involved in regulation of cell volume in neurons and muscles. We came to the similar conclusion by the study of magnetosensitive hydration of these both tissues.

As can be seen in Figure 13.17a-1 and b-3 in intact (control) animals, 10^{-11} M ouabain has dehydration in the brain cortex and hydration of heart muscle tissues. The SMF exposure in both tissues has dehydration effect. But in ouabain preinjected animals, the SMF exposure had opposite effects on cortex and heart tissue hydration: SMF exposure has hydration effect on both cortex and heart muscle tissue, while at 10^{-10} to 10^{-8} M concentration of ouabain, SMF had dehydration effect on cortex and hydration effect on heart muscle. Although the SMF exposure has different effects on cortex and muscle tissues, it has depressing effect on ouabain binding with α_3 receptors in both tissues (Figure 13.18a and b). It is extremely interesting that α_3 receptors are only magnetosensitive, while the ouabain bindings with α_2, α_1 receptor affinity are nonmagnetosensitive (Figure 13.18a and b). These data allow us to conclude that high-affinity ouabain receptors are a magnetosensor through which the modulation effect of SMF on cell hydration is realized.

However, the question is arising: whether the SMF-induced depression of ouabain binding with α_3 receptors is determined by the decrease of the number of functional active receptors in membrane (Ayrapetyan et al., 1984) or by the decrease of receptor affinity. It is obvious that by decreasing ouabain concentration in cell bathing medium, a number of mechanisms through which ouabain could modulate the cell volume will decrease. Therefore, it is suggested that the highest-affinity α_3 receptors could modulate the cell volume by activation of the mechanism that has the highest sensitivity to ouabain. As the functions of α_3 receptors closely correlate with the Na^+/Ca^{2+} exchange that has a key role in both intracellular Ca^{2+}-dependent receptor affinity and cell volume regulation, it can be considered as a potential mechanism responsible for determining the magnetosensitivity of cell hydration. To check this suggestion in the next series of our experiments—the dose-dependent ouabain effect on Ca^{2+} efflux and its magnetosensitivity on brain cortex and heart muscle were studied. Figure 13.19 demonstrates that the SMF exposure of animals has activation effect on Ca^{2+} efflux in both tissues as compared to its basal one in control. But the 10^{-11} M ouabain concentration has an inactivation effect on Ca^{2+} efflux in cells of brain cortex and activation effect on heart muscle. The SMF exposure on cortex at 10^{-11} M

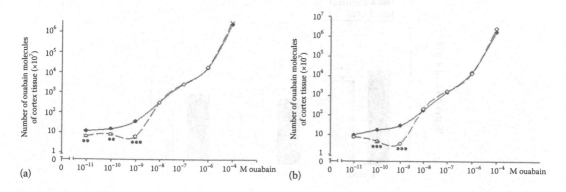

(a)

(b)

FIGURE 13.18 Dose-dependent ^{3}H-ouabain binding with cell membrane of adult rat's brain cortex (a) and heart muscle (b) tissues.

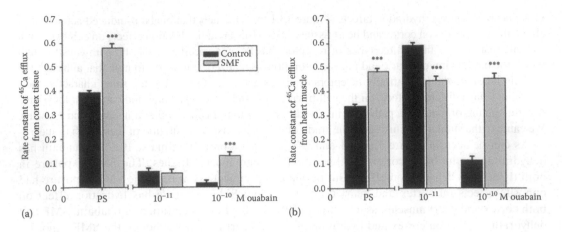

(a)   (b)

FIGURE 13.19 The effect of SMF on rate constant of $^{45}Ca^{2+}$ efflux from rat's brain cortex (a) and heart muscle (b) in ouabain-free, 10^{-11} and 10^{-10} M ouabain-containing PS. Error bars indicate the standard error of the mean (SEM) for three independent experiments. The symbol (***) indicates $P<0.001$. (From Ayrapetyan, S.N., *Environ. J.*, 32, 210, 2012, Figure 10.)

ouabain concentration had no significant effect on Ca^{2+} efflux in cells of brain cortex, while in the case of heart muscle, it had a strong activation effect on it (Figure 13.19). It seems strange that the SMF-induced decrease of Ca^{2+} efflux in the heart muscle at 10^{-11} M ouabain, which was accompanied by cell dehydration, was not accompanied by the decrease of ouabain binding with receptors (Figure 13.18). Probably, it can be explained by 10^{-11} M ouabain-induced modulation of the Ca^{2+} efflux by Ca^{2+}-pump or (and) intracellular Ca^{2+} storage changes. The final conclusion on it needs a more detailed investigation.

In order to elucidate the differences between direct MW radiation effect on brain and MW pretreated PS effect on cell hydration, the comparative study of direct MW exposure on brain and brain slices bathing in PS in vitro state as well as MW-treated PS injection animals on cortex hydration was carried out (Figure 13.20). As the presented data indicate in both cases, direct MW exposure on brain and brain slices leads to cell dehydration, while MV-treated PS intraperitoneal injection leads to the increase of cell cortex hydration. On the bases of these data, it can be suggested that both cell

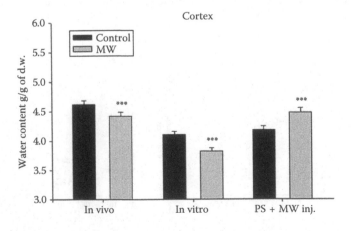

FIGURE 13.20 The effects of direct exposure on brain (in vivo) and brain slice bathing in PS (in vitro) and intraperitoneal injection of MW-treated PS on brain tissue hydration (PS + MW inj.).

bathing medium and intracellular water are targets through which nonthermal biological effects of MW on brain are realized.

In order to find out whether the NIR-induced hydration changes could have effect on behavioral activity of mammals including human beings, we have studied the NIR effect on pain threshold of rats by using *hot plate* method. By our previous study, the close correlation with neuronal hydration and its excitability and chemosensitivity was shown. The cell swelling leads to the increase of its sensitivity to physical and chemical factors, while the shrinkage has the opposite effect (Ayrapetyan, 1980; Ayrapetyan et al., 1984, 1988). On the bases of these data, the hypothesis that overhydration could bring to overexcitation serving as a nociceptive signal, which is transferred into the central neuronal system and generates pain sensation, was suggested (Ayrapetyan, 1998). By studying the correlation between cell hydration and pain threshold to *hot plate* and also the hypertonic solution (mannitol) of animal injection and the effect of anesthetic ketamine, the hypothesis that the tissue dehydration leads to increase of pain threshold was proved (Ayrapetyan et al., 2010; Heqimyan et al., 2012; Ayrapetyan et al., 2013). As demonstrated in Figure 13.21, both MV exposure and dehydration of mannitol injection–induced brain tissue lead to the increase of thermal pain threshold in rats (Figure 13.21).

Thus, the presented data on NIR effects on water structure as well as on different biological objects (plants, microbes, isolating neurons, and heart muscle of snails and rats) allow us to suggest the cell bathing medium as one of the primary targets and cell hydration as an extra sensitive and universal biomarker for biological effects of different types of NIR.

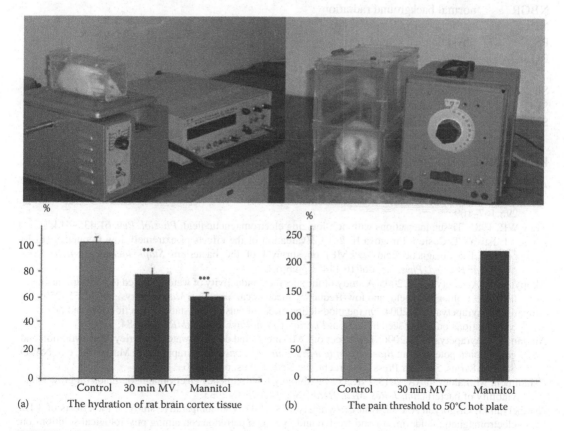

(a) The hydration of rat brain cortex tissue

(b) The pain threshold to 50°C hot plate

FIGURE 13.21 The effects of 4 Hz, 30 dB vibration and hypertonic solution (mannitol) injection on the hydration of brain cortex tissue (a) and pain threshold to hot plate (b).

ACKNOWLEDGMENTS

This work was supported by Armenian governmental grants, European Office of Aerospace Research and Development Research Grant Project No. A-803P and European Office of Aerospace Research and Development Research Grant Project No. 1592P. We express our gratitude to Anna Sargsyan and Ani Gyurjinyan from UNESCO Chair in Life Sciences for secretarial work.

ABBREVIATIONS

Ach	acetylcholine
ATP	adenosine triphosphate
cAMP	cyclic adenosine monophosphate
cGMP	cyclic guanosine monophosphate
CPM	calculated count per minute
dB	decibel
DNA	deoxyribonucleic acid
ELF EMF	extremely low-frequency electromagnetic field
EMFM	EMF-treated medium
ER	endoplasmic reticulum
LBGR	low background radiation
MPS	magnetized physiological solution
MV	mechanical vibration
MW	microwave
NBGR	normal background radiation
NIR	nonionizing radiation
PS	physiological solution
ROS	reactive oxygen spaces
SAR	specific absorption rate
SD	standard deviation
SEC	specific electrical conductance
SMF	static magnetic field

REFERENCES

Adams RJ, Schwartz A, Grupp G, Grupp IL, Lee S-W, Wallick ET, Powell T, Twist VW, Gathiram P. 1982. High-affinity ouabain binding site and low dose positive inotropic effect in rat myocardium. *Nature* 296:167–169.

Adey WR. 1981. Tissue interactions with non-ionizing electromagnetic field. *Physiol. Rev.* 61:435–514.

Ahmed I, Istivan T, Cosic I, Pirogova E. 2013. Evaluation of the effects of extremely low frequency (ELF) pulsed electromagnetic fields (PEMF) on survival of the bacterium *Staphylococcus aureus*. *EPJ Nonlinear Biomed. Phys.* 1:5. doi:10.1140/epjnbp12.

Akopyan SN, Ayrapetyan SN. 2005. A study of the specific conductivity of water exposed to constant magnetic field, electromagnetic field, and low-frequency mechanical vibration. *Mol. Biophys.* 50(2):255–259.

Amyan AM, Ayrapetyan SN. 2004. On the modulation effect of pulsing and static magnetic fields and mechanical vibrations on barley seed hydration. *Physiol. Chem. Phys. Med. NMR* 36:69–84.

Amyan AM, Ayrapetyan SN. 2006. The effect of EMF-pretreated distilled water on barley seed hydration and germination potential. In: *Bioelectromagnetics: Current Concepts*. Ayrapetyan S, Markov M, eds. NATO Science Series. Springer Press, Dordrecht, the Netherlands, pp. 65–86.

Aslanimehr M, Pahlevan AA. 2013. Effects of extremely low frequency electromagnetic fields on growth and viability of bacteria. *Int. J. Res. Med. Health Sci.* 1(2):2307–2083.

Ayrapetyan G, Grigoryan A, Dadasyan E, Ayrapetyan S. 2007. The comparative study of the effects of 4 Hz electromagnetic fields, infrasound-treated and hydrogen peroxide containing physiological solutions on Na pump-induced inhibition of heart muscle contractility. *The Environmentalist* 27:483–488.

Ayrapetyan GS, Papanyan AV, Hayrapetyan HV, Ayrapetyan SN. 2005. Metabolic pathway of magnetized fluid-induced relaxation effects on heart muscle. *Bioelectromagnetics* 26:624–630.

Ayrapetyan SN. 1980. On the physiological significance of pump induced cell volume changes. *Adv. Physiol. Sci.* 23:67–82.

Ayrapetyan SN. 1998. The application of the theory of metabolic regulation to pain. In: *Pain Mechanism and Management*. Ayrapeyan SN, Apkarian AV, eds. IOS Press, Amsterdam, the Netherlands, pp. 3–14.

Ayrapetyan SN. 2001. Na-K pump and Na:Ca exchanger as metabolic regulators and sensors for extraweak signals in neuromembrane. In: *Modern Problems of Cellular and Molecular Biophysics*. Ayrapetyan SN, North ACT, eds., Noyan Tapan, Yerevan, pp. 31–57.

Ayrapetyan SN. 2006. Cell aqua medium as a preliminary target for the effect of electromagnetic fields. In: *Bioelectromagnetics: Current Concepts*. Ayrapetyan S, Markov M, eds. NATO science series. Springer Press, Dordrecht, the Netherlands, pp. 31–64.

Ayrapetyan SN. 2012. Cell hydration as a universal marker for detection of environmental pollution. *Environ. J.* 32:210–221.

Ayrapetyan SN, Amyan AM, Ayrapetyan GS. 2006. The effect of static magnetic fields, low frequency electromagnetic fields and mechanical vibration on some physiochemical properties of water. In: *Water in Cell Biology*. Pollack G, Cameron I, Wheatley D, eds. Springer Press, Dordrecht, the Netherlands, pp. 151–164.

Ayrapetyan SN, Arvanov VL. 1979. On the mechanism of the electrogenic sodium pump dependence of membrane chemosensitivity. *Comp. Biochem. Physiol.* 64(A):601–604.

Ayrapetyan SN, Avanesian AS, Avetisian TH, Majinian SB. 1994. Physiological effects of magnetic fields may be mediated through actions on the state of calcium ions in solution. In: *Biological Effects of Electric and Magnetic Fields*, Vol. 1. Carpenter D, Ayrapetyan S, eds., Academic Press, New York, pp. 181–192.

Ayrapetyan SN, Dadasyan E, Ayrapetyan G, Hayrapetyan H, Baghdasaryan N, Mikaelyan E. 2009. The nonthermal effect of weak intensity millimeter waves on physicochemical properties of water solutions. *Electromagnet. Biol. Med.* 28:331–341.

Ayrapetyan SN, Heqimyan A, Deghoyan A. 2013. Cell dehydration as a mechanism of ketamine analgesic and anesthetic effects. *J. Bioequiv. Bioavailab.* 5:136–141.

Ayrapetyan SN, Heqimyan A, Nikoghosyan A. 2012. Age-dependent brain tissue hydration, Ca exchange and their dose-dependent ouabain sensitivity. *Bioequival. Bioavailab.* 4:060–068.

Ayrapetyan SN, Hunanian ASh, Hakobyan SN. 2004. The 4 Hz EMF-treated physiological solution depress Ach-induced neuromembrane current. *Bioelectromagnetics* 25(5):397–399.

Ayrapetyan SN, Musheghyan G, Deghoyan A. 2010. The brain tissue dehydration as a mechanism of analgesic effect of hypertonic physiological solution in rats. *J. Int. Dental Med. Res.* 3:93–98.

Ayrapetyan SN, Rychkov GY, Suleymanyan MA. 1988. Effects of water flow on transmembrane ionic currents in neurons of Helix pomatia and in squid giant axon. *Comp. Biochem. Physiol.* 89:179–186.

Ayrapetyan SN, Suleymanyan MA, Sagian AA, Dadalyan SS. 1984. Autoregulation of electrogenic sodium pump. *Cell. Mol. Neurobiol.* 4:367–384.

Azatian KV, White AR, Walker RJ, Ayrapetyan SN. 1998. Cellular and molecular mechanisms of nitric oxide-induced heart muscle relaxation. *Gen. Pharmacol.* 30:543–553.

Baghdasaryan NS, Mikayelyan YR, Barseghyan SV, Dadasyan EH, Ayrapetyan SN. 2012a. The density-dependency of dark and low background radiation effects on water and water solution properties. *Electromagnet. Biol. Med.* 31(1):87–100.

Baghdasaryan NS, Mikayelyan YR, Barseghyan S, Dadasyan EH, Ayrapetyan SN. 2012b. The modulating impact of illumination and background radiation on 8 Hz-induced infrasound effect on physicochemical properties of physiological solution. *Electromagnet. Biol. Med.* 31(4):310–319.

Baghdasaryan NS, Mikayelyan YR, Nikoghosyan AK, Ayrapetyan SN. 2013. The impact of background radiation, illumination and temperature on EMF-induced changes of aqua medium properties. *Electromagnet. Biol. Med.* 32(3):390–400.

Baker PF, Blaustein MP, Hodgkin AL, Steinhardt SA. 1969. The influence of Ca on Na efflux in squid axons. *J. Physiol.* 200:431–458.

Belyaev A. 2005. Non-thermal biological effects of microwaves. *Microwave Rev.* 11:13–29.

Belyaev A. 2012. *Evidence for Disruption by Modulation Role of Physical and Biological Variables in Bioeffects of Non-Thermal Microwaves for Reproducibility, Cancer Risk and Safety Standards*. BioInitiative Working Group 15.

Binhi VN. 2012. Two types of magnetic biological effects: Individual and batch effects. *Biophysics* 57:237–243.

Binhi VN, Rubin AB. 2007. The kT paradox and possible solutions. *Electromagnet. Biol. Med.* 26:45–62.

BioInitiative Report WHO: Working Group 2012. *A Rationale for Biologically-Based Public Exposure Standards for Electromagnetic Radiation*, Sage C, Carpenter DO, eds.

Blanco G. 2005. The Na/K-ATPase and its isozymes: What we have learned using the baculovirus expression system. *Front Biosci.* 10:2397–2411.

Blank M, Goodman RM. 2012. Electromagnetic fields and health: DNA-based dosimetry. *Electromagn. Biol. Med.* 31(4):243–249.

Blaustein MP, Lederer WJ. 1999. Na$^+$/Ca^{2+} exchange. Its physiological implications. *Physiol. Rev.* 79:763–854.

Blaustein MP, Mordecai P, Blaustein JZ, Hamlyn JM. 2009. The pump, the exchanger, and endogenous ouabain. *Hypertension* 53:291–298.

Bruskov VI, Chernikov AV, Gudkov SV, Masalimov ZHK. 2003. Activation of reducing properties of anions in sea water under the action of heat. *Biofizika* 48:1022–1029.

Chaplin MF. 2006. Information exchange within intracellular water. In *Water and the Cell*. Pollack, G. et al. eds. Springer, Netherlands, pp. 113–123.

Chemeris NK, Gapeyev AB, Sirota NP, Gudkova OYu, Kornienko NV, Tankanag AV, Konovalov IV, Buzoverya ME, Suvorov VG, Logunov VA. 2004. DNA damage in frog erythrocytes after in vitro exposure to a high peak-power pulsed electromagnetic field. *Mutat. Res.* 558:27–34.

Danielian AA, Ayrapetyan SN. 1999. Changes of hydration of rats' tissues after in vivo exposure to 0.2 Tesla steady magnetic field. *Bioelectromagnetics* 20(2):123–128.

Devyatkov ND. 1973. Effect of a SHF (mm-band) radiation on biological objects. *Uspekhi Fizicheskikh Nauk.* 110:453–454 (in Russian).

Dipolo R, Beaugé L. 2006. Na$^+$/Ca^{2+} exchanger: Influence of metabolic regulation on ion carrier interaction. *Physiol. Rev.* 86:155–203.

Domrachev GA, Rodigin YuL, Selivanovsky DA. 1992. Role of sound and liquid water as dynamically unstable polymeric system in mechano-chemically activated processes of oxygen production on Earth. *J. Phys. Chem.* 66:851–855.

Domrachev GA, Rodigin YuL, Selivanovsky DA. 1993. Mechano-chemically activated water dissociation in a liquid phase. *Proc. Russ. Acad. Sci.* 329:258–265.

Foster KR. 2006. The mechanisms paradox. In: *Biomagnetics*. Ayrapetyan SN, Markov MS, eds. Springer, Dordrecht, the Netherlands, pp. 17–29.

Gapeyev AB. 2011. The role of fatty acids in realization of anti-tumor effects of extremely high-frequency electromagnetic radiation. In: *2011 21st International Crimean Conference on Microwave and Telecommunication Technology (CriMiCo)*, Crimea.

Gapeyev AB, Lushnikov KV, Shumilina JuV, Sirota NP, Sadovnikov VB, Chemeris NK. 2003. Effects of low intensity extremely high frequency electromagnetic radiation on chromatin structure of lymphoid cells in vivo and in vitro. *Radiat. Biol. Radioecol.* 43:87–92 (in Russian).

Gapeyev AB, Mikhailik EN, Chemeris NK. 2009. Features of anti-inflammatory effects of modulated extremely high-frequency electromagnetic radiation. *Bioelectromagnetics* 30:454–461.

Gapeyev AB, Kulagina T, Alexander V. 2013. Exposure of tumor-bearing mice to extremely high-frequency electromagnetic radiation modifies the composition of fatty acids in thymocytes and tumor tissue. *Int. J. Radiat. Biol.* 89:602–610.

Gharibova LS, Avetisian TH, Ayrapetyan VE, Ayrapetyan SN. 1996. Effect of PMF on ^{45}Ca influx in excitable and unexcitable cells and proliferative activity of rat spleen cells. *Radioecology* N5:718–721.

Grigoryev Y. 2012. Evidence for effects on the immune system supplement. Immune system and EMF RF. In: BioInitiative Working Group 8. Bioinitiative Report WHO 2012. Sage C, Carpenter D, eds. www.bioinitiative.org.

Gromozova EN, Voychuk SI, Zelena LB, Gretskey IA. 2011. Microorganisms as a model system for studying the biological effects of electromagnetic non-ionizing radiation. Safety Engineering UDC 613.648.2:681.586. doi:10.7562/SE2012.2.02.06.

Gudkova OYu, Gudkov SV, Gapeev AB, Bruskov VI, Rubanik AV, Chemeris NK. 2005. The study of the mechanisms of formation of reactive oxygen species in aqueous solutions on exposure to high peak-power pulsed electromagnetic radiation of extremely high frequencies. *Biofizica* 50:773–779 (in Russian).

Heqimyan A, Narinyan L, Nikoghosyan A, Deghoyan A, Yeganyan A. 2012. Age dependency of high affinity ouabain receptors and their magneto sensitivity. *The Environmentalist* 32:228–235.

Hunanyan ASh, Ayrapetyan SN. 2007. The dose-dependent effect of hydrogen peroxide on neuromembrane chemosensitivity. *Electromagnet. Biol. Med.* 26:225–233.

Ilyin V, Batov A, Usanova N. 2010a. Impact of probiotic drugs, based on *Enterobacter faecium* autostrains, on human intestinal microflora in confined habitat. 38th COSPAR Scientific Assembly. July 15–18, 2010, Bremen, Germany, p.4.

Ilyin V, Solovieva Z, Panina J. 2010b. Operative control of human microflora in confined habitat. 38th COSPAR Scientific Assembly. July 15–18, 2010, Bremen, Germany, p.4.

Juhaszova M, Blaustein M. 1982. Na+ pump low and high ouabain affinity alpha subunit isoforms are differently distributed in cells. *Proc. Natl. Acad. Sci. USA* 94(5):1800–1805.

Kaczmarek LK. 2006. Non-conducting functions of ion channels. *Nat. Rev. Neurosci.* 7:761–771.

Klassen VI. 1982. *Magnetized Water Systems*. Chemistry Press, 296p. Moscow.

Lednev VV. 1991. Possible mechanism for the influence of weak magnetic field interactions with biological systems. *Bioelectromagnetics* 18:455–461.

Liboff AR. 1985. Geomagnetic cyclotron resonance in living cells. *J. Biol. Phys.* 13:99–102.

Lucchesi PA, Sweadner KJ. 1991. Postnatal changes in and skeletal muscle. *J. Biol. Chem.* 267:769–773.

Markov MS. 2004. Myosin light chain phosphorylation modification depending on magnetic fields I. *Theoret. Electomagnet. Biol. Med.* 23:55–74.

Markov MS. 2009. *Biological Effects of Electromagnetic Fields*. Springer, Dordrecht, the Netherlands (A special issue of *The Environmentalist*).

Markov MS, Todorov SI, Ratcheva MR. 1975. Biomagnetic effects of the constant magnetic field action on water and physiological activity. In: *Physical Bases of Biological Information Transfer*. Jensen K, Vassileva Yu, eds. Plenum Press, New York, pp. 441–445.

Martirosyan V, Baghdasaryan N, Ayrapetyan S. 2013a. The study of the effects of mechanical vibration at infrasound frequency on [3H]-thymidine incorporation into DNA of *E. coli* K-12. *Electromagnet. Biol. Med.* 32(1):40–47.

Martirosyan V, Baghdasaryan N, Ayrapetyan S. 2013b. Bidirectional frequency dependent effect of extremely low-frequency electromagnetic field on *E. coli* K-12. *Electromagnet. Biol. Med.* 32(3):291–300.

Narinyan L, Ayrapetyan G, Ayrapetyan S. 2012. Age-dependent magneto-sensitivity of heart muscle hydration. *Bioelectromagnetics* 33:452–458.

Narinyan L, Ayrapetyan G, Ayrapetyan S. May 2013. Age-dependent magnetosensitivity of heart muscle ouabain receptors. *Bioelectromagnetics* 34(4):312–322.

Narinyan LY, Ayrapetyan GS, Jaysankar de, Ayrapetyan SN. January 2014. Age-dependent increase in Ca^{2+} exchange magnetosensitivity in rat heart muscles. *Biochem. Biophys* 2:39–49.

Nawrotek P, Fijalkowski K, Struk M, Kordas M, Rakoczy R. 2014. Effects of 50 Hz rotating magnetic field on the viability of *Escherichia coli* and *Staphylococcus aureus*. *Electromagnet. Biol. Med.* 33(1):29–34.

Orecchini A, Paciaroni A, Petrillo C, Sebastiani F, De Francesco A, Sacchetti F. 2012b. Water dynamics as affected by interaction with biomolecules and change of thermodynamic state: A neutron scattering study. *J. Phys.: Condens. Matter* 24:064105. doi:10.1088/0953-8984/24/6/064105.

Orecchini A, Sebastiani F, Jasnin M, Paciaroni A, De Francesco A, Petrillo C, Moulin M, Haertlein M, Zaccai G, Sacchetti F. 2012a. Collective dynamics of intracellular water in living cells. *J. Phys. Condens. Matter* 24:6.

Parsegian VA, Rand RP, Ran DC. 2000. Osmotic stress crowding, preferential hydration and binding. A comparison of perspectives. *Proc. Natl. Acad. Sci. USA* 97:3987–3992.

Parton R, Simons K. 2007. The multiple faces of caveolae. *Nature* 8(3):185–194.

Rehman S et al. 2007. Raman spectroscopy of biological tissues. *Appl. Spectrosc. Rev.* 42:493–541.

Saghyan AA, Ayrapetyan SN, Carpenter DO. 1996. Low dose of ouabain stimulates the Na:Ca exchange in helix Pomatia neuros. *Mol. Neurobiol.* 16:180–185.

Schwartz A. 1989. Calcium antagonists: Review and perspective on mechanism of action. *Am. J. Cardiol.* 64(17):31–91.

Segatore B, Setacci D, Bennato F, Cardigno R, Amicosante G, Iorio R. 2012. Evaluations of the effects of extremely low-frequency electromagnetic fields on growth and antibiotic susceptibility of *Escherichia coli* and *Pseudomonas aeruginosa*. *Int J Microbiol.* 2012: 587293, Article ID 587293, 7 p. http://dx.doi.org/10.1155/2012/587293.

Stepanyan PS, Ayrapetyan GS, Margaryan GF, Arakelyan GA, Ayrapetyan SN. 2000. The effects of infrasound vibration on water and DNA solution properties. *Radiat. Biol. Radioecol.* 40:435–438 (in Russian).

Stepanyan RS, Ayrapetyan SN. 1999. The effect of mechanical vibration on the water conductivity. *Biophysics* 44(2):197–202 (in Russian).

Szent-Gyorgyi A. 1968. *Bioelectronics: A Study in Cellular Regulations, Defense and Cancer*. Academic Press, New York, pp. 54–56.

Takeuchi A, Tatsumi S, Sarai N, Terashima K, Matsuoka S, Noma A. 2006. Ionic mechanisms of cardiac cell swelling induced by blocking Na+/K+ pump as revealed by experiments and simulation. *J. Gen. Physiol.* 128(5):495–507.

Thomas RC. July 1972. Electrogenic sodium pump in nerve and muscle cells. *Physiol. Rev.* 52(3):563–594.

Trushin MV. 2003. The possibility role of electromagnetic fields in bacterial communication. *J. Microbiol. Immunol. Infect.* 153–160.

Wymore T, Deerfield DW, Hempel J. 2007. Mechanistic implications of the cysteine-nicotinamide adduct in aldehyde dehydrogenase based on quantum mechanical/molecular mechanical simulations. *Biochemistry* 46:9495–9506.

14 Age-Dependent Magnetic Sensitivity of Brain and Heart Tissues

Armenuhi Heqimyan, Lilia Narinyan,
Anna Nikoghosyan, and Sinerik Ayrapetyan

CONTENTS

Age-dependent magnetosensitivity is one of the modern problems of environmental sciences (Repacholi and Greenebaum, 1999; Repacholi et al., 2005; Kheifets et al., 2010). But the mechanism determining cell magnetosensitivity is still not clear. It is known that hypermagnetosensitivity appears in pathology, including aging (Adey, 1993, Miller et al., 1996; Ahlbom et al., 2004; Grigoriev, 2004; Gabriel, 2005; Kheifets and Shimkhada, 2005). However, the nature of the mechanism determining hypermagnetosensitivity in pathology stays also unclear.

Although the important physiological role of water in cell functional activity is widely accepted, its messenger role in the generation of various diseases (Ryu et al., 2008; Weinstein et al., 2010; Briones and Darwish, 2014), including aging-induced increase of neurological (Heneka et al., 2010; Nedergaard et al., 2010; Verkhratsky et al., 2012) and cardiovascular (Takeuchi et al., 2006; Sheydina et al., 2011) disorders, has not been clarified yet. However, it is known that the aforementioned pathalogical processes are accompanied by tissue dehydration, which is a reliable predictor of increasing frailty, progressive deterioration in cognitive function, and overall reduction in the quality of life (Warren et al., 1994; Miller et al., 1996; Wilson and Morley, 2003; D'Anci et al., 2006). Our weak knowledge on the metabolic mechanism(s) controlling cell hydration is the main barrier for understanding its role in norm and pathology.

Na⁺/K⁺ pump dysfunction, which is a common consequence of any pathology, including aging, has a crucial role in regulation of two fundamental cell parameters determining its functional activity such as cell hydration and intracellular Ca²⁺ homeostasis. It is revealed that tissue dehydration (Zglinicki and Schewe, 1995; Fraser and Arieff, 2001; Gabriel, 2005; Ayrapetyan et al., 2012; Heqimyan et al., 2012; Narinyan et al., 2012) and the increase of intracellular Ca²⁺ ions [Ca²⁺]ᵢ serve as markers for aging (Landfield, 1987; Khachaturian, 1989; Wang et al., 2000; Kostyuk and Lukyanetz, 2006; Weisleder et al., 2006; Thibault et al., 2007). However, the individual role of different catalytic isoforms of Na⁺/K⁺-ATPase (working molecule of pump) in regulation of cell hydration and Ca²⁺ homeostasis in aging as well as their magnetosensitivity is not investigated in the literature available for us. Therefore, these questions were the subject for our previous investigations (Ayrapetyan et al., 2012; Heqimyan et al., 2012; Narinyan et al., 2012, 2013).

14.1 AGE DEPENDENCY OF [^{3}H]-OUABAIN BINDING WITH CELL MEMBRANE AND ITS MAGNETOSENSITIVITY

As it is established (Juhaszova and Blaustein, 1997), Na$^+$/K$^+$-ATPase in brain cortex and heart muscle tissues has three identified isoforms (α_1, α_2, α_3) of α catalytic subunits. They are characterized by different affinities to cardiac glycoside ouabain (specific inhibitor for Na$^+$/K$^+$-ATPase). α_1 (with low affinity) and α_2 (with middle affinity) isoforms are involved in transportation of Na$^+$ and K$^+$ ions through cell membrane, while α_3 (with high affinity) is a gate for intracellular signaling system (Xie and Askari, 2002; Wu et al., 2013). Although the role of low-affinity (α_1) receptors is well established, the functional role of high-affinity ones (α_3) is still a matter of discussion (Baker and Willis, 1972; Lucchesi and Sweadner, 1991; Gao et al., 2002; Blaustein et al., 2009). It was shown that in spite of the fact that α_3 has no ionic transporting function its compact localization with Na$^+$/Ca^{2+} exchanger in plasma membrane and its crucial role in regulation of Na$^+$/Ca^{2+} exchange can be considered as a well-established fact (Adams et al., 1982; Therien et al., 1996; Juhaszova and Blaustein, 1997; James et al., 1999; Liu et al., 2000; Blaustein et al., 2009).

In our studies (Ayrapetyan et al., 1984, 2012; Heqimyan et al., 2012; Narinyan et al., 2012), the curve of [^{3}H]-ouabain dose-dependent binding with cell membrane in brain cortex (Figure 14.1a) and heart muscle (Figure 14.1b) tissues consisted of three components having different affinities (α_3 10^{-11}–10^{-9} M; α_2 10^{-9}–10^{-7} M; and α_1 10^{-7}–10^{-4} M). These data are in harmony with several experimental data in snail nerve ganglia (Ayrapetyan et al., 1984) and in mammal's muscles and neurons (Sweadner, 1989; Therien et al., 1996; Juhaszova and Blaustein, 1997). As can be seen in Figure 14.1, all the three zones of [^{3}H]-ouabain binding curves were more pronounced in brain cortex (Figure 14.1a) and heart muscle (Figure 14.1b) tissues of young rats compared to those in adult and old ones. As it is described in a number of studies (Lucchesi and Sweadner, 1991; Gao et al., 2002;

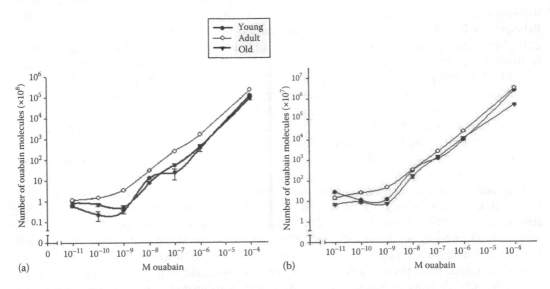

FIGURE 14.1 Number of ouabain molecules as a function of [^{3}H]-ouabain different concentrations in brain cortex (a) and heart muscle (b) tissues of three animal groups. Abscissas indicate [^{3}H]-ouabain concentrations; ordinates are logarithmic and define the number of ouabain molecules in tissues. Each point in the curve is the mean ± SD for three independent experiments. Error bars of each point are not detected because it blended with each other.

Blaustein et al., 2009), the expression of high-affinity receptors reaches at its maximum in adult animals, and then their number decreases depending on age. These data clearly indicate that α_3 receptors have more age-dependent characters than α_2 and α_1 ones.

To elucidate the magnetosensitivity of an individual family of the aforementioned ouabain receptors, [³H]-ouabain dose-dependent binding in brain cortex and heart muscle at three aged groups of rats was studied (Heqimyan et al., 2012; Narinyan et al., 2012). Our data showed that in brain cortex (Figure 14.2a) and heart muscle (Figure 14.3a) tissues of young animals, SMF exposure had more pronounced modulation effect on ouabain binding with α_2 and α_3 receptors, while in case of adults, SMF exposure caused a dose-dependent decrease in [³H]-ouabain binding only with α_3 receptors (Figures 14.2b and 14.3b). It is extremely interesting that in older animals, the magnetosensitivity of ouabain receptors disappeared (Figures 14.2c and 14.3c). Presented data on the similarity of age-dependent dynamics of brain cortex and heart muscle magneto-sensitivity suggest a common mechanism(s) determining the decrease of their age-dependent magnetosensitivity.

It is well known that ouabain binding is determined by both the number of cell functionally active receptors, depending on cell membrane surface (Ayrapetyan et al., 1984) and agonist's

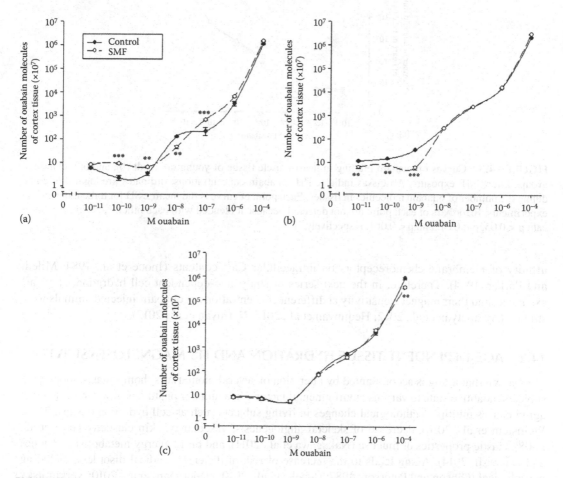

(a)   (b)

(c)

FIGURE 14.2   Curves of [³H]-ouabain binding in brain cortex tissue of young (a), adult (b), and old (c) animal groups after SMF exposure. Abscissas indicate [³H]-ouabain concentrations; ordinates are logarithmic and define the number of ouabain molecules in tissues. Each point in the curve is the mean ± SD for three independent experiments. Error bars of each point are not detected because it blended with it. ** and *** indicate $p < 0.01$ and $p < 0.001$, respectively.

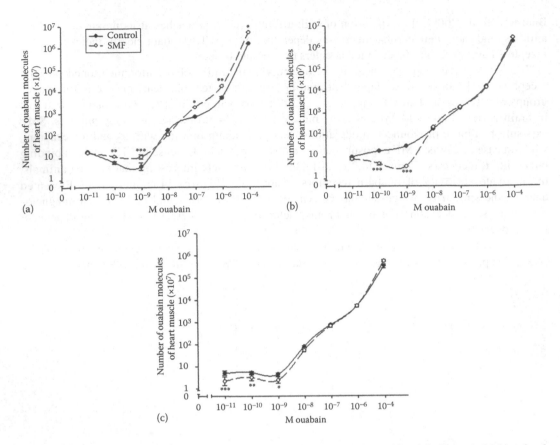

FIGURE 14.3 Curves of ouabain binding in heart muscle tissue of young (a), adult (b), and old (c) animal groups after SMF exposure. Abscissas indicate [^{3}H]-ouabain concentrations; ordinates are logarithmic and define the number of ouabain molecules in tissues. Each point of curve is the mean ± SD for three independent experiments. Error bars of each point are not detected because it blended with each other. *, **, and *** indicate $p < 0.05$, $p < 0.01$, and $p < 0.001$, respectively.

affinity of membrane chemoreceptors on intracellular Ca^{2+} contents (Inoue et al., 1984; Miledi and Parker, 1984). Therefore, in the next series of study, age-dependent cell hydration, Na^+/Ca^{2+} exchange, and their magnetosensitivity at different concentrations of ouabain-injected animals were studied (Ayrapetyan et al., 2012; Heqimyan et al., 2012; Narinyan et al., 2012).

14.2   AGE-DEPENDENT TISSUE HYDRATION AND ITS MAGNETOSENSITIVITY

It is known that aging is accompanied by alteration of several biomarkers' homeostasis, modifying subjects' adaptive state to various environmental factors. In a number of studies, it was revealed that aging causes multiple pathological changes in living subjects such as cell hydration (Evans, 2009; Weinstein et al., 2010), decrease of skeletal angiogenesis, vascularity, skin elasticity (Ryu et al., 2008), elastic properties of mucosa (Nakagawa et al., 2011), and brain energy metabolism (Briones and Darwish, 2014). Aging leads to the increase of risk of different medical disorders, including neurological (Gibson and Peterson, 1987; Heneka et al., 2010; Nedergaard et al., 2010; Verkhratsky et al., 2012) and cardiovascular diseases (Kaplan et al., 2007; see Sheydina et al., 2011).

At present, it is clear that cell hydration is one of the fundamental metabolic controlling cell parameters determining the functional activity of cell (Ayrapetyan, 1980, 2012; Haussinger, 1996a,b; Parsegian et al., 2000; Lang, 2007; Hoffman et al., 2009). According to *cell swelling* theory (Haussinger, 1996a; Lang, 2007), cell hydration can shift the pattern of cellular metabolism: cell swelling triggers proliferation, while cell shrinkage promotes apoptotic patterns of metabolism. Metabolic activity of cell hydration is determined by a number of its surface-dependent functionally active protein molecules in cell membrane (Ayrapetyan et al., 1984; Parton and Simons, 2007) and hydration-dependent regulation of intracellular macromolecules activity by *folding and unfolding mechanisms* (Parsegian et al., 2000).

Among the number of mechanisms involved in cell volume regulation, Na^+/K^+ pump has a fundamental role, and Na^+ gradient serves as a source of energy for a number of secondary ionic transporters, such as Na^+/Ca^{2+}, Na^+/H^+ exchangers, sugars, amino acids, and osmolytes (Hoffman et al., 2009). It is known that aging is accompanied by alteration of tissue dehydration as a result of Na^+/K^+ pump activity dysfunction (Zglinicki and Schewe, 1995; Gabriel, 2005; Ayrapetyan et al., 2012; Heqimyan et al., 2012; Narinyan et al., 2012). As can be seen in Figure 14.4, tissue hydration in brain cortex and heart muscle of young animals (Figure 14.4a) is higher than those in older ones (Figure 14.4b). It was shown that age-dependent cell hydration was a reason of Na^+/K^+ pump dysfunction (Ayrapetyan et al., 2012; Narinyan et al., 2012), but which of the three isoforms of Na^+/K^+ pump was responsible for age-dependent decrease of cell hydration was the subject for our study. For this purpose, dose-dependent ouabain effect on cell hydration in the tissues of young and older rats was studied (Ayrapetyan et al., 2012; Narinyan et al., 2012).

As can be seen from the data presented in Figure 14.5, [^3H]-ouabain dose-dependent curves of cell hydration for both brain cortex and heart muscle in young animals consist of minimum four zones (Figure 14.5a and b, dotted lines), which indicate that there are four mechanisms (with different ouabain sensitivities) involved in the regulation of cell hydration. In older animals, the curves of ouabain dose-dependent cell hydration in brain cortex had smooth character, that is, a number of mechanisms involved in cell volume regulation were depressed (Figure 14.5a), while the dose-dependent curve of cell hydration in heart muscle had different multicomponent character. Probably, it can be due to high content of actomyosin, the contractility of which has an essential role in regulation of muscle hydration. Therefore, it suggests that in regulation of cell hydration in brain cortex and heart muscle, different mechanisms are involved.

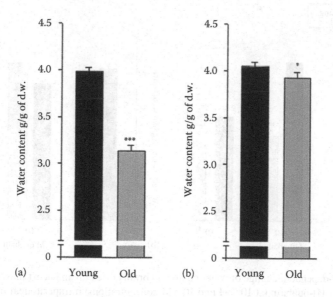

FIGURE 14.4 Initial level of water content in young and old rats' brain cortex (a) and heart muscle (b) tissues. Ordinates indicate the mean value of tissues' water content. As a control, the data of young animals were taken. Error bars indicate the standard error of the mean (SEM) for three independent experiments. * and *** indicate p < 0.05 and p < 0.001, respectively.

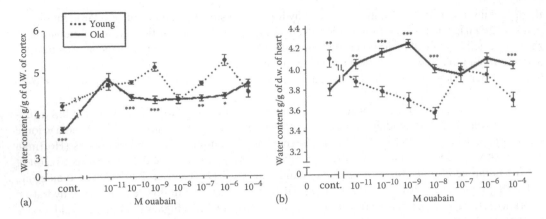

FIGURE 14.5 Curves of ouabain dose-dependent changes of water content in brain cortex (a) and heart muscle (b) tissues of two animal groups. Abscissas indicate [³H]-ouabain concentrations; ordinates show the mean value of tissues' water content. Each point in the curve is the mean ± SD for three independent experiments. Error bars of each point are not detected because it blended with it. *, **, and *** indicate $p < 0.05$, $p < 0.01$, and $p < 0.001$, respectively.

As it was shown in previous data, high-affinity (α_3) ouabain receptors had more pronounced age and magnetosensitivities (Figures 14.2 and 14.3). The study of [³H]-ouabain 10^{-9} M (α_3 receptor agonist) and 10^{-4} M (α_1 receptor agonist) effects on cell hydration in brain cortex and heart muscle tissues indicated that in brain cortex, the activation of α_3 receptors leads to dehydration while α_1 receptors to the increase of cell hydration (Figure 14.6a). In case of heart muscle, both mentioned [³H]-ouabain concentrations had overhydration effect on tissues, but at 10^{-9} M concentration, the effect was more pronounced than at 10^{-4} M (Figure 14.6b).

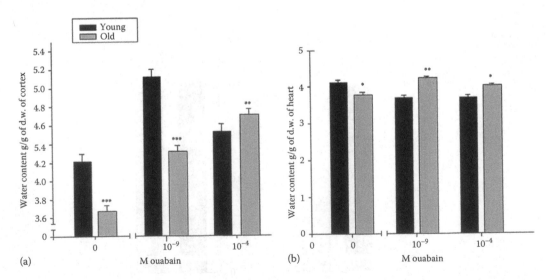

FIGURE 14.6 Age-dependent change of water content in brain cortex (a) and heart muscle (b) in two animal groups after PS and [³H]-ouabain (at 10^{-9} M and 10^{-4} M concentrations) intraperitoneal injections. Abscissas indicate ouabain doses (PS injection, i.e., 0 M ouabain, 10^{-9} M, 10^{-4} M). Ordinates indicate the mean value of tissues' water content. As a sham data, the mean values shown in Figure 14.5 were taken. Error bars indicate the SEM for three independent experiments.*, **, and *** indicate $p < 0.05$, $p < 0.01$, and $p < 0.001$, respectively.

Ouabain effects are traditionally explained by inactivation of Na$^+$/K$^+$ pump (Skou, 1957; Blaustein et al., 2009). On the one hand, different effects of [^{3}H]-ouabain at 10^{-9} M and 10^{-4} M in brain cortex and, on the other hand, more expressed overhydration effect at 10^{-9} M than at 10^{-4} M in heart muscle indicate nonreliability of the explanation of these two concentration effects on hydration by Na$^+$/K$^+$ pump inactivation.

As can be seen in Figure 14.7, SMF exposure on young animals led to cell dehydration of both brain cortex and heart muscle tissues (Figure 14.7a and b), while in older animals, it brought to over-hydration in cortex (Figure 14.7a) and had no effect on heart tissues (Figure 14.7b).

The individual role of α_3 and α_1 Na$^+$/K$^+$ ATPase isoforms in determining age-dependent magnetosensitivity of cell hydration under SMF exposure on intact and [^{3}H]-ouabain (10^{-9} M and 10^{-4} M)-poisoned two aged animal groups was studied (Figure 14.8).

As can be seen in Figure 14.8a, SMF exposure brought to significant dehydration of brain cortex in young intact and [^{3}H]-ouabain 10^{-9} M-poisoned animals, while at 10^{-4} M, it led to a significant overhydration (Figure 14.8a). In older animals, cell hydration of brain cortex was significantly increased at both 10^{-9} M and 10^{-4} M concentrations of ouabain, but after SMF exposure, the reverse (dehydration) effect was observed in brain cortex tissue (Figure 14.8b). In case of heart muscle tissue in young animals when both [^{3}H]-ouabain at 10^{-9} M and 10^{-4} M concentrations had slight dehydration effects, SMF exposure led to overhydration effect, which was more pronounced at [^{3}H]-ouabain 10^{-4} M than at [^{3}H]-ouabain 10^{-9} M (Figure 14.8c). In older animals, SMF exposure at [^{3}H]-ouabain 10^{-9} M had dehydration while at [^{3}H]-ouabain 10^{-4} M overhydration effects (Figure 14.8d).

The fact that SMF exposure-induced changes with membrane receptors (Figures 14.2 and 14.3) did not directly correlate with tissue hydration in animals of two aged groups (Figure 14.5) allows us to suggest that SMF effect on [^{3}H]-ouabain binding with corresponding receptors is implemented by intracellular mechanism regulating receptors' affinity to ouabain, which has also modulating effect on cell hydration.

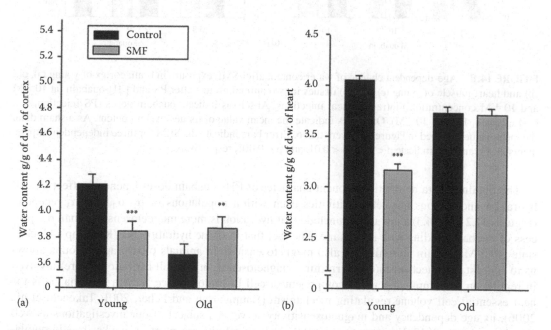

FIGURE 14.7 Change of water content in young and old rats' brain cortex (a) and heart muscle (b) tissues after SMF exposure. Ordinate indicates the mean value of tissues' water content. As a control, the data of initial water content were taken (Figure 14.5). Error bars indicate the SEM for three independent experiments. ** and *** indicate p < 0.01 and p < 0.001, respectively.

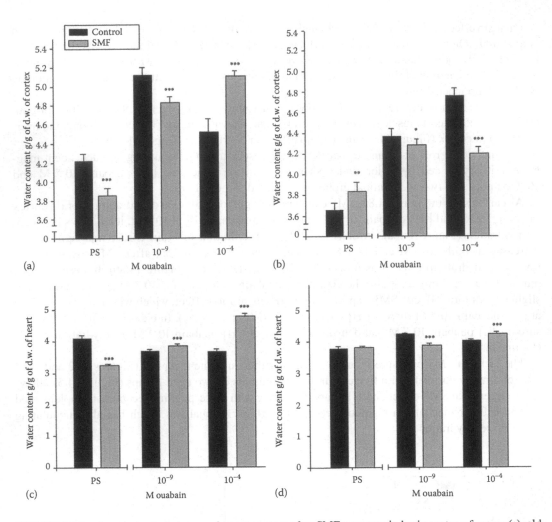

FIGURE 14.8 Age-dependent change of water content after SMF exposure in brain cortex of young (a), old (b) and heart muscle of young (c), old (d) tissues in two animal groups after PS and [³H]-ouabain (at 10^{-9} M and 10^{-4} M concentrations) intraperitoneal injections. Abscissas indicate ouabain doses (PS injection, i.e., 0 M ouabain, 10^{-9} M, 10^{-4} M). Ordinates indicate the mean value of tissues' water content. As a sham data, the mean values showed in Figure 14.5 were taken. Error bars indicate the SEM for three independent experiments. *, **, and *** indicate $p < 0.05$, $p < 0.01$, and $p < 0.001$, respectively.

The obtained data on multicomponent character of [³H]-ouabain dose-dependent effect on cell hydration and various magnetosensitivities even within populations of α_1, α_2, and α_3 receptors (Figures 14.2 and 14.3) allow us to conclude that hydration is more magnetosensitive than the process of ouabain binding with membrane. The fact that in tissue hydration Na⁺/K⁺ pump's blocked state (10^{-4} M ouabain poisoning) is also magnetosensitive in animals of two aged groups allows us to suggest that mechanism(s) determining magnetosensitivity of cell hydration is (are) involved in regulation of pump-independent component of cell hydration. Since Na⁺/Ca²⁺ exchange is the next essential cell volume regulating mechanism (Baumgarten and Feher, 2001; Takeuchi et al., 2006), its age dependency and magnetosensitivity serve as a subject of our investigations as well (Heqimyan et al., 2012; Narinyan et al., 2014). As high-affinity receptors have a key role in regulation of Na⁺/Ca²⁺ exchange and are characterized by high age and magnetosensitivities, they were suggested as one of the potential sensors determining the age-dependent magnetosensitivity of cell hydration. This suggestion was observed in our previous study of magnetosensitivity of Na⁺/Ca²⁺

exchange in neurons (Ayrapetyan et al., 1994) and heart muscle of snail (Ayrapetyan et al., 2005). As Na^+/Ca^{2+} exchange functions in stoichiometry of $3Na^+:1Ca^{2+}$ (Baker et al., 1969), its activation in forward mode could have hydration while in reverse mode dehydration effect on cells. Therefore, to find out the possible role of age-dependent increase of intracellular $[Ca^{2+}]_i$ (Khachaturian, 1989) in aging-induced dysfunction of α_3 receptors and depression of their magnetosensitivity, the dependency of Na^+/Ca^{2+} exchange and its magnetosensitivity were the subjects of our next study (Ayrapetyan et al., 2012, Heqimyan et al., 2012; Narinyan et al., 2014).

14.3 AGE-DEPENDENT NA⁺/CA²⁺ EXCHANGE AND ITS MAGNETOSENSITIVITY

As intracellular Ca^{2+} accumulation has feedback inhibitory effects on Na^+/K^+ pump activity (Skou, 1957) and switches on pathways leading to cytoskeletal phosphorylation and contraction (Markov, 2004; Tan et al., 2014), the degree and duration of cell shrinkage would depend upon the activity of mechanisms that removes intracellular Ca^{2+} ions from the cell. There are two mechanisms through which Ca^{2+} efflux from cell is realized: Na^+/Ca^{2+} exchange in forward mode and Ca^{2+} pump. As cytosolic gel condition determining the ratio of osmotic active and bounding water in cytoplasm highly depends on $[Ca^{2+}]_i$ (Pollack, 2008), cell hydration was used as a marker for $[Ca^{2+}]_i$ buffer system.

Data presented in Figure 14.9a and b indicate that $^{45}Ca^{2+}$ uptake is much higher in brain cortex and heart muscle tissues of young animals' control group (nonpoisoned animals) than in those tissues in group of older animals. It is known that in normal (with low $[Na]_i$) cells, Ca^{2+} uptake realized by Ca^{2+} channels is compensated by Ca^{2+} pump and Na^+/Ca^{2+} exchange in forward mode, that is, $^{45}Ca^{2+}$ uptake = Ca^{2+} efflux (Blaustein and Lederer, 1999). By the previously presented data, it is shown that the nanomolar (nM) concentration of ouabain is agonist for α_3 receptors, while 10^{-4} M for α_1 receptors (Narinyan et al., 2013). As can be seen in Figure 14.9, 10^{-9} M ouabain had activation effect on $^{45}Ca^{2+}$ uptake in brain cortex and heart tissues compared with $^{45}Ca^{2+}$ uptake in the same tissues of nonpoisoned animals, while $^{45}Ca^{2+}$ in brain cortex and heart muscle tissues of 10^{-4} M ouabain-poisoned animals also increased, but it was less pronounced than in case of

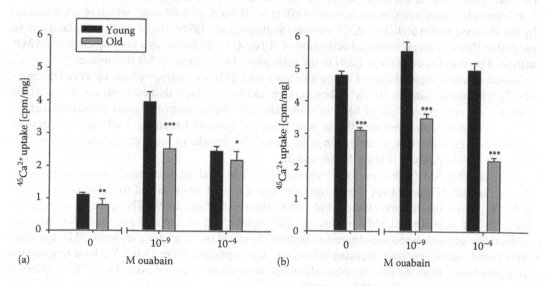

FIGURE 14.9 $^{45}Ca^{2+}$ uptake in brain cortex (a) and heart muscle (b) tissues after PS and nonradioactive ouabain (at 10^{-9} M and 10^{-4} M) intraperitoneal injections. Abscissas indicate ouabain doses (PS injection, i.e., 0 M ouabain, 10^{-9} M, 10^{-4} M). Ordinates indicate $^{45}Ca^{2+}$ uptake content counting per minute. Error bars indicate the SEM for three independent experiments. *, **, and *** indicate $p < 0.05$, $p < 0.01$, and $p < 0.001$, respectively.

10^{-9} M ouabain. The earlier presented data on age-dependent decrease of $^{45}Ca^{2+}$ uptake in 10^{-9} M and 10^{-4} M ouabain-poisoned rats can be explained by the age-dependent decrease of $Na^+/^{45}Ca^{2+}$ exchange in reverse (R) mode. The fact that, at 10^{-4} M ouabain, having strong inhibitory effect on Na^+/K^+ pump (Blaustein and Lederer, 1999) had no significant changes of $^{45}Ca^{2+}$ uptake in heart muscle of young rats but had inhibition effect in older ones indicates the existence of strong age-dependent decrease in the capacity of intracellular $[Ca^{2+}]_i$ buffer system. This suggestion is in close agreement with the literature data on aging-induced increase of $[Ca^{2+}]_i$ (Khachaturian, 1989; Kostyuk and Lukyanetz, 2006).

It is known that besides plasma membrane mechanisms, there are two cytoplasmic close-talking mechanisms determining $[Ca^{2+}]_i$ buffering properties: the thermodynamic activity of intracellular water (Pollack, 2008) and $[Ca^{2+}]_i$ sorption through cell intracellular structures (Brini and Carafoli, 2009).

Originally, it was thought that ouabain-induced increase in $[Ca^{2+}]_i$ is due to the activation of $R-Na^+/Ca^{2+}$ exchange, which is the consequence of Na^+/K^+ pump inactivation (Baker et al., 1969). However, earlier presented data (Figure 14.9) on the 10^{-9} M ouabain (which has no effect on Na^+/K^+ pump)-induced activation of $^{45}Ca^{2+}$ uptake significantly being higher than on 10^{-4} M ouabain (which has strong inhibitory effect on Na^+/K^+ pump) clearly indicate that this result cannot be explained only by Na^+/K^+ pump inactivation.

Since the difference between electrochemical gradients of Na^+ and Ca^{2+} ions ($E_{Na}-E_{Ca}$) serves as a source of energy for Na^+/Ca^{2+} exchange, the increase of $[Na^+]_i$ and decrease of $[Ca^{2+}]_i$ can bring to $R-Na^+/Ca^{2+}$ exchange activation (Baker et al., 1969). Therefore, 10^{-9} M ouabain-induced $R-Na^+/Ca^{2+}$ exchange activation can be considered as a result of $[Ca^{2+}]_i$ decrease leading to the increase in E_{Ca}. It is obvious that the latter could be done by the increase of $[Ca^{2+}]_i$ absorption with the help of intra-cellular structures. It is established that Ca^{2+} pump localized in membrane of ER transporting Ca^{2+} from cytoplasm to ER is activated by cyclic adenosine monophosphate (cAMP) (Brini and Carafoli, 2009). At the same time, there is a great number of literature data indicating that nM ouabain has elevation effect on the intracellular cAMP in different tissues, including dog renal cortex, goldfish intestinal mucosa, mouse pancreatic islets, murine epithelioid, fibroblastic cell lines, rat brain, and rat renal collecting tubule cells (Siegel, 1999). Earlier we have also shown that in snail neurons, low concentration of ouabain has activation effect on $R-Na^+/Ca^{2+}$ exchange, which is accompanied by the increase in intracellular cAMP content (Saghian et al., 1996). Based on these data, it can be suggested that α_3 receptor-induced activation of $R-Na^+/Ca^{2+}$ exchange as a consequence of cAMP-activated protein kinase activity leads to the activation of Ca^{2+} pump in ER membrane.

Based on the compared data of young animals under SMF exposure, we have observed decrease of $^{45}Ca^{2+}$ uptake at ouabain 10^{-9} M in brain cortex and heart muscle tissues. However, it is worth to note that the activation effect of SMF in young animals' brain cortex had more pronounced effect at 10^{-9} M ouabain compared with that in heart muscle (Figure 14.10a and c). SMF exposure in older animals had reversal, that is, activation effects on $^{45}Ca^{2+}$ uptake in brain cortex and heart muscle tissues at 10^{-9} M ouabain (Figure 14.10b and d).

The fact that SMF effect on $^{45}Ca^{2+}$ uptake at 10^{-9} M ouabain was more pronounced than in 10^{-4} M ouabain (Figure 14.10) clearly indicates that it cannot be explained from the point of view of Na^+/K^+ pump inactivation (Baker et al., 1969; Blaustein et al., 2009). Therefore, it is suggested that both nM ouabain and SMF modulate $^{45}Ca^{2+}$ uptake, but in opposite age-dependent manner, ouabain has activation and age-dependent decreasing effect on $^{45}Ca^{2+}$ uptake, while SMF has inac-tivation and age-dependent increasing effects on $^{45}Ca^{2+}$ uptake (Figure 14.10). We have reached the same conclusion from the data on the study of age-dependent magnetosensitivity of $^{45}Ca^{2+}$ efflux at different concentrations of ouabain in medium.

Thus, the study on age and magnetosensitivities of $^{45}Ca^{2+}$ uptake ($R-Na^+/Ca^{2+}$ exchange) has brought us to the conclusion that age-dependent decrease in Na^+/Ca^{2+} exchange activity in brain cor-tex and heart muscle tissues is a consequence of cGMP-dependent Ca^{2+} efflux dysfunction leading to $[Ca^{2+}]_i$ increase. The nM ouabain-induced activation of α_3 receptors leads to $R-Na^+/Ca^{2+}$ exchange

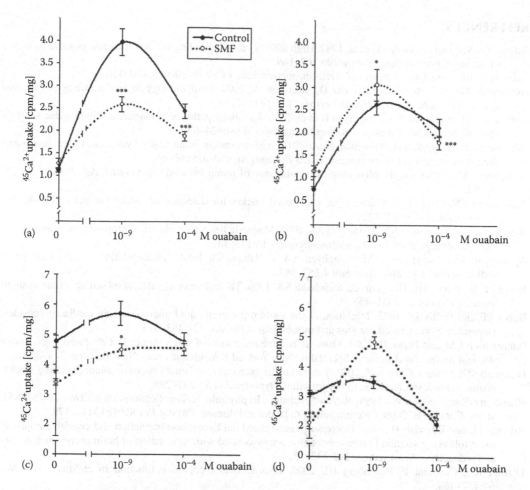

FIGURE 14.10 Curves of $^{45}Ca^{2+}$ uptake in brain cortex (a) and heart muscle (b) tissues after PS and nonradioactive ouabain (at 10^{-9} M and 10^{-4} M concentrations) intraperitoneal injections received after SMF exposure. Abscissas indicate ouabain doses (PS injection, i.e., 0 M ouabain, 10^{-9} M, 10^{-4} M). Ordinates indicate $^{45}Ca^{2+}$ uptake content counting per minute. Error bars indicate the SEM for three independent experiments. * and *** indicate $p < 0.05$ and $p < 0.001$, respectively.

as a result of cAMP-dependent Ca^{2+} pump activation in SR membrane (Brini and Carafoli, 2009). Age-dependent weakening of cGMP-dependent Ca^{2+} efflux from the cells through both Ca^{2+} pump and Na^+/Ca^{2+} exchange in forward mode can be considered as a consequence of dysfunction of cGMP formation process. SMF-induced elevation of $[cGMP]_i$ leads to the activation of Ca^{2+} efflux, which is more pronounced in cells of older rats' tissues that are rich in high $[Ca^{2+}]_i$ (Ayrapetyan et al., 1994, 2005).

Considering the presented data, the conclusion that intracellular Ca^{2+} has a multifunctional role in regulating the functional activity of cell, namely, in excitable tissues (brain cortex and heart muscle) and SMF-induced dramatic changes of intracellular Ca^{2+} in older animals, which have weakening intracellular Ca^{2+} buffer capacity, was made. This allows us to explain the hypermagnetosensitivity of living subjects in pathology, including aging, as a result of $[Ca^{2+}]_i$ increase. Therefore, it is suggested that high-affinity ouabain receptors, having a crucial role in regulation of intracellular Ca^{2+} homeostasis, can be considered as an age-dependent magnetosensitive gate of metabolic pathway through which the biological effect of magnetic field on the function of brain cortex and heart muscle tissues is realized.

REFERENCES

Adams RJ, Schwarts A, Grupp G et al. 1982. High affinity ouabain binding site and low-dose positive inotropic effect in rat myocardium. *Nature* 296:167–169.

Adey WR. 1993. Biological effects of electromagnetic fields. *J Cell Biochem* 5:410–416.

Ahlbom A, Green A, Kheifets L, Savitz D, Swerdlow A. 2004. Epidemiology of health effects of radiofrequency exposure. *Environ Health Perspect* 112:1741–1754.

Ayrapetyan G, Papanyan A, Hayrapetyan H et al. 2005. Metabolic pathway of magnetized fluid-induced relaxation effects on heart muscle. *Bioelectromagtnetics* 26(8):624–630.

Ayrapetyan S, Heqimyan A, Nikoghosyan A. 2012. Age-dependent brain tissue hydration, Ca exchange and their dose-dependent ouabain sensitivity. *J Bioequiv Availab* 4(5):60–68.

Ayrapetyan SN. 1980. On the physiological significance of pump induced cell volume. *Adv Physiol Sci* 23: 67–82.

Ayrapetyan SN. 2012. Cell hydration as a universal marker for detection of environmental pollution. *The Environmentalist* 32:210–221.

Ayrapetyan SN, Grigorian KV, Avanesyan AS. 1994. Magnetic fields alter electrical properties of solutions and their physiological effects. *Bioelectromagnetics* 15:133–142.

Ayrapetyan SN, Suleymanyan MA, Saghyan AA, Dadalyan SS. 1984. Autoregulation of the electrogenic sodium pump. *Cell Mol Neurobiol* 4:367–383.

Baker PF, Blaustein MP, Hodgkin AL, Steinhardt SA. 1969. The influence of calcium on sodium efflux in squid axons. *J Physiol* 200:431–458.

Baker PF and Willis JS. 1972. Inhibition of the sodium pump in squid giant axons by cardiac glycosides: Dependence on extracellular ions and metabolism. *J Physiol* 224:463–475.

Baumgarten CM and Feher JJ. 2001. Osmosis and the regulation of cell volume. In *Cell Physiology Source Book: A Molecular Approach*. Sperelakis (New York ed.), Academic Press, New York, pp. 319–355.

Blaustein MP, Zhang J, Chen et al. 2009. The pump, the exchanger, and endogenous ouabain: Signaling mechanisms that link salt retention to hypertension. *Hypertension* 53:291–298.

Blaustein NP and Lederer WJ. 1999. Na^+/Ca^{2+} exchange. Its physiological implications. *Physiol Rev* 79:763–854.

Brini M and Carafoli E. 2009. Calcium pumps in health and disease. *Physiol Rev* 89(4):1341–1378.

Briones TL and Darwish H. 2014. Decrease in age-related tau hyperphosphorylation and cognitive improvement following vitamin D supplementation are associated with modulation of brain energy metabolism and redox state. *Neuroscience* 262:143–55.

D'Anci KE, Constant F, Rosenberg IH. 2006. Hydration and cognitive function in children. *Nutr Rev* 64:457–464.

Evans D. 2009. *Osmotic and Ionic Regulation. Cells and Animals.*CRC Press, New York.

Fraser CL and Arieff AI. 2001. Na-K-ATPase activity decreases with aging in female rat brain synaptosomes. *Am J Physiol Renal Physiol* 281:F674–F678.

Gabriel C. 2005. Dielectric properties of biological tissue: Variation with age. *Bioelectromagnetics* 26:S12–S18.

Gao J, Wymore RS, Wang Y et al. 2002. Isoform-specific stimulation of cardiac Na/K pumps by nanomolar concentrations of glycosides. *J Gen Physiol* 119:297–312.

Gibson GE and Peterson C. 1987. Calcium and the aging nervous system. *Neurobiol Aging* 8:329–343.

Grigoriev Yu. 2004. Mobile phones and children: Is precaution warranted? *Bioelectromagnetics* 25:322–323.

Haussinger D. 1996a. Regulation of cell function by level of hydration. *Naturwissenschaften* 83:264–271.

Haussinger D. 1996b. The role of cellular hydration in regulation of cell function. *Biochem J* 313:697–710.

Heneka M, Rodriguez J, Verkhratsky A. 2010. Physiology of cell volume regulation in vertebrates. *Brain Res Rev* 63(1–2):189–211.

Heqimyan A, Narinyan L, Nikoghosyan A et al. 2012. Age dependency of high affinity ouabain receptors and their magneto sensitivity. *Environmentalist* 32:228–235.

Hoffman EK, Lambert IH, Pedersen SF. 2009. Physiology of cell volume regulation in vertebrates. *Physiol Rev* 89:193–277.

Inoue M, Oomura T, Yakushiji T, Akaike N. 1984. Intracellular calcium ions decrease the affinity of the GABA receptor. *Nature* 308:693–698.

James PF, Grupp IL, Grupp G et al. 1999. Identification of a specific role for the Na, K-ATPase a2 isoform as a regulator of calcium in the heart. *Mol Cell* 3:555–563.

Juhaszova M and Blaustein M. 1997. Na^+ pump low and high ouabain affinity alpha subunit isoforms are differently distributed in cells. *Proc Natl Acad Sci USA* 94(5):1800–1805.

Kaplan KP, Jurkovicova D, Babusikova E et al. 2007. Effect of aging on the expression of intracellular Ca^{2+} transport proteins in a rat heart. *Mol Cell Biochem* 301:219–226.

Khachaturian ZS. 1989. The role of calcium regulation in brain aging: Reexamination of a hypothesis. *Aging* 1:17–34.

Kheifets L, Ahlbom A, Crespi CM et al. 2010. Pooled analysis of recent studies of magnetic fields and childhood leukaemia. *Br J Cancer* 103:1128–1135.

Kheifets L. and Shimkhada R. 2005. Childhood leukemia and EMF: Review of the epidemiologic evidence. *Bioelectromagnetics* 26:S51–S59.

Kostyuk P, Lukyanetz E. 2006. Intracellular calcium signaling-basic mechanisms and possible alterations. In *Bioelectromagnetics: Current Concepts*. Ayrapetyan S and Markov M (eds.), NATO Science Series, Springer Press, Dordrecht, the Netherlands, pp. 87–122.

Landfield PW. 1987. "Increased calcium current" hypothesis of brain aging. *Neurobiol Aging* 8:346–347.

Lang F. 2007. Mechanisms and significance of cell volume regulation. *J Am Coll Nutr* 26:613S–623S.

Liu J, Tian J, Haas MZ et al. 2000. Ouabain interaction with cardiac Na^+, K^+-ATPase initiates signal cascades independent of changes in intracellular Na^+ and Ca^{2+} concentrations. *J Biol Chem* 275 (36):27838–27844.

Lucchesi PA and Sweadner KJ. 1991. Postnatal changes in Na,K-ATPase isoform expression in rat cardiac ventricle. Conservation of biphasic ouabain affinity. *J Biol Chem* 266:9327–9333.

Markov MS. 2004. Myosin light chain phosphorylation modification depending on magnetic fields. I. Theoretical. *Electromagnetic Biol Med* 23: 55–74.

Miledi R and Parker I. 1984. Chloride current induced by injection of calcium into Xenopus oocytes. *J Physiol* 357; 173–183.

Miller DH, Albert PS, Barkhof F et al. 1996. Guidelines for the use of magnetic resonance techniques in monitoring the treatment of multiple sclerosis. *Ann Neurol* 39:6–16.

Nakagawa K1, Sakurai K, Ueda-Kodaira Y, Ueda T. 2011. Age-related changes in elastic properties and moisture content of lower labial mucosa. *J Oral Rehabil* 38(4):235–241.

Narinyan L, Ayrapetyan G, Ayrapetyan S. 2012. Age-dependent magnetosensitivity of heart muscle hydration. *Bioelectromagnetics* 33:452–458.

Narinyan L, Ayrapetyan G, Ayrapetyan S. 2013. Age-dependent magnetosensitivity of heart muscle ouabain receptors. *Bioelectromagnetics* 34:312–322.

Narinyan L, Ayrapetyan G, De J, Ayrapetyan S. 2014. Age-dependent increase in Ca^{2+} exchange magnetosensitivity in rat heart muscles. *Biochem Biophys* 2:39–49.

Nedergaard M, Rodriguez J, Verkhratsky A. 2010. Glial calcium and diseases of the nervous system. *Cell Calcium* 47(2):140–149.

Parsegian VA, Rand RP, Rau DC. 2000. Osmotic stress, crowding, preferential hydration, and 3 binding: A comparison of perspectives. *Proc Natl Acad Sci USA* 97:3987–3992.

Parton RG and Simons K. 2007. The multiple faces of caveolae. *Nat Rev Mol Cell Biol* 8:185–194.

Pollack G. 2008. *Cells, Gels and the Engines of Life*. Ebner & Sons, Seatle, WA, pp. 170–178.

Repacholi M, Saunders R, van Deventer E, Kheifets L. 2005. Guest editors' introduction: Is EMF a potential environmental risk for children? *Bioelectromagnetics* 26(S7):S2–S4.

Repacholi MH and Greenebaum B. 1999. Interaction of static and extremely low frequency electric and magnetic fields with living systems: Health effects and research needs. *Bioelectromagnetics* 20 (3):133–160.

Ryu HS, Joo YH, Kim SO, Park KC, Youn SW. 2008. Influence of age and regional differences on skin elasticity as measured by the Cutometer. *Skin Res Technol* 14(3):354–358.

Saghian AA, Ayrapetyan SN, Carpenter D. 1996. Low dose of ouabain stimulates the Na/Ca exchange in neurons. *Cell Mol Neurobiol* 16:489–498.

Sheydina A, Riordon DR, Boheler KR. 2011. Molecular mechanisms of cardiomyocyte aging. *Clin Sci* 121:315–329.

Siegel GJ. 1999. *Basic Neurochemistry, 6th edition: Molecular, Cellular and Medical Aspects*. Lippincott-Raven, Philadelphia, PA, p. 1183.

Skou J. 1957. The influence of some cations on an adenosine triphosphatase from peripheral nerves. *Biochim Biophys Acta* 23:394–401.

Sweadner KJ. 1989. Isozymes of the Na^+/K^+-ATPase. *Biochim Biophys Acta* 988:185–220.

Takeuchi A, Tatsumi S, Sarai N et al. 2006. Ionic mechanisms of cardiac cell swelling induced by blocking Na+/K+ pump as revealed by experiments and simulation. *J Gen Physiol* 128:495–507.

Tan YX, Manz BN, Freedman TS et al. 2014. Increases in intracellular Ca^{2+} and Erk phosphorylation are restored by alteration of the actin cytoskeleton following inhibition of CskAS. *Nat Immunol* 15:186–194.

Therien AG, Nestor NB, Ball WJ, Blostein R. 1996. Tissue-specific versus isoform-specific differences in cation activation kinetics of the Na, K-ATPase. *Am Soc Biochem Mol Biol* 271:7104–7112.

Thibault O, Gant JC, Landfield PW. 2007. Expansion of the calcium hypothesis of brain aging and Alzheimer's disease: Minding the store. *Aging Cell* 6:307–317.

Verkhratsky A, Sofroniew M, Messing A et al. 2012. Neurological diseases as primary gliopathies: A reassessment of neurocentrism. *CAN Neuro* 4(3):e00082.

Wang ZM, Messi ML, Delbono O. 2000. L-type Ca^{2+} channel charge movement and intracellular Ca^{2+} in skeletal muscle fibers from aging mice. *Biophys J* 78:1947–1954.

Warren AJ, Colledge WH, Carlton MBL et al. 1994. The oncogenic cysteine-rich LIM domain protein rbtn2 is essential for erythroid development. *Cell* 78:45–58.

Weinstein RS, Wan C, Liu Q et al. 2010. Endogenous glucocorticoids decrease skeletal angiogenesis, vascularity, hydration, and strength in aged mice. *Aging Cell* 9(2):147–61.

Weisleder N, Brotto M, Komazaki Sh. 2006. Muscle aging is associated with compromised Ca^{2+} spark signaling and segregated intracellular Ca^{2+} release. *J Cell Biol* 174:639–645.

Wilson MM and Morley JE. 2003. Invited review: Aging and energy balance. *J Appl Physiol* 95(4):1728–1736.

Wu J, Akkuratov EE, Bai Y et al. 2013. Cell signaling associated with Na^+/K^+-ATPase: Activation of phosphatidylinositide 3-Kinase IA/Akt by Ouabain is independent of Src. *Biochemistry* 52:9059–9067.

Xie Z and Askari A. 2002. Na^+/K^+-ATPase as a signal transducer. *Eur J Biochem* 269:2434–2439.

Zglinicki T and Schewe C. 1995. Mitochondrial water loss and aging of cells. *Cell Biochem Funct* 13(3):181–187.

15 Electromagnetic Fields for Soft Tissue Wound Healing

Harvey N. Mayrovitz

CONTENTS

15.1 INTRODUCTION

This chapter deals with connections between wound healing and electromagnetic fields in the form of electrotherapy with electrodes (ET) or noncontact electromagnetic field (EMF) excitation. Either form is here referred to as electromagnetic field therapy (EMFT). Despite evidence of such connections, no mechanism can as yet account for such reported effects. A broad concept underlying EMFT in relation to soft tissue healing is that applied fields and currents beneficially affect functional aspects of cells and processes involved in tissue repair. Because standard treatments exist to deal with *typical* and uncomplicated wounds, EMFT is often reserved to treat wounds or ulcers that are chronic, non- or slow-healing, or recalcitrant to standard therapy.

One rationale for EMFT use for soft tissue healing initially derived from its efficacy in bone healing. Subsequent extensions to soft tissue healing have evolved with their own plausible rationales related to the body's natural bioelectric system (Becker, 1972, Nordenstrom, 1992) and from early observed relationships between electrical events and wound repair (Burr et al., 1938) and naturally occurring 10 μA current loops measured in human legs (Grimnes, 1984). Adding to these considerations is the fact that cellular function is largely determined by membrane electrical processes. A dermal wound will interrupt the normal epithelial cell potentials at the injury site causing an injury-related electric field and associated injury currents that are postulated to play an important role in the healing process (Barker et al., 1982, Foulds and Barker, 1983). The injury currents and associated electric fields arise because of the disruption of the normal transepidermal potential (TEP) that is maintained by Na^+ and Cl^- ion fluxes through their associated channels. In unbroken skin, this results in a TEP between stratum corneum (SC) and the basal layer of the epidermis of 20–50 mV with the SC relatively negative (Barker et al., 1982). When a wound is present, the TEP drives currents through the low electrical resistance of the wound, causing a lateral electric field (Nuccitelli, 2003), the magnitude of which decreases as the wound heals and which is smaller with advancing age (Nuccitelli et al., 2008, 2011). The injury current is reported as less than 1 mA with a lateral electric field of near 200 mV/mm at the wound site, which reduces to 0 at about 2 cm from the wound (Jaffe and Vanable, 1984). The injury current has been measured to be up to 22 nA/cm^2 at tips of accidentally amputated fingers in children (Illingworth and Barker, 1980). Because cells involved in healing are electrically charged, the endogenous bioelectricity facilitates cells to migrate to the wound (Vanable, 1989) to help the healing process.

EMFT may interact directly with wound currents or with related signal transduction processes (Lee et al., 1993), thereby restimulating retarded or arrested healing. Alternately, we argue that EMFT may mimic one or more intrinsic bioelectric effects and help trigger renewed healing. Accelerated healing by direct currents (200–800 µA) may be due to such a process (Carley and Wainapel, 1985). There is more recent evidence to support the notion that applied low-intensity currents in the range of the injury current can enhance wound healing (Balakatounis and Angoules, 2008). Another example may be that in which low-level currents of about 40 µA delivered via wearable devices reduced periwound edema from an initial depth of 1.6 mm to about 0.6 mm after 10 days of treatment (Young et al., 2011); however, additional research in this area is needed. Reported benefits of EMFT on cellular and other processes involved in wound repair are manifold and include edema reduction, blood flow changes and cellular proliferation, migration, differentiation, and upregulation of various cell functions as will be discussed subsequently. Although these and other factors are pieces of the puzzle, they suggest only a beneficial result. Verification needs human wound studies in a clinical setting. Such human studies are difficult due to the complexity of the wound healing process and by logistic and practical aspects of clinical wound research. In spite of these difficulties, clinical research with EMFT has progressed with promising findings and an increasing amount of direct and indirect evidence of benefits.

EMFT healing utility has mainly been tested in skin ulcers (arterial, venous, pressure, and diabetes). Some data support the concept that EMFT triggers healing of *stalled* ulcers with benefits ranging from those on a single subject with multiple wounds (Bentall, 1986) through a small number of randomized controlled trials (RCTs). For example, a meta-analysis showed that EMFT was associated with a 22% weekly healing rate compared with 9% for controls (Gardner et al., 1999). An analysis of 613 patients suggested a favorable EMFT effect (Akai and Hayashi, 2002). But not all clinical reports meet the rigorous inclusion criteria needed for the high confidence level to validate medical efficacy. Protocols that control and adequately characterize patient, wound, and treatment variables are logistically difficult and expensive, consuming both time and money. This issue has been emphasized (Vodovnik and Karba, 1992) and is now recognized. Animal experiments have also shown positive connections between EMFT and wound healing. But most animal studies use wound models quite different from human *chronic* wounds that are those in which repair is stalled and difficult to manage and the types most likely to benefit from adjunctive EMFT. Contrastingly, many EMFT-related effects on cells, tissues, and processes involved in tissue repair have been convincingly shown to occur as will be described.

It may be stated that the scientific case for an EMFT connection to soft tissue wound repair is not complete or fully validated. But based on clinical, experimental, and cellular observations, a connection between EMFT and wound healing is strongly suggested. In 2002, the accumulated information at that time led the US Centers for Medicare and Medicaid Services (CMS) to approve coverage for electrical stimulation as adjunctive therapy for various ulcers (stage III/IV pressure, arterial, venous, and diabetic) if the ulcers had not improved after 30 days of standard treatment. In 2003, CMS extended this acceptance to EMFT. More recently, some private insurers have adopted a similar policy for electrical stimulation (Aetna, 2014), but their policy clearly indicates that electrical stimulation is "electrical current via electrodes placed directly on the skin in close proximity to the ulcer," whereas they consider "high-frequency pulsed electromagnetic stimulation experimental and investigational." With respect to any potential EMFT–wound healing connection, we believe that much remains to be learned about the factors involved, mechanisms of action, specific targets, optimal dosing and patterns, and temporal strategies. These aspects need further focused research and exploration. The following sections are written with the hope that they will provide the needed background and framework to aid in this future process.

15.2 WOUND HEALING PROCESS SYNOPSIS

15.2.1 NORMAL HEALING

Normal wound healing has three broad phases: inflammation, proliferation, and remodeling. These occur in a well-ordered functionally overlapping sequence, the outcome of which depends upon interactions among many cell types, growth factors such as fibroblast growth factor (FGF) and vascular endothelial growth factor. Cells involved are vascular such as platelets, macrophages, mast, polymorphonuclear neutrophils (PMNs), monocytes, endothelial, and smooth muscle and dermal cells such as keratinocytes, melanocytes, Langerhans, fibroblasts, and myofibroblasts (Yamaguchi and Yoshikawa, 2001). As part of the repair process, cells release and/or interact with many components including structural proteins, growth factors, cytokines, chemokines, adhesion molecules, nitric oxide, trace elements, and proteases. Any of the broad array of cells and interactions could, in theory, be a target for adjunctive EMFT. Some of these as possible individual targets for EMFT are discussed in 5.0.

The initial inflammatory process helps limit blood loss (via clotting), promotes antibody and fibrin entry into interstitial spaces (by increased vascular permeability), and delivers needed blood flow via vasodilation. This early hyperemia increases O_2 delivery that supports antibacterial actions of accumulating neutrophils. Activated macrophages are attracted to the wound by chemotactic and/or galvanotactic signals causing them to release substances important for (1) angiogenesis, (2) granulation tissue maintenance, and (3) fibroblast and keratinocyte cell proliferation. Angiogenesis involves new wound capillary formation that is stimulated by angiogenic factors released from macrophages in response to low O_2 in the wound (Knighton et al., 1983) and by growth factors from fibroblasts and endothelial cells (ECs). Nitric oxide in the wound (Lee et al., 2001) also affects macrophage, fibroblast, and keratinocyte functions (Frank et al., 2002). Fibroblasts migrate to the wound and proliferate. Collagen synthesis is triggered by fibroblast-stimulating growth factors released from macrophages and continues at a rate dependent on the adequacy of blood flow to deliver O_2 and nutrients for protein synthesis. These nutrients include amino acids and, interestingly, ferrous iron. Epithelialization of open wounds depends on epithelial cell (1) migration triggered by epidermal growth factor released from macrophages and platelets, and (2) proliferation at the wound site. The epithelialization takes days to months depending on wound-related factors such as keratinocyte proliferation, migration, stratification, and differentiation and features of the extracellular matrix (O'Toole, 2001). Wounds close due to active contractile forces of myofibroblasts, which are provided with their energy substrates via blood flow. After wound closure, healing continues as wound remodeling continues for months to years. During remodeling, wound strength increases via collagen cross-linking, and excess collagen in the wound is eliminated, and many capillaries developed during early wound healing are resorbed.

15.2.2 Stalled Healing and Nonhealing Wounds

A wound is an added *metabolic organ*, and healing depends on the body's ability to meet the demands of this *temporary organ*. If healing delay exceeds 3 months, then it is often called *chronic, stalled, recalcitrant*, or *nonhealing*. Factors impeding healing may be local or systemic. Examples include infection and inadequate blood flow, O_2 delivery, or nutrient availability to support tissue building metabolic processes. Some conditions such as diabetes mellitus (DM) have further implications: hyperglycemia and impaired insulin signaling directly impair keratinocyte glucose utilization thereby altering proliferation and differentiation (Spravchikov et al., 2001). Inhibiting nitric oxide production reduces fibroblast and keratinocyte healing activities that delay healing (Akcay et al., 2000, Shi et al., 2001). Deficient wound concentrations of platelet-activating factor impair healing of venous ulcers (Stacey and Mata, 2000). Given the many potential causes for retarded healing, no *most important* target has been defined. But since blood supply plays a major role in healing, increases in blood flow and O_2 supply are often stated targets of EMFT and will be further discussed in Section 15.4.2.

Beyond blood flow, evidence (Costin et al., 2012, Pesce et al., 2013) suggests that EMFT can alter the wound bed cytokine profile moving it from a chronic proinflammatory profile to one that is anti-inflammatory, thus *kick-starting* stalled wounds. There is also evidence that PEMFT increases the tensile strength of experimental wounds (Strauch et al., 2007). A novel EMFT excitation pattern, based on a possible role of sensory nerves in healing, showed promise in treating a mixed etiology hard-to-heal wounds (Ricci and Afaragan, 2010). The pattern was a PEMF signal (4 ms, 4 Hz) embedded in stochastic noise. The noise component was used based on the assumption that it increases sensory nerve function via stochastic resonance (Kruglikov and Dertinger, 1994) and thereby improves healing. More study of this approach is warranted.

15.3 METHODS AND STRATEGIES FOR EMF-RELATED WOUND HEALING

EMFT may be applied at a wound site or remote to the wound, so therapy may use electric currents and fields in which the wound itself is directly or remotely treated. For ET, an electrode may be placed directly in the wound bed or the wound may be in the path of electrode pairs that straddle the wound. Alternately, electrostimulation may target nerves or tissue regions that functionally connect with, and potentially alter, wound site processes, either directly or via reflex effects with EMF or ET. EMF methodologies have been reviewed (Markov, 1995, McCulloch et al., 1995) and discussed in this volume. Since with EMF devices no contact electrodes are needed, they can produce effects in cases in which direct contact with skin is not advisable or possible such as when limbs and/or wounds are bandaged or otherwise covered. EMF devices use time-varying excitation that may modulate a carrier frequency historically at 27.12 MHz. EMF devices differ with respect to tissue heating effects with only some specifying device power but rarely giving energy delivered to target tissues. Tissue thermal effects can be reduced using low-duty cycles in which heat of a single high-power short pulse is dissipated during a much longer off-time between adjacent pulses. Pulse widths and shapes vary with some proprietary patterns claimed to be particularly effective with patents awarded for these claims. Wound treatment parameter variants include device power and magnetic field intensity, carrier frequency and pulse width, rate, and duty cycle. There are also variants regarding excitation pattern specifics, that is, whether stimulation is continuous or pulsed, galvanic or frequency modulated, biphasic or monophasic, symmetrical or asymmetrical, and sinusoidal or not and whether high-voltage or low-voltage stimulation is used (Markov, 1995, Kloth, 2005). This wide range of excitation parameters makes it difficult to correlate specific treatment parameters with wound healing efficacy. But PEMFT, with its inductive coupling to tissue, may provide a more uniform and predictable EMF signal in target tissues than is achieved with surface contact electrodes (Markov and Pilla, 1995), and tissue dose may be more reliably characterized. Further, the large spectral range of PEMF likely offers more chance for field coupling to produce effects in a wider range of possible (but as yet unspecified) biological processes.

15.4 CLINICAL AND RELATED FINDINGS RELEVANT TO WOUND TREATMENT USING EMFT

15.4.1 VENOUS LEG ULCERS

Venous leg ulcers (illustrated in Figure 15.1) are the most common chronic skin ulcer. They have a prevalence of near 1% (Nelzen, 2008), an open wound point prevalence of 0.1%–0.3%, which increases with age. Venous ulcer (VU) prevalence appears to be increasing (Lazarus et al., 2014). Venous reflux and venous hypertension due to incompetent venous valves and venous thrombosis are common findings in persons with VU. The genesis of the skin breakdown and ulceration is complex (Tassiopoulos et al., 2000). Its contributory factors include inflammation, upregulation of intercellular and vascular adhesion molecules (Peschen et al., 1999), protein-rich edema and leukocyte trapping (Smith, 2001), O_2 and microcirculatory deficits (Gschwandtner and Ehringer, 2001, Valencia et al., 2001), and PMN activation (McDaniel et al., 2013). Microvascular changes include dilated and tortuous capillaries with some capillary loss and increased capillary permeability with an increased transcapillary fluid efflux with tissue edema and altered microlymphatic function. Compression bandaging is the mainstay of standard treatment (Tang et al., 2012), which also helps normalize capillary numbers and size (Junger et al., 2000) and normalize the abnormally elevated leg blood flow (Mayrovitz and Larsen, 1994a).

In a prior review (Flemming and Cullum, 2001b), three eligible RCT studies were identified. A more recent review found no new additional eligible RCT studies (Aziz et al., 2013). In this latest review, the authors state that "there is no high quality evidence that electromagnetic therapy increases the rate of healing of venous leg ulcers, and further research is needed." Of the studies reviewed, one (Ieran et al., 1990) compared sham vs. PEMFT (75 Hz, peak field of 2.8 mT) in 37 patients (19 sham) for 90 days. VU were present for 30 months in the actively treated group vs. 23 months in controls. PEMFT was administered over the wound by patients at their homes for 3–4 h per day. After 90 days, 12 PEMF-treated ulcers were healed, compared to 6 in the sham group

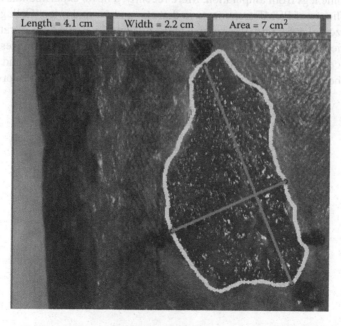

Length = 4.1 cm   Width = 2.2 cm   Area = 7 cm²

FIGURE 15.1 A typical VU located on the lateral calf. The perimeter of the VU is outlined so that its area can be determined and changes tracked and quantified to evaluate treatment effects. In this case, the area is 7 cm² determined by available software (www.clinsoft.org).

(p < 0.02). Granulation tissue, not present prior to treatment, was present in all PEMFT ulcers by day 15, whereas seven sham group ulcers showed new granulation tissue. The other RCT (Stiller et al., 1992) investigated 27 patients with recalcitrant VU. Patients were treated for 8 weeks at home for 3 h per day with a wearable portable EMF device that delivered a 22 Gauss bidirectional pulse of 3.5 ms at a 25% duty cycle. All received compression bandaging and daily 3 h leg elevation. At week 8, the PEMFT group (N = 17) had a 47.7% decrease in wound surface area vs. a 42.3% increase for sham (p < 0.0002). The investigators' global evaluations indicated that 50% of VU in the PEMFT group were healed or markedly improved vs. 0% in the sham group.

Other studies suggest benefits of various types of EMFT for VU healing. Twin 100 volt pulses (0.1 ms, 100 Hz) used directly on ulcers caused healing rates superior to standard therapy alone (Franek et al., 2000). In a unique study, EMF patterns were adjusted to interact with human monocytes and then used as the sole treatment for VU patients (Canedo-Dorantes et al., 2002). Results suggested a benefit of this treatment pattern. Improved VU healing of an ET pattern of frequency-modulated short pulse sequences has also been reported (Jankovic and Binic, 2008). A recent small study used a wearable EMF device to treat VU (Rawe and Vlahovic, 2012) with reported good results, while another wearable device used 40 µA currents that reduced periwound edema (Young et al., 2011) and indicated possible accelerated healing, but more study is needed.

15.4.2 ISCHEMIC ULCERS

A main predisposing condition for ischemic ulcers is advanced peripheral artery disease affecting arteries supplying the leg and foot. This ulcer type (example shown in Figure 15.2) can be painful and difficult to treat. Improving low blood flow is key, and therapies for which standard medical approaches have failed would be a most welcomed development. This ulcer type may be an important target of EMFT in the future. Pilot work (Goldman et al., 2002) using high-voltage pulses to treat ischemic ulcers in diabetic patients with very poor microcirculation raised periwound O_2 sufficiently to save some legs from amputation. More recently, a novel experimental wound model was used to test the efficacy of PEMF treatments in diabetic and normal mice (Callaghan et al., 2008). They used a trapezoidal pulse (200 µs rise time and 24 µs fall time) of 4.5 ms duration and 12 Gauss peak. It was applied at 15 Hz to full-thickness wounds on the mice's back. Results indicate that this PEMFT reduced healing time from about 24 to 18 days in diabetic mice and from 15 to about 11 days in nondiabetic mice. From other measurements, they concluded that improved healing was

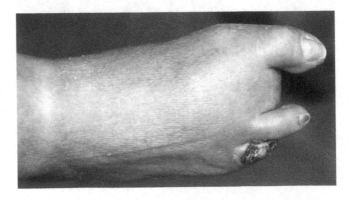

FIGURE 15.2 An ischemic ulcer on the small toe that has become necrotic. Ischemic ulcers arise due to inadequate blood flow often due to vascular disease. Their location varies but may be near bony surfaces exposed to external pressure. EMFT might be useful as a way to augment blood flow.

due to an upregulation of FGF that, via enhanced angiogenesis, improved circulation and thereby caused healing acceleration.

Low blood flow impedes ischemic ulcer healing; so for EMFT to be effective, it must improve the flow. Early work investigated flow augmentation to ischemic regions via reflex effects using pulsed radiofrequency excitation (PRF-EMF) at 27.12 MHz applied to epigastric regions in healthy persons (Erdman, 1960) and patients with peripheral arterial disease (Hedenius et al., 1966). A remote site instead of the foot or ulcer was used to produce a reflexive flow increase without imposing more metabolic demand on the distant (ischemic) region. Healthy subjects (N = 20) showed a dose-dependent increase in foot circulation judged by toe temperature and plethysmography (Erdman, 1960). Patients with intermittent claudication who received 12 PRF-FEMF treatments also had increased toe temperature (>3.0°C) (Hedenius et al., 1966), but elevations were short-lived after 20 min treatment. Cumulative effects appeared to be sustained, judged by increased pain-free walking distance.

More direct blood flow measures used laser Doppler (Mayrovitz and Larsen, 1994b,c 1996), which permits skin flow to be measured before, during, and after EMFT. PRF-EMF (65 µs, 600 pps, 1 G) was applied 1.5 cm above foot ulcers in diabetic patients. EMFT increased periulcer blood perfusion. Based on observed flow patterns, the authors judged the increase to be mainly due to increased number of capillaries with active blood flow. This suggestion was consistent with an EMF-related capillary recruitment process (Mayrovitz and Larsen, 1996) due to arteriolar vasodilation. Similar microcirculatory flow increases were reported for forearm skin of healthy persons (Mayrovitz and Larsen, 1992) and persons with post-mastectomy lymphedema (Mayrovitz et al., 2002a). Interestingly, no effects of static magnetic fields (500 G) were observed on hands (Mayrovitz et al., 2001) or forearms (Mayrovitz et al., 2002b) or in response to vasoconstricting maneuvers (Mayrovitz et al., 2005), but a slight *reduction* in finger resting skin blood flow was found (Mayrovitz and Groseclose, 2005).

A prior pilot study suggested that EMFT in the form of high-voltage pulses could increase peri-wound blood flow and increase ulcer healing (Goldman et al., 2004). More recently, high-voltage pulsed currents delivered via multiple pairs of electrodes on the leg and foot over multiple 2-week cycles resulted in an acceleration of the healing of arterial and mixed-factor ulcers as compared to standard treatments (Magnoni et al., 2013). This method uses a treatment protocol with a pseudo-random modulation of pulse sequence parameters during the course of about a 30 min treatment interval and is referred to as frequency modulated electromagnetic neural stimulation (FREMS). Prior use of this method indicated a potential to increase skin blood flow (Conti et al., 2009) perhaps suggesting at least one mechanism for its reported effectiveness in this study and previously (Jankovic and Binic, 2008). Along similar lines, pulsed EMF (15 Hz, 6 mT) applied to previously ischemic myocardium of mice caused an increase in both VGEF and capillary density with improved infarction size (Yuan et al., 2010) and indications of cardiomyogenesis (2 Hz, 2 ms, 4 nA) in cellular studies (Wen et al., 2013). The concept of a link between various forms of EMFT and blood flow is supported by other studies. For example, epidural spinal cord electrical stimulation (ESES) appears to benefit patients with severe lower extremity ischemia secondary to atherosclerotic disease. This approach, which uses implanted electrodes at the T10–T11 level and an implanted pulse generator, increased microscopically measured blood velocity in capillaries and the density of skin capillaries in the foot (Jacobs et al., 1988). In patients with rest pain and ischemic ulcers, this technique resulted in immediate pain reduction, and in most patients, it was accompanied by microscopically verified increases in capillary blood velocity and density and a significant increase in post-occlusive microvascular hyperemia (Jacobs et al., 1990). In more than half of these patients, the ulcers subsequently healed. Other studies using ESES have shown similar limb salvage rates and ulcer-healing potential (Graber and Lifson, 1987, Mingoli et al., 1993). In patients with and without ulcers, the amount of therapeutic success depends on increased transcutaneous oxygen tension (Claeys, 1997, Kumar et al., 1997, Fromy et al., 2002), which itself depends on an increased blood flow. Further work in assessing various EMFT methods and parameters targeting the treatment of arterial and ischemic ulcers is importantly needed.

15.4.3 DIABETES MELLITUS: RELATED ULCERS

Persons with DM are more susceptible to skin ulcers due to neuropathy, ischemia, and poor glycemic control. A higher likelihood of peripheral arterial disease and microvascular deficits increases chances for ischemia, tissue breakdown, and ulcer formation. DM skin ulcers may be difficult to heal for reasons that include reduced blood flow and wound O_2, functional deficits in cells normally involved in healing and infection. It takes much less local pressure to reduce skin blood flow near bony prominence in persons with DM (Fromy et al., 2002). If sensory neuropathy is present, normal pressure/pain signals are reduced or absent causing loss of warning sensations of impending tissue injury. Most of these ulcers are diabetic foot ulcers (DFUs) with plantar ulcers (Figure 15.3) a common type. For persons with DM, the prevalence of DFUs is reported as 4%–10% (Singh et al., 2005), an annual incidence of 2.5%–10.7% (Hunt, 2011) with the most common combination present reported as being neuropathy + minor foot trauma + foot deformity (Reiber et al., 1999) with edema accounting for 37% and ischemia for 35%. Persons with DM have a lifetime 10%–15% chance of getting a DFU (Gonzalez and Oley, 2000) with about half having vascular complications resulting in 0.5%–0.8% of them receiving amputations (Muller et al., 2002). Treatment includes foot offloading and standard wound care (Crawford and Fields-Varnado, 2013) with a variety of supplementary and adjunctive measures being reported as potentially efficacious (Brimson and Nigam, 2013, Holmes et al., 2013, Hamed et al., 2014, Houreld, 2014).

Regarding the utility of EMFT, an earlier study (Peters et al., 2001) used pulsed-galvanic electric stimulation (50 V, 100 µs), delivered through conductive stockings for 8 h every night to treat 40 DFUs. The EMFT was given for 12 weeks to half the patients with all patients receiving standard wound care and foot off-loading in this randomized, double-blind, placebo-controlled pilot study. Seventy-one percent of protocol-compliant patients receiving EMFT were healed compared with 29% in the sham treatment group (p = 0.038). The authors concluded that EMFT improved wound healing. A different EMFT approach was used to treat ulcers of a group of 80 diabetic patients. Daily treatment used a biphasic stimulation pattern with either asymmetric or symmetric square-wave pulses at amplitudes to activate intact peripheral nerves in skin. Controls consisted of groups that received either very low levels of stimulation current or no electrical stimulation. EMFT healing rates (ulcer perimeter changes) were significantly greater than controls only if asymmetric treatment was used (Baker et al., 1997). Further, a group of 64 diabetic patients with chronic ulcers, half of whom were treated with electrical nerve stimulation (80 Hz, 1 ms) sufficient to induce

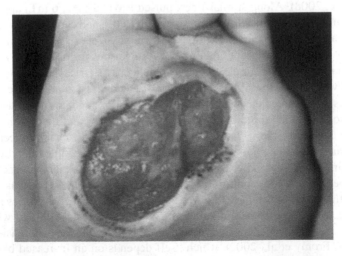

FIGURE 15.3 Diabetic plantar ulcer in a patient with sensory neuropathy. Though this ulcer appears on the bottom of the foot, this type can occur at other sites that are subject to unsensed sustained pressure.

paresthesia, were reported to have significantly reduced ulcer area and more healed ulcers after 12 weeks of twice-daily 20 min treatments (Lundeberg et al., 1992). More recently, in two interesting case studies, extremely recalcitrant ulcers were reported to be healed only after the introduction of 27.12 MHz pulsed EMFT (Larsen and Overstreet, 2008), and more recent case studies reported potential efficacy of micro-current-type EMFT (Lee et al., 2009, Ramadhinara and Poulas, 2013). The utility of PEMFT to heal wounds in diabetic mice with low-level currents has already been described (Callaghan et al., 2008) in Section 15.4.2. Despite the various reports citing positive outcomes of EMFT, it is felt by some that there is insufficient high-level evidence of efficacy as put forward in recent reviews (Game et al., 2012, Gottrup and Apelqvist, 2012). In some ways, plantar ulcers in persons with leprosy resemble diabetic ulcers. In a pilot, randomized, double-blind, controlled clinical trial (Sarma et al., 1997), 40 leprosy patients with plantar ulcers received standard treatment, and half of them also received pulsed sinusoidal magnetic fields (0.95–1.05 Hz, 2400 nT) for 4 weeks. Outcome measures changes in ulcer volume at treatment end. Control group ulcer volume of 2843 mm^3 was reduced to 1478 mm^3 at the end of the treatment; corresponding values in the EMFT group were 2428 and 337 mm^3. These data indicate that the EMFT caused significantly more rapid healing in these leprosy patients.

15.4.4 Pressure Ulcers

Pressure (decubitus) ulcers (PUs) result from sustained or inadequately relieved pressure, often on bony prominences including heel, trochanter, and sacral regions (Figure 15.4). The clinical stages of pressure ulceration range from nonblanching erythema (stage I) through full-thickness skin loss with extensive destruction and tissue necrosis involving muscle or bone (stage IV). These PUs are an important clinical, humanitarian, and economic problem with a prevalence ranging from about 5% to 50% depending on the type of care facility (Gottrup et al., 2013, Gunningberg et al., 2013, Aygor et al., 2014) and patient features that include such factors as age, nutritional status, mobility (Byers et al., 2000), coexisting morbidities (Amir et al., 2013, Moore et al., 2013, Scheel-Sailer et al., 2013), and general aspects of patient fragility and possibly ethnicity (Harms et al., 2014). Often, a final

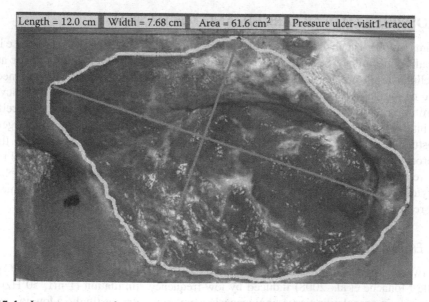

Length = 12.0 cm | Width = 7.68 cm | Area = 61.6 cm² | Pressure ulcer-visit1-traced

FIGURE 15.4 Large pressure ulcer located on the buttocks. Tracing outlines its perimeter and measured area is 61.6 cm². Pressure ulcers are often the result of blood flow deficits and tissue injury due to unrelieved pressure in the region of bony prominences.

common pathway is associated with blood flow changes within pressure-loaded tissue (Mayrovitz et al., 1997, 1998b, 1999, 2002d, Mayrovitz and Smith, 1998,1999, Mayrovitz and Sims, 2001). These concepts may have clinical utility in persons with spinal cord injuries (Jan et al., 2013, Smit et al., 2013, Sonenblum et al., 2014). Some experimental evidence suggests that both ischemia and ischemia–reperfusion injuries are involved (Peirce et al., 2000).

Based on an available RCT at the time (Sheffet et al., 2000, Flemming and Cullum, 2001a), EMFT for PU was judged to be insufficiently verified. In 2003, the CMS concluded that EMFT was an acceptable treatment modality for stage III and IV PU if healing had not progressed after 30 days of standard treatment. More recently, Smith et al. (2013) concluded that electrical stimulation was one of several adjunctive therapies that might have benefit. Some earlier studies, although not confirmatory, do provide interesting and suggestive findings of potential benefits of EMFT for PU. One small study (Comorosan et al., 1993) used PRF-EMFT (27.12 MHz) on patients with long-standing PU and found significant improvement over standard treatment alone. Another study (Salzberg et al., 1995) was randomized and double-blind and used PRF-EMFT or sham to treat 30 spinal cord–injured patients who had either stage II or III PU. Ulcers were treated for 30 min, twice daily, for 12 weeks, or until healed. The authors indicate that after controlling for the baseline status of the PU, PRF-EMFT was independently associated with a shorter median time to full healing. Use of the same PRF-EMFT method to treat patients with stage II or III long-standing PU also resulted in improved healing (Itoh et al., 1991). Similar positive results of pulsed current (300–600 µA) were reported in a double-blind placebo controlled study of long-standing stage II and III PU in which healing rates were improved with pulsed low-current treatment (Wood et al., 1993). Pulsed stimulation (200 V, 100 pps) used to treat PU in spinal cord–injured patients was reported to achieve a greater reduction in ulcer area as compared to a placebo group after 20 days of treatment (Griffin et al., 1991), and treatment of 150 persons with spinal cord injury using a pulsed biphasic stimulation (0.25 ms, 40 Hz, 15–25 mA) resulted in significantly faster healing (Stefanovska et al., 1993). Based on available clinical data, it appears to this author that there is a reasonable but not conclusive case for a beneficial effect of EMFT for treating PUs. Additional studies on the role of EMFT in PU treatment and its possible prevention appear warranted.

15.5 POSSIBLE CELLULAR TARGETS AND MECHANISMS

Mechanisms by which applied EMF alter cell properties and biological processes to cause improved wound healing are unknown. Theories describing how EMF interactions might occur at cellular and subcellular levels are discussed elsewhere in this volume. Whatever the specific mechanisms turn out to be in relation to wound healing, it is this author's opinion that clinical efficacy depends on determining the proper therapeutic parameters and timing to optimally modulate cell features and their interacting processes in the context of the wound healing cascade. Specific targets for any of the postulated mechanisms could theoretically be any cell or function involved in the wound healing process. A prior review (Mayrovitz, 2004) has described some of the potential targets for EMFT in relation to wound healing. These include ECs, fibroblasts, leukocytes, macrophages, and keratinocytes. The following summarizes key aspects of these and includes progress achieved since that prior review.

15.5.1 ENDOTHELIAL CELLS

A link between EMFT and EC possibly connected to wound healing is suggested by EC proliferation (Delle Monache et al., 2008) induced by low-frequency stimulation (1 mT, 50 Hz) although its specific role in granulation tissue angiogenesis is unknown. Interestingly, a low-intensity *static* magnetic field (120 µT) also increased EC proliferation (Martino et al., 2010). The involvement of a free radical mechanism was suggested (Martino, 2011). In vitro EC proliferation was described some years ago (Yen-Patton et al., 1988). Because ECs are intimately involved in the angiogenic process, the migration of endothelial progenitor cells in an electric field was carefully studied (Zhao et al., 2012). Results indicated that these cells move in the field's direction in association with VGEF signaling (Bai et al., 2011). This suggests a new way in which various EMFT signals may play a role. Because many effects of VGEF on EC depend on Ca^{++} signaling (Cheng et al., 2006), this may turn out to be another example of an EMF–Ca^{++} connection.

15.5.2 KERATINOCYTES

Wound healing needs epithelial cell proliferation, migration, and differentiation to re-epithelialize. Keratinocyte migration occurs during the proliferative phase and is guided by galvanotaxis whereby keratinocytes migrate to the wound and then differentiate. Reduced keratinocyte proliferation or migration can lead to impaired healing. Impaired migration may occur due to deficient lateral field gradients associated with the injury current previously described. The superimposed field strength of applied EMFT theoretically needed to overcome such a deficit would depend on multiple intrinsic factors. Recent work indicates that keratinocyte galvanotaxis direction and speed in the presence of combined alternating and direct electric fields depend on excitation frequency (Hart et al., 2013). Other patterns of applied EMF also indicate acceleration of keratinocyte migration in vitro (Huo et al., 2010). In addition to migration aspects, keratinocyte-released nitric oxide (NO), importantly involved in the wound healing process, is increased by keratinocyte exposure to EMFT (Patruno et al., 2010), proinflammatory chemokines are reduced (Vianale et al., 2008), and differentiation is enhanced (Arai et al., 2013), suggesting that each of these aspects may serve as targets in the EMFT–wound healing connection. In addition, there is now evidence for a Ca^{++} connection as suggested by a DC electric field–induced increase in Ca^{++} influx into undifferentiated keratinocytes (Dube et al., 2012).

15.5.3 FIBROBLASTS

Possible mechanisms whereby EMFT connects with fibroblasts in healing derive from interactions observed when fibroblasts are exposed to EMF of various types. An early suggestion was that fibroblast EMF exposure induces membrane currents that open voltage-controlled Ca^{++} channels (Biedebach, 1989) that cause changes in cell migration and proliferation. The impact of EMFT on Ca^{++} processes has been much studied. Sinusoidal EMF exposure (20 Hz, 8 mT) of human skin fibroblasts changed cellular Ca^{++} oscillation activity in a way that depended on the cell's differentiation state (Loschinger et al., 1999). Such changes in proliferation and differentiation could be triggered by transient increases in cAMP-dependent protein kinase activity (Thumm et al., 1999). Indeed, prior work showed that fibroblasts exposed to PRF-EMF (27.12 MHz) caused enhanced cell proliferation (George et al., 2002, Gilbert et al., 2002). Further, EMF stimulation (10–100 Hz, 0–130 $\mu A/cm^2$) of dermal fibroblasts in a collagen matrix showed an amplitude and frequency windowing process in cell proliferation in which an ion-interference mechanism was involved (Binhi and Goldman, 2000). The proposed ion-interference mechanism considers effects of induced electric gradients on protein-bound substrate ions. Other evidence of a Ca^{++} connection comes from studies in which fibroblasts exposed to 2 V/cm fields at 1 and 10 Hz (but not 100 Hz) caused a sixfold increase in internal Ca^{++} (Cho et al., 2002) likely due to more Ca^{++} influx via voltage-gated Ca^{++} channels. Such channel-gating processes can be started by membrane depolarization that could be due to forced vibration of free ions on either side of the membrane that causes membrane potential changes sufficient to open the voltage-gated channels (Panagopoulos et al., 2000). More recently (Sunkari et al., 2011), in vitro migration and proliferation were enhanced when fibroblasts were exposed to a low-intensity 1 GHz EMF. In addition, the release of fibroblast proinflammatory cytokines was reduced when exposed to pulsed EMF (Gomez-Ochoa et al., 2011). Increased migration and growth were also demonstrated when fibroblasts were exposed to fields ranging from

50 to 200 mV/mm in vitro (Rouabhia et al., 2013) with the possibility that low-intensity fields enhance fibroblast migration via a reactive O_2 species pathway (Tandon et al., 2014).

EMF–Ca^{++} connections in healing are ubiquitous and, in this author's view, are important research targets. A theoretical framework for a connection between imposed fields and Ca^{++} channel dynamics has recently been proposed (Sun et al., 2013). But it must be remembered that although Ca^{++} entry into fibroblasts is associated with fibroblast stimulation (Huang et al., 1999), Ca^{++} also affects blood vessels and blood flow, causing decreased flow if vascular smooth muscle is so exposed. To be effective in a wound healing sense, the timing of the application of EMFT-triggered Ca^{++} stimulations within the wound healing cycle should be considered. For now, little is known regarding optimal timing.

15.5.4 LEUKOCYTES AND MACROPHAGE INVOLVEMENT

Leukocyte involvement in wound healing occurs mainly when they are activated during the inflammatory phase. Activation occurs with a respiratory burst, a release of cytokines and O_2 radicals, and upregulation of cell surface receptors that cause increased leukocyte–endothelial adhesion. This adhesion was demonstrated to increase in vivo when exposed to sinusoidal magnetic fields (50 Hz) greater than about 30 mT (Ushiyama and Ohkubo, 2004). An additional factor that may be involved in this process is myeloperoxidase (MPO), which is positively charged and is subject to the effects of external fields. MPO plays a role in PMN recruitment (Klinke et al., 2011). PMNs normally move toward the wound via galvanotaxis (Rapp et al., 1988) where they are needed for their antibacterial actions. But their continued entry and presence may cause reduced local blood flow due to capillary plugging, abnormal vasoconstriction, and tissue damage associated with PMN continued enzyme release. Evidence of such impaired healing comes from studies on diabetic mice with a prolonged inflammatory phase and retarded wound healing (Wetzler et al., 2000). The prolonged inflammatory phase appeared to be due to the expression of inflammatory and leukocyte chemoattractant proteins released by keratinocytes that caused activated neutrophils and macrophages to be sustained within the wound. If the inflammatory phase is abnormally prolonged, EMFT effects on activated PMN could affect wound healing via modulations of cellular free Ca^{++} oscillations and membrane potentials that accompany PMN release of reactive O_2 metabolites. The intensity of these normal oscillations (0.05–0.1 Hz) could be increased if 20 ms pulses were delivered at the trough of these oscillations (Kindzelskii and Petty, 2000). Additionally, O_2 metabolites could be increased or decreased depending on the phase between applied field and the Ca^{++} oscillations. It is noteworthy that electrical stimulation enhanced PMN, monocyte, and macrophage migration in enhanced healing (Kloth and McCulloch, 1996).

15.6 BLOOD FLOW AND EDEMA AS EMF-RELATED WOUND HEALING PROCESSES AND TARGETS

15.6.1 BLOOD FLOW

As previously described, blood flow to and within the wound is an important target in EMFT-related wound healing especially in ischemic and diabetic wounds. EMF-related blood flow changes may be induced directly via EMF effects on vascular smooth muscle, ECs, or blood and indirectly via neural activation, as with transcutaneous electrical nerve stimulation (Kaada and Emru, 1988). Skin vessels are innervated by sensory, sympathetic, and parasympathetic fibers (Ruocco et al., 2002) and are each suitable EMF targets. Factors tending to reduce flow within the wound, such as trapped leukocytes, are also valid targets. Although blood flow deficits are involved in ischemic and in some diabetic ulcers, it is not necessarily true that more flow causes faster healing or that more tissue oxygenation is always best for the natural healing process. Effects of blood flow on healing on the timing of increases or decreases: If flow is initially too high, it may affect an angiogenesis trigger (low O_2), and if too high, it can increase edema. Alternatively, if flow is too low, it won't support wound metabolism and may cause sustained inflammation that inhibits healing. The need to reverse polarity of some EMFT forms to cause healing may reflect a need for different flow needs (Brown et al., 1989).

Blood flow in venous ulcers illustrates these points. Often, total leg flow is increased (Mayrovitz and Larsen, 1994a) as is periulcer skin microvascular flow (Mayrovitz and Larsen, 1994b, Malanin et al., 1999). But abnormally dilated arterioles are observed (Junger et al., 1996), and a maldistribution of flow between nutritive and nonnutritive pathways is present (Junger et al., 2000), possibly due to activated leukocytes plugging capillaries. If leukocytes are involved, then an EMF-related reduction in neutrophil activation and adherence might be beneficial in three ways: (1) reducing local ischemia in areas served by obstructed capillaries, (2) normalizing effects of enzymes and free radicals released by activated leukocytes, and (3) reducing edema caused by their activation. Further, in patients with venous ulcers, arteriolar vasoconstriction normally induced by standing is blunted (Belcaro et al., 1995) and likely contributes to microvascular hyper-perfusion that exacerbates hypertension in capillaries and venules thereby causing more tissue edema. Thus, a portrait emerges suggesting a plausible basis for delayed healing that includes an overall *hyperperfusion* with simultaneously reduced wound blood flow and localized tissue edema. This scenario suggests that an EMF-related selective *vasoconstriction* of non-nutrient circulation may be of benefit. Alternatively, an EMF-related increase in local nutritional wound blood flow, if it overcomes the relative ischemia without causing substantial edema, might favor wound healing. Normally, edema is controlled via compression bandaging, which, among other aspects, might redistribute microcirculation and thereby to normalize a deficient nutritional capillary network. EMFT therapy for venous ulcers should always be used in conjunction with compression bandaging.

It is noteworthy that patients with chronic venous insufficiency, a frequent forerunner of venous ulcers, have increased vasomotion frequency (Chittenden et al., 1992) causing spontaneous blood flow changes in the frequency range between 0.05 and 0.5 Hz (Mayrovitz, 1998a). This suggests that EMF-related effects on vasomotion (Xu et al., 1998) may impact wound flow and healing. Such EMF effects may work via effects on intracellular Ca^{++} oscillations and other Ca^{++}-signaling processes. Although not specifically studied in vascular smooth muscle cells, an EMF-related (50 Hz) reduction in total spectral power content of cytosolic Ca^{++} oscillations and specific changes in the low-frequency band ($0-10^{-3}$ Hz) have been demonstrated in human leukemia cells (Galvanovskis et al., 1996). An argument for the role of spectral power changes as a mode of cellular encoding has been made (Galvanovskis and Sandblom, 1998) although both amplitude and frequency may be involved. Such processes could be involved in EMF-related effects on arteriolar vasomotion and associate blood flow changes. Based on these and other considerations, it is the author's view that effectiveness of EMFT for altering blood flow to stimulate healing may be optimized by linking field/current parameters to rhythms of healing process via feedback that detects and accommodates naturally occurring physiological and vascular dynamics.

15.6.2 EDEMA

Interstitial fluid accumulation as edema or as a protein-rich lymphedema retards wound healing by decreasing blood flow, reducing O_2 diffusion to tissue (Hunt et al., 2000), and acting as a breeding ground for infection (Hunt et al., 1986). Initially, edema is due to capillary permeability changes in the inflammatory phase with damage or dysfunction of terminal lymphatics also probably involved. Edema presence is obvious under some conditions; sometimes, its presence is *silent*, as microedema in the wound environment and its effects on healing are often not considered. Further, the physical features of sustained edema may change over time due to a progressive increase in protein concentration and fibrin cross-linking. These changes further impact healing. Because of the well-established ability of EMFT to reduce gross edema, a question arises as to if EMF-related effects that may reduce microedema, either directly or by its effects on lymphatic pathways, play a role in possible favorable effects of EMFT on wound healing. PRF-EMF may affect lymphatic channels as they do blood vessels (Mayrovitz et al., 2002a). An impact of lymphatics is further suggested by the observation that lymphatic vessels near some ulcers are reduced in number and have partially destroyed endothelium (Eliska and Eliskova, 2001) and they have potential role in angiogenesis to support granulation tissue (Paavonen et al., 2000). Earlier results (Mayrovitz et al., 2002a) with PRF-EMFT (27.1 MHz) reduced edema in patients with postmastectomy lymphedema. Since in these patients the main deficit is reduced lymphatic pathways, reduced edema with EMFT may be achieved by the development of alternate lymphatic pathways. This suggests the possibility that a promising target for EMFT might be lymphatic vessels within and surrounding the wound area. New methods are now available to assess local tissue edema at almost any body site in a noninvasive manner (Mayrovitz, 2007, Mayrovitz et al., 2013a,b) and should aid in new research efforts to assess the possible connections between EMFT and wound-related edema changes.

15.7 CONCLUSIONS

The cumulative substantial evidence from cellular and animal experiments and from human studies indicates positive connections between forms of electromagnetic therapy and wound healing. The composite findings provide a useful framework and underlying basis for EMF therapy when used in a thoughtful and selective manner to treat certain chronic or recalcitrant wounds. But mechanisms are at best speculative with large gaps in our understanding of specific cellular and functional targets, therapeutic dose, and regimens to achieve *optimal* treatment of specific wound types.

The complexity of the wound healing process and the differential features of specific wound types require a selective approach to choose EMF therapy parameters, timing, and targets. This means that therapeutic EMF approaches should be based on physical and physiological considerations, which then are judged on therapeutic outcomes. The EMFT–wound connections described in this chapter may provide a basis for continued advances in this still-evolving adjunctive therapeutic modality.

REFERENCES

Aetna IC (2014) Clinical policy bulletin: Electrical stimulation for chronic ulcers (Number: 0680). vol. http://www.aetna.com/cpb/medical/data/600_699/0680.html. Accessed October 26, 2014.

Akai M, Hayashi K. Effect of electrical stimulation on musculoskeletal systems: A meta-analysis of controlled clinical trials. *Bioelectromagnetics* 23 (2002):132–143.

Akcay MN, Ozcan O, Gundogdu C, Akcay G, Balik A, Kose K, Oren D. Effect of nitric oxide synthase inhibitor on experimentally induced burn wounds. *J Trauma* 49 (2000):327–330.

Amir Y, Halfens RJ, Lohrmann C, Schols JM. Pressure ulcer prevalence and quality of care in stroke patients in an Indonesian hospital. *J Wound Care* 22 (2013):254, 256, 258–260.

Arai KY, Nakamura Y, Hachiya Y, Tsuchiya H, Akimoto R, Hosoki K, Kamiya S, Ichikawa H, Nishiyama T. Pulsed electric current induces the differentiation of human keratinocytes. *Mol Cell Biochem* 379 (2013):235–241.

Aygor HE, Sahin S, Sozen E, Baydal B, Aykar FS, Akcicek F. Features of pressure ulcers in hospitalized older adults. *Adv Skin Wound Care* 27 (2014):122–126.

Aziz Z, Cullum N, Flemming K. Electromagnetic therapy for treating venous leg ulcers. *Cochrane Database Syst Rev* 2 (2013):CD002933.

Bai H, Forrester JV, Zhao M. DC electric stimulation upregulates angiogenic factors in endothelial cells through activation of VEGF receptors. *Cytokine* 55 (2011):110–115.

Baker LL, Chambers R, DeMuth SK, Villar F. Effects of electrical stimulation on wound healing in patients with diabetic ulcers. *Diabetes Care* 20 (1997):405–412.

Balakatounis KC, Angoules AG. Low-intensity electrical stimulation in wound healing: Review of the efficacy of externally applied currents resembling the current of injury. *Eplasty* 8 (2008):e28.

Barker AT, Jaffe LF, Vanable JW, Jr. The glabrous epidermis of cavies contains a powerful battery. *Am J Physiol* 242 (1982):R358–R366.

Becker RO. Augmentation of regenerative healing in man. A possible alternative to prosthetic implantation. *Clin Orthop* 83 (1972):255–262.

Belcaro G, Laurora G, Cesarone MR, De Sanctis MT, Incandela L. Microcirculation in high perfusion micro-angiopathy. *J Cardiovasc Surg* (Torino) 36 (1995):393–398.

Bentall RHC. Low-level pulsed radiofrequency fields and the treatment of soft-tissue injuries. *Bioelec Bioenerg* 16 (1986):531–548.

Biedebach MC. Accelerated healing of skin ulcers by electrical stimulation and the intracellular physiological mechanisms involved. *Acupunct Electrother Res* 14 (1989):43–60.

Binhi VN, Goldman RJ. Ion-protein dissociation predicts 'windows' in electric field-induced wound-cell proliferation. *Biochim Biophys Acta* 1474 (2000):147–156.

Brimson CH, Nigam Y. The role of oxygen-associated therapies for the healing of chronic wounds, particularly in patients with diabetes. *J Eur Acad Dermatol Venereol* 27 (2013):411–418.

Brown M, McDonnell MK, Menton DN. Polarity effects on wound healing using electric stimulation in rabbits. *Arch Phys Med Rehabil* 70 (1989):624–627.

Burr HS, Harvey SC, Taffel M. Bio-electric correlates of wound healing. *Yale J Biol Med* 11 (1938):104–107.

Byers PH, Carta SG, Mayrovitz HN. Pressure ulcer research issues in surgical patients. *Adv Skin Wound Care* 13 (2000):115–121.

Callaghan MJ, Chang EI, Seiser N, Aarabi S, Ghali S, Kinnucan ER, Simon BJ, Gurtner GC. Pulsed electro-magnetic fields accelerate normal and diabetic wound healing by increasing endogenous FGF-2 release. *Plast Reconstr Surg* 121 (2008):130–141.

Canedo-Dorantes L, Garcia-Cantu R, Barrera R, Mendez-Ramirez I, Navarro VH, Serrano G. Healing of chronic arterial and venous leg ulcers with systemic electromagnetic fields. *Arch Med Res* 33 (2002):281–289.

Carley PJ, Wainapel SF. Electrotherapy for acceleration of wound healing: Low intensity direct current. *Arch Phys Med Rehabil* 66 (1985):443–446.

Cheng HW, James AF, Foster RR, Hancox JC, Bates DO. VEGF activates receptor-operated cation channels in human microvascular endothelial cells. *Arterioscler Thromb Vasc Biol* 26 (2006):1768–1776.

Chittenden SJ, Shami SK, Cheatle TR, Scurr JH, Coleridge Smith PD. Vasomotion in the leg skin of patients with chronic venous insufficiency. *Vasa* 21 (1992):138–142.

Cho MR, Marler JP, Thatte HS, Golan DE. Control of calcium entry in human fibroblasts by frequency-dependent electrical stimulation. *Front Biosci* 7 (2002):a1–a8.

Claeys LG. Improvement of microcirculatory blood flow under epidural spinal cord stimulation in patients with nonreconstructible peripheral arterial occlusive disease. *Artif Organs* 21 (1997):201–206.

Comorosan S, Vasilco R, Arghiropol M, Paslaru L, Jieanu V, Stelea S. The effect of diapulse therapy on the healing of decubitus ulcer. *Rom J Physiol* 30 (1993):41–45.

Conti M, Peretti E, Cazzetta G, Galimberti G, Vermigli C, Pola R, Scionti L, Bosi E. Frequency-modulated electromagnetic neural stimulation enhances cutaneous microvascular flow in patients with diabetic neuropathy. *J Diabetes Complications* 23 (2009):46–48.

Costin GE, Birlea SA, Norris DA. Trends in wound repair: Cellular and molecular basis of regenerative therapy using electromagnetic fields. *Curr Mol Med* 12 (2012):14–26.

Crawford PE, Fields-Varnado M. Guideline for the management of wounds in patients with lower-extremity neuropathic disease: An executive summary. *J Wound Ostomy Continence Nurs* 40 (2013):34–45.

Delle Monache S, Alessandro R, Iorio R, Gualtieri G, Colonna R. Extremely low frequency electromagnetic fields (ELF-EMFs) induce in vitro angiogenesis process in human endothelial cells. *Bioelectromagnetics* 29 (2008):640–648.

Dube J, Rochette-Drouin O, Levesque P, Gauvin R, Roberge CJ, Auger FA, Goulet D et al. Human keratinocytes respond to direct current stimulation by increasing intracellular calcium: Preferential response of poorly differentiated cells. *J Cell Physiol* 227 (2012):2660–2667.

Eliska O, Eliskova M. Morphology of lymphatics in human venous crural ulcers with lipodermatosclerosis. *Lymphology* 34 (2001):111–123.

Erdman WJ. Peripheral blood flow measurements during application of pulse high frequency currents. *Amer J Orthop* 2 (1960):196–197.

Flemming K, Cullum N. Electromagnetic therapy for the treatment of pressure sores. *Cochrane Database Syst Rev* 1 (2001a):CD002930.

Flemming K, Cullum N. Electromagnetic therapy for the treatment of venous leg ulcers. *Cochrane Database Syst Rev* 1 (2001b):CD002933.

Foulds IS, Barker AT. Human skin battery potentials and their possible role in wound healing. *Br J Dermatol* 109 (1983):515–522.

Franek A, Polak A, Kucharzewski M. Modern application of high voltage stimulation for enhanced healing of venous crural ulceration. *Med Eng Phys* 22 (2000):647–655.

Frank S, Kampfer H, Wetzler C, Pfeilschifter J. Nitric oxide drives skin repair: Novel functions of an established mediator. *Kidney Int* 61 (2002):882–888.

Fromy B, Abraham P, Bouvet C, Bouhanick B, Fressinaud P, Saumet JL. Early decrease of skin blood flow in response to locally applied pressure in diabetic subjects. *Diabetes* 51 (2002):1214–1217.

Galvanovskis J, Sandblom S. Periodic forcing of intracellular calcium oscillators: Theoretical studies of the effects of low frequency fields on the magnitude of oscillations. *Bioelec Bioenerg* 46 (1998):161–174.

Galvanovskis J, Sandblom J, Bergqvist B, Galt S, Hamnerius Y. The influence of 50-Hz magnetic fields on cytoplasmic Ca^{2+} oscillations in human leukemia T-cells. *Sci Total Environ* 180 (1996):19–33.

Game FL, Hinchliffe RJ, Apelqvist J, Armstrong DG, Bakker K, Hartemann A, Londahl M, Price PE, Jeffcoate WJ. A systematic review of interventions to enhance the healing of chronic ulcers of the foot in diabetes. *Diabetes Metab Res Rev* 28 Suppl 1 (2012):119–141.

Gardner SE, Frantz RA, Schmidt FL. Effect of electrical stimulation on chronic wound healing: A meta-analysis. *Wound Repair Regen* 7 (1999):495–503.

George FR, Lukas RJ, Moffett J, Ritz MC. In-vitro mechanisms of cell proliferation induction: A novel bioactive treatment for accelerating wound healing. *Wounds* 14 (2002):107–115.

Gilbert TL, Griffin N, Moffett J, Ritz MC, George FR. The provant wound closure system induces activation of p44/42 MAP kinase in normal cultured human fibroblasts. *Ann NY Acad Sci* 961 (2002):168–171.

Goldman R, Brewley B, Golden M. Electrotherapy reoxygenates inframalleolar ischemic wounds on diabetic patients. *Adv Skin Wound Care* 15 (2002):112–120.

Goldman R, Rosen M, Brewley B, Golden M. Electrotherapy promotes healing and microcirculation of infrapopliteal ischemic wounds: A prospective pilot study. *Adv Skin Wound Care* 17 (2004):284–294.

Gomez-Ochoa I, Gomez-Ochoa P, Gomez-Casal F, Cativiela E, Larrad-Mur L. Pulsed electromagnetic fields decrease proinflammatory cytokine secretion (IL-1beta and TNF-alpha) on human fibroblast-like cell culture. *Rheumatol Int* 31 (2011):1283–1289.

Gonzalez ER, Oley MA. The management of lower-extremity diabetic ulcers. *Manag Care Interface* 13 (2000):80–87.

Gottrup F, Apelqvist J. Present and new techniques and devices in the treatment of DFU: A critical review of evidence. *Diabetes Metab Res Rev* 28 Suppl 1 (2012):64–71.

Gottrup F, Henneberg E, Trangbaek R, Baekmark N, Zollner K, Sorensen J. Point prevalence of wounds and cost impact in the acute and community setting in Denmark. *J Wound Care* 22 (2013):413–414, 416, 418–422.

Graber JN, Lifson A. The use of spinal cord stimulation for severe limb-threatening ischemia: A preliminary report. *Ann Vasc Surg* 1 (1987):578–582.

Griffin JW, Tooms RE, Mendius RA, Clifft JK, Vander Zwaag R, el-Zeky F. Efficacy of high voltage pulsed current for healing of pressure ulcers in patients with spinal cord injury. *Phys Ther* 71 (1991):433–442; discussion 442–434.

Grimnes S. Pathways of ionic flow through human skin in vivo. *Acta Derm Venereol* 64 (1984):93–98.

Gschwandtner ME, Ehringer H. Microcirculation in chronic venous insufficiency. *Vasc Med* 6 (2001):169–179.

Gunningberg L, Hommel A, Baath C, Idvall E. The first national pressure ulcer prevalence survey in county council and municipality settings in Sweden. *J Eval Clin Pract* 19 (2013):862–867.

Hamed S, Bennett CL, Demiot C, Ullmann Y, Teot L, Desmouliere A. Erythropoietin, a novel repurposed drug: An innovative treatment for wound healing in patients with diabetes mellitus. *Wound Repair Regen* 22 (2014):23–33.

Harms S, Bliss DZ, Garrard J, Cunanan K, Savik K, Gurvich O, Mueller C, Wyman JF, Eberly L, Virnig B. Prevalence of pressure ulcers by race and ethnicity for older adults admitted to nursing homes. *J Gerontol Nurs* 40 (2014):20–26.

Hart FX, Laird M, Riding A, Pullar CE. Keratinocyte galvanotaxis in combined DC and AC electric fields supports an electromechanical transduction sensing mechanism. *Bioelectromagnetics* 34 (2013):85–94.

Hedenius P, Odeblad E, Wahlstrom L. Some preliminary investigations on the therapeutic effect of pulsed short waves in intermittent claudication. *Curr Therap Res* 8 (1966):317–321.

Holmes C, Wrobel JS, Maceachern MP, Boles BR. Collagen-based wound dressings for the treatment of diabetes-related foot ulcers: A systematic review. *Diabetes Metab Syndr Obes* 6 (2013):17–29.

Houreld NN. Shedding light on a new treatment for diabetic wound healing: A review on phototherapy. *Scientific World J* 2014 (2014):398412.

Huang JS, Mukherjee JJ, Chung T, Crilly KS, Kiss Z. Extracellular calcium stimulates DNA synthesis in synergism with zinc, insulin and insulin-like growth factor I in fibroblasts. *Eur J Biochem* 266 (1999):943–951.

Hunt DL. Diabetes: Foot ulcers and amputations. *Clin Evid* (Online) 2011 (2011):0602. Published online August 26, 2011.

Hunt TK, Hopf H, Hussain Z. Physiology of wound healing. *Adv Skin Wound Care* 13 (2000):6–11.

Hunt TK, Rabkin J, von Smitten K. Effects of edema and anemia on wound healing and infection. *Curr Stud Hematol Blood Transfus* 53 (1986):101–113.

Huo R, Ma Q, Wu JJ, Chin-Nuke K, Jing Y, Chen J, Miyar ME, Davis SC, Li J. Noninvasive electromagnetic fields on keratinocyte growth and migration. *J Surg Res* 162 (2010):299–307.

Ieran M, Zaffuto S, Bagnacani M, Annovi M, Moratti A, Cadossi R. Effect of low frequency pulsing electromagnetic fields on skin ulcers of venous origin in humans: A double-blind study. *J Orthop Res* 8 (1990):276–282.

Illingworth CM, Barker AT. Measurement of electrical currents emerging during the regeneration of amputated finger tips in children. *Clin Phys Physiol Meas* 1 (1980):87.

Itoh M, Montemayor JS, Jr., Matsumoto E, Eason A, Lee MH, Folk FS. Accelerated wound healing of pressure ulcers by pulsed high peak power electromagnetic energy (Diapulse). *Decubitus* 4 (1991):24–25, 29–34.

Jacobs MJ, Jorning PJ, Beckers RC, Ubbink DT, van Kleef M, Slaaf DW, Reneman RS. Foot salvage and improvement of microvascular blood flow as a result of epidural spinal cord electrical stimulation. *J Vasc Surg* 12 (1990):354–360.

Jacobs MJ, Jorning PJ, Joshi SR, Kitslaar PJ, Slaaf DW, Reneman RS. Epidural spinal cord electrical stimulation improves microvascular blood flow in severe limb ischemia. *Ann Surg* 207 (1988):179–183.

Jaffe LF, Vanable JW, Jr. Electric fields and wound healing. *Clin Dermatol* 2 (1984):34–44.

Jan YK, Liao F, Rice LA, Woods JA. Using reactive hyperemia to assess the efficacy of local cooling on reducing sacral skin ischemia under surface pressure in people with spinal cord injury: A preliminary report. *Arch Phys Med Rehabil* 94 (2013):1982–1989.

Jankovic A, Binic I. Frequency rhythmic electrical modulation system in the treatment of chronic painful leg ulcers. *Arch Dermatol Res* 300 (2008):377–383.

Junger M, Klyscz T, Hahn M, Rassner G. Disturbed blood flow regulation in venous leg ulcers. *Int J Microcirc Clin Exp* 16 (1996):259–265.

Junger M, Steins A, Hahn M, Hafner HM. Microcirculatory dysfunction in chronic venous insufficiency (CVI). *Microcirculation* 7 (2000):S3–S12.

Kaada B, Emru M. Promoted healing of leprous ulcers by transcutaneous nerve stimulation. *Acupunct Electrother Res* 13 (1988):165–176.

Kindzelskii AL, Petty HR. Extremely low frequency pulsed DC electric fields promote neutrophil extension, metabolic resonance and DNA damage when phase-matched with metabolic oscillators. *Biochim Biophys Acta* 1495 (2000):90–111.

Klinke A, Nussbaum C, Kubala L, Friedrichs K, Rudolph TK, Rudolph V, Paust HJ et al. Myeloperoxidase attracts neutrophils by physical forces. *Blood* 117 (2011):1350–1358.

Kloth LC. Electrical stimulation for wound healing: A review of evidence from in vitro studies, animal experiments, and clinical trials. *Int J Low Extrem Wounds* 4 (2005):23–44.

Kloth LC, McCulloch JM. Promotion of wound healing with electrical stimulation. *Adv Wound Care* 9 (1996):42–45.

Knighton DR, Hunt TK, Scheuenstuhl H, Halliday BJ, Werb Z, Banda MJ. Oxygen tension regulates the expression of angiogenesis factor by macrophages. *Science* 221 (1983):1283–1285.

Kruglikov IL, Dertinger H. Stochastic resonance as a possible mechanism of amplification of weak electric signals in living cells. *Bioelectromagnetics* 15 (1994):539–547.

Kumar K, Toth C, Nath RK, Verma AK, Burgess JJ. Improvement of limb circulation in peripheral vascular disease using epidural spinal cord stimulation: A prospective study. *J Neurosurg* 86 (1997):662–669.

Larsen JA, Overstreet J. Pulsed radio frequency energy in the treatment of complex diabetic foot wounds: Two cases. *J Wound Ostomy Continence Nurs* 35 (2008):523–527.

Lazarus G, Valle MF, Malas M, Qazi U, Maruthur NM, Doggett D, Fawole OA, Bass EB, Zenilman J. Chronic venous leg ulcer treatment: Future research needs. *Wound Repair Regen* 22 (2014):34–42.

Lee BY, Al-Waili N, Stubbs D, Wendell K, Butler G, Al-Waili T, Al-Waili A. Ultra-low microcurrent in the management of diabetes mellitus, hypertension and chronic wounds: Report of twelve cases and discussion of mechanism of action. *Int J Med Sci* 7 (2009):29–35.

Lee RC, Canaday DJ, Doong H. A review of the biophysical basis for the clinical application of electric fields in soft-tissue repair. *J Burn Care Rehabil* 14 (1993):319–335.

Lee RH, Efron D, Tantry U, Barbul A. Nitric oxide in the healing wound: A time-course study. *J Surg Res* 101 (2001):104–108.

Loschinger M, Thumm S, Hammerle H, Rodemann HP. Induction of intracellular calcium oscillations in human skin fibroblast populations by sinusoidal extremely low-frequency magnetic fields (20 Hz, 8 mT) is dependent on the differentiation state of the single cell. *Radiat Res* 151 (1999):195–200.

Lundeberg TC, Eriksson SV, Malm M. Electrical nerve stimulation improves healing of diabetic ulcers. *Ann Plast Surg* 29 (1992):328–331.

Magnoni C, Rossi E, Fiorentini C, Baggio A, Ferrari B, Alberto G. Electrical stimulation as adjuvant treatment for chronic leg ulcers of different aetiology: An RCT. *J Wound Care* 22 (2013):525–526, 528–533.

Malanin K, Kolari PJ, Havu VK. The role of low resistance blood flow pathways in the pathogenesis and healing of venous leg ulcers. *Acta Derm Venereol* 79 (1999):156–160.

Markov M. Electric current and electromagnetic field effects on soft tissue: Implications for wound healing. *Wounds* 7 (1995):94–110.

Markov M, Pilla AA. Electromagnetic stimulation of soft tissues: Pulsed radio frequency treatment of postoperative pain and edema. *Wounds* 7 (1995):143–151.

Martino CF. Static magnetic field sensitivity of endothelial cells. *Bioelectromagnetics* 32 (2011):506–508.

Martino CF, Perea H, Hopfner U, Ferguson VL, Wintermantel E. Effects of weak static magnetic fields on endothelial cells. *Bioelectromagnetics* 31 (2010):296–301.

Mayrovitz H, Sims N, Macdonald J. Effects of pulsed radio frequency diathermy on postmastectomy arm lymphedema and skin blood flow: A pilot investigation. *Lymphology* (2002a): 35(suppl):353–356.

Mayrovitz HN (1998a) Assessment of Human Microvascular Function. In: *Analysis of Cardiovascular Function*, Drzewiecki, G. and Li, J. (eds), pp. 248–273. New York: Springer.

Mayrovitz HN. Pressure and blood flow linkages and impacts on pressure ulcer development. *Adv Wound Care* 11 (1998b):4.

Mayrovitz HN (2004) *Electromagnetic Linkages in Soft Tissue Wound Healing*. In: Markov, M. (ed.), Chapter 30, Marcell-Dekker, New York, pp. 461–483.

Mayrovitz HN. Assessing local tissue edema in postmastectomy lymphedema. *Lymphology* 40 (2007):87–94.

Mayrovitz HN, Bernal M, Brlit F, Desfor R. Biophysical measures of skin tissue water: Variations within and among anatomical sites and correlations between measures. *Skin Res Technol* 19 (2013a):47–54.

Mayrovitz HN, Groseclose EE. Effects of a static magnetic field of either polarity on skin microcirculation. *Microvasc Res* 69 (2005):24–27.

Mayrovitz HN, Groseclose EE, King D. No effect of 85 mT permanent magnets on laser-Doppler measured blood flow response to inspiratory gasps. *Bioelectromagnetics* 26 (2005):331–335.

Mayrovitz HN, Groseclose EE, Markov M, Pilla AA. Effects of permanent magnets on resting skin blood perfusion in healthy persons assessed by laser Doppler flowmetry and imaging. *Bioelectromagnetics* 22 (2001):494–502.

Mayrovitz HN, Groseclose EE, Sims N. Assessment of the short-term effects of a permanent magnet on normal skin blood circulation via laser-Doppler flowmetry. *Sci Rev Altern Med* 6 (2002b):5–9.

Mayrovitz HN, Larsen PB. Effects of pulsed electromagnetic fields on skin microvascular blood perfusion. *Wounds* 4 (1992):197–202.

Mayrovitz HN, Larsen PB. Leg blood flow in patients with venous ulcers: Relationship to site and ulcer area. *Wounds* 6 (1994a):195–200.

Mayrovitz HN, Larsen PB. Periwound skin microcirculation of venous leg ulcers. *Microvasc Res* 48 (1994b):114–123.

Mayrovitz HN, Larsen PB. Standard and near-surface laser-Doppler perfusion in foot dorsum skin of diabetic and nondiabetic subjects with and without coexisting peripheral arterial disease. *Microvasc Res* 48 (1994c):338–348.

Mayrovitz HN, Larsen PB. Functional microcirculatory impairment: A possible source of reduced skin oxygen tension in human diabetes mellitus. *Microvasc Res* 52 (1996):115–126.

Mayrovitz HN, Macdonald J, Smith JR. Blood perfusion hyperaemia in response to graded loading of human heels assessed by laser-Doppler imaging. *Clin Physiol* 19 (1999):351–359.

Mayrovitz HN, McClymont A, Pandya N. Skin tissue water assessed via tissue dielectric constant measurements in persons with and without diabetes mellitus. *Diabetes Technol Ther* 15 (2013b):60–65.

Mayrovitz HN, Sims N. Biophysical effects of water and synthetic urine on skin. *Adv Skin Wound Care* 14 (2001):302–308.

Mayrovitz HN, Sims N, Taylor MC. Sacral skin blood perfusion: A factor in pressure ulcers? *Ostomy Wound Manage* 48 (2002d):34–38, 40–32.

Mayrovitz HN, Smith J. Heel-skin microvascular blood perfusion responses to sustained pressure loading and unloading. *Microcirculation* 5 (1998):227–233.

Mayrovitz HN, Smith J, Delgado M, Regan MB. Heel blood perfusion responses to pressure loading and unloading in women. *Ostomy Wound Manage* 43 (1997):16–20, 22, 24 passim.

Mayrovitz HN, Smith JR. Adaptive skin blood flow increases during hip-down lying in elderly women. *Adv Wound Care* 12 (1999):295–301.

McCulloch JM, Kloth LC, Feedar JA (1995) Wound healing alternatives in management. In: *Contemporary Perspectives in Rehabilitation*, Wolf, S.E. (ed.), pp. 275–310. Philadelphia, PA: F. A. Davis.

McDaniel JC, Roy S, Wilgus TA. Neutrophil activity in chronic venous leg ulcers—A target for therapy? *Wound Repair Regen* 21 (2013):339–351.

Mingoli A, Sciacca V, Tamorri M, Fiume D, Sapienza P. Clinical results of epidural spinal cord electrical stimulation in patients affected with limb-threatening chronic arterial obstructive disease. *Angiology* 44 (1993):21–25.

Moore Z, Johanssen E, van Etten M. A review of PU prevalence and incidence across Scandinavia, Iceland and Ireland (Part I). *J Wound Care* 22 (2013):361–362, 364–368.

Muller IS, de Grauw WJ, van Gerwen WH, Bartelink ML, van Den Hoogen HJ, Rutten GE. Foot ulceration and lower limb amputation in type 2 diabetic patients in dutch primary health care. *Diabetes Care* 25 (2002):570–574.

Nelzen O. Prevalence of venous leg ulcer: The importance of the data collection method. *Phlebolymphology* 15 (2008):143–150.

Nordenstrom BE. Impact of biologically closed electric circuits (BCEC) on structure and function. *Integr Physiol Behav Sci* 27 (1992):285–303.

Nuccitelli R. A role for endogenous electric fields in wound healing. *Curr Top Dev Biol* 58 (2003):1–26.

Nuccitelli R, Nuccitelli P, Li C, Narsing S, Pariser DM, Lui K. The electric field near human skin wounds declines with age and provides a noninvasive indicator of wound healing. *Wound Repair Regen* 19 (2011):645–655.

Nuccitelli R, Nuccitelli P, Ramlatchan S, Sanger R, Smith PJ. Imaging the electric field associated with mouse and human skin wounds. *Wound Repair Regen* 16 (2008):432–441.

O'Toole EA. Extracellular matrix and keratinocyte migration. *Clin Exp Dermatol* 26 (2001):525–530.

Paavonen K, Puolakkainen P, Jussila L, Jahkola T, Alitalo K. Vascular endothelial growth factor receptor-3 in lymphangiogenesis in wound healing. *Am J Pathol* 156 (2000):1499–1504.

Panagopoulos DJ, Messini N, Karabarbounis A, Philippetis AL, Margaritis LH. A mechanism for action of oscillating electric fields on cells. *Biochem Biophys Res Commun* 272 (2000):634–640.

Patruno A, Amerio P, Pesce M, Vianale G, Di Luzio S, Tulli A, Franceschelli S, Grilli A, Muraro R, Reale M. Extremely low frequency electromagnetic fields modulate expression of inducible nitric oxide synthase, endothelial nitric oxide synthase and cyclooxygenase-2 in the human keratinocyte cell line HaCat: Potential therapeutic effects in wound healing. *Br J Dermatol* 162 (2010):258–266.

Peirce SM, Skalak TC, Rodeheaver GT. Ischemia-reperfusion injury in chronic pressure ulcer formation: A skin model in the rat. *Wound Repair Regen* 8 (2000):68–76.

Pesce M, Patruno A, Speranza L, Reale M. Extremely low frequency electromagnetic field and wound healing: Implication of cytokines as biological mediators. *Eur Cytokine Netw* 24 (2013):1–10.

Peschen M, Lahaye T, Hennig B, Weyl A, Simon JC, Vanscheidt W. Expression of the adhesion molecules ICAM-1, VCAM-1, LFA-1 and VLA-4 in the skin is modulated in progressing stages of chronic venous insufficiency. *Acta Derm Venereol* 79 (1999):27–32.

Peters EJ, Lavery LA, Armstrong DG, Fleischli JG. Electric stimulation as an adjunct to heal diabetic foot ulcers: A randomized clinical trial. *Arch Phys Med Rehabil* 82 (2001):721–725.

Ramadhinara A, Poulas K. Use of wireless microcurrent stimulation for the treatment of diabetes-related wounds: 2 case reports. *Adv Skin Wound Care* 26 (2013):1–4.

Rapp B, de Boisfleury-Chevance A, Gruler H. Galvanotaxis of human granulocytes. Dose-response curve. *Eur Biophys J* 16 (1988):313–319.

Rawe IM, Vlahovic TC. The use of a portable, wearable form of pulsed radio frequency electromagnetic energy device for the healing of recalcitrant ulcers: A case report. *Int Wound J* 9 (2012):253–258.

Reiber GE, Vileikyte L, Boyko EJ, del Aguila M, Smith DG, Lavery LA, Boulton AJ. Causal pathways for incident lower-extremity ulcers in patients with diabetes from two settings. *Diabetes Care* 22 (1999):157–162.

Ricci E, Afaragan M. The effect of stochastic electrical noise on hard-to-heal wounds. *J Wound Care* 19 (2010):96–103.

Rouabhia M, Park H, Meng S, Derbali H, Zhang Z. Electrical stimulation promotes wound healing by enhancing dermal fibroblast activity and promoting myofibroblast transdifferentiation. *PLoS One* 8 (2013):e71660.

Ruocco I, Cuello AC, Parent A, Ribeiro-da-Silva A. Skin blood vessels are simultaneously innervated by sensory, sympathetic, and parasympathetic fibers. *J Comp Neurol* 448 (2002):323–336.

Salzberg CA, Cooper-Vastola SA, Perez F, Viehbeck MG, Byrne DW. The effects of non-thermal pulsed electromagnetic energy on wound healing of pressure ulcers in spinal cord-injured patients: A randomized, double-blind study. *Ostomy Wound Manage* 41 (1995):42–44, 46, 48 passim.

Sarma GR, Subrahmanyam S, Deenabandhu A, Babu CR, Madhivathanan S, Kesavaraj N. Exposure to pulsed magnetic fields in the treatment of plantar ulcers in leprosy patients—A pilot, randomized, double-blind, controlled clinical trial. *Indian J Lepr* 69 (1997):241–250.

Scheel-Sailer A, Wyss A, Boldt C, Post MW, Lay V. Prevalence, location, grade of pressure ulcers and association with specific patient characteristics in adult spinal cord injury patients during the hospital stay: A prospective cohort study. *Spinal Cord* 51 (2013):828–833.

Sheffet A, Cytryn AS, Louria DB. Applying electric and electromagnetic energy as adjuvant treatment for pressure ulcers: A critical review. *Ostomy Wound Manage* 46 (2000):28–33, 36–40, 42–24.

Shi HP, Most D, Efron DT, Tantry U, Fischel MH, Barbul A. The role of iNOS in wound healing. *Surgery* 130 (2001):225–229.

Singh N, Armstrong DG, Lipsky BA. Preventing foot ulcers in patients with diabetes. *JAMA* 293 (2005):217–228.

Smit CA, Zwinkels M, van Dijk T, de Groot S, Stolwijk-Swuste JM, Janssen TW. Gluteal blood flow and oxygenation during electrical stimulation-induced muscle activation versus pressure relief movements in wheelchair users with a spinal cord injury. *Spinal Cord* 51 (2013):694–699.

Smith ME, Totten A, Hickam DH, Fu R, Wasson N, Rahman B, Motu'apuaka M, Saha S. Pressure ulcer treatment strategies: A systematic comparative effectiveness review. *Ann Intern Med* 159 (2013):39–50.

Smith PD. Update on chronic-venous-insufficiency-induced inflammatory processes. *Angiology* 52 Suppl 1 (2001):S35–S42.

Sonenblum SE, Vonk TE, Janssen TW, Sprigle SH. Effects of wheelchair cushions and pressure relief maneuvers on ischial interface pressure and blood flow in people with spinal cord injury. *Arch Phys Med Rehabil* (2014).

Spravchikov N, Sizyakov G, Gartsbein M, Accili D, Tennenbaum T, Wertheimer E. Glucose effects on skin keratinocytes: Implications for diabetes skin complications. *Diabetes* 50 (2001):1627–1635.

Stacey MC, Mata SD. Lower levels of PAI-2 may contribute to impaired healing in venous ulcers—A preliminary study. *Cardiovasc Surg* 8 (2000):381–385.

Stefanovska A, Vodovnik L, Benko H, Turk R. Treatment of chronic wounds by means of electric and electromagnetic fields. Part 2. Value of FES parameters for pressure sore treatment. *Med Biol Eng Comput* 31 (1993):213–220.

Stiller MJ, Pak GH, Shupack JL, Thaler S, Kenny C, Jondreau L. A portable pulsed electromagnetic field (PEMF) device to enhance healing of recalcitrant venous ulcers: A double-blind, placebo-controlled clinical trial. *Br J Dermatol* 127 (1992):147–154.

Strauch B, Patel MK, Navarro JA, Berdichevsky M, Yu HL, Pilla AA. Pulsed magnetic fields accelerate cutaneous wound healing in rats. *Plast Reconstr Surg* 120 (2007):425–430.

Sun GX, Wang LJ, Xiang C, Qin KR. A dynamic model for intracellular calcium response in fibroblasts induced by electrical stimulation. *Math Biosci* 244 (2013):47–57.

Sunkari VG, Aranovitch B, Portwood N, Nikoshkov A. Effects of a low-intensity electromagnetic field on fibroblast migration and proliferation. *Electromagn Biol Med* 30 (2011):80–85.

Tandon N, Cimetta E, Villasante A, Kupferstein N, Southall MD, Fassih A, Xie J, Sun Y, Vunjak-Novakovic G. Galvanic microparticles increase migration of human dermal fibroblasts in a wound-healing model via reactive oxygen species pathway. *Exp Cell Res* 320 (2014):79–91.

Tang JC, Marston WA, Kirsner RS. Wound Healing Society (WHS) venous ulcer treatment guidelines: What's new in five years? *Wound Repair Regen* 20 (2012):619–637.

Tassiopoulos AK, Golts E, Oh DS, Labropoulos N. Current concepts in chronic venous ulceration. *Eur J Vasc Endovasc Surg* 20 (2000):227–232.

Thumm S, Loschinger M, Glock S, Hammerle H, Rodemann HP. Induction of cAMP-dependent protein kinase A activity in human skin fibroblasts and rat osteoblasts by extremely low-frequency electromagnetic fields. *Radiat Environ Biophys* 38 (1999):195–199.

Ushiyama A, Ohkubo C. Acute effects of low-frequency electromagnetic fields on leukocyte-endothelial interactions in vivo. *In vivo* 18 (2004):125–132.

Valencia IC, Falabella A, Kirsner RS, Eaglstein WH. Chronic venous insufficiency and venous leg ulceration. *J Am Acad Dermatol* 44 (2001):401–421; quiz 422–404.

Vanable JW, Jr. (1989) Integumentary potentials and wound healing. In: *Electric Fields in Vertebrate Repair*, Ed. Borgens RB, pp. 171–224. Alan R. Liss, Inc, New York.

Vianale G, Reale M, Amerio P, Stefanachi M, Di Luzio S, Muraro R. Extremely low frequency electromagnetic field enhances human keratinocyte cell growth and decreases proinflammatory chemokine production. *Br J Dermatol* 158 (2008):1189–1196.

Vodovnik L, Karba R. Treatment of chronic wounds by means of electric and electromagnetic fields. Part 1. Literature review. *Med Biol Eng Comput* 30 (1992):257–266.

Wen L, Zhang C, Nong Y, Yao Q, Song Z. Mild electrical pulse current stimulation upregulates S100A4 and promotes cardiogenesis in MSC and cardiac myocytes coculture monolayer. *Cell Biochem Biophys* 65 (2013):43–55.

Wetzler C, Kampfer H, Stallmeyer B, Pfeilschifter J, Frank S. Large and sustained induction of chemokines during impaired wound healing in the genetically diabetic mouse: Prolonged persistence of neutrophils and macrophages during the late phase of repair. *J Invest Dermatol* 115 (2000):245–253.

Wood JM, Evans PE, 3rd, Schallreuter KU, Jacobson WE, Sufit R, Newman J, White C, Jacobson M. A multicenter study on the use of pulsed low-intensity direct current for healing chronic stage II and stage III decubitus ulcers. *Arch Dermatol* 129 (1993):999–1009.

Xu S, Okano H, Ohkubo C. Subchronic effects of static magnetic fields on cutaneous microcirculation in rabbits. *In vivo* 12 (1998):383–389.

Yamaguchi Y, Yoshikawa K. Cutaneous wound healing: An update. *J Dermatol* 28 (2001):521–534.

Yen-Patton GP, Patton WF, Beer DM, Jacobson BS. Endothelial cell response to pulsed electromagnetic fields: Stimulation of growth rate and angiogenesis in vitro. *J Cell Physiol* 134 (1988):37–46.

Young S, Hampton S, Tadej M. Study to evaluate the effect of low-intensity pulsed electrical currents on levels of oedema in chronic non-healing wounds. *J Wound Care* 20 (2011):368, 370–363.

Yuan Y, Wei L, Li F, Guo W, Li W, Luan R, Lv A, Wang H. Pulsed magnetic field induces angiogenesis and improves cardiac function of surgically induced infarcted myocardium in Sprague-Dawley rats. *Cardiology* 117 (2010):57–63.

Zhao Z, Qin L, Reid B, Pu J, Hara T, Zhao M. Directing migration of endothelial progenitor cells with applied DC electric fields. *Stem Cell Res* 8 (2012):38–48.

16 Physical Regulation in Cartilage and Bone Repair

Ruggero Cadossi, Matteo Cadossi, and Stefania Setti

CONTENTS

16.1 INTRODUCTION

The therapeutic possibilities at the physician's disposal foresee the use of both chemical and physical energies. While the use of chemicals (drugs) and ionizing electromagnetic energy for disease treatment has been well defined in different branches of medicine and surgery, this has not been the case for nonionizing electromagnetic energies. Clinical biophysics is the branch of medical science that studies the action process and the effects of nonionizing physical stimuli utilized for therapeutic purposes. The principles underpinning clinical biophysics entail the recognizability and specificity of the physical energy applied on the skeletal tissue, that is, biophysical stimulation (BS).

The BS of bone and cartilage repair is employed in many countries in the orthopedic field to promote and reactivate the formation of bone tissue, to control joint inflammation, and to stimulate anabolic activities of cartilage. The scientific origins of the BS techniques for bone healing are acknowledged to lie in the by now classic studies performed first by Fukada and Yasuda [1], then by Bassett and Becker [2]; study of the effect of BS techniques for cartilage healing is more recent [3].

16.2 BIOPHYSICAL STIMULATION OF BONE REPAIR

The aforementioned studies performed in the 1950s and 1960s highlighted the relation between bone tissue mechanical deformation and electric potentials. Bone generates two types of electric signal: one in response to mechanical deformation and the other in the absence of deformation.

The signal induced by structural deformation following the application of a load is present in bone, not necessarily vital, and can be ascribed to a dual origin: (1) to the direct piezoelectric effect and (2) to the electrokinetic phenomenon of the flow potential [4–10]. The electrical signal induced by the mechanical deformation has been considered to be the transducer of a physical force in a cell response and has been taken to be the mechanism that determines the continuous adaptation of the mechanical competence of bone to variations in load, according to the well-known Wolff's law [11].

In the absence of mechanical stress, vital bone generates an electrical signal detectable in vivo as surface stationary bioelectric potential and ex vivo as stationary electric (ionic) current that can be measured [12–20].

On the basis of these premises, biophysical enhancement of osteogenesis with different stimuli has been developed: pulsed electromagnetic field (PEMF, inductive system), capacitive-coupling electric field (CCEF, capacitive system), and the use of low-intensity pulsed ultrasound (LIPU, ultrasound system) (Figure 16.1).

16.2.1 MECHANISM OF ACTION OF BONE STIMULATION AND PRECLINICAL STUDIES

Biophysical interactions of physical stimuli at the cell membrane are not well understood, but a lot of studies have shown that BS stimulates the synthesis of extracellular matrix molecules and may be useful in clinically stimulating the repair of fractures and nonunions [21]. Various authors agree on the fact that the cell membrane plays the fundamental role in recognizing and transferring the physical stimulus to the various metabolic pathways of the cell; by this mechanism of action, a cell recognizes a physical stimulus and thus modifies its functions. The PEMF stimulation causes the liberation of calcium ions (Ca^{++}) from the smooth endoplasmic reticulum, whereas with the CCEF stimulation, an increase in Ca^{++} transport across voltage-gated channels is observed; for LIPU stimulation, ideal candidates for performing this function would seem to be mechanosensitive ion channels, although a genuine mechanoreceptor has not yet been identified [21,22] (Figure 16.2). The intracellular increase of the Ca^{++} determines a series of enzyme responses with resulting gene transcription (several bone morphogenetic proteins [BMPs], transforming growth factor-beta [TGF-β1], and collagen) and cell proliferation [22]. Upregulation of TGF-β1 mRNA expression has been reported in mechanically loaded bones; different groups have demonstrated increases in proliferation, differentiation, and transcription of mRNA for several

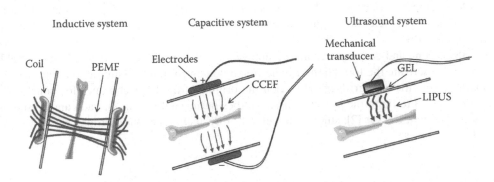

FIGURE 16.1 The techniques of biophysical stimulation of bone tissue.

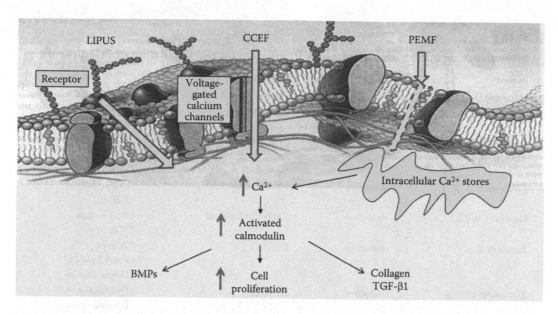

FIGURE 16.2 Schematic drawing of the signal transduction pathways of inductive, capacitive, and ultrasound system.

BMPs and TGF-βs in skeletal tissue with all biophysical system exposure [23–41] (Table 16.1). Physical agents may be synergistic with endogenously synthesized or exogenously applied growth factors in tissue repair; the application of physical stimuli results in changes in gene expression for signaling proteins. The interactions between growth factors and physical stimuli are a very fertile area for investigation.

The US Food and Drug Administration (FDA) has recognized in the inductive system the ability to enhance the healing process by an increased production of different growth factors, including multiple BMPs and other osteopromotive growth factors necessary to facilitate the healing of fractures and fusions. BS, in comparison with drug administration, is able to produce a local concentration of growth factor synthesis, without any systemic side effects. Nevertheless, it is important to keep in mind that, as with a drug, the dosage of physical stimulus is fundamental if positive effects on osteogenesis are to be produced. The biological effects of BS depend not only on the length of treatment time but also on the signal characteristics: intensity, waveform, frequency.

In vitro studies have shown that the physical stimuli increase the synthesis of bone matrix and favor the proliferation and differentiation of the osteoblast-like primary cells [42,43]. Fassina et al. investigated the effect of PEMF on SAOS-2 human osteoblast proliferation and on calcified matrix production over a polyurethane porous scaffold and showed a higher cell proliferation and a greater expression of decorin, fibronectin, osteocalcin, osteopontin, TGF-β1, type I collagen, and type III collagen in PEMF-stimulated culture than in controls [28]. Hartig et al. demonstrated that the exposure to CCEF of osteoblast-like primary cells increases the synthesis of bone matrix and favors their proliferation and differentiation [35]. Ryaby et al. reported that LIPU increased calcium incorporation in bone-cell cultures, reflecting a change in cell metabolism [44]. This increase in second messenger activity was paralleled by the modulation of adenylate cyclase activity and TGF-β1 synthesis in osteoblastic cells.

In vivo, the authors have observed an increase in the formation of bone tissue [45] and a shorter healing time of experimental fractures and/or bone lesions treated with inductive system [46,47]. Studies of newly formed bone tissue performed with tetracycline labeling have demonstrated

TABLE 16.1

Inductive, Capacitive, and Ultrasound Stimulation on the Regulation of TGF-β/BMPs

Authors	Physical System	In Vitro Models	Results
Nagai and Ota [23]	Inductive	Chick embryonic calvaria	↑BMP-2,4 mRNA
Bodamyali et al. [24]	Inductive	Rat calvarial osteoblasts	↑Proliferation ↑BMP-2,4 mRNA
Guerkov et al. [25]	Inductive	Human nonunion cells	↑TGF-β1
Aaron et al. [26]	Inductive	Endochondral ossification in vivo	↑Differentiation ↑TGF-β1mRNA protein
Lohmann et al. [27]	Inductive	MG63 osteoblasts	↑Differentiation ↑TGF-β1
Fassina et al. [28]	Inductive	SAOS-2 osteoblasts	↑Proliferation ↑TGF-β1
Jansen et al. [29]	Inductive	hBMSCs	↑TGF-β1 ↑BMP-2mRNA ↑Differentiation
Esposito et al. [30]	Inductive	hBMSCs	↑Proliferation ↑Differentiation
Ceccarelli et al. [31]	Inductive	hBMSCs	↑Proliferation ↑ECM deposition
Lim et al. [32]	Inductive	Human alveolar BMSCs	↑Proliferation ↑Differentiation
Zhou et al. [33]	Inductive	Rat calvarial osteoblasts	↑Proliferation
Zhuang et al. [34]	Capacitive	Osteoblastic cells (MC3T3-E1)	↑Proliferation ↑TGF-β1mRNA
Hartig et al. [35]	Capacitive	Osteoblast from periosteum explants	↑Proliferation ↑Differentiation
Wang et al. [36]	Capacitive	Osteoblastic cells (MC3T3-E1)	↑BMP-2,4,5,6,7mRNA
Bisceglia et al. [37]	Capacitive	Osteoblast-like cell lines (SAOS-2)	↑Proliferation
Clark et al. [38]	Capacitive	Human calvarial osteoblasts	↑BMP-2,4 mRNA ↑TGF-β1, β2, β3mRNA ↑FGF-2
Hauser et al. [39]	Ultrasound	Osteoblast-like cell lines (SAOS-2)	↑Proliferation
Fassina et al. [40]	Ultrasound	SAOS-2 human osteoblasts	↑Proliferation ↑ECM deposition
Xue et al. [41]	Ultrasound	Alveolar bone in vivo	↑BMP-2 mRNA

that, following exposure to PEMF, the ability of the osteoblast activity to lay down bone tissue (mineral apposition rate), that is to form trabeculae, is doubled [48]. Fini et al. demonstrated a significant increase in bone microhardness and in osteointegration at the bone interface of a hydroxyapatite cylinder implanted in trabecular bone of distal femur of rabbit stimulated with PEMF [49]. In experimental fractures produced on rabbit fibula, a significant shortening of healing time has been observed with the use of CCEF [50]. In the presence of osteoporosis induced by castration of rats, CCEF has been capable of inhibiting the onset of this pathology [51]. Positive effects on osteogenesis have been reported also with the use of LIPU in several animal studies. Wang et al. studied the healing of bilateral closed femoral shaft fractures in rats [52]. They reported a 67% increase in stiffness in the group treated with LIPU, significantly greater than the increase in the controls (p < 0.02).

16.2.2 RATIONALE FOR CLINICAL USE OF BIOPHYSICAL STIMULATION ON BONE REPAIR

BS was approved for clinical use by the FDA 40 years ago. Every year since then, tens of thousands of patients have undergone treatment throughout the world [53]; however, only products whose clinical effectiveness is well documented in the literature should be used.

Over the last few years since BS of osteogenesis has been in clinical use, a great number of clinical trials have been performed using the appropriate double-blind or control group protocols. These study designs were dictated by the need to discriminate effectively between the effects of BS and other possible associated orthopedic approaches, and to quantify the efficacy of the treatments in humans to favor bone consolidation.

In orthopedic practice, osteogenetic activity aimed at consolidation of a fracture continually comes up against problems of mechanical and biological kinds [54]. Among the factors that may jeopardize a repair process at bone tissue level, primary consideration is usually accorded to the mechanical aspects, on which orthopedic research has successfully focused for many years. More recently, it has been observed that failed consolidation may be ascribed to an insufficient osteogenetic response at the level of the fracture site, rather than to inadequate immobilization. Just as stimulation of a fracture with evident problems of mobility or diastasis between the stumps is contraindicated, so it appears unusual to operate on a patient with a satisfactory mechanical stability of the lesion when the problem can be attributed to an impaired osteogenetic response.

BS should be started only if the mechanical stability and the alignment of the fracture are guaranteed and if the gap extension does not exceed half of the diameter of the fractured bone [55].

These observations represent the rationale for indication of BS treatment; bearing in mind these principles, the rate of success obtained with BS exceeds 90%.

We now report the main results of trials in different clinical settings with the use of the BS techniques described earlier.

16.2.3 RISK FRACTURES

Some fracture patterns that occur in specific bones may compromise the blood supply (femoral neck fractures, talus) or may severely involve the surrounding soft tissue (tibial shaft). These fractures should be considered at risk because they are more likely to end up in a delayed union or a nonunion. BS has been shown to be able to accelerate healing of risk fractures treated with plaster cast or external fixation or complex fractures with serious damage to the soft tissues and exposure of the bone tissue [56,57]. In all cases, BS succeeded in shortening the average time of healing (25%–38%). Fractures that consolidate in 70–80 days from trauma do not benefit from BS. However, in those cases where the site and morphology of fracture, type of exposure, or conditions of the patient presage difficulties in the repair process, BS is rightly indicated [58]. Risk fractures amount to 20% of all fractures. The probability that they may evolve in nonunion (i.e., 5%–10% of fractures in the United States) is huge (20%–25%) and justifies the application of BS. Fontanesi et al., in a controlled study of 40 tibia fresh fractures treated with plaster, remarked how the effect of stimulation is evidenced through a reduction in average healing times ($p < 0.005$) [59]. Donkerwolcke Coussaert too noted a shortening in the time to union of stimulated tibial recent fractures treated with external fixation [60]. Faldini et al. reported the effects of PEMF in patients with femoral neck fractures treated with screw fixation in a randomized double-blind study [61]. Fracture healing was achieved in 94% of active patients compared with 69% of the placebo group. Benazzo et al. [62], using CCEF technique, observed earlier recovery in athletes with stress fracture. Heckman et al. [63] performed a randomized, double-blind, placebo-controlled trial of 67 closed or grade-I open tibial fractures to evaluate the effect of LIPU on the healing of cortical fractures. LIPU treatment led to a significant (24%) reduction in the time to clinical healing as well as to a 38% decrease in the time to overall (clinical and radiographic) healing.

BS increased fracture healing rate and reduced healing time, leading to a better quality of life in patients suffering from risk fractures.

16.2.4 OSTEOTOMIES

Three Italian double-blind studies have been performed on osteotomies: human femoral intertrochanteric osteotomies [64], tibial osteotomies [65], and osteotomies in patients undergoing massive bone graft [66]. The osteotomies of tibia and femur showed how the application of PEMF stimulation favors rapid healing of the osteotomic line and, in the case of femur osteotomy, an early mineralization of the bone callus demonstrated by computer analysis of the x-ray films. As regards the effects on massive bone grafts, a significant shortening of the healing time (29%, p<0.05) from 9 to 6 months was observed for patients not undergoing chemotherapy after the surgery.

16.2.5 NONUNIONS

In the international literature, there is abundant clinical evidence on the effectiveness of BS on nonunions; the authors have reported a success rate in the treatment of nonunions above 73%–85% [67,68] (Figure 16.3).

A clinical trial has demonstrated how the percentage of union obtainable with surgery (used to correct inadequate mechanical conditions) or with BS (when the failed union can be attributed to a biological deficiency only) is exactly the same, 87%, for noninfected nonunions. In the presence of infection, the percentage of success of surgery falls to 40%, whereas infection does not impair the good result of BS [69]. It is clear that in the presence of infection and unsatisfactory mechanical

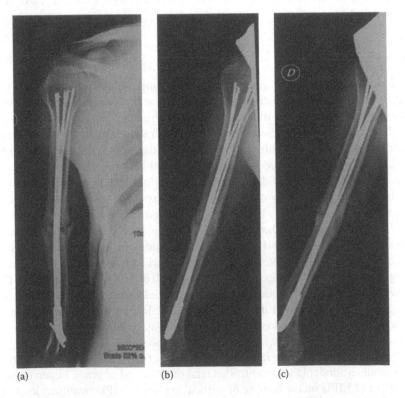

(a)   (b)   (c)

FIGURE 16.3 A 23-year-old male: (a) 5 months from trauma starts PEMF treatment; (b) complete healing after 3 months of PEMF treatment; (c) 3 months' follow-up.

conditions the association of surgery and stimulation can offer the best results. The union rate in patients suffering from tibial nonunions and surgically treated was 91% for patients who underwent PEMF stimulation compared to only 83% of the control group; the average radiological healing time was 3.3 months with PEMF and 4.9 months without [70]. Scott and King [71], in a prospective double-blind study of patients suffering from nonunion, reported a success healing rate of 60% in an CCEF active group, while none of the patients healed in the placebo group (p=0.004). Eighty-four percent healing was achieved in patients suffering from nonunion and stimulated with CCEF [72]. Romanò et al. included 49 patients affected by septic nonunions in a prospective study treated with LIPU and antibiotic therapy [73]. Bone healing was achieved in 39 patients (85.1%); 7 were considered failures, while 3 patients decided to discontinue the treatment. The authors conclude that LIPU is a conservative treatment that may avoid the need for additional complex surgery.

16.2.6 PROSTHESIS

The activation of the osteogenetic activity immediately after the insertion of an implant favors its integration and, most importantly, guarantees implant stability in the long term. BS has the strong potentiality to enhance orthopedic implant fixation on the basis of the assumption that the processes of bone healing around implants involve the activation of osteogenetic processes, similar to those of the bone healing of fractures and defects, at least in terms of initial host response.

Rispoli et al. [74] reported successful treatment with PEMF in 76% of patients with painful uncemented hip prostheses. More recently, Kennedy et al. [75] performed a double-blind study on 37 patients with femoral component loosening; 53% of patients were treated successfully with BS versus 11% of control patients. In 1995, Steinberg [76] reported a case of a 44-year-old patient in whom osteolytic changes that developed around the distal end of the femoral prosthesis appeared to reverse with the combined use of anti-inflammatory drugs and PEMF. A study on the effect of PEMF treatment in 24 patients with aseptic loosening of hip prostheses was performed also by Konrad et al. [77]. After 6 months of treatment and 1 year later, pain and hip function improved significantly, and there was also a significant improvement in both isotope scans and ultrasonography. Dallari et al. investigated the effects of PEMFs in a randomized, double-blind study on 30 patients undergoing hip revision [78]. Ninety days after surgery, the clinical outcome was significantly improved in stimulated patients compared to the control group, as well as bone mineral density. The authors conclude that PEMF treatment aids clinical recovery and bone stock restoration after total hip arthroplasty revision surgery.

In any case, there are no contraindications for the use of BS due to the electrolysis of metals with consequent production of toxic substances; thus, BS is an effective treatment to encourage osteointegration in the presence of biomaterials and to avoid complications deriving from implant failure.

16.2.7 AVASCULAR NECROSIS OF THE FEMORAL HEAD

Osteonecrosis of the femoral head represents the final phase of a pathological process involving the cancellous bone marrow edema, subchondral bone and cartilage damage, culminating in osteoarthrosis of the joint, with consequent arthroplasty in the majority of cases. It is therefore necessary to act in the early phases with a treatment that can prevent or slow the progression of the osteonecrosis. In patients with osteonecrosis of the hip, a broad comparative study of the various orthopedic treatments for the joint (core decompression and PEMF), compared with the indication for non-weight-bearing, demonstrated that BS achieved better long-term results [79]. These observations have been confirmed by independent study [80]; moreover, BS was revealed as the preferred treatment for avascular necrosis of the femoral head in the early stages of the condition, Ficat I and II [81].

The effects of BS are already evident in the short term, if we consider the capacity of the physical stimulus to exert a strong anabolic effect on articular cartilage and to reabsorb subchondral bone marrow edema, a condition that can seriously damage the joint. The long-term effects of BS can be summarized as a capacity to promote strong osteogenetic activity in the necrotic area and to prevent trabecular fractures and the consequent collapse of the subchondral bone. This noninvasive therapy is well tolerated by patients, is effective, and does not preclude surgical solutions.

16.2.8 SPINAL FUSION OF VERTEBRAL FRACTURES

Spinal fusion has been the most popular treatment for various spinal disorders. A common challenge to spinal fusion procedures is a postoperative failure to fuse, known as a nonunion or pseudarthrosis. However, the reported pseudarthrosis rate was not low. It was estimated that 25%–81% of lumbar fusion procedures failed to unite [82]. The earliest reported use of electrical stimulation to improve the efficacy of spinal fusion was over a quarter century ago. Dwyer et al. [83] were the first to show that adjunctive electrical stimulation improved the fusion rate of patients undergoing both anterior and posterior spinal fusion. By 1985, increasing interest in the use of electrical stimulation to improve spinal fusion outcomes resulted in a continually growing number of clinical and basic scientific studies, which have validated the clinical and scientific utilities of electrical stimulation to enhance the success of spinal fusion. In Goodwin's study [84], the overall fusion success rates were found to be statistically higher in the CCEF-stimulated group (85%) compared to the control group (65%). In a level I trial, Rossini reported excellent results when using CCEF for pain relief in patients with osteoporotic vertebral fractures and chronic pain [85].

16.3 BIOPHYSICAL STIMULATION OF CARTILAGE REPAIR

The most advanced frontier in orthopedics and traumatology is represented by treatment modalities that aim to accelerate cartilage tissue repair processes and to reconstruct the functional properties of the damaged tissues, especially joint tissues for those with inflammatory and degenerative diseases. Continually integrating knowledge of the fundamentals in new research with orthopedic practices creates strong options to resolve the biggest clinical problems [86]. Modern treatment modalities aim to perfect the techniques of tissue repair and regeneration by promoting significant anabolic tissue activity.

Currently, the advanced surgical treatments used in joint repairs include various different strategies, including implanting into the damaged site engineered tissue, mesenchymal stem cells, differentiated cells, and growth factors [87]. So far, these treatment methods have not yet offered successes that can be reproduced and integrated over time [88]. More specifically, long-term clinical results have proven unsatisfactory because in many cases fibrous tissue forms, and there are inferior mechanical properties with limited durability over time [89]. Furthermore, patients often continue over time to refer persistent symptoms, such as pain or intra-articular effusion [90].

International literature outlines how the results of these modern treatment modalities are incredibly wide ranging [88]. The success or failure of a surgical intervention to repair tissue is completely dependent on the type of implant, the dexterity demanded for the positioning of the engineered tissue, and the quality of the articular environment where it is being placed. Surgical treatments, even those minimally invasive, such as arthroscopy, still provoke local damage that in turn generates its own inflammatory reaction of varying degree, which is difficult to control over time and which may alter the cellular homeostasis.

Inflammation represents a serious deterioration for articular tissues, cartilage, ligaments, meniscus, and soft tissues, and must be controlled within the shortest time possible. When there are proinflammatory cytokines, such as interleukin-1beta (IL-1β) and tumor necrosis factor-alpha (TNF-α) within the articular environment, metalloproteinase activities are stimulated, and prostaglandin E_2 (PGE$_2$) is released leading to inhibition of the extracellular matrix synthesis with its own accompanying enzymatic destruction so that the reparative activity on the cartilaginous tissue is fibrous [91–93]. The inflammatory cell activity and the release of the proinflammatory cytokines into the synovial fluid are responsible for the catabolic effects on the cartilaginous matrix that degenerates, subsequently also leading to the loss of the mechanical functionality [94,95].

The process of cartilage degeneration can be described as the alteration of two opposite metabolic activities that, under physiological conditions, are normally balanced. On one hand, there is the catabolic function that tends to damage cartilage tissue and on the other, the anabolic function that maintains and protects the cartilage. Owing to its low reparative capacity, it is fundamentally important to maintain the health of all articular cartilage, in all of its components, cells, and extracellular matrix, even including the subchondral bone, and this is done by stimulating the functional activity of the chondrocyte and inhibiting damage caused by inflammation.

To this end, the concept of chondroprotection identifies the options of various treatments including pharmacological, physical, and surgical, whether individually or combined, that allow conservation of the cartilage itself or that aim to limit damage due to degenerative, pathological, and traumatic processes and to inflammatory reactions.

Physiologically, the human body controls inflammation through activating numerous cellular processes, such as involving adenosine receptors, especially A_{2A} and A_3 [96,97]. In one animal study on septic arthrosis, it was found that a specific A_{2A} adenosine receptor agonist significantly reduced the cartilage damage, the synovial inflammation, and infiltration of leukocytes [98,99]. Notwithstanding, drugs with A_{2A} adenosine receptor agonist action, although considered chondroprotective, are still under trial owing to the development of possible side effects.

Therefore, it is fundamentally important to develop local treatment methods that limit and prevent cartilage tissue degeneration without undervaluing the intrinsic possibilities of the damaged tissue itself and its reparative or regenerative capabilities which in turn may affect spontaneous healing depending on the characteristics of the injury and the patient. Among these new chondroprotective methods, BS is playing a fundamental role.

16.3.1 Mechanism of Action of Cartilage Stimulation and Preclinical Studies

The use of BS for joint treatment must respond to the need to treat articular cartilage in its entire extension and depth as well as involve the articular structures, such as the meniscus, ligaments, synovial membrane, and all the way to the subchondral bone. These problems have now been resolved exclusively through the use of specific PEMF, called I-ONE therapy, which has been well demonstrated by the Cartilage Repair and Electromagnetic Stimulation (CRES) study group during a large translational research focused on the chondroprotective effects of stimulation. The phases are summarized here as follows [100].

In vitro analysis of BS has documented considerable results on various cellular models. In two studies on human neutrophils, BS creates a strong adenosine-agonist effect specific for the A_{2A} and A_3 receptors [101,102]. This effect, mediated by the increase in the number of receptors themselves, determines a significant increase in the cyclical AMP (cAMP) and a reduction in the release of superoxide anion (O_2^-) suggesting an anti-inflammatory effect. This mechanism of action has been confirmed in chondrocyte and synoviocyte cultures [103,104]. Previous studies have shown that BS was able to stimulate human adult articular chondrocytes cultured in monolayer [105,106]. Furthermore, similar to some drugs, bovine synovial fibroblasts, cultivated in the presence of inflammatory stimulants, such as TNF-α or lipopolysaccharides, that were exposed to BS resulted in a decrease in the release of PGE_2, a molecule involved in the inflammatory process and an important pain reducer, and a decrease in the expression of cyclooxygenase-2 (COX-2), the key enzyme that determines the production of PGE_2 [107]. In human synovial fibroblasts taken from osteoarthritis patients, Ongaro et al. demonstrated that BS reduces the synthesis of inflammatory mediators such as PGE_2 and proinflammatory cytokines (IL-6 and IL-8), while stimulating the release of interleukin-10 (IL-10), an anti-inflammatory cytokine [108]. A greater increase in gene expression and a greater proliferation of porcine chondrocytes, cultivated on a collagen scaffold, and a lower level of gene expression for

MMP-13 in articular bovine chondrocytes seeded into scaffold were measured in cultures treated with BS compared with the controls [109–111].

More recently, Ongaro et al. studied the effect of BS on mesenchymal stem cells during chondrogenic differentiation in the presence of IL-1β showing a significant role of PEMF in counteracting the IL-1β-induced inhibition of chondrogenesis and suggesting BS as a therapeutic strategy for improving the clinical outcome of cartilage engineering repair procedures [112].

These observations suggest that BS must be administered to control inflammation, protect articular cartilage, and, in a final analysis, block cellular apoptosis.

Dose–response curves, run ex vivo on full-thickness bovine articular cartilage, per exposure time (h/day), per field peak value (mT), and per frequency (Hz), made it possible to identify the most effective therapeutic parameters and conditions to use for in vivo and clinical trials: 1.5 mT, 75 Hz, 4 h/day (I-ONE therapy, IGEA, Italy) [113] (Figure 16.4). In regard to this, it is important to underscore that, as with drugs, treatment dosage of BS is fundamental to achieve effective therapy.

The ex vivo results found with the I-ONE therapy indicate that this specific BS blocks the progression of the cartilage matrix degeneration induced by the proinflammatory cytokines, such as IL-1β, and encourages anabolic chondrocyte activity thereby resulting in an increase in proteoglycan synthesis [114]. This effect has been observed even when the anabolic activity growth factors were present, such as insulin growth factor-1 (IGF-1) [115]. Recently, it was found that I-ONE therapy induces a strong anabolic action, in combination with the IGF-1, also in human cartilage explants taken from patients with knee osteoarthritis [116].

The authors conclude that I-ONE therapy has assumed an important role in the treatment to prevent progression of initial osteoarthritis by exercising a strong anti-inflammatory and chondroprotective effect.

Based on the in vitro and ex vivo results, studies were conducted on small and large animals to evaluate the I-ONE therapy effects, such as (1) preventing spontaneous osteoarthritic degeneration of the knee in Dunkin Hartley guinea pigs [117,118], (2) repairing damaged tissue following osteochondral implants in adult sheep [119].

In the guinea pigs, the I-ONE therapy was found to be able to block progression of the osteoarthritic cartilage damage, limit cartilage fibrillation, conserve cartilage depth, and prevent sclerosis phenomena in the subchondral bone. These results are in accord with those of other authors demonstrating how in animals that underwent BS there was an increase in the TGF-β synthesis and inhibition of the TNF-α synthesis with a clear anabolic and trophic effect on the articular cartilage [120].

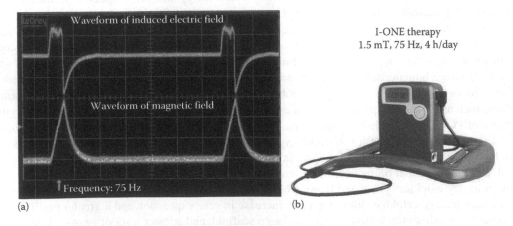

(a)   (b)

FIGURE 16.4   (a) Signal waveform of I-ONE therapy. (b) Dosage of I-ONE therapy and clinical device.

In adult sheep, a significant osteointegration of autologous osteochondral implants was seen as well as an improved cartilaginous integration in the animals treated with the I-ONE therapy. The synovial liquid, removed at the end of the treatment from the control animals, contained higher levels of IL-1β and TNF-α and a lower concentration of TGF-β compared to the animals treated with the I-ONE therapy.

With this, the antidegenerative, reparative, and anti-inflammatory effects of treatment with I-ONE therapy can be accounted for also in vivo.

16.3.2 CLINICAL USE OF BIOPHYSICAL STIMULATION ON ARTICULAR CARTILAGE REPAIR

Activation of the receptor involved in the inflammatory process, inhibition of the catabolic cytokine effect, decrease in the release of the pro-apoptotic molecule, and the strong anabolic action with the growth factors highlight how specific dosage of PEMF is antidegenerative and joins in with controlling the inflammatory phenomena and the reparative processes (Table 16.2).

I-ONE therapy can therefore be used proactively as (1) *postsurgical treatment* with the objective of quickly controlling local inflammatory response due to the surgical operation and, over the long term, to maintain the mechanical and biological properties of the cartilage or engineered tissue by means of an effective chondroprotective effect; (2) *post-arthroplasty treatment* to inhibit the inflammatory processes that affect the periarticular tissues and to avoid the development of chronic

TABLE 16.2
PEMF Effect on Different In Vitro Models for Cartilage Healing

Authors	In Vitro Models	PEMF Effect
Vincenzi et al. [104], Pezzetti et al. [105], and De Mattei et al. [106]	Human chondrocytes	Increases cell proliferation
Varani et al. [101,102]	Human neutrophils	Upregulation of A_{2A} and A_3 adenosine receptor
		Inhibited the generation of O_2^-
De Mattei et al. [114,115]	Bovine articular cartilage explants	Inhibits the effect of IL-1β on PG synthesis
		Acts in concert with IGF-I in stimulating PG synthesis
Varani et al. [103] and De Mattei et al. [107]	Bovine chondrocytes and fibroblast-like synoviocytes	Upregulation of A_{2A} and A_3 adenosine receptors
		Inhibits PGE_2 release
Chang et al. [109,110]	Porcine chondrocyte cells embedded in collagen constructs	Cell proliferation
		Promotes GAG production and gene expression
		Promotes GAG production
		Increase in type II collagen
Ongaro et al. [116]	Human cartilage explants	Augments cartilage anabolic activities
		Counteract the catabolic activity of IL-1β
		Act in concert with IGF-I in stimulating PG synthesis
Ongaro et al. [108]	Human osteoarthritic synovial fibroblasts	Inhibits the release of PGE_2 and IL-1β, IL-6, and IL-8
		Stimulates the release of IL-10
Ongaro et al. [112]	Bovine mesenchymal stem cells	Counteract the IL-1β-induced inhibition of chondrogenesis
Hilz et al. [111]	Primary articular bovine chondrocytes seeded into scaffolds	Highest level of GAG/DNA and gene expression for type II/type I collagen ratio
		Lower level of gene expression for MMP-13

pain and functional limitations; and (3) *conservative treatment* to limit the progression of a degenerative process such as osteoarthritis that comes with age and is accelerated by inflammatory and/or traumatic events.

16.3.2.1 I-ONE Therapy in Postsurgical Treatment

It is already well known that a trauma or simple arthroscopy can release proinflammatory cytokines into the joint environment that cause degeneration of the surrounding tissues. For example, injuries to the anterior cruciate ligament (ACL) will result in a significant functional deficit displayed as major instability, with progressive secondary damage to other articular structures and with the progressive development of precocious degenerative alterations in the articular environment (cartilage, meniscus). In particular, rupture of the ACL causes an increase in proinflammatory cytokines during the acute phase (IL-6 and IL-8) that alter the cartilage metabolism favoring catabolic phenomena leading to the development of osteoarthritis [121]. Consider that only 41% and 33% of patients undergoing patellar and gracilis semitendinosus tendon reconstruction respectively reported a normal score level on the IKDC evaluation [122]. Furthermore, Berg et al. underscore how the continued exposure of the ligament to the synovial liquid enzymes, cytokines, and growth factor inhibitors may interfere with the spontaneous healing of the ligament itself and may offer a nonmechanical explanation for the enlargement of the bone tunnel [123].

The clinical experiences gained in the CRES clinical study using the I-ONE therapy included patients with damage to the knee joint enrolled in randomized clinical protocols with a control group (evidence level I). The protocols included patients suffering from cartilage injuries and treated with microfracture [124], with autologous chondrocyte implant [125], and with reconstruction of the ACL, mostly associated with meniscectomy [126].

Independently of the injury treated, statistical analysis of the group showed that the percentage of patients assuming nonsteroidal anti-inflammatory drugs (NSAIDs) is significantly lower in those patients treated with I-ONE therapy compared to the control group ($p < 0.05$) and that the recovery times for functionality of the knee were significantly reduced in the short term and conserved even long after the intervention. Benazzo et al. demonstrated that 2 years after the reconstruction of ACL, a complete functional recovery was achieved by 86% of the patients in the active group compared with 75% of the patients in the placebo group [126]. Zorzi et al., after 3 years of follow-up, found that 87.5% of patients treated with microfractures in the placebo group were unable to return to full activities compared to 37.5% in the active group ($p < 0.05$) [124].

These clinical evidence level I studies highlight how early treatment of the joint immediately after surgery makes it possible to achieve improved functional recovery, or a full return to normal activities, and to conserve them over the long term.

16.3.2.2 I-ONE Therapy in Post-Arthroplasty Treatment

Total knee arthroplasty (TKA) operations are quite frequent, and their number is continually increasing. The interventions are complex, and the periarticular tissues are significantly affected. The success of TKA is linked to surgical factors (type of arthroplasty, prosthetic design, prosthetic material, surgical technique) and to *biological*-type factors (inflammation, pain, edema, impingement). So much so that it is already well known that important functional limitations may be primarily caused by an excessive reaction that affects the periarticular tissues. In the days following TKA, the presence of a heavy local inflammatory component is associated with pain and functional limitation, which normally resolve in a few months. At times, however, it may require longer periods and may develop into chronic inflammation, often discrete, that lingers for years. In these patients, the functional conditions, although better compared to the original status, cannot be considered completely satisfactory even though there are no alterations in the stability or positioning of prosthesis. Baker in fact affirms that 19.8% of patients report pain even 1 year after TKA [127]. Beswick et al. report that many patients (10%–34%) continue to have significant pain and functional limitations even years later after TKA [128]. The inflammatory response is primarily due to the

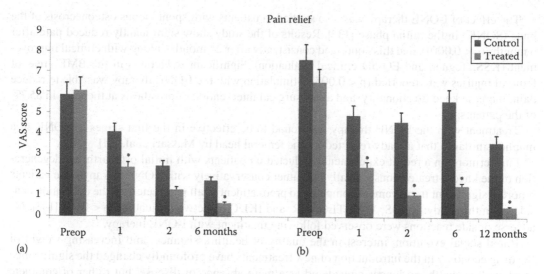

FIGURE 16.5 Visual analogue score (VAS) pain evaluation of patients at different follow-ups in Adravanti et al. study (a) and Moretti et al. study (b). *p < 0.05 for the treated group vs control group.

presence of proinflammatory cytokines in the synovia. In fact, an inverse relationship has been demonstrated between the quantity of IL-6 proinflammatory cytokines in the joint and the functional recovery of the patient [129].

In the two clinical trials, patients suffering from advanced knee osteoarthrosis who had undergone cemented posterior stabilized TKA were enrolled into a randomized and controlled study [130,131]. I-ONE therapy was administered 4 h/day for 60 days. Both studies showed that there was an immediate and significant resolution of the pain even just 1 month after TKA, and this was maintained at all follow-ups with a significant difference compared to the control group (Figure 16.5). An early resolution of the joint swelling and a lower assumption of NSAIDS were reported in the stimulated group compared to the control one, thereby significantly improving the patients' quality of life.

The clinical efficacy of the I-ONE therapy was also evaluated over the long term. At an average of 3 years after TKA, a phone call follow-up was done. There were positive results with significant differences (p < 0.05) across all parameters analyzed (pain relief and walking aid) reported by the group stimulated with I-ONE therapy compared to the control group.

The studies on TKA indicate how the clinical results and actual results affecting the patients' quality of life, over the short and long terms, can be improved by adopting a local control therapy for the joint environment, especially where it will reduce the duration and the depth of the inflammatory response developing after TKA. The use of the I-ONE therapy makes it possible to act in a specific and focused method aiming the effects exclusively on the joint and on the soft tissues, without any side effects that unfortunately are frequently associated with drug treatments.

16.3.2.3 I-ONE Therapy in Conservative Treatment

Edema, algodystrophy, and osteonecrosis are pathologies that share the same initial pathogenic event: ischemic damage to the bone. In edema and primarily in algodystrophy, there is an inflammatory process in a localized osteoporosis zone with pain. In the initial stages, the surviving bone tissue almost always is able to guarantee total recovery and return of the mechanical and biological characteristics of the healthy bone. This does not occur with necrosis, and the devascularized tissues progressively accumulate structural alterations based on the progression of the pathology.

The effect of I-ONE therapy was also studied in patients with spontaneous osteonecrosis of the knee (SONK) in the initial phase [132]. Results of the study show significantly reduced pain after 6 months (p < 0.0001), and this continued consistently after 24 months, along with clinical improvement (KSS, Tegner, and EQ-5D clinical evaluation). Significant reductions in the BME areas of femoral injuries were reported (p < 0.005). Stimulation with the I-ONE therapy was able to reduce pain, increase knee functionality, and avoid surgical intervention of prosthesis at the knee in 85.7% of the patients.

Treatment with the I-ONE therapy was found to be effective in the first stages of SONK in a manner similar to that already reported for the femoral head by Massari et al. [81].

Furthermore, in a recent experiment conducted on patients with initial osteoarthritic degeneration of the knee—treated nonsurgically but rather conservatively with I-ONE therapy—the patients reported significant improvement compared to pretreatment in all parameters of the evaluation considered in this study: KOOS, VAS, TEGNER, and IKDC objective evaluation (p < 0.05) [133]. No adverse or side reactions were observed following treatment with I-ONE therapy.

Rapid social evolution, interest in the quality of health assistance, and increasing weight of the drug economy in the introduction of new treatments have profoundly changed the significance attributed to health, no longer considered as a mere absence of disease, but rather of complete physical, mental, and social well-being. In this view, various conservative treatment options for cartilage damage have been made available over the past few years, and many studies are currently underway to clarify some still unanswered questions regarding the long-term duration of these procedures.

Treatment with the I-ONE therapy represents a valid therapeutic approach for initial osteoarthritis or localized cartilage injury. It is an easy-to-use treatment and eliminates any possible side effects typical in drug treatments. Its use is well tolerated by patients.

16.4 CONCLUSIONS

During the last century, treatment with different physical energy has been introduced to favor tissue healing, and its clinical use is now based on sound scientific evidence. Undoubtedly, the orthopedic community has played a central role in the development and understanding of the importance of the physical stimuli to control biological activities, having studied the clinical importance of BS of endogenous bone repair in risk fractures, osteotomies, nonunions, avascular necrosis, and in the presence of prosthesis. Recently, the results of a double-blind study show that BS favors patient recovery, enhances the healing process up to 12 months after stimulation, and has a positive effect on chronic pain in patients with osteoporotic vertebral fractures, inducing a reduction in the use of analgesic drugs. In consideration of the sensitivity of cartilage tissue to physical stimuli, the orthopedic community has now focused its interest on the joint to prevent cartilage degeneration, to enhance cartilage repair, and to favor patients' function recovery. Today, several new areas of medicine are interested in the possibility to utilize nonchemical means to treat different pathologies (acute cerebral ischemia, Alzheimer's disease). Some applications are in the opening or preliminary phases of research; however, everything suggests that these therapeutic possibilities will be more and more utilized.

BS, when compared to the pharmacological treatments, has the advantage of being easy to administer. Being a local therapy, it can reach maximum *concentration* at the treatment site and thus maximum therapeutic efficacy, without general negative side effects. BS seems to be suitable for protracted treatments in the presence of chronic degenerative diseases, whereas it does not seem to be able to treat systemic disorders.

Further development of the clinical application of physical energy introduces many and complex questions; however, the possibility to recognize and define a therapeutic development area as clinical biophysics represents a fundamental moment of synthesis necessary to create a common ground of reference for researchers in different fields.

In Europe, there is still an important open issue that may significantly limit the development and use of this new therapeutic approach that involves physics, engineering, biology, and medicine. Unlike the United States, where clearance to the market is regulated by the FDA, in Europe, there is no requirement of approval by a competent agency. In the United States, safety and efficacy must be proven, whereas in Europe, at present, only *electrical safety* (CE mark) is required.

While this may favor the development of new technologies, it has also meant the proliferation of systems of treating patients with no scientific basis or study demonstrating their effectiveness. This exposes patients to the risk of being treated with devices whose efficacy has not been proven or, worse, that may actually be harmful. This deficiency will certainly need to be remedied by the responsible authorities in the near future.

BS is an important area of clinical biophysics applied to human pathology. It requires care and precision in use if it is to ensure the success expected of it by physicians and patients.

REFERENCES

1. Fukada E, Yasuda I. On the piezoelectric effect of bone. *J Phys Soc Jpn*. 1957;12:121–128.
2. Bassett CA, Becker RO. Generation of electric potentials in bone in response to mechanical stress. *Science*. 1962;137:1063–1064.
3. Xu J, Wang W, Clark CC, Brighton CT. Signal transduction in electrically stimulated articular chondrocytes involves translocation of extracellular calcium through voltage-gated channels. *Osteoarthritis Cart*. 2009;17(3):397–405.
4. Galvani L. *De viribus electricitatis in motu musculari commentarius*. Bologna, Italy: Ex typographia Instituti Scientiarum, 1791.
5. Black J. *Electrical Stimulation*. New York: Praeger, 1987.
6. Guzelsu N. Piezoelectric and electrokinetic effects in bone tissue. *Electro Magnet Biol*. 1993;12(1):51–82. Review.
7. Green J, Kleeman CR. Role of bone in regulation of systemic acid-base balance. *Kidney Int*. 1991;39:9–26.
8. Otter MW, Vincent R, Palmieri VR, Dadong DWu, Seiz KG, Mac Ginitie LA, Cochran GVB. A comparative analysis of streaming potentials in vivo and in vitro. *J Orthopaedic Res*. 1992;10:710–719.
9. Pollack SR. Bioelectrical properties of bone. Endogenous electrical signals. *Orthop Clin North Am*. 1984;15:3–14.
10. Behari J. Electrostimulation and bone fracture healing. *Biomed Eng*. 1992;18:235–254.
11. Wolff J. *Das Gesetz der Transformation der Knochen*. Berlin, Germany: Hirschwald, 1892.
12. Friedenberg ZB, Brighton CT. Bioelectric potentials in bone. *J Bone Joint Surg Am*. 1966;48(5):915–923.
13. Friedenberg ZB, Dyer R, Brighton CT. Electro-osteograms of long bones of immature rabbits. *J Dent Res*. 1971;50(3):635–639.
14. Friedenberg ZB, Harlow MC, Heppenstall R, Brighton CT. The cellular origin of bioelectric potentials in bone. *Calcif Tissue Res*. 1973;13(1):53–62.
15. Rubinacci A, Brigatti L, Tessari L. A reference curve for axial bioelectric potentials in adult rabbit tibia. *Bioelectromagnetics*. 1984;5(2):193–202.
16. Chakkalakal DA, Wilson RF, Connolly JF. Epidermal and endosteal sources of endogenous electricity in injured canine limbs. *IEEE Trans Biomed Eng*. 1988;35(1):19–30.
17. Lokietek W, Pawluk RF, Bassett CA. Muscle injury potentials: A source of voltage in the undeformed rabbit tibia. *J Bone Joint Surg Br*. 1974;56(2):361–369.
18. Borgens RB. Endogenous ionic currents traverse intact and damaged bone. *Science*. 1984;225(4661):478–482.
19. De Ponti A, Villa I, Boniforti F, Rubinacci A. Ionic currents at the growth plate of intact bone: occurrence and ionic dependence. *Electro- Magneto Biol*. 1996;15(1):37–48.
20. Rubinacci A, De Ponti A, Shipley A, Samaja M, Karplus E, Jaffe LF. Bicarbonate dependence of ion current in damaged bone. *Calcif Tissue Int*. 1996;58:423–428.
21. Aaron RK, Boyan BD, Ciombor DM, Schwartz Z, Simon BJ. Stimulation of growth factor synthesis by electric and electromagnetic fields. *Clin Orthop Relat Res*. 2004;(419):30–37. Review.
22. Brighton CT, Wang W, Seldes R, Zhang G, Pollack SR. Signal transduction in electrically stimulated bone cells. *J Bone Joint Surg Am*. 2001;83-A(10):1514–1523.
23. Nagai M, Ota M. Pulsating electromagnetic field stimulates mRNA expression of bone morphogenetic protein -2 and -4. *J Dental Res*. 1994;73:1601–1605.

24. Bodamyali T, Bhatt B, Hughes FJ, Winrow VR, Kanczler JM, Simon B, Abbott J, Blake DR, Stevens CR. Pulsed electromagnetic fields simultaneously induce osteogenesis and upregulate transcription of bone morphogenetic proteins 2 and 4 in rat osteoblasts in vitro. *Biochem Biophys Res Commun.* 1998;250(2):458–461.

25. Guerkov HH1, Lohmann CH, Liu Y, Dean DD, Simon BJ, Heckman JD, Schwartz Z, Boyan BD. Pulsed electromagnetic fields increase growth factor release by nonunion cells. *Clin Orthop Relat Res.* 2001;(384):265–279.

26. Aaron RK, Ciombor DM, Keeping H, Wang S, Capuano A, Polk C. Power frequency fields promote cell differentiation coincident with an increase in transforming growth factor-beta(1) expression. *Bioelectromagnetics.* 1999;20(7):453–458. Erratum in: *Bioelectromagnetics.* 2000;21(1):73.

27. Lohmann CH, Schwartz Z, Liu Y, Guerkov H, Dean DD, Simon B, Boyan BD. Pulsed electromagnetic field stimulation of MG63 osteoblast-like cells affects differentiation and local factor production. *J Orthop Res.* 2000;18(4):637–646.

28. Fassina L, Visai L, Benazzo F, Benedetti L, Calligaro A, De Angelis MG, Farina A, Maliardi V, Magenes G. Effects of electromagnetic stimulation on calcified matrix production by SAOS-2 cells over a polyurethane porous scaffold. *Tissue Eng.* 2006;12(7):1985–1999.

29. Jansen JH, van der Jagt OP, Punt BJ, Verhaar JA, van Leeuwen JP, Weinans H, Jahr H. Stimulation of osteogenic differentiation in human osteoprogenitor cells by pulsed electromagnetic fields: An in vitro study. *BMC Musculoskelet Disord.* 2010;11:188.

30. Esposito M, Lucariello A, Riccio I, Riccio V, Esposito V, Riccardi G. Differentiation of human osteoprogenitor cells increases after treatment with pulsed electromagnetic fields. *In Vivo.* 2012;26(2):299–304.

31. Ceccarelli G, Bloise N, Mantelli M, Gastaldi G, Fassina L, De Angelis MG, Ferrari D, Imbriani M, Visai L. A comparative analysis of the in vitro effects of pulsed electromagnetic field treatment on osteogenic differentiation of two different mesenchymal cell lineages. *Biores Open Access.* 2013;2(4):283–294.

32. Lim K, Hexiu J, Kim J, Seonwoo H, Cho WJ, Choung PH, Chung JH. Effects of electromagnetic fields on osteogenesis of human alveolar bone-derived mesenchymal stem cells. *Biomed Res Int.* 2013;2013:296019. doi: 10.1155/2013/296019. Epub 2013 Jun 19.

33. Zhou J, Wang JQ, Ge BF, Ma XN, Ma HP, Xian CJ, Chen KM. Different electromagnetic field waveforms have different effects on proliferation, differentiation and mineralization of osteoblasts in vitro. *Bioelectromagnetics.* 2013. doi: 10.1002/bem.21794.

34. Zhuang H, Wang W, Seldes RM, Tahernia AD, Fan H, Brighton CT. Electrical stimulation induces the level of TGF-β1 mRNA in osteoblastic cells by a mechanism involving calcium/calmodulin pathway. *Biochem Biophys Res Comm.* 1997;237:225–229.

35. Hartig M, Joos U, Wiesmann HP. Capacitively coupled electric fields accelerate proliferation of osteoblast-like primary cells and increase bone extracellular matrix formation in vitro. *Eur Biophys J.* 2000;29(7):499–506.

36. Wang Z, Clark CC, Brighton CT. Up-regulation of bone morphogenetic proteins in cultured murine bone cells with use of specific electric fields. *J Bone Joint Surg Am.* 2006;88(5):1053–1065.

37. Bisceglia B, Zirpoli H, Caputo M, Chiadini F, Scaglione A, Tecce MF. Induction of alkaline phosphatase activity by exposure of human cell lines to a low-frequency electric field from apparatuses used in clinical therapies. *Bioelectromagnetics.* 2011;32(2):113–119.

38. Clark CC, Wang W, Brighton CT. Up-regulation of expression of selected genes in human bone cells with specific capacitively coupled electric fields. *J Orthop Res.* July 2014; 32(7):894–903.

39. Hauser J, Hauser M, Muhr G, Esenwein S. Ultrasound-induced modifications of cytoskeletal components in osteoblast-like SAOS-2 cells. *J Orthop Res.* 2009;27(3):286–294.

40. Fassina L, Saino E, De Angelis MG, Magenes G, Benazzo F, Visai L. Low-power ultrasounds as a tool to culture human osteoblasts inside cancellous hydroxyapatite. *Bioinorg Chem Appl.* 2010:456240. doi: 10.1155/2010/456240. Epub 2010 Mar 31.

41. Xue H, Zheng J, Cui Z et al. Low-intensity pulsed ultrasound accelerates tooth movement via activation of the BMP-2 signaling pathway. *PLoS ONE.* 2013;8(7):e68926. doi: 10.1371/journal.pone.0068926. Print 2013.

42. Aaron RK, Ciombor DM. Acceleration of experimental endochondral ossification by biophysical stimulation of the progenitor cell pool. *J Orthop Res.* 1996;14(4):582–589.

43. Ciombor DM, Aaron RK. The role of electrical stimulation in bone repair. *Foot Ankle Clin.* 2005;10(4): 579–593. Review.

44. Ryaby JT, Mathew J, Duarte-Alves P. Low intensity pulsed ultrasound affects adenylate cyclase activity and TGF-β synthesis in osteoblastic cells. *Trans Orthop Res Soc.* 1992;7:590.

45. Aaron RK, Ciombor DM, Jolly G. Stimulation of experimental endochondral ossification by low-energy pulsing electromagnetic fields. *J Bone Miner Res*. 1989;4(2):227–233.

46. Bassett CAL, Pawluk RJ, Pilla AA. Augmentation of bone repair by inductively coupled electromagnetic fields. *Science*. 1974;3;184(136):575–577.

47. Bassett C, Valdes M, Hernandez E. Modification of fracture repair with selected pulsing electromagnetic fields. *J Bone Joint Surg*. 1982;64A:888–895.

48. Canè V, Botti P, Farneti D, Soana S. Electromagnetic stimulation of bone repair: A histomorphometric study. *J Orthop Res*. 1991;9(6):908–917.

49. Fini M, Giavaresi G, Giardino R, Cavani F, Cadossi R. Histomorphometric and mechanical analysis of the hydroxyapatite-bone interface after electromagnetic stimulation: An experimental study in rabbits. *J Bone Joint Surg Br*. 2006;88(1):123–128.

50. Brighton CT, Hozack WJ, Brager MD, Windsor RE, Pollack SR, Vreslovic EJ, Kotwick JE. Fracture healing in the rabbit fibula when subjected to various capacitively coupled electrical fields. *J Orthop Res*. 1985;3(3):331–340.

51. Brighton CT, Luessenhop CP, Pollack SR, Steinberg DR, Petrik ME, Kaplan FS. Treatment of castration-induced osteoporosis by a capacitively coupled electrical signal in rat vertebrae. *J Bone Joint Surg Am*. 1989;71(2):228–236.

52. Wang SJ, Lewallen DG, Bolander ME, Chao EY, Ilstrup DM, Greenleaf JF. Low intensity ultrasound treatment increases strength in a rat femoral fracture model. *J Orthop Res*. 1994;12:40–47.

53. Bassett CAL. Therapeutic uses of electric and magnetic fields in orthopedics. In: *Biological Effects of Electric and Magnetic Fields*, D.O. Carpenter and S. Ayrapetyan (eds.). San Diego, CA: Academic Press, Vol. II, pp. 13–48;1994.

54. Frost HM. The biology of fracture healing. An overview for clinicians. Part I, II. *Clin Orthop Relat Res*. 1989;(248):283–309. Review.

55. Cadossi R, Traina CG, Massari L. Electric and magnetic stimulation of bone repair: Review of the European experience. In: *Symposium of Physical Regulation of Skeletal Repair. American Academy of Orthopaedic Surgeons*, R.K. Aaron, M.E. Bolander (eds.). Rosemont, IL, pp. 37–51;2005.

56. Hinsenkamp M, Bourgois R, Bassett CA, Chiabrera A, Burny F, Ryaby J. Electromagnetic stimulation of fracture repair. Influence on healing of fresh fracture. *Acta Orthop Belg*. 1978;44(5):671–698.

57. Verardi V, Bottai M, Mazzeo M, Eminente A. I CEMP nella stimolazione dei tessuti molli. In: *Impiego dei campi elettromagnetici pulsati in ortopedia e traumatologia*, G.C. Traina, L. Romanini, L. Massari, C. Villani, R. Cadossi (eds.). Roma, Italy, Vol. I, pp. 205–210;1995.

58. Chiabrera, A., Cadossi, R., Bersani, F., Franceschi, C., Bianco, B. Electric and magnetic field effects on the immune system. In: *Biological Effects of Electric and Magnetic Fields*, D.O. Carpenter and S. Ayrapetyan (eds.). London, U.K.: Academic Press, Vol. II, pp. 121–145;1994.

59. Fontanesi G, Traina GC, Giancecchi F, Tartaglia I, Rotini R, Virgili B, Cadossi R, Ceccherelli G, Marino AA. La lenta evoluzione del processo riparativo di una frattura può essere prevenuta? *G.I.O.T.* 1986;XII(3):389–404.

60. Donkerwolcke Coussaert M, Hinsenkamp M, Burny F. Electromagnetic stimulation of fresh fractures treated with Hoffmann external fixation. *Orthopedics* 1984;7:411–416.

61. Faldini C, Cadossi M, Luciani D, Betti E, Chiarello E, Giannini S. Electromagnetic bone growth stimulation in patients with femoral neck fractures treated with screws: Prospective randomized double-blind study. *Current Orthop Practice*. 2010;282–287.

62. Benazzo F, Mosconi M, Beccarisi G, Galli U. Use of capacitive coupled electric fields in stress fractures in athletes. *Clin Orthop Relat Res*. 1995;(310):145–149.

63. Heckman JD1, Ryaby JP, McCabe J, Frey JJ, Kilcoyne RF. Acceleration of tibial fracture-healing by non-invasive, low-intensity pulsed ultrasound. *J Bone Joint Surg Am*. 1994;76(1):26–34.

64. Borsalino G, Bagnacani M, Bettati E, Fornaciari G, Rocchi R, Uluhogian S, Ceccherelli G, Cadossi R, Traina G. Electrical stimulation of human femoral intertrochanteric osteotomies: Double blind study. *Clin Orthop Relat Res*. 1988;(237):256–263.

65. Mammi GI, Rocchi R, Cadossi R, Traina GC. The electrical stimulation of tibial osteotomies. Double-blind study. *Clin Orthop Relat Res*. 1993;(288):246–253.

66. Capanna R, Donati D, Masetti C, Manfrini M, Panozzo A, Cadossi R, Campanacci M. Effect of electromagnetic fields on patients undergoing massive bone graft following bone tumor resection: A double-blind study. *Clin Orthop Relat Res*. 1994;(306):213–221.

67. Schmidt-Rohlfing B, Silny J, Gavenis K, Heussen N. Electromagnetic fields, electric current and bone healing—What is the evidence? *Z Orthop Unfall*. 2011;149(3):265–270.

68. Assiotis A, Sachinis NP, Chalidis BE. Pulsed electromagnetic fields for the treatment of tibial delayed unions and nonunions. A prospective clinical study and review of the literature. *J Orthop Surg Res.* 2012;7:24.

69. Traina GC, Fontanesi G, Costa P, Mammi GI, Pisano F, Giancecchi F, Adravanti P. Effect of electromagnetic stimulation on patients suffering from nonunion. A retrospective study with a control group. *J Bioelectricity.* 1991;10:101–117.

70. Cebriàn, JL, Gallego P, Francès A, Sànchez P, Manrique E, Marco F, Lòpez-Duràn L. Comparative study of the use of electromagnetic fields in patients with pseudoarthrosis of tibia treated by intramedullary nailing. *Int Orthop.* 2010;34:437–440.

71. Scott G, King JB. A prospective double blind trial of electrical capacitive coupling in the treatment of non-union of long bones. *J Bone Joint Surg Am.* 1994;76(6):820–826.

72. Impagliazzo A, Mattei A, Spurio Pompili GF, Setti S, Cadossi R. Treatment of nonunited fractures with capacitively coupled electric field. *J Orthop Traumatol.* 2006;7:16–22.

73. Romanò CL, Romanò D, Logoluso N. Low-intensity pulsed ultrasound for the treatment of bone delayed union or nonunion: A review. *Ultrasound Med Biol.* 2009;35(4):529–536.

74. Rispoli FP, Corolla FM, Mussner R. The use of low frequency pulsing electromagnetic fields in patients with painful hip prosthesis. *J Bioelectricity.* 1988;7:181.

75. Kennedy WF, Roberts CG, Zuege RC, Dicus WT. Use of pulsed electromagnetic fields in treatment of loosened cemented hip prostheses. A double-blind trial. *Clin Orthop.* 1993;286:198–205.

76. Steinberg GG. Reversible osteolysis. *J Arthroplasty.* 1995;10:556–559.

77. Konrad K, Sevcic K, Foldes K, Piroska E, Molnar E. Therapy with pulsed electromagnetic fields in aseptic loosening of total hip prostheses: A prospective study. *Clin Rheumatol.* 1996;15:325–328.

78. Dallari D, Fini M, Giavaresi G, Del Piccolo N, Stagni C, Amendola L, Rani N, Gnudi S, Giardino R. Effects of pulsed electromagnetic stimulation on patients undergoing hip revision prostheses: A randomized prospective double-blind study. *Bioelectromagnetics.* 2009;30(6):423–430.

79. Aaron RK, Lennox D, Bunce GE, Ebert T. The conservative treatment of osteonecrosis of the femoral head. A comparison of core decompression and pulsing electromagnetic fields. *Clin Orthop Relat Res.* 1989;(249):209–218.

80. Musso ES, Mitchell SN, Schink-Ascani M, Bassett CA. Results of conservative management of osteonecrosis of the femoral head. A retrospective review. *Clin Orthop Relat Res.* 1986; (207):209–215.

81. Massari L, Fini M, Cadossi R, Setti S, Traina GC. Biophysical stimulation with pulsed electromagnetic fields in osteonecrosis of the femoral head. *J Bone Joint Surg Am.* 2006;88(Suppl 3):56–60.

82. Meril AJ. Direct current stimulation of allograft in anterior and posterior lumbar interbody fusion. *Spine* 1994;19:2393–2398.

83. Dwyer AF, Yau AC, Jefcoat KW. Use of direct current in spine fusion. *J Bone Joint Surg. (Am).* 1974;56:442.

84. Goodwin CB, Brighton CT, Guyer RD, Johnson JR, Light KI, Yuan HA. A double blind study of capacitively coupled electric stimulation as a adjunct to lumbar spinal fusion. *Spine.* 1999;24(13):1349–1356.

85. Rossini M, Viapiana O, Gatti D, de Terlizzi F, Adami S. Capacitively coupled electric field for pain relief in patients with vertebral fractures and chronic pain. *Clin Orthop Relat Res.* 2010;468(3):735–740.

86. Moran CJ, Shannon FJ, Barry FP et al. Translation of science to surgery: Linking emerging concepts in biological cartilage repair to surgical intervention. *J Bone Joint Surg Br.* 2010;92(9):1195–1202. Review.

87. Getgood A, Brooks R, Fortier L, Rushton N. Articular cartilage tissue engineering: Today's research, tomorrow's practice? *J Bone Joint Surg Br.* 2009;91(5):565–576. Review.

88. Harris JD, Siston RA, Pan X, Flanigan DC. Autologous chondrocyte implantation: A systematic review. *J Bone Joint Surg Am.* 2010;92(12):2220–2233. Review.

89. Pelttari K, Wixmerten A, Martin I. Do we really need cartilage tissue engineering? *Swiss Med Wkly.* 2009;139(41–42):602–609. Review.

90. Brun P, Dickinson SC, Zavan B, Cortivo R, Hollander AP, Abatangelo G. Characteristics of repair tissue in second-look and third-look biopsies from patients treated with engineered cartilage: Relationship to symptomatology and time after implantation. *Arthritis Res Ther.* 2008;10(6):R132.

91. Guilak F, Fermor B, Keefe FJ, Kraus VB, Olson SA, Pisetsky DS, Setton LA, Weinberg JB. The role of biomechanics and inflammation in cartilage injury and repair. *Clin Orthop Relat Res.* 2004;(423):17–26. Review.

92. Kuettner KE, Cole AA. Cartilage degeneration in different human joints. *Osteoarthritis Cartilage.* 2005;13(2):93–103. Review.

93. Ortiz LA, Dutreil M, Fattman C, Pandey AC, Torres G, Go K, Phinney DG. Interleukin 1 receptor antagonist mediates the antiinflammatory and antifibrotic effect of mesenchymal stem cells during lung injury. *Proc Natl Acad Sci USA.* 2007;104(26):11002–11007.

94. Schuerwegh AJ, Dombrecht EJ, Stevens WJ, Van Offel JF, Bridts CH, De Clerck LS. Influence of pro-inflammatory (IL-1 alpha, IL-6, TNF-alpha, IFN-gamma) and anti-inflammatory (IL-4) cytokines on chondrocyte function. *Osteoarthritis Cartilage.* 2003;11(9):681–687.

95. Goldring SR, Goldring MB. The role of cytokines in cartilage matrix degeneration in osteoarthritis. *Clin Orthop Relat Res.* 2004;427S:S27–S36.

96. Tesch AM, MacDonald MH, Kollias-Baker C, Benton HP. Chondrocytes respond to adenosine via A(2) receptors and activity is potentiated by an adenosine deaminase inhibitor and a phosphodiesterase inhibitor. *Osteoarthritis Cartilage.* 2002;10(1):34–43.

97. Borea PA, Gessi S, Bar-Yehuda S Fishman P. A3 adenosine receptor: Pharmacology and role in disease. *Handb Exp Pharmacol.* 2009;(193):297–327. Review.

98. Cohen SB, Gill SS, Baer GS, Leo BM, Scheld WM, Diduch DR. Reducing joint destruction due to septic arthrosis using an adenosine2A receptor agonist. *J Orthop Res.* 2004;22(2):427–435.

99. Cohen SB, Leo BM, Baer GS, Turner MA, Beck G, Diduch DR. An adenosine A2A receptor agonist reduces interleukin-8 expression and glycosaminoglycan loss following septic arthrosis. *J Orthop Res.* 2005;23 (5):1172–1178.

100. Massari L, Benazzo F, De Mattei M, Setti S, Fini M; CRES Study Group. Effects of electrical physical stimuli on articular cartilage. *J Bone Joint Surg Am.* 2007;89(Suppl 3):152–161. Review. No abstract available. Erratum in: *J Bone Joint Surg Am.* 2007;89(11):2498. CRES Study Group [added].

101. Varani K, Gessi S, Merighi S, Iannotta V, Cattabriga E, Spisani S, Cadossi R, Borea PA. Effect of low frequency electromagnetic fields on A_{2A} adenosine receptors in human neutrophils. *Br J Pharmacol.* 2002;136(1):57–66.

102. Varani K, Gessi S, Merighi S, Iannotta V, Cattabriga E, Pancaldi C, Cadossi R, Borea PA. Alteration of A_3 adenosine receptors in human neutrophils and low frequency electromagnetic fields. *Biochem Pharmacol.* 2003;66(10):1897–1906.

103. Varani K, De Mattei M, Vincenzi F, Gessi S, Merighi S, Pellati A, Ongaro A, Caruso A, Cadossi R, Borea PA. Characterization of adenosine receptors in bovine chondrocytes and fibroblast-like synoviocytes exposed to low frequency low energy pulsed electromagnetic fields. *Osteoarthritis Cartilage.* 2008;16(3):292–304.

104. Vincenzi F, Targa M, Corciulo C, Gessi S, Merighi S, Setti S, Cadossi R, Goldring MB, Borea PA, Varani K. Pulsed electromagnetic fields increased the anti-inflammatory effect of A_{2A} and A_3 adenosine receptors in human T/C-28a2 chondrocytes and hFOB 1.19 osteoblasts. *PLoS ONE.* 2013;8(5):e65561. doi: 10.1371/journal.pone.0065561. Print 2013.

105. Pezzetti F, De Mattei M, Caruso A, Cadossi R, Zucchini P, Carinci F, Traina GC, Sollazzo V. Effects of pulsed electromagnetic fields on human chondrocytes: An in vitro study. *Calcif Tissue Int.* 1999;65(5):396–401.

106. De Mattei M, Caruso A, Pezzetti F, Pellati A, Stabellini G, Sollazzo V, Traina GC. Effects of pulsed electromagnetic fields on human articular chondrocyte proliferation. *Connect Tissue Res.* 2001;42(4):269–279.

107. De Mattei M, Varani K, Masieri FF, Pellati A, Ongaro A, Fini M, Cadossi R, Vincenzi F, Borea PA, Caruso A. Adenosine analogs and electromagnetic fields inhibit prostaglandin E(2) release in bovine synovial fibroblasts. *Osteoarthritis Cart.* 2009;17(2):252–262.

108. Ongaro A, Varani K, Masieri FF et al. Electromagnetic fields (EMFs) and adenosine receptors modulate prostaglandin E(2) and cytokine release in human osteoarthritic synovial fibroblasts. *J Cell Physiol.* 2012;227(6):2461–2469.

109. Chang CH, Loo ST, Liu HL, Fang HW, Lin HY. Can low frequency electromagnetic field help cartilage tissue engineering? *J Biomed Mater Res A.* 2010;92(3):843–851.

110. Chang SH, Hsiao YW, Lin HY. Low-frequency electromagnetic field exposure accelerates chondrocytic phenotype expression on chitosan substrate. *Orthopedics.* 2011;34(1):20.

111. Hilz FM, Ahrens P, Grad S et al. Influence of extremely low frequency, low energy electromagnetic fields and combined mechanical stimulation on chondrocytes in 3-D constructs for cartilage tissue engineering. *Bioelectromagnetics.* 2014;35(2):116–128.

112. Ongaro A, Pellati A, Setti S, Masieri FF, Aquila G, Fini M, Caruso A, De Mattei M. Electromagnetic fields counteract IL-1β activity during chondrogenesis of bovine mesenchymal stem cells. *J Tissue Eng Regen Med.* 2012. doi: 10.1002/term.1671. [Epub ahead of print].

113. De Mattei M, Fini M, Setti S, Ongaro A, Gemmati D, Stabellini G, Pellati A, Caruso A. Proteoglycan synthesis in bovine articular cartilage explants exposed to different low-frequency low-energy pulsed electromagnetic fields. *Osteoarthritis Cart.* 2007;15(2):163–168.

114. De Mattei M, Pasello M, Pellati A, Stabellini G, Massari L, Gemmati D, Caruso A. Effects of electromagnetic fields on proteoglycan metabolism of bovine articular cartilage explants. *Connect Tissue Res.* 2003;44(3–4):154–159.

115. De Mattei M, Pellati A, Pasello M, Ongaro A, Setti S, Massari L, Gemmati D, Caruso A. Effects of physical stimulation with electromagnetic field and insulin growth factor-I treatment on proteoglycan synthesis of bovine articular cartilage. *Osteoarthritis Cart.* 2004;12(10):793–800.

116. Ongaro A, Pellati A, Masieri FF, Caruso A, Setti S, Cadossi R, Biscione R, Massari L, Fini M, De Mattei M. Chondroprotective effects of pulsed electromagnetic fields on human cartilage explants. *Bioelectromagnetics.* 2011;32(7):543–551.

117. Fini M, Giavaresi G, Torricelli P, Cavani F, Setti S, Cane V, Giardino R. Pulsed electromagnetic fields reduce knee osteoarthritic lesion progression in the aged Dunkin Hartley guinea pig. *J Orthop Res.* 2005;23(4):899–908.

118. Fini M, Torricelli P, Giavaresi G, Aldini NN, Cavani F, Setti S, Nicolini A, Carpi A, Giardino R. Effect of pulsed electromagnetic field stimulation on knee cartilage, subchondral and epiphyseal trabecular bone of aged Dunkin Hartley guinea pigs. *Biomed Pharmacother.* 2008;62(10):709–715.

119. Benazzo F, Cadossi M, Cavani F, Fini M, Giavaresi G, Setti S, Cadossi R, Giardino R. Cartilage repair with osteochondral autografts in sheep: Effect of biophysical stimulation with pulsed electromagnetic fields. *J Orthop Res.* 2008;26(5):631–642.

120. Ciombor DM, Aaron RK, Wang S, Simon B. Modification of osteoarthritis by pulsed electromagnetic field—A morphological study. *Osteoarthritis Cart.* 2003;11(6):455–462.

121. Bigoni M, Sacerdote P, Turati M et al. Acute and late changes in intraarticular cytokine levels following anterior cruciate ligament injury. *Orthop Res.* 2013;31(2):315–321.

122. Biau DJ, Tournoux C, Katsahian S, Schranz P, Nizard R. ACL reconstruction: A meta-analysis of functional scores. *Clin Orthop Relat Res.* 2007;458:180–187.

123. Berg EE, Pollard ME, Kang Q. Interarticular bone tunnel healing. *Arthroscopy.* 2001;17(2):189–195.

124. Zorzi C, Dall'oca C, Cadossi R, Setti S. Effects of pulsed electromagnetic fields on patients' recovery after arthroscopic surgery: Prospective, randomized and double-blind study. *Knee Surgery, Sports Traumatol Arthrosc.* 2007;15(7):830–834.

125. Collarile M, Cadossi M, Chiarello E, Setti S, Zorzi C. L'efficacia della stimolazione biofisica nell'accelerare il recupero funzionale dopo trapianto di condrociti autologhi: Studio prospettico, randomizzato, controllato. 97° Congresso Nazionale della Società Italiana di Ortopedia e Traumatologia, Roma 2012.

126. Benazzo F, Zanon G, Pederzini L et al. Effects of biophysical stimulation in patients undergoing arthroscopic reconstruction of anterior cruciate ligament: Prospective, randomized and double blind study. *Knee Surg Sports Traumatol Arthrosc.* 2008;16(6):595–601.

127. Baker PN, van der Meulen JH, Lewsey J, Gregg PJ. National joint registry for England and Wales. The role of pain and function in determining patient satisfaction after total knee replacement. Data from the national joint registry for England and Wales. *J Bone Joint Surg Br.* 2007;89(7):893–900.

128. Beswick AD, Wylde V, Gooberman-Hill R, Blom A, Dieppe P. What proportion of patients report long-term pain after total hip or knee replacement for osteoarthritis? A systematic review of prospective studies in unselected patients. *BMJ.* 2012;Open 2:e000435.

129. Ugraş AA, Kural C, Kural A, Demirez F, Koldaş M, Cetinus E. Which is more important after total knee arthroplasty: Local inflammatory response or systemic inflammatory response? *Knee.* 2011;18:113–116.

130. Moretti B, Notarnicola A, Moretti L, Setti S, De Terlizzi F, Pesce V, Patella V. I-ONE therapy in patients undergoing total knee arthroplasty: A prospective, randomized and controlled study. *BMC Musculoskelet Disord.* 2012;13:88.

131. Adravanti P, Nicoletti S, Setti S, Ampollini A, de Girolamo L. Effect of pulsed electromagnetic field therapy in patients undergoing total knee arthroplasty: A randomised controlled trial. *Int Orthop.* 2014;38(2):397–403.

132. Marcheggiani Muccioli GM, Grassi A, Setti S, Filardo G, Zambelli L, Bonanzinga T, Rimondi E, Busacca M, Zaffagnini S. Conservative treatment of spontaneous osteonecrosis of the knee in the early stage: Pulsed electromagnetic fields therapy. *Eur J Radiol.* 2013;82(3):530–537.

133. Gobbi A, Lad D, Petrera M, Karnatzikos G. Symptomatic early osteoarthritis of the knee treated with pulsed electromagnetic fields: Two-year follow-up. *Cartilage.* 2014;5(2):76–83.

17 Magnetic Fields for Pain Control

William Pawluk

CONTENTS

17.1 GENERAL

The issue of pain treatment is an extremely urgent health and socioeconomic problem. Pain, in acute, recurrent, and chronic forms, is prevalent across age, cultural background, and sex and costs North American adults an estimated $10,000–$15,000 per person annually. At least one in four adults in North America is suffering from some form of pain at any given moment. These estimates do not include the nearly 30,000 people that die in North America each year due to nonsteroidal anti-inflammatory drug-induced gastric lesions (Thomas and Prato, 2002).

This large population of people in pain relies heavily upon the medical community for the provision of pharmacological treatment. Many physicians are now referring chronic pain sufferers to non-drug-based therapies, that is, complementary and alternative medicine (CAM) in order to reduce drug dependencies, risks of invasive procedures, and/or side effects. More than a third of American adults report using some form of CAM, with total visits to CAM providers each year now exceeding those to primary care physicians. Annual out-of-pocket costs for CAM are estimated to exceed $27 billion (Institute of Medicine, 2005).

The ability to relieve pain is very variable and unpredictable, depending on the source or location of pain and whether it is acute or chronic. Pain mechanisms are complex and have peripheral and central nervous system aspects. Therapies should be tailored to the specifics of the pain process in the individual patient. Most effective pain management strategies require multiple concurrent approaches, especially for chronic pain. It is rare that a single modality, including magnetic field (MF) therapy, solves the problem completely.

The level of evidence for the value of pulsed electromagnetic fields (PEMFs) for managing pain varies significantly. From an insurance coverage and public policy perspective, the standard for the quality of evidence even in randomized controlled trials is very demanding. From the perspective of the clinician, who lives and functions with a high degree of ambiguity, reasonable levels of evidence are all that is necessary for the clinician to make decisions and recommendations. For most clinicians, the level of evidence, relative to managing the patient's pain, is typically observational since there is very little value in more objective measures of pain and since the pain under management tends to be in multiple locations and in a very wide variety of types of patients. Rarely do clinicians have sufficient useful biological (e.g., from imaging studies) or biochemical (from available laboratory studies) data to drive decision-making for pain management. Most decision-making at the clinical level resolves to physical examination of patients or the patient's own self-report.

While biologists and physicists attempt to discern and discover the mechanisms of biological processes that may contribute to pain, measurements of these phenomena are not typically accessible to the clinician. While pharmacologists may determine that a particular biological or cellular physiological mechanism is a desired goal for a pharmacologic intervention, the clinician cannot typically measure this level of aspect of the problem. Therefore, the clinician relies on observable phenomena, such as physical examination, blood pressures, cholesterol levels, and C-reactive protein levels, which are not direct surrogates for the mechanism of a given medication. So, to the extent that controlled clinical trials exist, despite some of their methodological limitations, for the clinician, this is a better level of information for decision-making than mechanistic studies. Studies that report on observable physiological parameters may be of the greatest value in the practical clinical management of pain.

Static EMFs have been used for centuries to control pain and other biological problems. This review explores the value of magnetic therapy in managing pain, presenting the scientific basis supporting these modalities. This includes the use of MFs, produced by both static (permanent) and time-varied (most commonly, pulsed) magnetic fields (PEMFs). Fields of various strengths, waveforms, and frequencies have been evaluated, depending on the clinical conditions or aspects selected for the study. There is as yet no *gold standard*. And, it is doubtful that there ever will be because of the very nature of the complexities involved. After thousands of patient-years of use globally, very little risk has been found to be associated with MF therapies (Markov, 2004). Standards and guidelines for safety have been promulgated and published (ICNIRP, 2010). The primary precautions or contraindications relate to implanted electrical devices, pregnancy (because of lack of data), and seizures with certain kinds of frequency patterns in seizure-prone individuals.

MFs affect pain perception in many different ways. These actions are both direct and indirect. Direct effects of MFs are neuron firing, calcium ion movement, membrane potentials, endorphin levels, nitric oxide, dopamine levels, acupuncture actions, and nerve regeneration. Indirect benefits of MFs on physiological function are on circulation, muscle, edema, tissue oxygen, inflammation, healing, prostaglandins, cellular metabolism, and cell energy levels (Jerabek and Pawluk, 1996).

Pain relief mechanisms vary by the type of stimulus used (Takeshige and Sato, 1996). For example, needling to the pain-producing muscle, application of a static MF or external qigong, or needling to an acupuncture point all reduce pain by different mechanisms. In guinea pigs, pain could be induced by reduction of circulation in the muscle (ischemia) and reduced by recovery of circulation. Muscle pain relief is induced by recovery of circulation due to the enhanced release of acetylcholine as a result of activation of the cholinergic vasodilator nerve endings innervated to the muscle artery (Takeshige and Sato, 1996).

17.2   ALTERNATIVES FOR PAIN MANAGEMENT

In clinical practice, it is rare to manage pain with one simple approach. Usually, many approaches are used simultaneously. These include supplements and herbs (Eriksen et al., 1996; Arnold and Thornbrough, 1999; Randall et al., 2000), acupuncture (Thomas et al., 1992; Wong and

Rapson, 1999), and chiropractic adjustments or manipulations (Haldeman and Rubinstein, 1993), among others.

Both static (permanent) and time-varied (pulsed) EMFs have been studied for the management of a myriad of health conditions, including pain. Most of this review will focus on PEMFs.

Since the turn of this century, a number of electrotherapeutic, magneto-therapeutic, and electromagnetic medical devices have emerged for treating a broad spectrum of trauma, tumors, and infections with static and pulsed EMFs (Jerabek and Pawluk, 1996). The current scientific literature indicates that short-term periodic exposure to PEMFs has emerged as one of the most effective and safe forms of therapy for many conditions, including pain.

An assessment of the difference in the effectiveness of pain relief between pulsed radiofrequency (PRF) and electro-acupuncture (Lin et al., 2010) stimulation in patients with chronic low-back pain was made. Visual analog scale (VAS) pain score, the Oswestry disability index (ODI), and Short Form 36 (SF-36) were used to assess pain relief and functional improvement effect of PRF and electro-acupuncture, in a randomized controlled trial. PRF therapy was significantly better after 1 month of treatment. But electro-acupuncture also showed functional improvement in the lumbar spine based on the ODI scores.

17.3 STATIC MAGNETIC FIELDS FOR PAIN RELIEF

While most of the evidence presented in this review relates to PEMFs, static MFs have also been found in various studies to have benefits in pain management. This review of necessity is limited in the number of studies regarding static MF benefits, and the ones presented are somewhat unique in their approach to addressing the pain problem.

Several studies found value in the use of static magnets. A review of magnets applied to acupuncture points shows that this application has variable usefulness (Colbert et al., 2008). However, an acupuncture-like action may be at least one of the explanations for the benefit of locally applied magnets. Magnetic wrist bracelets have been found helpful to decrease the pain from osteoarthritis (OA) of the hip and knee (Harlow et al., 2004). In this study, commercial neodymium magnets were tested on 194 individuals in a randomized, placebo-controlled trial for pain control in OA of the hip and knee. The magnets in group A were 170–200 mT; in group B, they were 21–30 mT; and group C used nonmagnetic steel washers. The Western Ontario and McMaster universities osteoarthritis index (WOMAC A) was measured at entry, 4 and 12 weeks. Secondary outcomes included changes in WOMAC B and C scales and a VAS for pain. Mean pain scores were reduced more in the standard magnet group (group A) than in the sham control group (mean difference 1.3 points). The intriguing result is that even with nonlocal application of a magnetic therapy for pain from OA of the hip and knee, pain in these remote joints decreases when wearing magnetic wrist bracelets. A lower-intensity magnet group (group B) was used as a low *dose* control. The assumption here is that the lower-intensity magnetic therapy is expected to be more likely to be comparable to the sham group than the higher-intensity magnetic therapy. However, in this study, even the weaker magnets produced a measurable benefit, although it was not statistically different from the stronger magnets. Still, there was a trend for the stronger magnets to produce better results than the weaker magnets. The mean reduction in WOMAC A scores in the intervention group of 2.9 (27% change from baseline score) and the difference above placebo (1.3 points) is similar to that found in trials of conventional medical OA treatments.

A uniquely designed static magnetic system was also studied in rheumatoid arthritis (RA) and knee pain (Segal et al., 2001) in a randomized, double-blind, controlled, multisite clinical trial in 64 patients. Four static magnets enclosed in a molded plastic circular case (group A) or a control magnetic device with only one magnet (group B) was taped to the knee of each subject to be worn continually for 1 week. A greater reduction in reported pain in the A group was sustained through 1-week follow-up (40.4% versus 25.9%), supported by twice-daily pain diary recordings, p<0.0001 for each treatment period versus baseline. Between-group comparisons found no significant difference

($p < 0.23$), which shows no difference in pain reduction results between the stronger four-magnet unit versus the one-magnet unit enclosures. Subjects in group A reported an average decrease in their global assessment of disease activity of 33% over 1 week, as compared with a 2% decrease in the control group ($p < 0.01$). After 1 week, 68% of the treatment group reported feeling better or much better, compared with 27% of the control group, and 29% and 65%, respectively, reported feeling the same as before treatment ($p < 0.01$). Despite what appear to be clinically relevant differences between the stronger and weaker magnetic groups, the lack of statistical significance may well be due to an underpowered study or the treatment time was not carried on long enough. Because even weaker magnets may still show significant clinical benefits, studies using them as controls are likely to need larger numbers of subjects in order to detect meaningful differences. A similar issue occurred in the Harlow study (Harlow et al., 2004).

Several studies reported less positive results. These are instructive for potential methodological considerations for future static magnet studies. A static magnetic foil placed in a molded insole for the relief of heel pain was used for 4 weeks to treat heel pain (Caselli et al., 1997). Sixty percent of patients in the treatment and sham groups reported improvement. The magnetic foil offered no advantage over the plain insole. This study, like others with low numbers of patients, may have had significant design limitations, including not having had a large enough sample to show differences, especially when considering dosimetry at the target tissue. Flexible static magnetic shoe insert foils produce fairly weak fields. Since MFs, especially from static magnets, drop off in strength very rapidly from the surface of the applicator (Pilla, 1998), the MF *dose* delivered to the target tissue is usually much lower than assumed or theoretical values. The depth of penetration of the desired MF intensity is critical to achieve clinically meaningful results (Markov, 2004). As a result, desired outcomes may never be seen for the magnetic device, and conditions studied or much longer treatment times may be necessary to achieve even any meaningful results.

Chronic pain frequently presented by post-polio patients can be relieved by application of static MFs applied directly over trigger points using 300–500 G static magnets for 45 min (Vallbona and Richards, 1999).

Treatment with a flexible permanent magnetic pad for 21 days reduced chronic muscular low-back pain six times more than placebo (Preszler, 2000). This has also been effective for herniated lumbar disks, spondylosis, radiculopathy, sciatica, and arthritis. Pain relief is sometimes experienced as early as 10 min or in some cases takes as long as 14 days.

17.4 PEMFs AND PAIN MANAGEMENT

Several authors have reviewed the experience with PEMFs in Eastern Europe (Jerabek and Pawluk, 1996) and elsewhere (Trock, 2000) and provided a synthesis of the typical physiological findings of practical use to clinicians, resulting from magnetic therapies. These include, at a minimum, reduction in edema and muscle spasm/contraction, improved circulation, enhanced tissue repair, and natural antinociception. These are the fundamentals of the repair of cell injury. PEMFs have been used extensively in many conditions and medical disciplines, being most effective in treating rheumatic or musculoskeletal disorders. PEMFs produced significant reduction of pain, improvement of spinal functions, and reduction of paravertebral spasms. In clinical practice, PEMFs have been found to be an aid in the therapy of orthopedic and trauma problems (Borg et al., 1996).

The ability of PEMFs to affect pain is at least in part dependent on the ability of PEMFs to positively affect human physiological or anatomic systems. The human nervous system is strongly affected by therapeutic PEMFs (Prato et al., 2001). Animals exposed to static and extremely low-frequency (ELF) MFs are also affected by the presence of light, which strengthens the effects of PEMFs (Prato et al., 1999).

One of the most reproducible results of weak ELF MF exposure is an effect upon neurologic pain signal processing (Thomas and Prato, 2002). This evidence suggests that PEMFs would also be an effective complement for treating patients suffering from both chronic and acute pain.

The placebo response may explain as much as 40% of an analgesia response from any pain treatment (Colloca et al., 2013), and needs to be accounted for in research design to assure adequate sample sizes. However, aside from this aspect of accounting for the placebo effect, the central nervous system mechanisms responsible for the placebo response, that is, central cognitive and behavioral processes, can be addressed directly in managing pain and include medications, hypnosis, mindfulness meditation, and psychotherapy. In addition, these placebo response–related central processes appear to be an appropriate target with magnetic therapies for managing pain. Amplifying MF manipulation of cognitive and behavioral processes has been evaluated in animal behavior studies and in humans, affecting at the very least opiate receptors (Del Seppia et al., 2007). Therefore, amplifying the placebo response with centrally focused MFs would generally be expected to be additive to pain management using MF therapies elsewhere on the body. One particular application considering this approach is rapid transcranial magnetic stimulation (rTMS). rTMS is being increasingly evaluated for this purpose and already found effective in reducing centrally and peripherally caused pain (Lefaucheur et al., 2004). A review of the rTMS technology is available at Aleman (2013).

Current transcranial magnetic coil stimulators can activate brain neural structures without deep electrode placement and the discomfort associated with transcutaneous electrical stimulation used in pain control. The possibility of reducing pain in patients with localized musculoskeletal processes is by applying repetitive magnetic stimulation (rMS) on tender noncentral body regions (Pujol et al., 1998). Thirty patients were randomized to receive 40 min of real or sham magnetic stimulation. After a single session, real magnetic stimulation significantly exceeded the sham effect, decreasing VAS scores, using a 100 mm scale, by 59% in the treated group and 14% in sham-treated patients (p = 0.001). Pain relief regularly persisted several days. Results indicate that powerful magnetic coil stimulation can efficiently reduce pain originating from localized musculoskeletal injuries.

In the opposite situation of the placebo response, not all circumstances involving pain respond equally well to any management approach, including PEMFs. The time it takes to heal the underlying cause contributing to pain is an important variable. To elucidate one possible mechanism of pain reduction with PEMFs, a randomized, double-blinded, placebo-controlled, crossover experiment was conducted with human volunteers (Fernandez et al., 2007). Pain was caused by infusion of hypertonic 5% saline into the forearm. Subjects received active or sham PEMF. There were no significant differences in mean VAS pain scores between the two machines. There are many possible explanations for this lack of apparent benefit in this experimental model, including inadequate PEMF dosing.

17.5 POTENTIAL MECHANISMS OF MAGNETIC FIELD EFFECTS ON PAIN

Cell injury itself involves multiple processes (Kumar, 2007), which, if mitigated, can be expected to reduce the perception of pain and limit the results of the cell injury. Therefore, this is the goal of clinical management. If the cause of pain cannot be reduced or eliminated, then the goals of pain management shift to reducing the perception of pain or blocking the pain signal traffic otherwise.

Research on the use of PEMFs for pain management focuses on the multiple mechanisms of the production of pain. The primary mechanisms of the production of pain in local tissue in response to cell injury include, to varying degrees, edema, apoptosis or necrosis, diminished vascular supply, reduced cellular energy production, and impaired repair processes. PEMF therapies address many of these different aspects of cell injury (Jerabek and Pawluk, 1996).

Magnetic therapy increases the threshold of pain sensitivity (Thomas and Prato, 2002) and activates the anticoagulation system (Khamaganova et al., 1993), which increases circulation to tissue. PEMF treatment stimulates production of opioid peptides, activates mast cells and increases electric capacity of muscular fibers, helps with edema and pain before or after a surgical operation (Pilla, 2013), increases amino acid uptake (De Loecker et al., 1990), and induces changes in transmembrane energy transport enzymes, allowing energy coupling and increased biological chemical transport work.

Healthy humans normally have reduced pain perception and decreased pain-related brain signals (Prato et al., 2001). Biochemical changes in the blood of treated patients are found that support the pain reduction benefit. PEMFs cause a significant improvement in normal standing balance in adult humans (Thomas et al., 2001). PEMFs couple with muscular processing or upper-body nervous tissue functions, which indicate CNS sensitivity that likely improves central pain processing.

Various kinds of PEMFs have been found to reduce pain. For example, various MFs applied to the head or to an extremity, for 1–60 min, with intervals between exposures from several minutes to several hours, randomly sequenced with sham exposures allowed the study of brain reactions by various objective measures (Kholodov, 1998). EEGs showed increased low-frequency rhythms. Low-frequency EEG rhythms may explain the common perception of relaxation and sleepiness with ELF EMFs. Even weak AC MFs affect pain perception and pain-related EEG changes in humans (Sartucci et al., 1997). A 2 h exposure to 0.2–0.7 G ELF MFs caused a significant positive change in pain-related EEG patterns.

The benefits of PEMF use may last considerably longer than the time of use. This is a common clinical observation. In rats, a single exposure produces pain reduction both immediately after treatment and even at 24 h after treatment (Cieslar et al., 1994). The analgesic effect is still observed at the 7th and 14th day of repeated treatment and even up to 14 days after the last treatment. Repeated presentation of painful stimuli in rats can significantly elevate the threshold of response to painful stimuli. One group (Fleming et al., 1994) investigated the ability of magnetic pulse stimuli to produce increases in pain thresholds, simulating thalamic pain syndrome. Exposure to the PEMFs increased the pain threshold progressively over 3 days. Pain suppression was maintained on the second and third days relative to other treatments. The pain threshold following the third MF exposure was significantly greater than those associated with morphine and other treatments. Brain-injured and normal rats both showed a 63% increase in mean pain threshold. The mechanism may involve endorphins, having important implications for clinical practice and the potential for a reduction in reliance on habit-forming medications.

PEMFs promote healing of soft tissue injuries by reducing edema and increasing resorption of hematomas (Markov and Pilla, 1995), thereby reducing pain. Low-frequency PEMFs reduce edema primarily during treatment sessions. PEMFs at very high frequencies applied for 20–30 min cause decreases in edema lasting several hours following an exposure session. PRFs induce vasoconstriction at the injury site, probably a primary mechanism in the anti-edema effect. PRFs also displace negatively charged plasma proteins found in the traumatized tissue. This is expected to increase lymphatic flow, an additional factor contributing to edema reduction.

PEMF signals induce maximum electric fields in the mV/cm range at frequencies below 5 kHz. PRF fields are typically shortwave band, 13–40 MHz range, carrier waves. Modulation, for example, consisting of modulated sinusoidal waves of lower frequencies, will prevent heating in target tissues. Higher-intensity PRFs have many applications involving tissue ablation, through heating destruction of tissue. PRFs considered in this review do not include the ablative PRFs.

Chronic pain often occurs from aberrant small neural networks with self-perpetuated neurogenic inflammation. It is thought that high-intensity pulsed magnetic stimulation (HIPMS) non-invasively depolarizes neurons and can facilitate recovery following injury (Ellis, 1993). HIPMS, intensity up to 1.17 T, was used to study recovery after injury in patients with posttraumatic/postoperative low-back pain, reflex sympathetic dystrophy (RSD), neuropathy, thoracic outlet syndrome, and endometriosis. The outcome VAS difference was 0.4–5.2 with sham treatments versus 0–0.5 for active treatments. The author proposed that the pain reduction was likely due to induced eddy currents.

In normal subjects, a magnetic stimulus over the cerebellum reduces the size of responses evoked by cortical electrical stimulation (Ugawa et al., 1997). Magnetic stimulation over the cerebellum produces the same effect as electrical stimulation, even in ataxic patients, and may be useful for the pain associated even with peripheral muscle spasticity. Direct electrical stimulation of the brain is an accepted clinical procedure (Levy et al., 2010). Since electrical stimulation is generally uncomfortable and often invasive, PEMF stimulation, whether with lower-intensity PEMF systems or higher-intensity rTMS (Lefaucheur et al., 2004), is being seen to be a safer and equally effective alternative.

Even when magnetic field stimulation (MFS) of high enough intensity is used to cause quadriceps muscle contractions (Han et al., 2006), it appears to cause less pain at similar peak muscle torque levels than neuromuscular electrical stimulation (NMES). The VAS was compared at the same peak torque reached by each method of stimulation. The mean tolerable maximum peak torque was higher, almost double, with MFS versus NMES. So, magnetic stimulation produced less pain at the same isometric peak torque. MFS may even be effective in reducing pain in the presence of high-intensity muscle contractions.

Effects on the tissues of the body and the symptoms of pain have been found across a wide spectrum of electromagnetic frequencies, including high-frequency PEMFs. For example, significant reductions in pain were found in individuals with acute whiplash injuries using 27.12 MHz PEMF stimulation (Foley-Nolan et al., 1992). The same group (Foley-Nolan et al., 1990) had previously found that individuals with persistent neck pain lasting greater than 8 weeks had statistically significantly greater improvement in their pain compared to controls. The controls were then crossed over onto PEMF treatment and had similar results.

For more detailed discussion of the potential mechanisms of action of MFs to treat pain, see Markov (2004). The author discusses some of the parameters that may be necessary to properly choose a therapeutic MF with respect to the target tissue to be stimulated. The research literature on magnetic therapies for pain management is very variable in describing the particular parameters of the magnetic therapy apparatus being studied. This leaves the clinician at a significant disadvantage in determining which MFs produce the best results for the given condition being treated. Further, the author states, "during the past 25 years more than 2 million patients have been treated worldwide for a large variety of injuries, pathologies and diseases. This large number of patients exhibited a success rate of approximately 80%, with virtually no reported complications." The author goes on to describe a number of mechanisms of cellular action of EMFs that may be deemed responsible for the therapeutic benefit in improving pain.

In another study, Shupak et al. (2004) looked at possible mechanisms or influencing factors for the effects of PEMFs on pain, especially on sensory and pain perception thresholds. It appears that MF exposure does not affect temperature perception but can increase pain thresholds, indicating an analgesic effect.

Based on the review by Del Seppia et al. (2007), it appears that at least one of the mechanisms involved in PEMF effects on pain and nociception is the opiate receptor. Another study in rats (Fleming et al., 1994) found that there was an analgesic effect comparable to more noxious tactile stimulation, that is, stress-induced analgesia. There was an approximately 50% increase in the pain threshold in response to electrical current stimulation.

In a study to gain a better understanding of pain perception (Robertson et al., 2010), a functional magnetic resonance imaging study was done to assess how the neuromodulation effect of MFs influences the processing of acute thermal pain in normal volunteers. ELF MFs (from DC to 300 Hz) have been shown to affect pain sensitivity in snails, rodents, and humans. Because of this research, it is unlikely that a pure placebo response is involved. This neuroimaging study found changes in specific areas of the brain with pain stimuli that are definitely modified by low-intensity PEMF exposure.

Chronic pain is often accompanied with or results from decreased circulation or perfusion to the affected tissues, for example, cardiac angina or intermittent claudication. PEMFs have been shown to improve circulation (Guseo, 1992). Pain syndromes due to muscle tension and neuralgias improve.

17.6 NEUROPATHY

Peripheral neuropathy can be an extremely painful condition that is very challenging to manage. Two randomized controlled studies failed to show significant results in diabetic peripheral neuropathy (DPN) (Wróbel et al., 2008; Weintraub et al., 2009). Another two studies showed significant improvements in DPN (Cieslar et al., 1995; Graak et al., 2009). There were significant methodological differences among the studies.

A large study (Weintraub et al., 2009) was conducted to determine whether repetitive and cumulative exposure to low-frequency PEMF to the feet can reduce neuropathic pain (NP) and influence nerve regeneration. Two-hundred and twenty-five patients with DPN stage II or III were randomized in a double-blind, placebo-controlled parallel study, across 16 academic and clinical sites in 13 states to PEMF or sham (placebo) devices. They applied their treatments 2 h per day to their feet for 3 months. Pain reduction scores were measured using a VAS, the neuropathy pain scale (NPS), and the patient's global impression of change (PGIC). A subset of subjects underwent serial 3 mm punch skin biopsies from three standard lower-limb sites for epidermal nerve fiber density (ENFD) quantification. There was a significant dropout rate of 13.8%. The PEMF versus sham group had reductions in DPN symptoms on the PGIC (44% versus 31%; $p = 0.04$). There were no significant differences in the NP intensity on NPS or VAS. Of the 27 patients who completed serial biopsies, 29% of the PEMF group had an increase in the distal leg ENFD of at least 0.5 SDs, while none did in the sham group ($p = 0.04$). Those with increases in distal thigh ENFD had significant decreases in pain scores. The conclusion was that PEMF at this dose was not effective specifically in reducing NP. However, neurobiological effects on ENFD, PGIC, and reduced itching scores were hopeful and suggest that future studies should be attempted with higher PEMF intensities 3000–5000 G, longer duration of exposure, and a larger biopsy cohort. Since most of the therapeutic approaches to DPN have poor success rates, relying mostly on the suppression of pain with medications, this study is encouraging in actually demonstrating potential nerve regeneration improvements.

Another randomized, placebo-controlled, double-blind study (Wróbel et al., 2008) was conducted to assess an ELF PEMF effect on pain intensity, quality of life and sleep, and glycemic control in patients with painful diabetic polyneuropathy. Sixty-one patients were randomized into a study group of 32 patients exposed to a low-frequency, low-intensity MF or a sham control group of 29 patients. Pain durations were greater than 2 years in both groups. Treatments were for 3 weeks, 20 min a day, 5 days a week. Questionnaires, completed at the beginning, after 1–3 and 5 weeks, included SFMPQ-VAS (pain evaluation), EuroQol EQ-5D, and MOS Sleep Scale. Significant reductions in pain intensity were seen in both the study group VAS 73 mm at baseline versus 33 mm after 3 weeks and controls, VAS 69 mm at baseline versus 41 mm after 3 weeks. The extent of pain reduction did not differ significantly between the groups at any time. The conclusion was that this low-intensity ELF PEMF, used for only 3 weeks, had no advantage over sham exposure in reducing pain intensity. In the Weintraub study, patients were treated for 3 months, providing a longer opportunity to produce sustainable changes in the tissues. Since neuropathy is a very stubborn problem to treat, it is likely that both of these neuropathy studies were too short for the severity of neuropathy present, treatment protocols, measures, and equipment used.

In another study (Graak et al., 2009) on NP, using low-power, low-frequency PEMF of 600 and 800 Hz, 30 patients, 40–68 years of age with DPN stages N1a, N1b, N2a, were randomly allocated to three groups of 10 in each. Groups 1 and 2 were treated with low-power 600 and 800 Hz PEMF for 30 min for 12 consecutive days. Group 3 served as control on usual medical treatment. Pain and motor nerve conduction parameters (distal latency, amplitude, nerve conduction velocity) were assessed before and after treatment. They found significant reduction in pain and statistically significant ($p < 0.05$) improvement in distal latency and nerve conduction velocity in experimental Groups 1 and 2. Using this particular protocol, low-frequency PEMF was seen to reduce NP as well as for retarding the progression of neuropathy even when applied even for a short span of time. What could happen with longer-term treatment remains to be determined.

Thirty-one patients with diabetes mellitus (type I and II), with intense symptoms of neuropathy, were treated (Cieslar et al., 1995). They had 20 exposures to variable sinusoidal PEMF, 40 Hz, 15 mT, every day for 12 min. Reduction of pain and paresthesias, vibration sensation, and improved muscle strength was seen in 85% of patients, all significantly better than sham controls.

Carpal tunnel syndrome is another form of neuropathy, affecting the median nerve at the wrist. There are many different approaches to the treatment of carpal tunnel syndrome, including surgery, with varying success. In a randomized, double-blinded, placebo-controlled trial (Weintraub and Cole, 2008), a commonly commercially available combination of simultaneous static and dynamic, rotating time-varying dynamic MFS was used to treat the wrist. There was a significant reduction of *deep* pain. Ten months of active PEMF resulted in improvement in nerve conduction and subjective improvement on examination (40%), pain scores (50%), and a global symptom scale (70%).

The neuropathy of postherpetic neuralgia, a very common and painful condition, often medically resistant, responded to PEMF (Kusaka et al., 1995). A combination of static and pulsed MF device was placed on the pain/paresthesia areas or over the spinal column or limbs. Treatments continued until symptoms improved or adverse side effects occurred. Therapy was effective in 80%. This treatment approach shows that treatment for pain problems may either be localized to the area of pain or over the spinal column or limbs, away from the pain. Treatment over the appropriate related spinal segment offers the opportunity to interrupt the afferent pain signal traffic to the brain. This approach has been frequently used with success in Eastern European studies (Jerabek and Pawluk, 1996). Another author reported a more general clinical series in postherpetic pain in which better results happened in patients simultaneously suffering from neck and low-back pain (Di Massa et al., 1989).

Posttraumatic, late-stage RSD or now called regional complex pain syndrome, a form of neuropathy, is very painful and largely untreatable by standard medical approaches. In one report, ten 30 min PEMF sessions of 50 Hz followed by a further 10 sessions at 100 Hz plus physiotherapy and medication reduced edema and pain at 10 days (Saveriano and Ricci, 1989). There was no further improvement at 20 days. The author had a personal case treated with a 27.12 MHz PEMF signal, in a nurse who was almost completely disabled in her left upper extremity. She used her device for about an hour a day. Within about 1 month, she had about 70% recovery, and within 2 months, she had essentially normal function with no further sensitivity to touch, changes in temperature, etc. She maintained her recovery with continued treatments in the home setting.

rTMS has been reported for the treatment of RSD in a sham-controlled trial (Picarelli et al., 2010) and one case series (Pleger et al., 2004). In the controlled trial, the active treatment group of five patients, during treatment, had a significant reduction in pain intensities. Reduction in the mean VAS scores was 4.65 cm (50.9%) against 2.18 cm (24.7%) in the s-rTMS group. The highest reduction occurred at the 10th session. In the earlier reported case series, patients were treated in the motor cortex contralateral to the affected side. Seven out of ten patients reported decreased pain intensities. Pain relief occurred 30 s after stimulation, with a maximum effect at 15 min later. Pain re-intensified up to 45 min after rTMS. Sham rTMS did not alter pain perception. Both of these studies appear to indicate that the benefit of even rTMS is limited to the time of treatment. This may well relate back to an earlier point regarding the need to heal the underlying cause in order to achieve sustainable results. Short-term rTMS treatments are unlikely to more durably impact the underlying cellular and physiological dysfunctions but, in some cases, may have a definite value in short-term management.

17.7 ORTHOPEDIC OR MUSCULOSKELETAL USES

Musculoskeletal conditions, especially with related pain, are most frequently treated with MF therapies. Among these, one of the most common conditions is lumbar arthritis, as a cause of back pain. Chronic low-back pain affects approximately 15% of the US population during their lifetime (Preszler, 2000). Given the current treatment options available through conventional medical therapy, with their attendant risks, there is a large unmet need for safe and effective alternative therapies (Institute of Medicine, 2005).

PEMFs of 35–40 mT give relief or elimination of pain about 90%–95% of the time for lumbar OA, improve results from other rehabilitation therapies, and secondarily, additionally improve related neurologic symptoms (Mitbreit et al., 1986). Even PEMFs of 5–15 G used at the site of pain and related trigger points also help (Rauscher and Van Bise, 2001). Some patients remained pain free 6 months after treatment.

Using peripherally applied ELF high-intensity magnetic stimulation (rMS), with an rTMS device, benefits were found in musculoskeletal pain from painful shoulder with abnormal supraspinatus tendon, tennis elbow, ulnar compression syndrome, carpal tunnel syndrome, semilunar bone injury, traumatic amputation neuroma of the median nerve, persistent muscle spasm of the upper and lower back, inner hamstring tendinitis, patellofemoral arthritis, osteochondral lesions of the heel, posterior tibial tendinitis, upper back muscle spasms, and rotator cuff injury (Pujol et al., 1998).

In a series of 240 patients treated in an orthopedic practice with PEMFs, patients had decreased pain (Schroter, 1976) from rheumatic illnesses, delayed healing process in bones, and pseudoarthrosis, including those with infections, fractures, aseptic necrosis, venous and arterial circulation, RSD (all stages), osteochondritis dissecans, osteomyelitis, and sprains and strains and bruises. The clinically determined success rate approached 80%. About 60% of loosened hip prostheses have subjective relief of pain and walk better, without a cane. Even so, x-ray evidence of improvement was seen periodically, as evidenced by cartilage/bone reformation, including the joint margin. If the goal in pain management is to heal the underlying tissue, not just manage symptoms, evidence, typically from imaging studies, can drive the duration of treatment to obtain the most long-lasting and more permanent results.

To further expand on these points, in both research and clinical settings, a determination always needs to be made regarding the length of the course of treatment. This needs to take consideration of the objectives of the treatment. Patients are often happy simply by having a reduction in their symptoms and improvement in their function. On the other hand, the clinician may be aware that better and more long-standing results can be obtained by extending the treatment program. In the case of arthritis, an ideal situation could include resolution or improvement of the bony or cartilage changes, so the pain does not recur. This can best be determined through imaging studies. In the case of fractures, clinically, the person is considered released from care when there is a sufficient callus formed, allowing immobilization casting to be removed and full rehabilitation initiated. On the other hand, these fractures are at risk of breaking down full healing, leading to various levels of nonunion or delayed union, not uncommon problems in orthopedics. Therefore, continued x-ray evaluation of the bone may be necessary to know how long the course of magnetic treatment should be extended to prevent the possibility of union issues. Similar situations apply to other clinical conditions. So, in the earlier situations, simply managing the pain or other symptoms without adequate consideration of the ultimate goals would potentially lead to less than optimal results. The constraints of the research and clinical settings and other practical considerations most often drive the decisions to terminate treatment or the research. Because of these constraints, we do not commonly have adequate information about the proper length of treatment courses to guide clinical practice. If patients had unlimited access to PEMF systems, preferably in the home setting, better information would be able to be obtained to determine optimal courses of treatment.

The use of PEMFs is rapidly increasing and extending to soft tissue from its first applications to hard tissue (Pilla, 2013). EMF in current orthopedic clinical practice is frequently used to treat delayed and nonunion fractures, rotator cuff tendinitis, spinal fusions, and avascular necrosis, all of which can be very painful. Clinically relevant response to the PEMF is generally not always immediate, requiring daily treatment for upward of a year in the case of nonunion fractures. PRF applications appear to be best for the reduction of pain and edema. The acute tissue inflammation that accompanies the majority of traumatic and chronic injuries is essential to the healing process; however, the body often over-responds in the chronic lesion situation, and the resulting edema causes delayed healing and chronic pain. Edema reduction is an important target for PRF and PEMF applications.

Double-blind clinical studies have now been reported for chronic wound repair, acute ankle sprains, and acute whiplash injuries. PRFs have been found to accelerate reduction of edema in acute ankle sprains by up to fivefold (Pennington et al., 1993). Some of the best responses to PRFs appear to be during or immediately after the treatment of acute soft tissue injuries. For bone repair, while not commonly used for this purpose, limited experience shows that responses to PRFs are significantly slower. The voltage changes induced by PRF at binding sites in macromolecules affect ion binding kinetics with resultant modulation of biochemical cascades relevant to the inflammatory stages of tissue repair.

Even chronic musculoskeletal pain treated with MFs for only 3 days, once per day, can eliminate and/or maintain chronic musculoskeletal pain (Stewart and Stewart, 1989). Small, battery-operated PEMF devices with very weak field strengths have been found to benefit musculoskeletal disorders (Fischer, 2002). Because of the low strength used, treatment at the site of pain may need to last between 11 and 132 days, between two times per week, 4 h each, and, if needed, continuous use. Use at night could be near the head, for example, beneath the pillow, to facilitate sleep. Pain scale scores are significantly better in the majority of cases. Conditions that can be considered for treatment are arthritis, lupus erythematosus, chronic neck pain, epicondylitis, patellofemoral degeneration, fracture of the lower leg, and CRPS.

Back pain or whiplash syndrome treated with a very low-intensity PEMF twice a day for 2 weeks along with usual pain medications relieves pain in 8 days in the PEMF group versus 12 days in the controls (Thuile and Walzl, 2002). Headache is halved in the PEMF group, and neck and shoulder/arm pain improved by one-third versus medications alone.

Cervical spine-related nerve pain affects approximately 1 in 1000 adults per year. A prospective audit done initially on the effect of PRF treatment of the cervical dorsal root ganglion found satisfactory pain relief for a mean period of 9.2 months. The research group then went on to do a randomized sham-controlled trial (Van Zundert et al., 2007). At 3 months, the PRF group showed a significantly better outcome with regard to the global perceived effect (>50% improvement) and VAS (20-point pain reduction). The need for pain medication was significantly reduced in the active group after 6 months. Another application of PRF, using low field amplitude pulsed short-wave 27.12 Hz diathermy, has successfully treated persistent neck pain and improved mobility (Foley-Nolan et al., 1990). This system used a miniaturized, 9 V battery-operated, diathermy generator fitted into a soft cervical collar, for 3–6 weeks, 8 h daily. Seventy-five percent of improved range of motion (ROM) and pain were seen within 3 weeks of treatment.

Other PEMFs have been found (Kjellman et al., 1999) to have more benefit in the treatment of neck pain in some research, compared to physical therapy, for both pain and mobility.

A blinded randomized study was conducted to compare European spa therapy (ST) with PEMF therapy in chronic neck pain (Forestier et al., 2007a). There was significantly greater improvement in the PEMF group than the ST group (p = 0.02). As part of the earlier study, the authors also did a cost–benefit analysis (Forestier et al., 2007b). It is rare to find cost–benefit research on the use of PEMFs. The main outcome measure of the cost–benefit part of the research was the cost required to achieve an increase in health dimension scores on the MOS SF-36 comparing care in the 6 months preceding to 6 months after the start of the study. They found that the overall health care costs were less for the PEMF group than the ST and control groups. A gain of one physical MOS SF-36 unit over 1 year cost €3,400 for the PEMF group as a whole, €29,000 for the ST group, and €95,076 for the control group. It appears that the cost–benefit to society of the PEMF treatment group compared to ST or standard therapy produces a substantial cost savings.

One group evaluated pain and swelling after distal radius fractures after an immobilization period of 6 weeks (Cheing et al., 2005). Eighty-three patients were randomly allocated to receive 30 min of either ice plus PEMF (group A), ice plus sham PEMF (group B), PEMF alone (group C), or sham PEMF for 5 consecutive days (group D). All had a standard home exercise program. Outcome measures included a VAS for recording pain, volume displacement for measuring the swelling of the forearm, and a handheld goniometer for measuring the range of wrist motions. They were assessed, before treatment, and on days 1, 3, and 5 during treatment. At day 5, a significantly greater cumulative reduction in VAS as well as improved ulnar deviation ROM was found in group A than the other three groups. For volumetric measurement and pronation, participants in group A performed better than subjects in group D but not those in group B. The end result was that the addition of PEMF to ice therapy produces better overall treatment outcomes than ice alone, or PEMF alone, in pain reduction and ulnar ROM. This study points out the cumulative benefit of using both PEMFs and standard therapy, at least in radial fractures.

Treatment of lateral epicondylitis (tennis elbow) can be frustrating and challenging. Many therapeutic approaches have been used, including local steroid injection and surgery. This condition tends to recur regardless of the therapies used. Steroid injections carry their own risks, and so alternative methods of therapy that are less invasive and potentially harmful need to be developed. PEMFs have been used as a useful and safe candidate therapy. One group tested the efficacy of PEMF compared to sham PEMF and local steroid injection (Uzunca et al., 2007). Sixty patients with lateral epicondylitis were randomly and equally distributed into three groups as follows: group I received PEMF, group II sham PEMF, and group III a corticosteroid + anesthetic agent injection. Pain levels during rest, activity, nighttime, resisted wrist dorsiflexion, and forearm supination were investigated with VAS and algometer. All patients were evaluated before treatment, at the third week, and the third month. VAS values during activity and pain levels during resisted wrist dorsiflexion were significantly lower in group III than group I at the third week. Group I patients had lower pain during rest, activity, and nighttime than group III at the third month. PEMF appears to reduce lateral epicondylitis pain better than sham PEMF. Corticosteroid and anesthetic agent injections can be used in patients for rapid return to activities, along with PEMFs to produce a longer-standing benefit.

Another randomized sham-controlled study (Devereaux et al., 1985) on lateral humeral epicondylitis (tennis elbow) involved 30 patients with both clinical and thermographic evidence of tennis elbow. PEMF treatment, consisted of 15 Hz, delivering 13.5 mV and using a figure of eight coil with the loops over each epicondyle for 8 h a day in one or two sessions, for a minimum period of 8 weeks. They were significant improvements in grip strength at 6 weeks, with a slight decrease in difference at 8 weeks. There was little difference in the first 4 weeks. Since there were only 15 subjects in each treatment group, this study was probably underpowered for most of the other measurement indices used.

17.8 OSTEOARTHRITIS PAIN

OA affects about 40 million people in the United States. OA of the knee is a leading cause of disability in the elderly. Medical management is often ineffective and creates additional side-effect risks. Many patients with OA of the knee/s undergo many soft tissue and intra-articular injections, physical therapy, and many, eventually, arthroscopies or joint replacements.

An ELF sawtooth wave, 50 μT, whole-body and pillow applicator system has been in use for about 20 years in Europe. In one study using the system, applied 8 min twice a day for 6 weeks, it was shown to improve knee function and walking ability significantly (Pawluk et al., 2002). Pain, general condition, and well-being also improved. Medication use decreased. Plasma fibrinogen, C-reactive protein (a sign of inflammation), and the sedimentation rate all decreased by 14%, 35%, and 19% respectively. Sleep disturbances often contribute to increased pain perception. It was found to improve sleep, with 68% reporting good/very good results. Even after 1 year follow-up, 85% claim a continuing benefit in pain reduction. Medication consumption decreases from 39% at 8 weeks to 88% after 8 weeks.

A randomized double-blind controlled study on early knee OA (Nelson et al., 2013) found an almost 60% reduction in mean VAS pain scores within the first 5 or so days for the active treatment group. This improvement held and persisted for 42 days of the study. The sham group mean VAS scores were not significant at any time point. The mean VAS score change for the active treatment group was about threefold better than the sham group. A portable 6.8 MHz sinusoidal signal with a peak-induced electric field of 34 V/m was applied for 15 min twice a day for 14 days. The majority of the pain relief happened in the first 5 days and maintained stable for 42 days, till the end of the study. Pain relief persisted after the 14th day when the PEMF was discontinued. Pain relief decreased from a mean baseline VAS score of 6.85 to just under 4, after day 5.

A major limitation of magnetic therapy research studies is that they are often terminated too soon. This is not the case typically in clinical practice, when these therapies are often applied for extended periods of time, typically when symptoms have improved to a goal target. My own clinical practice experience has been that when individuals discontinue their therapies prematurely, for OA in particular, their pain eventually returns. So, they should either continue therapy over extended periods of time or get periodic retreatments. Other than clinical practice experience, we do not have good research-based data to guide clinical practice. By necessity, this study also did not evaluate clinical measures of function, effusion, and inflammatory markers or have evidence by physical imaging, such as MRI, of actual physical changes to the joint. Other knee arthritis research, using other ELF low-intensity signals (Pawluk et al., 2002), found functional improvements in objective knee scores. We continue to await imaging evidence for longer-term changes to the structure of the knee itself.

In another randomized, placebo-controlled study (Ay and Evcik, 2009), PEMF of 50 Hz, 105 µT, applied for 30 min, was used in 55 patients with grade 3 OA for only 3 weeks for pain relief and enhancing functional capacity of patients with knee OA. Pain improved significantly in both groups relatively equally ($p < 0.000$). However, there was significant improvement in morning stiffness and activities of daily living (ADL) compared to the control group. They did not find a beneficial symptomatic effect of PEMF in the treatment of knee OA in all patients.

In a rheumatology clinic study of knee OA (Pipitone and Scott, 2001), 75 patients received active PEMF treatment by a unipolar magnetic device or placebo for 6 weeks. The 9 V battery-operated device was <0.5 gauss with a low-frequency coil of 2 kHz plus harmonics up to 50 kHz modulated on a 3, 7.8, or 20 Hz base frequency and an ultrahigh frequency coil with a 250 MHz modulated frequency plus harmonics of the same modulation as the LF coil. Patients were instructed to use the magnetic devices three times a day. The 7.8 Hz modulation frequency was prescribed for the morning and afternoon treatments, while the 3 Hz modulation frequency was prescribed for the evening. Baseline assessments showed that the treatment groups were equally matched. Analysis at follow-up showed greater between group improvements in global scores of health status. Paired analysis showed significant improvements in the actively treated group in objective function, pain, disability, and quality of life at study end compared to baseline. These differences were not seen in the placebo-treated group.

In another randomized, double-blind, placebo-controlled clinical trial of knee OA in Denmark (Thamsborg et al., 2005), 83 patients had two 2 h of daily treatment, 5 days per week for 6 weeks. They were reevaluated at 2 and 6 weeks after treatment. Again, objective standardized measures were used. There was a significant improvement in ADL, stiffness, and pain in the PEMF-treated group. In the control group, there was no effect on ADL after 2 weeks and a weak change in ADL after 6 and 12 weeks. Even the control group had significant reductions in pain at all evaluations and in stiffness after 6 and 12 weeks. There were no between-group differences in pain over time. ADL score improvements for the PEMF-treated group appeared to be less with increasing age. When groups were compared, those <65 years of age had significant reduction in stiffness. While this tended to be a negative study, when looking at between-group comparisons, there were indications of improvement in ADLs and stiffness, especially in individuals younger than 65.

Twenty-seven OA patients treated with PEMF in a tube-like coil device for 18 half-hour exposures over 1 month had an average improvement of 23%–61% compared to 2%–18% in the placebo group (Trock et al., 1993). They were evaluated at baseline, midpoint of therapy, end of treatment, and 1 month later. The active treatment group had decreased pain and improved functional performance. Another study reported by the same group (Trock et al., 1994), including 86 patients with OA of the knee and cervical spine, showed significant changes from baseline for the treated patients at the end of treatment and at 1-month follow-up. Placebo patients also showed improvement but with less statistical significance at the end of treatment and had lost significance for most variables at 1-month follow-up. The study patients showed improvements in pain, pain on motion, patient overall assessment, and physician global assessment.

Using PEMFs to prevent the development or progression of OA may be another target for PEMFs. Meniscal tears of the knee frequently, whether operated or not, lead to arthritic changes, which can take place over long periods of time. In a longer-term 21-year follow-up study (Roos et al., 1998) of post-meniscectomy patients looking for x-ray signs of OA, mild x-ray changes were found in 71% of the knees, while more advanced changes were seen in 48%. The corresponding prevalence values in the control group were 18% and 7%, respectively. Knee symptoms were reported twice as often in the study group as in the controls. Surgical removal of a meniscus following knee injury is still a significant risk factor for knee OA development on x-ray, with a relative risk of 14.0 after 21 years. A 5-year x-ray follow-up study (Covall and Wasilewski, 1992) was done post arthroscopic meniscectomy. In this follow-up study, some 61% of the knee joints showed changes postoperatively, with about 15% showing significant progression. When these joints were compared to the nonoperated joint (serving as a self-control), only 40% of the operative knees showed relative progression and only 4% had significant progression.

Because of the relatively high frequency of OA in patients with meniscal tears or occurring post-meniscectomy, it appears that PEMFs could be indicated for routine use to prevent onset or progression. Of course, it remains to be proven that PEMFs can in fact slow down progression. At the moment at least, the evidence seems to indicate that PEMFs may be helpful in the function and pain aspects of existing OA of the knee. The ability to do longer-term comparisons is becoming contaminated by the significant increases in the numbers of people getting early joint replacements. We may well have to rely on clinicians using therapies like PEMFs to provide us with reports of the long-term benefits of MF therapies on preventing OA, until more formal and well-funded studies are available.

One study (Sutbeyaz et al., 2006) looked at the effect of PEMFs on pain, ROM, and functional status in patients with cervical osteoarthritis (COA). Thirty-four patients were included in a randomized double-blind study. PEMF was administrated to the whole body using a 1.8×0.6 m size whole body mat. They were on the mat for 30 min per session, twice a day for 3 weeks. Pain levels in the PEMF treatment group decreased significantly after therapy ($p < 0.001$), with no change in the sham group. Active ROM, neck muscle spasm, and disability (NPDS) scores also improved significantly after PEMF therapy ($p < 0.001$). No change was seen in the sham group. This study shows that PEMFs can give significant pain reduction in neck arthritis and can be used alone or with other therapies to give even greater benefits.

A 50 Hz pulsed sinusoidal MF, 35 mT field PEMF for 15 min, 15 treatment sessions, improves hip arthritis pain in 86% of patients. Average mobility without pain improved markedly (Rehacek et al., 1982). Forty-seven patients with periarthritis of the shoulder who were receiving outpatient physical therapy were randomized using a controlled triple-blind study design to conventional physical therapy or conventional physical therapy with pulsed MF therapy (Leclaire and Bourgouin, 1991). They received treatments three times a week for a maximum of 3 months. PEMF therapy was applied 30 min at a time at three different frequencies 10/15/30 Hz with matched intensities of 30/40/60 G over the course of the therapy program. This study showed no statistically significant benefit from magnetotherapy in the pain score, ROM, or improvement of functional status in patients with periarthritis of the shoulder. There appeared to be a trend toward slightly worse baseline function of the magnetic therapy group. This would therefore suggest that treatment was not carried out for a sufficient time. An improvement in the design of the study would have been to follow the individuals until they had achieved either goal recovery or full recovery, as would happen in clinical practice. Another possibility for the lack of benefit for the pulsed magnetic therapy group is that the frequencies and intensities used are not optimized for this particular condition, given the length and the frequency of treatments per week.

17.9 FIBROMYALGIA

Fibromyalgia (FM) is a complex syndrome, primarily affecting women. There is still no adequate standard conventional medical approach to this problem. While there is an approved medication for this condition, it is not always effective, because it is not dealt with the underlying cause and has significant side effects. PEMFs can frequently be very helpful. In one study (Sutbeyaz et al., 2009), 56 women with FM, aged 18–60 years, were randomly assigned to either PEMF or sham therapy, 30 min per session, twice a day for 3 weeks. Treatment outcomes were assessed after treatment and at 4 weeks, showing significant improvements in test scores at the end of therapy and at 4-week follow-up. The sham group also showed improvement at this time on all outcome measures except the specific FM questionnaire. So, low-frequency PEMF therapy can improve at least some general FM symptoms.

A low-intensity PEMF (400 µT) in a portable device fitted to their head was found to help FM. In a randomized, double-blind, sham-controlled clinical trial (Thomas et al., 2007), patients with either chronic generalized pain from FM (n = 17) or chronic localized musculoskeletal or inflammatory pain (n = 15) were exposed in treatments twice daily for 40 min over 7 days. A VAS scale was used. There was a positive difference with PEMF over sham treatment with FM, although not quite reaching statistical significance (p = 0.06). The same level of benefit was not seen in those without FM. In patients with other causes of chronic, nonmalignant pain, either longer periods of exposure are necessary or other approaches need to be considered.

The effect of specific PEMF exposure on pain and anxiety ratings was investigated in two patient populations (Shupak et al., 2006). A double-blind, randomized, placebo-controlled parallel design was used on the effects of an acute 30 min MF exposure (less than or equal to 400 µT; less than 3 kHz) on VAS-assessed pain and anxiety ratings in female RA and FM patients who received either the PEMF or sham exposure treatment. A significant pre–post effect was present for the FM patients, p < 0.01. There was no significant reduction in VAS anxiety ratings pre- to post-exposure.

Lying on a mattress pad embedded with static magnets at night for 16 weeks reduced pain in FM patients (Colbert et al., 1999). This was a randomized double-blind study, with 25 females sleeping on *magnetic* mattress pads or a nonmagnetized pad. Each pad had 270 comparable looking domino size (2 × 4.5 cm) ceramic pieces that were either magnetized or not. The magnets were measured to be about 1100 G on their surface. However, having in mind compression properties of the foam in the mattress pad, the actual MF at the body surface was in the range of 500–600 G depending on the patient body mass. The women sleeping on the experimental mattress pad experienced a statistically significant decrease in total myalgic pain of 12%, improved physical functioning of 30%, an average pain score decrease of 38% on VAS, and improvement in sleep of 37%. The control group had a 1% decrease in total myalgic score, 3% decrease in physical functioning, 8% decrease in pain score, and 6% improvement in sleep. It certainly appears that field intensity with static MFs can be very important in achieving adequate results, with higher field intensities likely producing better results. On the other hand, this FM study also indicates that MF density, that is, the number of MFs applied to the whole body simultaneously, may be a significant factor as well and may be the preferred approach given that it is often not clear where the, actual versus assumed, pain generators are.

17.10 POSTOPERATIVE PAIN

Postoperative pain is expected, with variable severity depending on the patient and the type of surgery. Surgeons seek new methods of pain control to reduce side effects and speed postoperative recovery. Several studies were found evaluating the value of MFs of postoperative pain.

An in vivo study of PEMFs was done in dogs postoperatively after ablation of ovaries and uterus to see how pain is affected and interacts with postoperative morphine analgesia. Sixteen healthy dogs were examined within 6 h postoperative at eight different time points. There were four groups: (1) control group (NaCl administration), (2) postoperative PEMF exposure (NaCl administration), (3) postoperative morphine application, and (4) postoperative morphine application plus PEMF exposure. The PEMF was 0.5 Hz, exposure intermittent, 20 min field on/20 min field off for 6 h, whole-body exposure. At 30 min, the total pain score for group 4 was significantly less than for the control group, but not significantly different from group 2 or 3. The results suggest that PEMF may augment morphine analgesia or be used separately postoperatively after invasive abdominal procedures.

After breast augmentation surgery, patients (Hedén and Pilla, 2008) applied a portable and disposable noninvasive, high-frequency and low-intensity PEMF device in a double-blind, randomized, placebo-controlled study. Healthy females undergoing breast augmentation for aesthetic reasons were separated into three cohorts: (n = 14) receiving bilateral PEMF treatment, (n = 14) receiving bilateral sham devices, and (n = 14) an active device to one breast and a sham device to the other breast. Pain levels were measured twice daily through the seventh day after surgery (POD 7), and postoperative analgesic use was also tracked. VAS scores decreased in the active cohort by almost three times the sham cohort by POD 3 ($p < 0.001$) and persisted at this level to POD 7. Postoperative pain medication use decreased nearly three times faster in the active versus the sham cohorts by POD 3 ($p < 0.001$). These results can be extended to include certainly the use of this form of PEMF for the control of almost any situation of postoperative pain, especially involving surgery on superficial physical structures.

In another surgical study, this time post breast reduction for symptomatic macromastia, PEMFs were studied not only on their results on postoperative pain but also on potential mechanisms, including cytokines and angiogenic factors in the wound bed (Rodhe et al., 2010). Twenty-four patients were randomized in a double-blind, placebo-controlled, randomized fashion to a sham control or a low-intensity 27.12 Hz PEMF configured to modulate the calmodulin-dependent nitric oxide signaling pathway. Pain levels were measured by VAS, and narcotic use was recorded. The PEMF used produced a 57% decrease in mean pain scores at 1 h ($p < 0.01$) and a 300% decrease at 5 h ($p < 0.001$), persisting to 48 h postoperatively in the active versus the control group, along with a concomitant 2.2-fold reduction in narcotic use in active patients ($p = 0.002$). Mean IL-1β in wound exudates was 275% lower ($p < 0.001$), suggesting fairly rapid reductions in acute posttraumatic inflammation.

On the other hand, some research has found a lack of benefit of PEMFs postoperatively. Pain after elective inguinal hernia repair was evaluated in a double-blind randomized, non-PEMF controlled trial using a high-frequency low-intensity portable PEMF device (Reed et al., 1987). The device had an output rate of 320 Hz, pulse width of 60 μs, and maximum power output of 1 W. Treatment was 15 min twice a day, over and under the thigh. VAS at 24 and 48 h postoperatively showed no difference between treated and untreated groups. This study most likely used treatment times that were too short for the intensities used, and the electrodes were placed remote to the actual wound, not over the surgical site.

Severe joint inflammation following trauma, arthroscopic surgery, or infection can damage articular cartilage; thus, every effort should be made to protect cartilage from the catabolic effects of proinflammatory cytokines and stimulate cartilage anabolic activities. A pilot, randomized, prospective, and double-blind study (Zorzi et al., 2007) was done to evaluate the effects of PEMFs (75 Hz, rectangular) after arthroscopic treatment of knee cartilage. Patients with knee pain were recruited and treated by arthroscopy with chondroabrasion and/or perforations and/or radiofrequencies. There were two groups: lower-intensity control (MF at 0.05 mT) and active (MF of 1.5 mT). PEMFs were used for 90 days, 6 h per day. Objective measures were used before arthroscopy, and after 45 and 90 days, the use of anti-inflammatories (NSAIDs) were recorded. Three-year follow-up interviews were also used (n = 31). Knee score values at 45 and 90 days were higher in the active group at

90 days (p < 0.05). NSAID use was 26% in the active group and 75% in the control group (p = 0.015). At 3-year follow-up, those completely recovered was higher in the active group (p < 0.05).

Anterior cruciate ligament reconstruction, now common surgical procedure, is usually performed by a minimally invasive arthroscopic procedure. Even so, arthroscopy may elicit an inflammatory joint reaction detrimental to articular cartilage. PEMFs would be expected to mitigate some of these drug reactions. To study this possibility, a prospective, randomized, and double-blind study was done on 69 patients with a 75 Hz, 1.5 mT device, 4 h per day for 60 days versus sham device (Benazzo et al., 2008). At follow-up, active treatment patients showed a statistically significant faster recovery (p < 0.05). The use of anti-inflammatories was less frequent (p < 0.05). Joint swelling and return to normal ROM occurred faster (p < 0.05). The 2-year follow-up did not show statistically significant difference between the two groups. In addition, a subset analysis of 29 patients (15 in the active group; 14 in the placebo group) who concurrently had meniscectomy, function scores between the two groups were even larger than observed in the whole study. So this particular PEMF signal is expected to shorten postoperative recovery time and limit joint inflammation.

17.11 SHOULDER PAIN

Shoulder pain is the third most common musculoskeletal problem and accounts for 5% of visits in primary care. Although many treatments are described, there is no consensus on optimal treatment, and up to 40% of patients still have pain 12 months after initially seeking help for pain. Previously, the effect of transcutaneous pulsed radiofrequency treatment was evaluated in a retrospective audit (Taverner and Loughnan, 2014) that showed good pain relief for a mean 395 days and justified this randomized sham-controlled trial. In this study, 51 patients had evaluations at 4 and 12 weeks by a blinded observer and compared with baseline. There were reductions in pain at night, pain with activity, and functional improvement with active but not sham. Active treatment showed significant reductions of 24/100 in pain at night and 20/100 with activity at 4 weeks and 18/100 and 19/100, respectively, at 12 weeks from baseline. Pain at both rest and shoulder elevation were not improved by active treatment.

Shoulder overuse, subacromial impingement syndrome, is a frequent and commonly disabling type of shoulder pain. A double-blinded, PEMF randomized, and controlled study was done along with other standard conservative treatment modalities in acute rehabilitation (Atkas et al., 2007). Forty-six patients received a standard program for 3 weeks of pendulum exercises and cold packs five times a day, restriction of shoulder extension, daily activities, and meloxicam 15 mg daily. One group was given PEMF of 50 Hz, 30 G with a U-shaped applicator, 25 min per session, 5 days per week for 3 weeks. The other group was given sham PEMF. A VAS, total constant shoulder function score, and ADL were measured, before and after treatment. When compared with baseline, significant improvements were seen in all these variables at the end of the treatment in both groups (p < 0.05). The active PEMF group had a higher baseline resting VAS with rest pain and activity pain and a 0.8 VAS difference from baseline to posttreatment assessments, with significantly higher standard deviations. So, it appears that this study was underpowered to be able to detect differences using the treatment protocols applied. Because both groups received standard therapies as well, including ice packs and medication, it would be expected to be harder to find a statistically significant difference with this small number of study patients, for the duration of the study, for the additional benefit of PEMFs. It is quite probable that for short-term studies like this, a much higher-intensity PEMF system would be needed to demonstrate significant differences in benefit.

17.12 PELVIC PAIN SYNDROME

Noninflammatory chronic pelvic pain syndrome (CPPS) can be quite disabling in both men and women, frequently with no adequate treatment options. A study (Leippold et al., 2005) was designed to prospectively evaluate sacral magnetic high-frequency stimulation as a treatment option for patients with noninflammatory CPPS (CPPS, category IIIB). Fourteen men were treated with sacral magnetic stimulation, 10 treatment sessions once a week for 30 min at a frequency of 50 Hz. Twelve of fourteen men reported improvement but only during the time of stimulation. Inventory scores before and after treatment did not change. There was no sustained effect beyond the time of stimulation on the mean scores for pain, micturition complaints, or quality of life. Sacral magnetic stimulation in patients with CPPS IIIB reduces pain only during stimulation. The fact the pain relief is obtained during treatment is notable and valuable. Because this level of frequency of treatments is less likely to induce healing in the tissues causing the pain syndrome, it may be reasonable to expect only a reduction in pain during the treatment course and not a more enduring benefit. While this treatment approach does not appear to be useful, it remains to be seen whether a change in the protocol may produce more enduring results.

Gynecologic pelvic pain may also benefit from PEMFs. A high-voltage, high-frequency system (Jorgensen et al., 1994) was used in the setting of ruptured ovarian cysts, postoperative pelvic hematomas, chronic urinary tract infection, uterine fibrosis, dyspareunia, endometriosis, and dysmenorrhea. Ninety percent of patients experienced marked rapid relief from pain, with pain subsiding within 1–3 days after PEMF treatment, eliminating supplementary analgesics. Unfortunately, longer-term data are not available to determine the durability of the therapeutic response.

17.13   MISCELLANEOUS PAIN APPLICATIONS

In dentistry, periodontal disease may cause bone resorption severe enough to require bone grafting. Grafting is followed by moderate pain peaking several hours afterward. Repeated PEMF exposure for 2 weeks eliminates pain within a week. Even single PEMF exposure to the face for 30 min of a 5 mT field and related conservative treatment produce much lower pain scores versus controls (Tesic et al., 1999).

Results of PRF PEMF in a case series either eliminates or improves, even at 2 weeks following therapy, pain in 80% of patients with pelvic inflammatory disease, 89% with back pain, 40% with endometriosis, 80% with postoperative pain, and 83% with lower abdominal pain of unknown cause (Punnonen et al., 1980).

PEMFs have been found to be helpful in headaches. For migraine headaches, high-frequency PEMFs applied to the inner thighs for at least 2 weeks are effective short-term therapy (Sherman et al., 1999). Longer exposures lead to greater reduction of headache activity. One month after a treatment course, 73% of patients report decreased headache activity versus 50% of placebo treatment. Another 2 weeks of treatment after the 1-month follow-up gives an additional 88% decrease in headache activity. Patients with headache treated with a PEMF for 15 days after failing acupuncture and medications get effective relief of migraine, tension, and cervical headaches at about 1 month after treatment (Prusinski et al., 1987). They have at least a 50% reduction in frequency or intensity of the headaches and reduction in analgesic drug use. Cluster and posttraumatic headaches do not respond as well.

17.14   SUMMARY

PEMFs of various kinds, strengths, and frequencies included have been found to have good results in a wide array of painful conditions. There is little risk when compared to the potential invasiveness of other therapies and the risk of toxicity, addiction, and complications from medications. This creates an ideal setup for clinicians to attempt PEMFs before other more potentially harmful treatments are attempted, especially for long-term treatment of chronic pain conditions. Clearly, more research is needed to elaborate mechanisms and optimal treatment parameters and the best MF systems for given applications. While it is the goal of the researcher to find the optimal system for a particular indication, the goal is opposite for clinicians, that is, to have the most generalizable system for the most problems. This latter situation is expected to be the best, most cost-effective approach, not only for patients but also for the health care system.

Most magnetic systems used in the research setting have never been commercially available. On the other hand, a large percentage of commercially available systems have been inadequately studied. Most PEMF systems typically available to the clinician are the commercial systems. Most clinicians using PEMF systems rely on the general body of knowledge of PEMF effects on biology as being clinically useful, bolstered when possible with evidence from clinical trials. Ultimately, as with any other therapy, including pharmaceuticals, clinical experience will guide variations in treatment protocols, whether only one PEMF system or multiple systems are used. Because of the range of effects of different PEMF systems, it may be best clinical practice to have several different PEMF systems to achieve desired results for the widest majority of clinical presentations. These could include, but are not limited to, for example, high-intensity local systems; lower-intensity, whole-body complex signal systems for health maintenance or very sensitive individuals; medium-intensity local and/or whole-body systems that allow selection of individual frequencies; or systems with extremely low intensities that provide a broad spectrum of frequencies.

Many studies that have been reported in this review have been controlled trials, and many have been randomized double-blind placebo. Many studies have small sample sizes and therefore are often underpowered, often yielding negative or conflicting results. Negative results do not necessarily indicate the lack of potential for the technology and are simply the probability of an inadequate research design. This reflects the very complex nature of clinical practice across a large spectrum and variety of pain conditions.

There are very few studies comparing different PEMF systems. There is no consistency in the device design even for the same clinical category, such as OA of the knee. Even though a device may have been found to have statistically significant benefits in a given condition, it is not predictable that it would be equally effective in other situations. What works well in vitro may not work well in vivo. The reverse is also true that what might not work well in vitro may work well in in vivo in living systems, because of harmonics, biological amplification, and much more complex physiological systems. Also, what works well in animals may not work well in humans.

The reasons for apparent inconsistency of results are that the issues being treated are by themselves complex and at the very least include different demands and responses for biological windows for frequencies, intensities, and waveforms. This is not even to consider the effects of the durations of individual treatments or courses of treatment. Understanding the dose delivered at the target tissue is critical and is one of the main reasons that static MFs have not been more widely accepted, that is, often being of insufficient intensities to reach deeper tissues. Clinical experience has shown that deeper problems, or those that are chronically more severe, respond better with higher-intensity MFs, regardless of whether static or time varied, independent of considering waveform and frequency.

Medical practitioners are becoming gradually aware of the potential of MFs to successfully treat or significantly benefit the myriad of problems presented to them, reducing risks of conventional medical approaches when feasible. A major decision for clinicians remains whether to do in-office treatment or encourage patients to purchase their own systems for home use. Experience demonstrates that longer-term, daily, in-home use produces the most durable results. In-office treatments can provide some indication of responsiveness of the condition to magnetic therapy. In some circumstances, a course of treatments in the office setting may resolve the pain for a considerable period of time. After 25 years of clinical use of MF therapies, it has been my experience that long-term home use works best, especially for chronic pain, almost regardless of the system purchased.

While very few PEMF systems have been US Food and Drug Administration (FDA) approved, it is even rare for a system to be FDA approved specifically for pain and then become covered by Medicare or other health insurance. So, regardless of whether treatments are in-office or home-based, they are highly unlikely to be covered by insurance.

It may be reasonable enough to conclude that the body of evidence for the use of PEMFs for pain management is sufficiently robust to provide some degree of credibility for the clinician to be able to consider this technology in the management of their pain patients. Knowing which systems to use, understanding how to adjust or select stimulation parameters, understanding the clinical dimensions of the problems needed to be treated, and selecting appropriate conditions for pain management using PEMFs will provide the clinician and the patient the most useful results.

REFERENCES

Aktas I, Akgun K, Cakmak B. (August 2007) Therapeutic effect of pulsed electromagnetic field in conservative treatment of subacromial impingement syndrome. *Clin Rheumatol* 26(8): 1234–1239.

Aleman A. (August 2013) Use of repetitive transcranial magnetic stimulation for treatment in psychiatry. *Clin Psychopharmacol Neurosci* 11(2): 53–59.

Arnold MD, Thornbrough LM. (August 1999) Treatment of musculoskeletal pain with traditional Chinese herbal medicine. *Phys Med Rehabil Clin N Am* 10(3): 663–671, ix–x.

Ay S, Evcik D. (April 2009) The effects of pulsed electromagnetic fields in the treatment of knee osteoarthritis: A randomized, placebo-controlled trial. *Rheumatol Int* 29(6): 663–666.

Benazzo F, Zanon G, Pederzini L et al. (June 2008) Effects of biophysical stimulation in patients undergoing arthroscopic reconstruction of anterior cruciate ligament: Prospective, randomized and double blind study. *Knee Surg Sports Traumatol Arthrosc* 16(6): 595–601.

Borg MJ, Marcuccio F, Poerio AM et al. (October 1996) Magnetic fields in physical therapy. Experience in orthopedics and traumatology rehabilitation. *Minerva Med* 87(10): 495–497.

Caselli MA, Clark N, Lazarus S, Velez Z et al. (January 1997) Evaluation of magnetic foil and PPT insoles in the treatment of heel pain. *J Am Podiatr Med Assoc* 87(1): 11–16.

Cheing GL, Wan JW, Kai Lo S. (November 2005) Ice and pulsed electromagnetic field to reduce pain and swelling after distal radius fractures. *J Rehabil Med* 37(6): 372–377.

Cieslar G, Mrowiec J, Sieron A et al. (1994) The reactivity to thermal pain stimulus in rats exposed to variable magnetic field. *Balneol Pol* 36(3–4): 24–28.

Cieslar G, Sieron A, Radelli J. (1995) The estimation of therapeutic effect of variable magnetic fields in patients with diabetic neuropathy including vibratory sensibility. *Balneol Pol* 37(1): 23–27.

Colbert AP, Cleaver J, Brown KA et al. (September 2008) Magnets applied to acupuncture points as therapy— A literature review. *Acupunct Med* 26(3): 160–170.

Colbert AP, Markov MS, Banerji M et al. (1999) Magnetic mattress pad use in patients with fibromyalgia: A randomized double-blind pilot study. *J Back Musculoskelet Rehabil* 13: 19–31.

Colloca L, Klinger R, Flor H et al. (April 2013) Placebo analgesia: Psychological and neurobiological mechanisms. *Pain* 154(4): 511–514.

Covall DJ, Wasilewski SA. (1992) Roentgenographic changes after arthroscopic meniscectomy: Five-year follow-up in patients more than 45 years old. *Arthroscopy* 8(2): 242–246.

De Loecker W, Cheng N, Delport PH. Effects of pulsed electromagnetic fields on membrane transport. In *Emerging Electromagnetic Medicine*, O'Connor ME, Bentall RHC, Monahan JC (eds.), New York: Springer-Verlag, 1990, pp. 45–59.

Del Seppia C, Ghione S, Luschi P et al. (2007) Pain perception and electromagnetic fields. *Neurosci Biobehav Rev* 31(4): 619–642.

Devereaux MD, Hazleman BL, Thomas PP. (October–December 1985) Chronic lateral humeral epicondylitis— A double-blind controlled assessment of pulsed electromagnetic field therapy. *Clin Exp Rheumatol* 3(4): 333–336.

Di Massa A, Misuriello I, Olivieri MC et al. (1989) Pulsed magnetic fields. Observations in 353 patients suffering from chronic pain. *Minerva Anestesiol* 55(7–8): 295–299.

Ellis WV. (1993) Pain control using high-intensity pulsed magnetic stimulation. *Bioelectromagnetics* 14(6): 553–556.

Eriksen W, Sandvik L, Bruusgaard D. (October 1996) Does dietary supplementation of cod liver oil mitigate musculoskeletal pain? *Eur J Clin Nutr* 50(10): 689–693.

Feine JS, Lund JP. (May 1997) An assessment of the efficacy of physical therapy and physical modalities for the control of chronic musculoskeletal pain. *Pain* 71(1): 5–23.

Fernandez MI, Watson PJ, Rowbotham DJ. (August 2007) Effect of pulsed magnetic field therapy on pain reported by human volunteers in a laboratory model of acute pain. *Br J Anaesth* 99(2): 266–269.

Fischer G. (2002) Relieving pain in diseases of the musculoskeletal system with small apparatuses that produce magnetic fields, Personal communication.

Fleming JL, Persinger MA, Koren SA. (1994) Magnetic pulses elevate nociceptive thresholds: Comparisons with opiate receptor compounds in normal and seizure-induced brain-damaged rats. *Electro Magnetobiol.* 13(1): 67–75.

Foley-Nolan D, Barry C, Coughlan RJ et al. (1990) Pulsed high frequency (27 MHz) electromagnetic therapy for persistent neck pain. A double blind, placebo-controlled study of 20 patients. *Orthopedics* 13(4): 445–451.

Foley-Nolan D, Moore K, Codd M et al. (1992) Low energy high frequency pulsed electromagnetic therapy for acute whiplash injuries. A double blind randomized controlled study. *Scand J Rehabil Med* 24(1): 51–59.

Forestier R, Françon A, Saint-Arromand F et al. (April 2007a) Are SPA therapy and pulsed electromagnetic field therapy effective for chronic neck pain? Randomised clinical trial. First part: Clinical evaluation. *Ann Readapt Med Phys* 50(3): 140–147.

Forestier R, Françon A, Saint Arroman F et al. (April 2007b) Are SPA therapy and pulsed electromagnetic field therapy effective for chronic neck pain? Randomised clinical trial. Second part: Medicoeconomic approach. *Ann Readapt Med Phys* 50(3): 148–153.

Graak V, Chaudhary S, Bal BS et al. (April 2009) Evaluation of the efficacy of pulsed electromagnetic field in the management of patients with diabetic polyneuropathy. *Int J Diabetes Dev Ctries* 29(2): 56–61.

Guseo A. Physiological effects of pulsing electromagnetic field. In *First Congress of European Bioelectromagnetics Association (EBEA)*, Brussels, Belgium, January 1992, s.31.

Haldeman S, Rubinstein SM. (January 1993) The precipitation or aggravation of musculoskeletal pain in patients receiving spinal manipulative therapy. *J Manipulat Physiol Ther* 16(1): 47–50.

Han TR, Shin HI, Kim IS. (July 2006) Magnetic stimulation of the quadriceps femoris muscle: Comparison of pain with electrical stimulation. *Am J Phys Med Rehabil* 85(7): 593–599.

Harlow T, Greaves C, White A, Brown L, Hart A, Ernst E. (2004) Randomised controlled trial of magnetic bracelets for relieving pain in osteoarthritis of the hip and knee. *BMJ* 329: 1450–1454.

Hedén P, Pilla AA. (July 2008) Effects of pulsed electromagnetic fields on postoperative pain: A double-blind randomized pilot study in breast augmentation patients. *Aesthetic Plast Surg* 32(4): 660–666.

ICNIRP (2010) Guidelines for limiting exposure to time-varying electric and magnetic fields (1 Hz–100 kHz). *Health Phys* 99(6): 818–836.

Institute of Medicine (IOM) of the National Academies. *Complementary and Alternative Medicine in the United States.* Washington, DC: The National Academies Press, 2005, p. 1.

Jerabek J, Pawluk W. *Magnetic Therapy in Eastern Europe: A Review of 30 Years of Research.* Chicago, IL: Advanced Magnetic Research of the Delaware Valley, 1996.

Jorgensen WA, Frome BM, Wallach C. (1994) Electrochemical therapy of pelvic pain: Effects of pulsed electromagnetic fields (PEMF) on tissue trauma. *Eur J Surg* 160(574 Suppl): 83–86.

Khamaganova IV, Boinich ZV, Arutiunova ES. (1993) Clinical aspects of the use of a pulsed magnetic field. *Fizicheskaia Meditzina* 3(1–2): 35–37.

Kholodov YA. A non-specific initial response of brain to various electromagnetic fields. In *International Meeting of Electromagnetic Fields: Biological Effects and Hygienic Standards*, Moscow, Russia, May 1998.

Kjellman GV, Skargren EI, Oberg BE. (1999) A critical analysis of randomised clinical trials on neck pain and treatment efficacy. A review of the literature. *Scand J Rehabil Med* 31(3): 139–152.

Kumar V. Cell injury, cell death and adaptations. In *Robbins and Cotran Pathologic Basis of Disease, Professional Edition*, Kumar V, Abbas AK, and Aster JC (eds.), 8th edn. Elsevier, Philadelphia, PA, 2007, Chapter 1.

Kusaka C, Seto A, Nagata T et al. (1995) Pulse magnetic treatment and whole-body, alternating current magnetic treatment for post-herpetic neuralgia. *J Jpn Biomagnet Bioelectromagnet Soc* 8(2): 29–38.

Leclaire R, Bourgouin J. (April 1991) Electromagnetic treatment of shoulder periarthritis: A randomized controlled trial of the efficiency and tolerance of magnetotherapy. *Arch Phys Med Rehabil* 72(5): 284–287.

Lefaucheur JP, Drouot X, Menard-Lefaucheur I et al. (April 2004) Neurogenic pain relief by repetitive transcranial magnetic cortical stimulation depends on the origin and the site of pain. *J Neurol Neurosurg Psychiatr* 75(4): 612–616.

Leippold T, Strebel RT, Huwyler M et al. (2005) Sacral magnetic stimulation in non-inflammatory chronic pelvic pain syndrome. *BJU Int* 95: 838–841.

Levy R, Deer TR, Henderson J. (March–April 2010) Intracranial neurostimulation for pain control: A review. *Pain Phys* 13(2): 157–165.

Lin ML, Lin MH, Fen JJ et al. (2010) A comparison between pulsed radiofrequency and electro-acupuncture for relieving pain in patients with chronic low back pain. *Acupunct Electrother Res* 35(3–4): 133–146.

Markov MS. Magnetic and electromagnetic field therapy: Basic principles of application for pain relief. In *Bioelectromagnetic Medicine,* Rosch PJ and Markov MS (eds.), New York, Marcel Dekker, 2004, pp. 251–264.

Markov MS, Pilla AA. (1995) Electromagnetic field stimulation of soft tissue: Pulsed radiofrequency treatment of post-operative pain and edema. *Wounds* 7(4): 143–151.

Mitbreit IM, Savchenko AG, Volkova LP et al. (1986) Low-frequency magnetic field in the complex treatment of patients with lumbar osteochondrosis. *Ortop Travmatol Protez* (10): 24–27.

Nelson FR, Zvirbulis R, Pilla AA. (August 2013) Non-invasive electromagnetic field therapy produces rapid and substantial pain reduction in early knee osteoarthritis: A randomized double-blind pilot study. *Rheumatol Int* 33(8): 2169–2173.

Pawluk W, Turk Z, Fischer G, Kobinger W. Treatment of osteoarthritis with a new broadband PEMF signal. *Presentation in the 24th Annual Meeting of Bioelectromagnetics Society*, Quebec City, Quebec, Canada, June 2002.

Pennington GM, Danley DL, Sumko MH et al. (February 1993) Pulsed, non-thermal, high-frequency electromagnetic energy (DIAPULSE) in the treatment of grade I and grade II ankle sprains. *Mil Med* 158(2): 101–104.

Picarelli H, Teixeira MJ, de Andrade DC et al. (November 2010) Repetitive transcranial magnetic stimulation is efficacious as an add-on to pharmacological therapy in complex regional pain syndrome (CRPS) type I. *J Pain* 11(11): 1203–1210.

Pilla AA. Electromagnetic therapeutics: State-of-the-art in hard and soft tissue applications. *Presentation in the Fourth International Congress of European Bioelectromagnetics Assoc. (EBEA)*, Zagreb, Croatia, November 1998.

Pilla AA. (June 2013) Nonthermal electromagnetic fields: From first messenger to therapeutic applications. *Electromagnet Biol Med* 32(2): 123–136.

Pipitone N, Scott DL. (2001) Magnetic pulse treatment for knee osteoarthritis: A randomised, double-blind, placebo-controlled study. *Curr Med Res Opin* 17(3): 190–196.

Pleger B, Janssen F, Schwenkreis P et al. (February 12, 2004) Repetitive transcranial magnetic stimulation of the motor cortex attenuates pain perception in complex regional pain syndrome type I. *Neurosci Lett* 356(2): 87–90.

Prato FS, Del Seppia C, Kavaliers M et al. Stress-induced analgesia in house mice and deer mice is reduced by application of various magnetic fields conditions. *21st Annual Meeting of Bioelectromagnetics Society*, Long Beach, CA, June 1999, Abstract 6–3:38.

Prato FS, Thomas AW, Cook CM. (2001) Human standing balance is affected by exposure to pulsed ELF magnetic fields: Light intensity-dependent effects. *Neuroreport* 12(7): 1501–1505.

Preszler RR. A non-invasive complementary method of reducing chronic muscular low back pain using permanent magnetic therapy. Master thesis, Physician Assistant Studies, University of Nebraska School of Medicine, Physician Assistant Program, Lincoln, Omaha, NE, 2000.

Prusinski A, Wielka J, Durko A. (1987) Pulsating electromagnetic field in the therapy of headache. In *Second Symposium on Magnetotherapy*, Szekesfehervar, Hungary, May 1987. *J Bioelectr* 7(1): 127–128.

Pujol J, Pascual-Leone A, Dolz C et al. (1998) The effect of repetitive magnetic stimulation on localized musculoskeletal pain. *Neuroreport* 9(8): 1745–1748.

Punnonen R, Gronroos M, Luikko P et al. (1980) The use of pulsed high-frequency therapy (Curapuls) in gynecology and obstetrics. *Acta Obstet Gynecol Scand* 59(2): 187–188.

Randall C, Randall H, Dobbs F et al. (June 2000) Randomized controlled trial of nettle sting for treatment of base-of-thumb pain. *J Royal Soc Med* 93(6): 305–309.

Rauscher E, Van Bise WL. Pulsed magnetic field treatment of chronic back pain. *Presentation in the 23rd Annual Meeting of Bioelectromagnetics Society*, St. Paul, MN, June 2001.

Reed MW, Bickerstaff DR, Hayne CR, Wyman A, Davies J. (June 1987) Pain relief after inguinal herniorrhaphy. Ineffectiveness of pulsed electromagnetic energy. *Br J Clin Pract* 41(6): 782–784.

Rehacek J, Straub J, Benova H. (1982) The effect of magnetic fields on coxarthroses. *Fysiatr Revmatol Vestn* 60(2): 66–68.

Robertson JA, Théberge J, Weller J et al. (March 6, 2010) Low-frequency pulsed electromagnetic field exposure can alter neuroprocessing in humans. *J Royal Soc Interface* 7(44): 467–473.

Rohde C, Chiang A, Adipoju O et al. (June 2010) Effects of pulsed electromagnetic fields on interleukin-1 beta and postoperative pain: A double-blind, placebo-controlled, pilot study in breast reduction patients. *Plast Reconstr Surg* 125(6): 1620–1629.

Roos H, Laurén M, Adalberth T et al. (April 1998) Knee osteoarthritis after meniscectomy: Prevalence of radiographic changes after twenty-one years, compared with matched controls. *Arthritis Rheumatol* 41(4): 687–693.

Sartucci F, Bonfiglio L, Del Seppia C et al. (1997) Changes in pain perception and pain-related somatosensory evoked potentials in humans produced by exposure to oscillating magnetic fields. *Brain Res* 769(2): 362–366.

Saveriano G, Ricci S. (April 1989) Experiences in treating secondary post-traumatic algodystrophy with low-frequency PEMFs in conjunction with functional rehabilitation. In *International Symposium in Honor of Luigi Galvani*, Bologna, Italy. *J Bioelectr* 8(2): 320.

Schroter M. (March/April 1976) Conservative treatment of 240 patients with magnetic field therapy. *Medizinisch-Orthopadische Technik* (2): 78.

Segal NA, Toda Y, Huston J et al. (2001) Two configurations of static magnetic fields for treating rheumatoid arthritis of the knee: A double-blind clinical trial. *Arch Phys Med Rehabil* 82(10): 1453–1460.

Sherman RA, Acosta NM, Robson L. (1999) Treatment of migraine with pulsing electromagnetic fields: A double-blind, placebo-controlled study. *Headache* 39(8): 567–575.

Shupak NM, McKay JC, Nielson WR et al. (Summer 2006) Exposure to a specific pulsed low-frequency magnetic field: A double-blind placebo-controlled study of effects on pain ratings in rheumatoid arthritis and fibromyalgia patients. *Pain Res Manage* 11(2): 85–90.

Shupak NM, Prato FS, Thomas AW. (June 10, 2004) Human exposure to a specific pulsed magnetic field: Effects on thermal sensory and pain thresholds. *Neurosci Lett* 363(2): 157–162.

Stewart DJ, Stewart JE. (1989) The destabilization of an abnormal physiological balanced situation, chronic musculoskeletal pain, utilizing magnetic biological device. *Acta Med Hung* 46(4): 323–337.

Sutbeyaz ST, Sezer N, Koseoglu BF. (February 2006) The effect of pulsed electromagnetic fields in the treatment of cervical osteoarthritis: A randomized, double-blind, sham-controlled trial. *Rheumatol Int* 26(4): 320–324.

Sutbeyaz ST, Sezer N, Koseoglu F et al. (October 2009) Low-frequency pulsed electromagnetic field therapy in fibromyalgia: A randomized, double-blind, sham-controlled clinical study. *Clin J Pain* 25(8): 722–728.

Takeshige C, Sato M. (April–June 1996) Comparisons of pain relief mechanisms between needling to the muscle, static magnetic field, external qigong and needling to the acupuncture point. *Acupunct Electrother Res* 21(2): 119–131.

Taverner M, Loughnan T. (February 2014) Transcutaneous pulsed radiofrequency treatment for patients with shoulder pain booked for surgery: A double-blind, randomized controlled trial. *Pain Pract* 14(2): 101–108.

Tesic D, Djuric M, Pekaric-Nadj N et al. PEMF aided pain reduction in stomatology. *Presentation in the 21st Annual Meeting of Bioelectromagnetics Society*, Long Beach, CA, Abstract P-141, June 1999.

Thamsborg G, Florescu A, Oturai P et al. (July 2005) Treatment of knee osteoarthritis with pulsed electromagnetic fields: A randomized, double-blind, placebo-controlled study. *Osteoarthr Cartil* 13(7): 575–581.

Thomas AW, Drost DJ, Prato FS. (2001) Human subjects exposed to a specific pulsed (200 uT) magnetic field: Effects on normal standing balance. *Neurosci Lett* 297(2): 121–124.

Thomas AW, Graham K, Prato FS et al. (Winter 2007) A randomized, double-blind, placebo-controlled clinical trial using a low-frequency magnetic field in the treatment of musculoskeletal chronic pain. *Pain Res Manage* 12(4): 249–258.

Thomas AW, Prato FS. Magnetic field based pain therapeutics and diagnostics. *Presentation in the 24th Annual Meeting of Bioelectromagnetics Society*, Quebec City, Quebec, Canada, June 2002.

Thomas AW, White KP, Drost DJ et al. (August 17, 2001) A comparison of rheumatoid arthritis and fibromyalgia patients and healthy controls exposed to a pulsed (200 microT) magnetic field: Effects on normal standing balance. *Neurosci Lett* 309(1): 17–20.

Thomas D, Collins S, Strauss S. (March 1992) Somatic sympathetic vasomotor changes documented by medical thermographic imaging during acupuncture analgesia. *Clin Rheumatol* 11(1): 55–59.

Thuile C, Walzl M. (2002) Evaluation of electromagnetic fields in the treatment of pain in patients with lumbar radiculopathy or the whiplash syndrome. *Neuro Rehabil* 17: 63–67.

Trock DH. (February 2000) Electromagnetic fields and magnets. Investigational treatment for musculoskeletal disorders. *Rheum Dis Clin N Am* 26(1): 51–62, viii.

Trock DH, Bollet AJ, Dyer RH Jr et al. (March 1993) A double-blind trial of the clinical effects of pulsed electromagnetic fields in osteoarthritis. *J Rheumatol* 20(3): 456–460.

Trock DH, Bollet AJ, Markoll R. (October 1994) The effect of pulsed electromagnetic fields in the treatment of osteoarthritis of the knee and cervical spine. Report of randomized, double blind, placebo controlled trials. *J Rheumatol* 21(10): 1903–1911.

Ugawa Y, Terao Y, Hanajima R et al. (September 1997) Magnetic stimulation over the cerebellum in patients with ataxia. *Electroencephalogr Clin Neurophysiol* 104(5): 453–458.

Uzunca K, Birtane M, Taştekin N. (January 2007) Effectiveness of pulsed electromagnetic field therapy in lateral epicondylitis. *Clin Rheumatol* 26(1): 69–74.

Vallbona C, Richards T. (August 1999) Evolution of magnetic therapy from alternative to traditional medicine. *Phys Med Rehabil Clin N Am* 10(3): 729–754.

Van Zundert J, Patijn J, Kessels A et al. (January 2007) Pulsed radiofrequency adjacent to the cervical dorsal root ganglion in chronic cervical radicular pain: A double blind sham controlled randomized clinical trial. *Pain* 127(1–2): 173–182.

Weintraub MI, Cole SP. (July–August 2008) A randomized controlled trial of the effects of a combination of static and dynamic magnetic fields on carpal tunnel syndrome. *Pain Med* 9(5): 493–504.

Weintraub MI, Herrmann DN, Smith AG et al. (July 2009) Pulsed electromagnetic fields to reduce diabetic neuropathic pain and stimulate neuronal repair: A randomized controlled trial. *Arch Phys Med Rehabil* 90(7): 1102–1109.

Wong JY, Rapson LM. (August 1999) Acupuncture in the management of pain of musculoskeletal and neurologic origin. *Phys Med Rehabil Clin N Am* 10(3): 531–545, vii–viii.

Wróbel MP, Szymborska-Kajanek A, Wystrychowski G et al. (September 2008) Impact of low frequency pulsed magnetic fields on pain intensity, quality of life and sleep disturbances in patients with painful diabetic polyneuropathy. *Diabetes Metab* 34(4 Pt 1): 349–354.

Zorzi C, Dall'Oca C, Cadossi R et al. (July 2007) Effects of pulsed electromagnetic fields on patients' recovery after arthroscopic surgery: Prospective, randomized and double-blind study. *Knee Surg Sports Traumatol Arthrosc* 15(7): 830–834.

18 Electromagnetic Fields in Plastic Surgery

Application to Plastic and Reconstructive Surgical Procedures

Christine H. Rohde, Erin M. Taylor, and Arthur A. Pilla

CONTENTS

18.1 INTRODUCTION

Acute postoperative pain is a common complaint, estimated to occur in 80% of surgical patients (Apfelbaum et al., 2003), despite current available treatment. Postoperative pain leads to increased morbidity, number of hospital days, and health care costs. The current standard of treatment for postoperative pain is opioid narcotics, which can have undesirable side effects and the potential for addiction and abuse. Additional pain modalities have included pain pumps that deliver local anesthetics to the targeted site. Pain pumps are invasive, require additional procedures for placement and removal, and have varying efficacy (Liu et al., 2006; Kazmier et al., 2008). Furthermore, local anesthetics and opioids have primary analgesic, rather than anti-inflammatory effects.

Pulsed electromagnetic fields (PEMFs) are cleared by regulatory bodies worldwide for use as adjunctive therapy to accelerate the repair of delayed and nonunion fractures and chronic wounds, as well as to accelerate the reduction of postoperative pain and inflammation (Pilla, 2007). As the mechanisms of PEMF bioeffects have become better understood, PEMF devices have become portable, disposable, and very economical. This has allowed the scope of application of PEMF to be readily expanded to the management of postoperative pain and inflammation. In the clinical setting, several double-blind, placebo-controlled, randomized clinical studies, including breast reduction (BR; Rohde et al., 2010), breast augmentation (Heden and Pilla, 2008; Rawe et al., 2012), and autologous flap breast reconstruction (Rohde et al., 2012) procedures, have reported that nonthermal pulsed radio frequency fields significantly accelerate postoperative pain and inflammation reduction, and, concomitantly, postoperative narcotic requirements. Two of these studies examined the levels of interleukin-1beta (IL-1β), an inflammatory cytokine involved in pain hypersensitivity, in wound exudates, and found significantly decreased IL-1β production and wound exudate volume in the first 24 h postoperatively (Rohde et al., 2010, 2012). Taken together, this body of results suggests that PEMF has the potential to be an essential tool for the surgeon to accelerate the reduction of postsurgical pain and inflammation, decrease patient morbidity, and enhance surgical outcomes.

18.2 BACKGROUND

For the vast majority of surgical procedures, pain mechanisms involve increased sensitivity of nociceptors due to increased presence of proinflammatory cytokines in the wound milieu. Thus, neutrophils and macrophages, the first cellular responders in the inflammatory phase of wound repair, produce a major proinflammatory cytokine, interleukin-1beta (IL-1β), immediately after injury, which in turn causes high, proinflammatory, amounts of nitric oxide (NO) to be released into the wound bed (LaPointe and Isenović, 1999). Extended exposure to high concentrations of NO induces cyclooxygenase-2 (COX-2) and increases levels of prostaglandins (PGEs), unnecessarily extending the inflammatory phase of wound repair, which can lead to pain, fibrosis, and delayed healing (Broughton et al., 2006). Alternative approaches to decrease postoperative pain involve slowing the appearance of proinflammatory agents at the surgical site (Binshtok et al., 2008).

PEMF therapy, using a variety of signal configurations, has long been reported to reduce pain and inflammation (Ross and Harrison, 2013b). In addition to the many clinical reports reviewed by Ross and others, which demonstrate that PEMF can produce clinically and physiologically significant acute and chronic pain relief, Cadossi and colleagues have provided strong evidence that PEMF increases the anti-inflammatory effect of adenosine receptors (De Mattei et al., 2009; Ongaro et al., 2012; Vincenzi et al., 2013), which is proposed as a mechanism for postoperative pain relief in knee replacement surgery (Adravanti et al., 2014).

Although PEMF signals with a vast range of waveform parameters have been reported to reduce acute pain and enhance healing, a common unifying mechanism, which involves Ca^{2+}-dependent NO signaling, has been proposed (Pilla et al., 1999, 2011; Pilla, 2007, 2012, 2013). Enhanced levels of transient NO in a challenged cell or tissue produce enhanced levels of cyclic guanosine monophosphate (cGMP), which in turn can rapidly decrease the rate of release of inflammatory cytokines (Ren and Torres, 2009) and increase the release of growth factors and blood flow (Werner and Grose, 2003). A PEMF signal can enhance Ca^{2+}-dependent NO release, provided it is configured to satisfy its asymmetrical binding kinetics to calmodulin (CaM) and is detectable above baseline electrical noise in the Ca/CaM binding pathway (Pilla, 2007; Pilla et al., 2011). Radio frequency PEMF signals can easily be configured to meet this criterion, with the added benefit that the generators and coil applicators can be made to be portable and disposable for the ease of clinical use, particularly in the postoperative setting.

This review considers PEMF with a pulse-modulated 27.12 MHz radio frequency carrier, which can be configured to modulate the NO/cGMP signaling pathway. The proposed mechanism is that PEMF accelerates the binding of cytosolic Ca^{2+} to CaM in a cell, which has been challenged physically or chemically, leading to a transient increase in NO, which in turn enhances a transient increase in cGMP release. PEMF modulation of CaM/NO/cGMP signaling can lead to more rapid resolution of inflammation. This proposed mechanism is outlined in Figure 18.1. Several peer-reviewed studies at the cellular level support this mechanism (Pilla et al., 1999, 2011; Pilla, 2007, 2011, 2012, 2013). One study examined the effects of PEMF on NO release in real time from neuronal cells during an inflammatory challenge with LPS (Pilla, 2012). PEMF produced an immediate threefold increase in NO and had no effect on NO without the LPS challenge. This study confirmed that PEMF affects only challenged tissues, supporting the consistent reports of no adverse effects from this PEMF signal (Heden and Pilla, 2008; Rohde et al., 2010, 2012). Other studies showed that

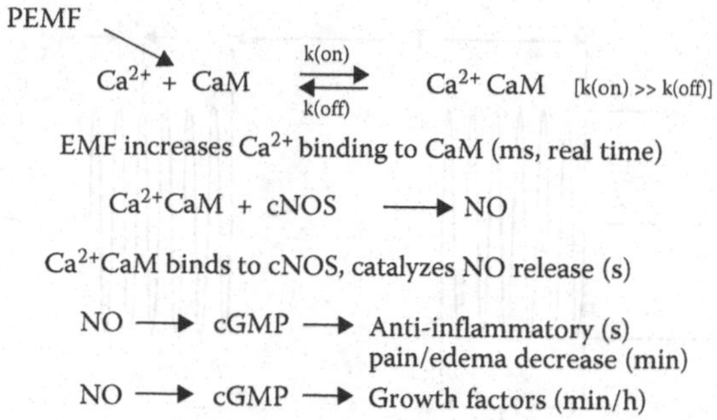

PEMF

$$Ca^{2+} + CaM \underset{k(off)}{\overset{k(on)}{\rightleftharpoons}} Ca^{2+} CaM \quad [k(on) \gg k(off)]$$

EMF increases Ca^{2+} binding to CaM (ms, real time)

$$Ca^{2+}CaM + cNOS \longrightarrow NO$$

$Ca^{2+}CaM$ binds to cNOS, catalyzes NO release (s)

$$NO \longrightarrow cGMP \longrightarrow \text{Anti-inflammatory (s)} \\ \text{pain/edema decrease (min)}$$

$$NO \longrightarrow cGMP \longrightarrow \text{Growth factors (min/h)}$$

FIGURE 18.1 Proposed PEMF mechanism. PEMF accelerates the binding of cytosolic Ca^{2+} to CaM, which activates cNOS leading to increased NO with downstream anti-inflammatory effects and decreased pain and edema. (Adapted from Pilla, A. et al., *Biochim. Biophys. Acta.*, 1810, 1236, 2011; Pilla, A.A., *Electromagn. Biol. Med.*, 32, 123, 2013.)

CaM antagonists block the PEMF effect on NO and cGMP in fibroblasts and endothelial cells (Pilla et al., 2011; Pilla, 2013), further supporting the proposed mechanism. Radio frequency PEMF has been reported to accelerate wound (Strauch et al., 2007) and tendon repair (Strauch et al., 2006) in animal models and angiogenesis in a thermal myocardial injury model (Strauch et al., 2009; Pilla, 2013) and downregulate IL-1β in fibroblasts (Moffett et al., 2012), in rats after posttraumatic brain injury (Rasouli et al., 2012), and in a mouse cerebral ischemia model (Pena-Philippides et al., 2014). Taken together, these results provide strong support for an anti-inflammatory mechanism based upon PEMF modulation of CaM/NO/cGMP signaling in challenged cells and tissues.

18.3 CLINICAL EVIDENCE OF PEMF EFFECT

18.3.1 NONTHERMAL RADIO FREQUENCY PEMF SIGNAL

The radio frequency PEMF signal considered in this review was configured a priori to modulate CaM-dependent enzymes and was first tested on myosin light chain (MLC) kinase in a cell-free enzyme assay for MLC phosphorylation (Pilla et al., 1999; Pilla, 2007). These early studies confirmed that PEMF accelerated the binding kinetics of Ca^{2+} to CaM. Subsequent studies showed that PEMF enhanced the activity of CaM-dependent constitutive NO synthase (cNOS = neuronal nNOS or endothelial eNOS) in the NO signaling pathway (Pilla, 2013). Extensive blinded testing on cutaneous wound repair in a rat model led to a signal configuration that inductively deposits a physiologically meaningful dose in the tissue (Strauch et al., 2007). The resulting PEMF signal in current clinical use consists of repetitive bursts of a 27.12 MHz radio frequency carrier. This signal takes advantage of the asymmetrical voltage-dependent kinetics of Ca^{2+} binding to CaM. Thus, Ca/CaM binding occurs in 1–10 ms, whereas Ca/CaM dissociation requires the better part of a second. Therefore, any inductively coupled PEMF signal for which the opposite polarity duration is well below 1 s, for example, 27.12 MHz, will always produce a net increase in Ca^{2+} bound to CaM.

Specifically, the PEMF signal used in the studies reviewed here consists of a shortwave radio frequency carrier at 27.12 MHz, a frequency that the Federal Communications Commission designates as a medical therapy frequency (Figure 18.2). The RF carrier is modulated with 2 ms bursts (within the kinetics of Ca/CaM binding) repeating at 2 bursts/s. The peak induced electric field is adjusted to be in the 3–10 V/m range. The maximum specific absorption rate (SAR), a measure of peak power deposited in a kilogram of tissue, for this signal has been measured to be 40 mW/kg, which is well below the level at which heat could be detected in a cell or tissue target (Pilla, 2013; Panagopoulos et al., 2013). The PEMF signal produces no sensation and is detectable only with specialized laboratory equipment, ensuring complete double-blinding in randomized, placebo-controlled clinical studies.

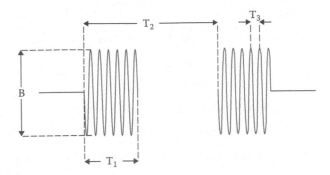

FIGURE 18.2 Radio frequency PEMF signal. $1/T_3$: 27.12 MHz, shortwave RF; T_1: 2 ms burst; $1/T_2$: 2 bursts/s; B: peak induced magnetic field 2–5 μT; peak induced electric field: 3–10 V/m; SAR: 40 ± 10 mW/kg. This signal produces no sensation and is measureable only with specialized laboratory equipment.

The PEMF devices employed in the clinical studies discussed in this review are manufactured by Ivivi Health Sciences, LLC, San Francisco, CA, and donated for each study. The device, known as the SofPulse, has been cleared by the Food and Drug Administration for pain and edema relief and is reimbursed by Medicare for chronic wound repair. The devices are noninvasive, nonpharmacologic, have no reported adverse side effects, and are configured with an automatic treatment regimen.

The PEMF signal is inductively applied with a single or dual 19 cm single turn coil, chosen to best fit the location of tissue injury. The dual coil unit has been used postoperatively for patients undergoing BR or breast augmentation (Figure 18.3), and the single coil unit has been used postoperatively for patients undergoing the more complex surgery of autologous breast reconstruction with transverse rectus abdominus myocutaneous (TRAM) flaps (Figure 18.6). The PEMF device weighs only 2.4 ounces, fits comfortably in a postsurgical bra or dressing, and requires no further intervention once activated in the operating room at the completion of the surgery. The devices add approximately $200 to the cost of the surgical procedure. This cost is similar to that of wound catheter equipment (Liu et al., 2006), with the advantages that the PEMF device is noninvasive, nonpharmacologic, and does not require additional invasive intervention.

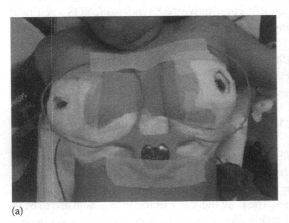 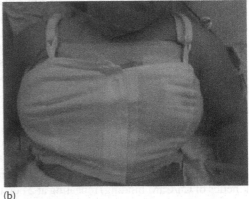

(a)   (b)

FIGURE 18.3 Dual coil PEMF device (SofPulse Duo, Ivivi Health Sciences, San Francisco, CA) on breast reduction patient. The PEMF device is soft, flexible, easily placed over dressings, and comfortable to be used within a postsurgical bra. PEMF is applied to each breast alternately during each pulse cycle using a multiplexing technique.

18.3.2 PEMF in Breast Augmentation Surgery

This double-blind, randomized, and placebo-controlled study included 42 healthy women under-going breast augmentation (Heden and Pilla, 2008), randomly divided into three cohorts: the first group with bilateral PEMF treatment (n = 14), the second group with bilateral sham devices (n = 14), and the third group with an active device on one breast and a sham device on the other breast (active/sham cohort) (n = 14), providing data from 80 breasts for analysis. The active PEMF devices deliv-ered a 30 min treatment regimen every 4 h for the first 3 days postoperatively, then every 8 h for the next 3 days, and finally every 12 h until the follow-up visit around postoperative day 7. Pain levels were assessed by self-evaluation with a visual analog scale (VAS), previously validated for post-surgical pain (Bodian et al., 2001; Coll et al., 2004). VAS scores were recorded twice daily through postoperative day 7, and postoperative narcotic use was followed.

The results showed that both the active cohort and the active/sham cohort experienced a near threefold decrease in VAS pain scores by postoperative day 3 versus the sham cohort (P < 0.001), which persisted at this magnitude through postoperative day 7. Concomitantly, postoperative nar-cotic use decreased almost threefold faster in the active and active/sham versus sham cohorts by postoperative day 3 (P < 0.001). The reason that PEMF was as effective in the active/sham and active cohorts is because measurement showed that the contralateral sham breast still experienced about 60% of the electric field amplitude delivered to the contralateral active breast.

It is of interest to note the results from a second breast augmentation study using a radio fre-quency signal with the same carrier frequency but substantially different pulse modulation and lower amplitude (Rawe et al., 2012). Those results showed that pain and narcotic use decreased only about 1.5-fold faster in the active versus sham cohorts, reflecting that a lower dose of PEMF was employed. Taken together, these studies demonstrated that PEMF therapy can reduce postoperative pain and narcotic medication use when used in combination with standard postoperative care in breast augmentation patients.

18.3.3 PEMF in Breast Reduction Surgery

This study examined the effects of PEMF on reducing postoperative pain in BR patients through a double-blind, placebo-controlled, randomized study (Rohde et al., 2010). PEMF was applied imme-diately postoperative with an automatic treatment regimen of 20 min every 4 h (Q4). The Q4 treat-ment regimen was based on the standard every 4 h narcotic dosing regimen for postoperative pain relief. The study included 24 patients who underwent BR for symptomatic macromastia. PEMF devices were placed on all patients, half of whom were randomly assigned to active treatment regi-mens and half of whom were assigned to sham regimens. Endpoints included VAS scores, pain medication use in Percocet equivalents, and cytokine levels in the wound exudates.

PEMF therapy with a Q4 hour treatment regimen significantly decreased postoperative pain and narcotic use in the immediate period after surgery in BR patients. PEMF produced a 57% decrease in mean pain scores at 1 h (P < 0.01) and a threefold decrease at 5 h (P < 0.001), which persisted to 48 h postoperatively, in the active versus sham groups. Concomitantly, the use of narcotic pain medication decreased more than twofold in active versus sham patients (P = 0.002). IL-1β and wound exudate volume results will be discussed separately later. The results for pain and narcotic use, summarized in Figure 18.4, provide level 1 therapeutic evidence that PEMF can significantly reduce postoperative pain and narcotic use in the immediate postoperative period after BR surgery.

18.3.3.1 Effect of PEMF Dosing Regimen

The effect of PEMF treatment regimen on postoperative pain and narcotic use was examined in an extension of the BR study summarized earlier. Results from rat traumatic brain injury research reported that the initial inflammatory response, as measured by IL-1β, was reduced

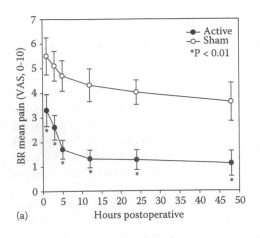
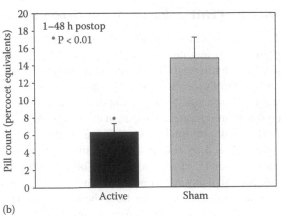

FIGURE 18.4 PEMF effect on postoperative pain and narcotic requirements in breast reduction patients. (a) PEMF produced a 57% decrease in mean pain scores at 1 h (P<0.01) and a threefold decrease at 5 h (P<0.001), which persisted to 48 h postoperatively, in the active versus sham groups. (b) Use of narcotic pain medication decreased more than twofold in active patients (P=0.002). (Adapted from Rohde, C. et al., *Plast Reconstr. Surg.*, 125, 1620, 2010.)

about 10-fold with a more frequent PEMF regimen of 5 min every 20 min. The PEMF effect on IL-1β was significantly larger than that on the physiological endpoint of pain in the initial BR study; however, there was no physiological endpoint, for example, cognition or behavior in the rat study. Therefore, a prospective cohort study with two additional treatment regimens was incorporated into the previous double-blind, placebo-controlled PEMF study in BR patients (Taylor et al., 2014). The two additional cohorts included a 5 min PEMF every 20 min (5/20) group and a 15 min PEMF every 2 h (Q2) group, which were compared with the historical active 20 min PEMF every 4 h (Q4 active) and sham (Q4 sham) groups. Data from 50 patients (13 in the 5/20 group, 13 in the Q2 group, and 12 in the Q4 [active] and 12 in the Q4 [sham] groups) were available for analysis. Visual analog scores and pain levels in Percocet equivalents were recorded as in the original study.

Results for Q4 (active) and Q2 were similar and showed a clinically significant decrease in postoperative pain between 1 and 14 h postoperatively, normalized to the 1 h values. VAS scores at 14 h postoperative were 47% and 37% of pain at 1 h in the Q4 (active) and Q2 cohorts, respectively (P<0.01). However, the 5/20 group showed no effect compared with the sham group (P=0.271) and was greater than twofold higher than that for the Q2 and Q4 (active) groups (P<0.01). The rate of pain decrease to 14 h postop in Q4 (active) and Q2 cohorts was more than threefold faster than that in the 5/20 and Q4 (sham) cohorts (P<0.01). Concomitantly, the data for pain medication use showed a clinically significant decrease for Q4 (active) and Q2 regimens (P<0.02). Again, narcotic usage was not significantly different between the 5/20 regimen and Q4 (sham) groups (P=0.439) and twofold higher than the Q4 (active) and Q2 cohorts (P<0.02). The results for Q2 provide further validation for the Q4 data and show that PEMF can provide significant postoperative pain relief for BR patients. The results are summarized in Figure 18.5.

However, the 5/20 regimen was not effective. The following explanation, based upon the possible effect of PEMF on CaM-dependent phosphodiesterase (PDE), was suggested. The PEMF signal used in this study produces an instantaneous burst of NO during each burst (Pilla, 2012). The amount of increased NO is directly dependent upon induced voltage amplitude, repetition rate, and total exposure time. Each of these parameters can modify the instantaneous amount of Ca^{2+}-dependent increase in NO in tissues produced by PEMF. In this study, PEMF signal parameters, including repetition rate, were identical for all patients in active cohorts. The dosing change was

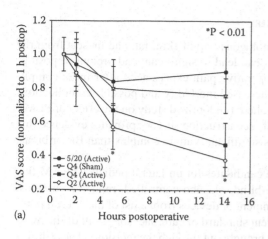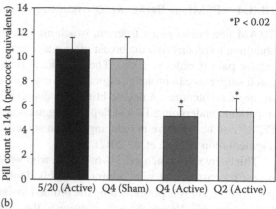

FIGURE 18.5 Effect of PEMF regimen on postoperative pain and narcotic requirements in breast reduction patients. (a) VAS scores at 14 h postoperative were 47% and 37% of pain at 1 h in the Q4 (active) and Q2 cohorts, respectively (P<0.01). The 5/20 cohort showed no effect compared with the sham group (P=0.271) and was greater than twofold higher than that for the Q2 and Q4 (active) groups (P<0.01). (b) Postoperative pain medication use showed a clinically significant decrease for Q4 (active) and Q2 regimens (P<0.02). Narcotic usage was not significantly different between the 5/20 regimen and Q4 (sham) groups (P=0.439) and twofold higher than the Q4 (active) and Q2 cohorts (P<0.02). (Adapted from Taylor, E. et al., *J. Surg. Res.*, 2014 [Epub ahead of print].)

signal regimen. The rate of increase in NO from PEMF in tissue for the 5/20 regimen is nearly 2.5-fold faster than that for Q2 and fourfold faster than that for Q4 (active).

The dynamics of Ca^{2+}-dependent NO release are tightly regulated through a negative feedback mechanism that involves PDE isoenzymes (Mo et al., 2004; Miller et al., 2009; Batchelor et al., 2010). These studies show that the rate of Ca^{2+}-dependent NO production in challenged tissue is not only modulated by PEMF, but also by PDE activity, which, itself, is also Ca^{2+} dependent. Thus, PEMF can enhance both the activity of cNOS, which increases NO release, and the activity of PDE, which decreases NO release in the tissue. Thus, if PEMF increases NO too rapidly, PDE isoenzyme activity may also be enhanced sufficiently to reduce or block the effects of PEMF on NO signaling. It is therefore suggested that the ineffectiveness of the 5/20 PEMF regimen on postoperative pain is because NO was increased too rapidly in the wound bed, further accelerating the activation of CaM-dependent PDE. The end result was that PEMF-enhanced NO release in the wound bed was limited or blocked by its effect on the dynamics of NO/cGMP/PDE signaling. This led to no significant persistent difference in the rate of pain decrease post-BR surgery in patients in the 5/20 cohort compared with those in the Q4 (sham) cohort.

PEMF effect on the dynamics of NO/cGMP/PDE signaling may also be the explanation for results in two recent publications, which show that PEMF effects depend upon signal configuration. The first showed that the PEMF effect on breast cancer cell apoptosis was significant when the same waveform, applied for the same exposure time, repeated at 20 Hz, but not at 50 Hz (Crocetti et al., 2013). A second study showed that the expression of inflammatory markers, tumor necrosis factor (TNF) and nuclear factor-kappa beta (NF-κB), in challenged macrophages was reduced by PEMF when the same waveform, applied for the same exposure time, was repeated at 5 Hz, but not at 15 or 30 Hz (Ross and Harrison, 2013a). It is of interest to note that Ca^{2+}-dependent NO/cGMP signaling modulates the expressions of these inflammatory markers (Ha et al., 2003; Yurdagul et al., 2013), suggesting that the effect of increased repetition rate is consistent with increased production of NO by PEMF at a rate high enough for negative feedback from PDE isoforms to predominate, thus blocking the PEMF effect on NO signaling. This effect is similar to the increase in NO in tissue expected from the 5/20 PEMF regimen in the BR study described earlier.

18.3.4 PEMF in Breast Reconstruction Surgery

TRAM flap breast reconstruction, which uses an autologous flap of skin, fat, and muscle from the abdomen to reconstruct the breast after mastectomy, can lead to significant and prolonged postoperative pain (Cordeiro, 2008). The significant postoperative pain experienced after more complicated surgeries can prolong surgical recovery with increased morbidity and possible complications in surgical outcomes. A double-blind, placebo-controlled, randomized study on the effect of PEMF in patients undergoing TRAM flap autologous breast reconstruction surgery provides evidence that PEMF can be effective in reducing pain in surgeries significantly more complex than BR or breast augmentation (Rohde et al., 2012).

Thirty-two women, aged 34–65 years, who were candidates for unilateral pedicled TRAM flap breast reconstruction were admitted to this double-blind, placebo-controlled, randomized study. Randomization was performed by the blinded assignment of devices from a list of their serial numbers. Use of PEMF was the only addition to the current standard of care. Jackson–Pratt drains were placed into the breast and abdominal donor sites and brought out through the incisions. These drains were left in place while the patient was in hospital. This permitted the collection of wound exudates in the immediate postoperative stages of healing. Exudates were collected and stored at −80°C for subsequent analysis.

Patients were randomly assigned two disposable PEMF devices placed within the surgical dressings at the breast flap and abdominal donor sites (Figure 18.6). PEMF therapy was identical for both sites and was programmed to continuously apply PEMF for 15 min every 2 h. Devices were activated on transfer to the recovery stretcher.

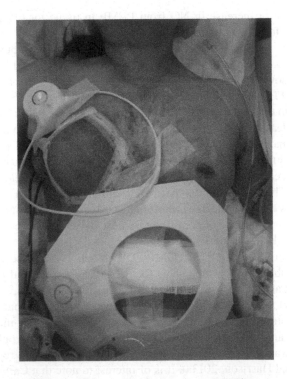

FIGURE 18.6 Single-coil PEMF device on transverse rectus abdominus myocutaneous (TRAM) flap patient. For this more complex breast reconstruction surgery, one PEMF single-coil unit is placed over the donor abdominal site, and one PEMF single-coil unit is placed on the recipient breast site. The abdominal single unit has a more rigid case to prevent excessive folding of the device, while maintaining appropriate placement. (Adapted from Rohde, C. et al., *Plast Reconstr. Surg.*, 130, 91, 2012.)

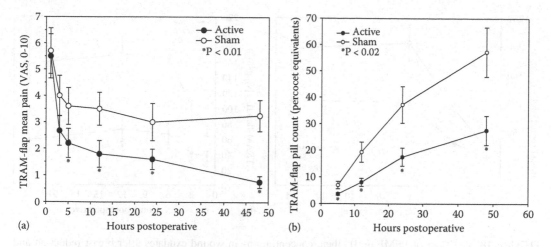

FIGURE 18.7 Effect of PEMF on the reduction of postoperative pain and narcotic use in TRAM flap patients. (a) Overall VAS results: Pain in the active cohort at 48 h is only 14% of pain at 1 h (P < 0.001), whereas it is 60% of pain at 1 h in the sham cohort (NS, P = 0.195). Pain decreases in the active cohort nearly threefold faster than in the sham cohort (P < 0.001). (b) Sham patients required a mean of about twofold more narcotics than active patients and had an increased narcotic use at more than 2.5-fold the rate of active patients. (Adapted from Rohde, C. et al., *Plast Reconstr. Surg.*, 130, 91, 2012.)

The primary outcome measure in this study was the rate of decrease in postsurgical pain. Secondary outcomes were cytokine concentration in the wound bed, wound exudate volume, and postop narcotic requirements. Pain levels were assessed by self-evaluation with a VAS. VAS data were obtained at intervals starting at hour 1 postop and at specified intervals thereafter for 72 h. Use of narcotic pain medication (oxycodone/acetaminophen) over the first 48 h was assessed by comparing pill counts for each group. Wound exudate was collected hourly starting at 1 h postop for the first 6 h and at 6–8 h intervals thereafter.

The results showed that PEMF therapy led to significantly decreased pain and narcotic use in patients after TRAM flap surgery. Patients in the active cohort had pain scores at 48 h at only 14% of pain at 1 h (P < 0.001), whereas patients in the sham cohort had pain scores at 48 h at 60% of pain at 1 h (P = 0.195). Pain decreased in the active cohort nearly threefold faster than in the sham cohort (P < 0.001). Patients in the active cohort were essentially pain-free by 48 h postoperatively, whereas sham patients still experienced significant enough pain to require medication. Sham patients required a mean of about twofold more narcotics than active patients and increased narcotic use at more than 2.5-fold the rate of active patients. These results, summarized in Figure 18.7, suggest that nonthermal PEMF therapy can produce rapid reductions in postoperative pain and narcotic use in complex surgical procedures.

18.4 PEMF IS AN ANTI-INFLAMMATORY

Interleukin-1beta plays a large role in mediating postoperative pain. Produced by macrophages, IL-1β is a proinflammatory cytokine that induces COX-2 and pain hypersensitivity. The severity of surgical trauma has correlated with greater levels of IL-1β, with increasing levels of systemic IL-1β corresponding to increasing severity of surgical insult (Baigrie et al., 1992).

Two studies on the effects of PEMF in BR patients and TRAM flap reconstruction patients examined the levels of interleukin-1beta (IL-1β) in wound exudates (Rohde et al., 2010, 2012). Both studies demonstrated significantly decreased IL-1β production and wound exudate volume in the first 24 h postoperatively. In the BR study, mean IL-1β concentration was significantly

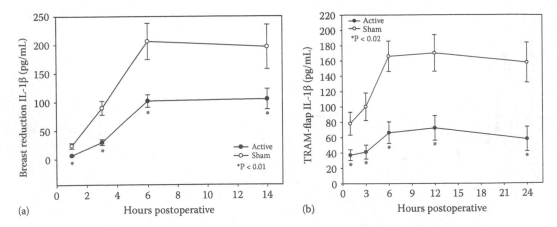

FIGURE 18.8 Effect of PEMF on IL-1beta concentrations in wound exudates after breast reduction and TRAM flap breast reconstruction. (a) In the breast reduction study, mean IL-1β concentration was significantly lower by more than threefold at 1 h postoperatively (P < 0.001), twofold at 6 h (P < 0.001), and twofold at 14 h (P < 0.01) in the active versus sham groups. (b) In the TRAM flap study, mean IL-1β concentration was significantly lower starting at 1 h postoperative (P < 0.02). IL-1β increases nearly threefold faster over the first 6 h for sham versus PEMF-treated patients (P < 0.01). IL-1β is more than threefold higher in the sham cohort by 24 h postoperative (P < 0.02). (Adapted from Rohde, C. et al., *Plast Reconstr. Surg.*, 125, 1620, 2010; Rohde, C. et al., *Plast Reconstr. Surg.*, 130, 91, 2012.)

lower by threefold at 1 h postoperatively (P < 0.001), twofold at 6 h (P < 0.001), and twofold at 18 h (P < 0.01) in the active versus sham groups (Rohde et al., 2010). Similarly, in the TRAM flap study (Rohde et al., 2012), mean IL-1β concentration was significantly lower starting at 1 h postoperative and increased nearly threefold faster over the first 6 h for sham versus PEMF-treated patients (P < 0.01). IL-1β was nearly threefold higher in the sham cohort by 6 h postoperative. These results for the BR and TRAM-flap studies are summarized in Figure 18.8. The decrease in IL-1β correlates well with the decrease in mean VAS scores over the same postoperative period in both of these studies. The significant reduction of IL-1β in wound exudates in patients treated with PEMF is consistent with a PEMF effect on the NO-signaling pathway, modulating the dynamics of an inflammatory cytokine involved in pain hypersensitivity.

Cumulative wound exudate volume is related to the amount of inflammation present in the wound bed. Edema in inflammatory processes results from the increased permeability of endothelial intracellular gap junctions, which allows proteins to move from the blood stream to the site of inflammation. In the BR and TRAM flap studies, cumulative wound volume in the first 24 h postoperative was nearly twofold lower in the active versus sham cohorts (P < 0.001). This correlates with the decrease in IL-1β, pain levels, and narcotic use in both studies. These results, summarized in Figure 18.9, suggest that inflammation was significantly lower in patients treated with active PEMF therapy devices. These data lend further support that PEMF affects the inflammatory response in tissues under stress.

18.5 PEMF BENEFIT OVER OTHER PAIN REDUCTION MODALITIES

Although multiple treatment modalities are available to reduce pain, many have adverse side effects or are contraindicated after certain types of surgery. Pharmacologic therapies to treat postoperative pain include opioids, nonsteroidal anti-inflammatory drugs (NSAIDS), local anesthetics, and NMDA receptor antagonists. As the most frequently used agent for postoperative pain relief, opioids block postsynaptic receptors of afferent neurons or activate inhibitory pathways for analgesic effects, but are not anti-inflammatory. In addition, opioids have undesirable side effects, including

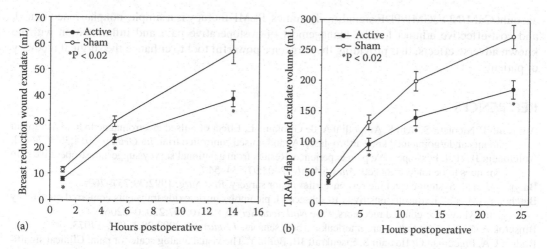

FIGURE 18.9 Effect of PEMF on cumulative wound exudate volume (inflammation) in (a) breast reduction and (b) TRAM flap breast reconstruction. Cumulative wound exudate volume increase from 14 to 24 h postoperative was nearly twofold higher in the sham versus active cohorts (P < 0.001).

nausea, vomiting, constipation, and addictive potential (Madadi et al., 2013). NSAIDs inhibit cyclooxygenases, including COX-1 and COX-2, thus inhibiting the formation of PGEs, prostacyclins, and thromboxanes from arachidonic acid. However, NSAIDS are rarely used postoperatively due to increased risk of hemorrhage (Swanton, Shorten, 2003) and decreased utility for severe pain. Local anesthetics, such as lidocaine and bupivacaine, block the sodium ion channels on the neuronal cell membrane, stabilizing the membrane and decreasing the rate of depolarization and repolarization to achieve analgesia (Ohsaka et al., 1994). Wound catheter equipment that deliver local anesthetic to the surgical site are similar in price to PEMF devices, but the devices are invasive, require an additional procedure, and do not directly target the inflammatory pathway (Kelly et al., 2001). Liposomal bupivacaine, or Exparel (Pacira Pharmaceuticals, Inc., Parsippany, NJ), is a long-lasting local anesthetic typically injected at the surgical site during surgery; however, Exparel generally costs more than PEMF devices, is useful only in the area where it is injected, lasts only up to 72 h, and does not affect the inflammatory cycle (Sinclair et al., 1993). NMDA receptor antagonists, such as ketamine, reduce the resulting amplification and prolongation of the pain pathway through noncompetitive blockade of the NMDA receptor antagonist, but ketamine is known for its hallucinogenic and disorienting effects, thus making it rare in the use of postoperative pain relief (Loix et al., 2011).

In contrast to all of the earlier points, PEMF provides significant postoperative pain relief through a noninvasive, nonpharmacologic, nonaddictive cost-effective means with no known adverse effects. Furthermore, PEMF reduces inflammation, which may facilitate faster tissue healing and postoperative recovery. Both economical and disposable, radio frequency PEMF devices can be easily incorporated in the postsurgical treatment of patients for lower morbidity, reduced inflammation, improved surgical outcomes, and decreased cost of health care.

18.6 CONCLUSIONS

The recent double-blind, placebo-controlled, randomized studies reviewed here provide strong evidence that nonthermal radio frequency PEMF therapy can produce rapid reductions in postoperative pain and markers of inflammation for both simple and complex surgical procedures. These effects lead to decreased postoperative narcotic use and morbidity, as well as enhanced surgical outcomes. The results from these studies support a PEMF effect on the NO signaling pathway, which takes into account CaM/NO/cGMP/PDE signaling dynamics. PEMF therapy is a simple, nonpharmacological, and cost-effective adjunct for the management of postoperative pain and inflammation with no known adverse effects, thus providing the surgeon a powerful tool to enhance the surgical recovery of patients.

REFERENCES

Adravanti P, Nicoletti S, Setti S, Ampollini A, de Girolamo L. Effect of pulsed electromagnetic field therapy in patients undergoing total knee arthroplasty: A randomised controlled trial. *Int Orthop.* 2014;38:397–403.

Apfelbaum JL et al. Post-operative pain experience: Results from a national survey suggest post-operative pain continues to be undermanaged. *Anesth Analg.* 2003;97:534–540.

Baigrie RJ et al. Systemic cytokine response after major surgery. *Br J Surg.* 1992;79:757–760.

Batchelor AM et al. Exquisite sensitivity to subsecond, picomolar nitric oxide transients conferred on cells by guanylyl cyclase-coupled receptors. *Proc Natl Acad Sci USA* 2010;107:22060–22065.

Binshtok AM et al. Nociceptors are interleukin-1beta sensors. *J Neurosci.* 2008;28:14062–14073.

Bodian CA, Freedman G, Hossain S, Eisenkraft JB, Beilin Y. The visual analog scale for pain: Clinical significance in post-operative patients. *Anesthesiology.* 2001;95:1356–1361.

Broughton G 2nd, Janis JE, Attinger CE. Wound healing: An overview. *Plast Reconstr Surg.* 2006;117 (7 Suppl):1e-S–32e-S.

Coll AM, Ameen JR, Mead D. Post-operative pain assessment tools in day surgery: Literature review. *J Adv Nurs.* 2004;46:124–133.

Cordeiro P. Breast reconstruction after surgery for breast cancer. *N Engl J Med.* 2008;359:1590–1601.

Crocetti S, Beyer C, Schade G, Egli M, Fröhlich J, Franco-Obregón A. Low intensity and frequency pulsed electromagnetic fields selectively impair breast cancer cell viability. *PLoS ONE.* 2013;8(9):e72944.

De Mattei M, Varani K, Masieri FF, Pellati A, Ongaro A, Fini M, Cadossi R, Vincenzi F, Borea PA, Caruso A. Adenosine analogs and electromagnetic fields inhibit prostaglandin E2 release in bovine synovial fibroblasts. *Osteoarthritis Cart.* 2009;17(2):252–262.

Ha KS, Kim KM, Kwon YG et al. Nitric oxide prevents 6-hydroxydopamine-induced apoptosis in PC12 cells through cGMP-dependent PI3 kinase/Akt activation. *FASEB J.* 2003;17:1036–1047.

Heden P, Pilla AA. Effects of pulsed electromagnetic fields on post-operative pain: A double-blind randomized pilot study in breast augmentation patients. *Aesthetic Plast Surg.* 2008;32:660–666.

Kazmier FR, Henry SL, Christiansen D, Puckett CL. A prospective, randomized, double-blind, controlled trial of continuous local anesthetic infusion in cosmetic breast augmentation. *Plast Reconstr Surg.* 2008;121:711–715.

Kelly DA, Ahmad M, Brull SJ. Preemptive analgesia I: Physiological pathways and pharmacological modalities. *Can J Anesth.* 2001;48:1000–1010.

LaPointe MC, Isenović E. Interleukin-1beta regulation of inducible nitric oxide synthase and cyclooxygenase-2 involves the p42/44 and p38 MAPK signaling pathways in cardiac myocytes. *Hypertension.* 1999;33 (1 Pt 2):276–282.

Liu SS, Richman JM, Thirlby RC, Wu CL. Efficacy of continuous wound catheters delivering local anesthetic for post-operative analgesia: A quantitative and qualitative systematic review of randomized controlled trials. *J Am Coll Surg.* 2006;203:914–932.

Loix S, De Kock M, Henin P. The anti-inflammatory effects of ketamine: State of the art. *Acta Anaesthesiolog Belgica.* 2011;62:47–58.

Madadi P, Sistonen J, Silverman G, Gladdy R, Ross CJ, Carleton BC, Carvalho JC, Hayden MR, Koren G. Life-threatening adverse events following therapeutic opioid administration in adults: Is pharmacogenetic analysis useful? *Pain Res Manag.* 2013;18:133–136.

Miller CL, Oikawa M, Cai Y et al. Role of Ca^{2+}/calmodulin-stimulated cyclic nucleotide phosphodiesterase 1 in mediating cardiomyocyte hypertrophy. *Circ Res.* 2009;105:956–964.

Mo E, Amin H, Bianco IH, Garthwaite J. Kinetics of a cellular nitric oxide/cGMP/phosphodiesterase-5 pathway. *J Biol Chem.* 2004;279:26149–26158.

Moffett J, Fray LM, Kubat NJ. Activation of endogenous opioid gene expression in human keratinocytes and fibroblasts by pulsed radiofrequency energy fields. *J Pain Res.* 2012;5:347–357.

Ohsaka A, Saionji K, Sato N, Igari J. Local anesthetic lidocaine inhibits the effect of granulocyte colony-stimulating factor on human neutrophil functions. *Experimental Hematol.* 1994;22:460–466.

Ongaro A et al. Electromagnetic fields (EMFs) and adenosine receptors modulate prostaglandin E(2) and cytokine release in human osteoarthritic synovial fibroblasts. *J Cell Physiol.* 2012;227:2461–2469.

Panagopoulos DJ, Johansson O, Carlo GL. Evaluation of specific absorption rate as a dosimetric quantity for electromagnetic fields bioeffects. *PLoS ONE*. 2013;8:e62663.

Pena-Philippides JC, Yang Y, Bragina O, Hagberg S, Nemoto E, Roitbak T. Effect of pulsed electromagnetic field (PEMF) on infarct size and inflammation after cerebral ischemia in mice. *Transl Stroke Res*. 2014;5:491–500.

Pilla AA. Mechanisms and therapeutic applications of time varying and static magnetic fields. In *Biological and Medical Aspects of Electromagnetic Fields*, Barnes F, Greenebaum B, (eds.). CRC Press: Boca Raton, FL. 2007; pp. 351–411.

Pilla AA. Electromagnetic fields instantaneously modulate nitric oxide signaling in challenged biological systems. *Biochem Biophys Res Commun*. 2012;426:330–333.

Pilla AA. Nonthermal electromagnetic fields: From first messenger to therapeutic applications. *Electromagn Biol Med*. 2013;32:123–136.

Pilla A, Fitzsimmons R, Muehsam D, Wu J, Rohde C, Casper D. Electromagnetic fields as first messenger in biological signaling: Application to calmodulin-dependent signaling in tissue repair. *Biochim Biophys Acta*. 2011;1810:1236–1245.

Pilla AA, Muehsam DJ, Markov MS, Sisken BF. EMF signals and ion/ligand binding kinetics: Prediction of bioeffective waveform parameters. *Bioelectrochem Bioenerg*. 1999;48:27–34.

Rasouli J et al. Attenuation of interleukin-1beta by pulsed electromagnetic fields after traumatic brain injury. *Neurosci Lett*. 2012;519:4–8.

Rawe IM et al. Control of post-operative pain with a wearable continuously operating pulsed radiofrequency energy device: A preliminary study. *Aesthet Plast Surg*. 2012;36:458–463.

Ren K, Torres R. Role of interleukin-1beta during pain and inflammation. *Brain Res Rev*. 2009;60:57–64.

Rohde C et al. Effects of pulsed electromagnetic fields on interleukin-1 beta and post-operative pain: A double-blind, placebo-controlled, pilot study in breast reduction patients. *Plast Reconstr Surg*. 2010;125:1620–1629.

Rohde C, Hardy K, Asherman J, Taylor E, Pilla AA. PEMF therapy rapidly reduces post-operative pain in TRAM flap patients. *Plast Reconstr Surg*. 2012;130(5S-1):91–92.

Ross CL, Harrison BS. Effect of pulsed electromagnetic field on inflammatory pathway markers in RAW 264.7 murine macrophages. *J Inflamm Res*. 2013a;6:45–51.

Ross CL, Harrison BS. The use of magnetic field for the reduction of inflammation: A review of the history and therapeutic results. *Altern Ther Health Med*. 2013b;19(2):47–54.

Sinclair R, Eriksson AS, Gretzer C, Cassuto J, Thomsen P. Inhibitory effects of amide local anaesthetics on stimulus-induced human leukocyte metabolic activation, ltb4 release and il-1 secretion in vitro. *Acta Anaesthesiolog. Scand*. 1993;37:159–165.

Strauch B et al. Pulsed magnetic field therapy increases tensile strength in a rat Achilles' tendon repair model. *J Hand Surg Am*. 2006;31:1131–1135.

Strauch B et al. Pulsed magnetic fields accelerate cutaneous wound healing in rats. *Plast Reconstr Surg*. 2007;120:425–430.

Strauch B et al. Evidence-based use of pulsed electromagnetic field therapy in clinical plastic surgery. *Aesthet Surg J*. 2009;29:135–143.

Swanton BJ, Shorten GD. Anti-inflammatory effects of local anesthetic agents. *Int Anesthesiol Clin*. 2003;41:1–19.

Taylor E, Hardy K, Alonso A, Pilla A, Rohde C. Pulsed electromagnetic field (PEMF) dosing regimen impacts pain control in breast reduction patients. *J Surg Res*. 2014 [Epub ahead of print].

Vincenzi F, Targa M, Corciulo C, Gessi S, Merighi S, Setti S, Cadossi R, Goldring MB, Borea PA, Varani K. Pulsed electromagnetic fields increased the anti-inflammatory effect of A_2A and A_3 adenosine receptors in human T/C-28a2 chondrocytes and hFOB 1.19 osteoblasts. *PLoS ONE*. 2013;8(5):e65561.

Werner S, Grose R. Regulation of wound healing by growth factors and cytokines. *Physiol Rev*. 2003;83:835–870.

Yurdagul A Jr, Chen J, Funk SD, Albert P, Kevil CG, Orr AW. Altered nitric oxide production mediates matrix-specific PAK2 and NF-κB activation by flow. *Mol Biol Cell*. 2013;24:398–408.

19 Daily Exposure to a Pulsed Electromagnetic Field for Inhibition of Cancer Growth
Therapeutic Implications

Ivan L. Cameron, Marko S. Markov, and W. Elaine Hardman

CONTENTS

19.1 INTRODUCTION

This chapter addresses four key questions on the effects of pulsed electromagnetic field (PEMF) as a potential therapy for treatment of tumorous cancers.

19.1.1 Does PEMF Therapy Slow Cancerous Tumor Growth or Cause Shrinking of Tumor Size?

A number of published reports on the growth of cancerous tumors of various types exposed to PEMF are summarized in Table 19.1. All of these reports demonstrated significant slowing of tumor growth when exposed to PEMF in the 0.8–120 Hz range using a variety of exposure conditions. None of these reports gave evidence of tumor shrinking (regression). Thus, the answer is that PEMF can be used to significantly slow tumor growth, but there is no evidence that PEMF, used as the sole therapy, causes tumor shrinkage or regression.

19.1.2 What Is the Most Effective PEMF Cancer Therapy?

Tumor growth has been used as an indicator of PEMF therapy effectiveness. Determining the most effective PEMF therapy calls for consideration of other measures including host survival, host side effects (i.e., body weight, cell proliferation in host tissues, organ weights, histopathology, and blood cell counts), metastatic incidence and the extent of tumor vascularization, and viable and necrotic volume density.

TABLE 19.1

Sample of Literature Reports on Effects of Electromagnetic Field Types on the Growth of Cancerous Tumors in Animal Hosts[a]

Type of Tumor	Frequency Hz and (pps)	Intensity Tesla	Exposure min, (h), or s/day	Significant Growth Retardation Yes or No	References
Melanoma	25	2–5 mT	3 h/day	Yes	Hu et al. (2010)
Hepatoma	100	0.7 mT	1 h 3×/day	Yes	Wen et al. (2011)
Colon	50	2.5 and 5.5 mT	70 min/day	Yes	Tofani et al. (2002)
Mammary	12 and 460	9 mT	10 min on alternate days	Yes	Bellossi and Desplaces (1991)
MX-1	50	15–20 mT	3 h/day	Yes	Berg et al. (2010)
Carcinogen induced	0.8	100 mT	8 h/day	Yes	Seze et al. (2000)
Mammary	120[b]	10–20 mT	10 min/day	Yes	Williams et al. (2001)
Mammary	120[b]	10 and 20 mT	3–80 min/day	Yes	Cameron et al. (2014)
Mammary	120[b]	15 mT	10 min/day	Yes	Cameron et al. (2005a,b)
Mammary	1	100 mT	60–180 min/day	No	Tatarov et al. (2011)
	1		360 min/day	Yes	
Sarcoma	0.16–1.3	0.6–2.0 T	15 min/day	Yes	Zhang et al. (2002)
Melanoma	50	5.5 mT	70 min/day	Yes	Tofani et al. (2003)
Sarcoma	50	250 mT	80 s/day	Yes	Yamaguchi et al. (2006)

[a] This sample is not comprehensive but is reasonably judged to be a representative.
[b] A semi-sine wave was used in the 120 Hz studies; other studies used a sine wave signal.

The choice of the most effective PEMF therapy conditions includes signal frequency (Hz), wave form, intensity, exposure duration, and frequency. However, few of the studies have been designed to answer the question of what is the most effective PEMF to treat cancer? Results of experiments designed to help answer this question are described in the next section.

19.1.3 How Do PEMF Therapy and Other Cancer Therapies Work?

The analysis of experimental data (Cameron et al., 2005a,b, 2014) has focused on the effect of a PEMF therapy on tumor growth and vascularity, host survival, host side effects, and metastasis using an implanted murine 16/c mammary adenocarcinoma in C3H/HeJ mice. Seven days after cancer implantation, tumor size reached 100 mm³, and the PEMF treatment has begun. Treatment PEMF ranges from 0 to 20 mT were given for 3–40 min once or twice a day. A semi-sine wave signal of 120 pulses/s was used. All of the seven treated groups had significantly slower tumor growth compared to the sham control group. Exposure to 20 mT for 10 min twice a day resulted in the greatest tumor growth suppression. PEMF-exposed mice had higher survival at 20 days post-tumor inoculation, and all PEMF groups had lower incidence of tumor cell metastasis than the control group. Pathological examination of host tissues showed no evidence of PEMF-related pathology.

Because tumor angiogenesis is essential for tumor growth, it was decided to study the extent of blood vessels in the tumors. PEMF had been shown to inhibit tumor angiogenesis (Markov et al., 2004) leading to an increase in the extent of tumor necrosis and a decrease in viable tumor area and in overall growth. Figure 19.1 is a photomicrograph of a 12 µm thick immunohistochemical slide section of the murine 16/c mammary adenocarcinoma tumor. Similar slide sections were used to

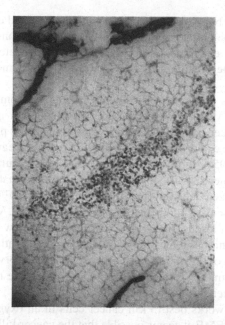

FIGURE 19.1 This photomicrograph is of a 12 µm thick cryosection of 16/C murine mammary adenocarcinoma. The tumor section was subjected to immunohistochemistry to detect the presence of CD 31 antigen, a marker for blood vessels. Blood vessels are visualized by the presence of the dark stain for CD 31 (examples at arrows). Areas of viable tissue are seen adjacent to the blood vessels. Necrotic (#) area is located even further from the blood vessels and contains condensed nuclei and cell fragments.

determine the extent of vascularization (dark black vascular regions CD 31 positive for blood vessels) and the surrounding viable tumor tissue. Areas further than about 100–150 µm away from a blood vessel were recognized as necrotic. Simple morphometric analysis of such slide sections, using either bright-field or phase contrast microscopy and a calibrated ocular grid, were used to determine the volume density of the vascular, the viable, and the nonviable necrotic tumor tissue.

Treatment with 15 mT for 10 min/day was found to give a maximum PEMF antiangiogenic therapy window. This PEMF therapy window, aimed at maximum suppression of tumor vascularization, was therefore selected as a safe adjuvant therapy to increase the effectiveness of conventional cancer therapy as reported next.

Knowing that a semi-sine wave signal of 15 mT at 120 pulses/s given for 10 min/day gave a maximum suppression of tumor vascularization antiangiogenic effect (Williams et al., 2001, Cameron et al., 2014), to test this PEMF exposure condition alone and in combination with gamma irradiation (IR) therapy in nude mice bearing a human MDA MB231 breast cancer xenograft (Cameron et al., 2005a) was decided. The cancer cells used in the study were transfected with a green fluorescence gene to help identify tumor cell metastasis. These cells were injected into the mammary fat pad of young female mice. Six weeks after injection, the tumorous mice were randomly assigned to one of four groups: sham control, 10 min daily PEMF, 200 Gy of IR every other day for a total of 800 Gy using a 137 Cs gamma cell-40 irradiator (this dose and exposure schedule are commonly used for radiotherapy in humans), and IR plus PEMF. Some mice were euthanized a day after the end of IR treatment, while others were euthanized 3 weeks later. Cross sections of the tumor were stained with CD 31 (an immunohistochemical marker of endothelial cells that line blood vessels) or with PAS (a marker of vascular endothelial cells). Sections were also stained for hypoxia-inducible factor 1 α (HIF).

Tumors less than 33 mm³ were colorless (not pink) when viewed through the skin of the nude mice, but turned pink as the tumor grew larger. Histology of the colorless tumors showed no evidence of a tumor capsule, but pink tumors had a well-vascularized capsule and had cancer cells but scant vascularization. The tumor regions greater than 100 μm from the capsule (the subcortex) had more blood vessels and endothelial cell pseudopods. It was determined that IR and PEMF therapy decreased the volume density of blood vessels, but the volume density of endothelial cell pseudopods increased.

Linear regression analysis of the subcortical zone vascularity from all of the treatment groups revealed that as the blood vessel area decreased, the pseudopod area significantly increased ($p < 0.005$). Most of the cancer cells in this subcortical zone stained positive for HIF, indicating a hypoxic tissue condition. The extent of cell pseudopods therefore appears to be an indicator of tissue hypoxia (Cameron et al., 2005a). In the tumor region located even further away from the capsule, a region of dying cells and necrosis was observed. It is proposed that the hypoxic tumor cells in the subcortical zone are oxygen deprived (hypoxic) and produce HIF leading to the production of growth factors, including vascular endothelial growth factor (VEGF). VEGF stimulates endothelial cells to form pseudopods that invade the hypoxic region and form new blood vessels (neoangiogenesis). It appears that PEMF inhibits a hypoxia-driven angiogenesis pathway.

The antiangiogenic action of PEMF may also help explain the observed decrease in the incidence of metastasis by PEMF treatment. To metastasize, blood vessels are needed for the movement of cancer from the tumor to other sites in the host body.

It is well known that IR works best to kill cancer cells in an oxygen-rich environment. Thus, when coadministered with PEMF, it seems possible that the cancer-killing potential of IR could be reduced due to the antiangiogenic action of the PEMF. PEMF is proposed to work not by a direct cell-killing process but by the suppression of hypoxia-driven angiogenesis. It was also observed that tumor vascularization of IR-treated mice returned to control levels by 21 days after the last IR treatment. Neoangiogenesis will eventually return the tissue to a normoxic state. Just how fast revascularization returns to the untreated control level after IR is an important but unanswered therapeutic question (Cameron et al., 2005a).

An experiment was designed, using zebrafish embryonic development to test if the 15 mT signal given for 10 min once a day specifically inhibits the hypoxia-driven angiogenic pathway or not. The embryos were cultured in shallow water in oxygen-permeable plastic Petri dishes. No evidence of tissue hypoxia was observed in tissue sections of the embryos (Cameron et al., 2005a). Newly fertilized zebrafish eggs were given daily exposures of 15 mT for 10 min. Developmental stages of treated and untreated (sham control) embryos were scored (Kimmel et al., 1995). Figure 19.2 summarizes the results. There was no difference in the rate of development due to PEMF treatment. Neither cell proliferation during cleavage nor angiogenesis at prime stages showed evidence of retardation of angiogenesis or of the rate of cleavage. Since the embryos were normoxic, it was concluded that PEMF is antiangiogenic only under hypoxic conditions. Another experiment was done to determine whether this same PEMF treatment retarded tail fin regeneration in adult zebrafish. The regeneration process was followed using microscopic observations at 35–100× magnification. Both PEMF-treated and nontreated fish regenerated their tail at the same rate. Markers of regeneration included length, number of fin ray bones, and blood vessel development. The thickness of the tail fin ranged from 250 to 500 μm. Thus, oxygen diffusion from the fin surface seems likely. Taken together, the results of these two zebrafish studies, using the same PEMF treatment as used in the tumor growth study in mice, showed that PEMF did not slow or retard either cell proliferation or angiogenesis during fish development or tail regeneration.

It is concluded that PEMF works to retard tumor growth and angiogenesis via a hypoxia-driven pathway but does not retard angiogenesis under the nonhypoxic conditions of the fish.

19.1.4 What Are the Prospects for PEMF Treatment of Cancer?

Previous studies have mainly dealt with the growth of transplantable cancer tumors in animal models. Studies on the effect of PEMF on blood-borne cancers are sparse. What can be said is that a

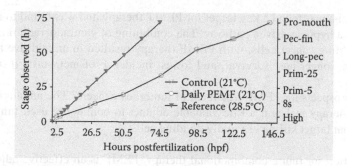

FIGURE 19.2 Temperature and 10 min daily exposure to PEMF on zebrafish development at 21°C. The staging and 28.5°C data are from Kimmel et al., *Dev. Dynam.*, 203, 253, 1995. Both the stage of development (h) and morphology are listed on the *y*-axis. Stage abbreviations are 8 somite (8s) and protruding mouth (pro-mouth). Time is listed on the *x*-axis. The slopes of the PEMF-treated and untreated embryos were not significantly different, but development was slower at 21°C than at 28.5°C.

PEMF that slowed tumor growth did so without harmful side effects. TEMF has not however been shown to shrink tumor size. This indicates that PEMF, used as sole cancer treatment modality, is not likely to cure cancer.

Future PEMF experiments need to take into account not only the PEMF exposure condition needed to explain optimum slowing of tumor growth but also other parameters like incidence of metastasis, host side effects, what is happening within the tumor (vascular, viable, and necrotic volume), and host tissue responses. This chapter also examines how PEMF and other cancer therapies work. The commonly used cancer therapies, like gamma irradiation and chemotherapy, work not only by killing rapidly dividing tumor cells but also by killing rapidly dividing normal host cell populations with harmful side effects. PEMF, on the other hand, has no known harmful side effects and appears to be a completely safe therapy.

Experimental results reveal that PEMF has an antiangiogenic effect that works via a tumor hypoxia-driven pathway but does not appear to have an antiangiogenic effect in the absence of the hypoxic state.

Use of PEMF has proved an effective adjunctive therapy when combined with gamma irradiation therapy (Cameron et al., 2005a). Future studies call for cessation of PEMF therapy for some time prior to a second round of IR treatment.

It is concluded that PEMF therapy of tumorous cancers is a safe effective adjunct that can be successfully used in combination with other common therapies and can enhance their effectiveness.

19.2 SUMMARY

Four questions were addressed in this chapter.

1. Does PEMF therapy slow cancerous tumor growth or cause shrinking of tumor size? Review of literature indicated that it does slow tumor growth, but it does not cause tumor shrinking.
2. What is the most effective PEMF cancer therapy? There have been little published data on this question, and only a few studies have measured more than just tumor growth. A summary of reports to determine the most effective PEMF for breast cancer therapy is presented.
3. How do PEMF therapy and other cancer therapies work? Gamma irradiation and most chemotherapy target killing of rapidly dividing tumor cells. Tumor vascularization

(angiogenesis) was found a key target for PEMT therapy and was found to inhibit angiogenesis via a hypoxia-driven pathway. The combining of gamma irradiation, which kills rapidly dividing cancer cells, with PEMF therapy resulted in an additive tumor growth suppression, longer host survival, and lowest incidence of metastasis (Cameron et al., 2005a,b).

4. What are prospects for PEMF treatment of tumorous cancers? The antiangiogenic action of PEMF therapy makes it a safe valuable adjunct to commonly used tumorous cancer therapies that target killing of rapidly dividing cells.

Here is a suggestion for future combinational therapy. PEMF is an effective adjunct therapy following IR therapy, but the continued daily PEMF therapy should be stopped sometime (perhaps 2–4 days) prior to a second round of IR therapy. This should allow resumption of angiogenesis, a decrease in hypoxic areas, and an increase in cell proliferation in well-oxygenated areas within the tumor, all of which are needed for an effective second round of IR treatment (Cameron et al., 2005a).

ACKNOWLEDGMENTS

This chapter is dedicated to the late C. Douglas William instigator and sponsor of much of our PEMF research. This work was supported in part by a contract from EMFT Therapeutics Inc. and in part by NIH grant CA75253. The expert assistance of Kerri Glaspie with manuscript preparation is gratefully acknowledged. We thank Stephen Scogin for his help with illustrations. We also thank Dr. N. Short for doing much of the zebrafish research.

REFERENCES

Bellossi A, Desplaces A. 1991. Effect of a 9 mT pulsed magnetic field on C3H/Bi female mice with mammary carcinoma: A comparison between the 12 Hz and the 460 Hz frequencies. *In Vivo*. 5:39–40.

Berg H, Gaoenther B, Hilger I, Radeva M, Traitcheva N, Wollweber L. 2010. Bioelectromagnetic field effects on cancer cells and mice tumors. *Electromagn Biol Med*. 29:132–143.

Cameron IL, Markov MS, Harman WE. 2014. Optimization of a pulsed electromagnetic field to retard breast cancer tumor growth and vascularity. *Cancer Cell Int*. In press.

Cameron IL, Sun LZ, Short N, Hardman WE, Williams CD. 2005a. Therapeutic electromagnetic field (TEMF) and gamma irradiation on human breast cancer xenograft growth, angiogenesis and metastasis. *Cancer Cell Int*. 5:23–28.

Cameron IL, Sun LZ, Short N, Hardman WE, Williams CD. 2005b. Daily pulsed electromagnetic field inhabits tumor angiogenesis via the hypoxia driven pathway. *Proc Am Assoc Cancer Res*. 46:1236.

Hu JH, St. Pierre LS, Buckner CD, Lafrenie RM, Persinger MA. 2010. Growth of injected melanoma cells is suppressed by whole body exposure to specific spatial-temporal configuration of weak intensity magnetic fields. *Int J Radiat Biol*. 86:79–88.

Kimmel CD, Ballard WW, Kimmel SR, Ullman B, Schilling TF. 1995. Stages of embryonic development of the Zebrafish. *Dev Dynam*. 203:253–310.

Markov MS, Williams CD, Cameron IL, Hardman WE, Salvatore JR. 2004. Can magnetic field inhibit angiogenesis and tumor growth. In: Rosch PJ and Markov MS (eds.), *Bioelectromagnetic Medicine*, Marcel Dekker, New York, pp. 625–636.

Seze R, Tuffet S, Moreau J, Veyret B. 2000. Effects of 100 mT time varying magnetic fields on the growth of tumors in mice. *Bioelectronics*. 21:107–111.

Tatarov I, Panda A et al. 2011. Effect of magnetic fields on tumor growth and viability. *Comp Med*. 61:339–345.

Tofani S, Barone D, Berardelli M, Berno E, Cintorino M, Foglia L, Ossola P, Ronchetto F, Toso E, Eandi M. 2003. Static and ELF magnetic fields enhance the in vivo anti-tumor efficacy of cisplatin against lewis lung carcinoma, but not of cyclophosphamide against B16 melanotic melanoma. *Pharmacol Res*. 48:83–90.

Tofani S, Cintorino M et al. 2002. Increased mouse survival, tumor growth inhibition and decreased immuno-activity p 53 after exposure to magnetic fields. *Bioelectromagnetics*. 23:230–238.

Wen J, Jiang S, Chen B. 2011. The effect of 100 Hz magnetic field combined with x-ray on hepatoma-implantation in mice. *Bioelectromagnetics*. 32:322–324.

Williams CD, Markov MS, Hardman WE, Cameron IL. 2001. Therapeutic electromagnetic field effects on angiogenesis and tumor growth. *Anticancer Res*. 21:3887–3892.

Yamaguchi S, Ogiue-Ikeda M, Sekino M, Ueno S. 2006. Effects of pulsed magnetic stimulation on tumor development and immune functions in mice. *Bioelectromagnetics*. 27:64–72.

Zhang X, Zhang H, Zhan X, Xiong W. 2002. Extremely low frequency ELF pulsed-gradient magnetic fields inhibit malignant tumor growth at different biological levels. *Cell Biol Int*. 26:599–603.

20 Nonionizing Radiation
Exposure Assessment and Risk

Michel Israel, M. Ivanova, V. Zaryabova, and T. Shalamanova

CONTENTS

20.1 INTRODUCTION

Wireless technologies are implemented in the whole life of the people during the last years. They enter everywhere in industry, medicine, transport, communication, media, data processing, etc. One of the main applications of wireless technology is the communication service. Following the development of mobile telephony, the technology passed through analog (1G) to digital system—from 2G (Global System for Mobile Communication [GSM]), 2.5 G (General Packet Radio Service [GPRS]), 3G (Universal Mobile Telecommunication System [UMTS]) to 4G (IP-based network) technology. Meanwhile, wireless local area network (WLAN); Worldwide Interoperability for Microwave Access (WiMAX); Bluetooth; digital enhanced cordless telecommunication (DECT) telephones; the ultrawide band imaging data processing technology in medicine, science, and technology ultra wide band (UWB); and many other applications are also in development and in practice.

The improvement of the wireless communication brought to an increased number of base stations to achieve a better coverage, good quality of signal, especially in residential areas.

The rapid introduction of mobile technology in modern life raises population's fears and concern for possible health effects of electromagnetic field's (EMF) exposure both to mobile phones users and to the general population exposed 24 h to EMF from base stations. Moreover, there are not proven evidences for any adverse effects of EMF long-term exposure. What is more, science does not yet provide precise clear responses to many questions in this field, as the risk for brain tumors, glioma and acoustic neuroma, changes in EEG, cognitive functions, for physiological changes.

Furthermore, little attention was paid to the risk of radiofrequency (RF) exposure on service staff of base stations and mobile communication antennae. Nevertheless, that there are special requirements for switching off the power during service close to the antennae, in most cases this is not the real situation: workers perform their activities during irradiation. Mild[12] reports similar practice among the service staff of base stations and antennae in Sweden. In many cases, nevertheless that the particular antenna is switched off, the technical operators are exposed to other sources as radiobroadcasting antennae, also other base stations.

There are still big differences between the exposure limits for EMF (especially for RF) implemented in European countries and worldwide. The reasons for it are the different approaches and rationales for the evaluation of the exposure. Many countries accept the International Commission on Non-Ionizing Radiation Protection (ICNIRP) Guidelines[7] but apply precautionary approaches or other philosophies to implement more conservative limits than those proposed by WHO, ICNIRP, or other international organizations.

Now, in Bulgaria, there are three licensed mobile operators, working in the GSM or CDMA, and UMTS systems (frequencies about 900, 1800, and 2100 MHz). They cover more than 98% of the country with RF signal for wireless mobile communication.

A large group of occupations are exposed to ultraviolet (UV) radiation in the working environment, some of them to natural/solar radiation—working outdoors and others with activities related to artificial sources. The risk of exposure for some professional groups applying UV sources in their work is very low because of the reason that sources are shielded or placed in a protective enclosure (food industry), or because of the short duration of the exposure (banks). For other professional groups, exposure limit values are exceeded in several seconds, or even a risk of exposure to working places in the vicinity of the sources exists—welding, dermatology, and physiotherapy. Factors having influence on exposure to sources of UV radiation can be summarized as follows: spectrum of radiation source, power, and duration of exposure.

Meanwhile, many studies have proven the harmful effects of UV radiation on human health. Target organs are mainly skin and eyes. The main identified adverse effects are as follows:

- On the eye: photokeratoconjunctivitis, cataracts, pterygium, pingueculum, droplet keratitis, etc.
- On the skin: erythema, burning, phototoxicity and photoallergy, premature aging, skin cancer (basal cell carcinoma, squamous cell carcinoma, malignant melanoma)[8,16]

UV radiation, encompassing UVC, UVB, and UVA, solar radiation, and use of UV-emitting tanning devices are recognized as carcinogenic to humans and included in the list of carcinogens as Group 1 by IARC.[8]

By 2010, there was no legal basis for the control of optical radiation in the working environment in Bulgaria. In 2010, Directive 2006/25/EC[3] was transposed in the national legislation introducing exposure limit values for coherent and noncoherent optical radiation and obligations of employers, training requirements, and health surveillance of workers.

There is still no legal basis for control of UV-emitting tanning devices.

20.2 REGULATIONS CONCERNING NIR (RF AND OPTICAL) PROTECTION OF WORKERS AND GENERAL POPULATION IN BULGARIA: CURRENT SITUATION

In 2014, the following regulations concerning human exposure to RF and optical radiation in Bulgaria are in practice:

20.2.1 RF FIELDS

20.2.1.1 General Public

For *EMF exposure to general population* from communication systems in residential areas in the frequency range 30 kHz to 30 GHz: *Ordinance No. 9, State Gazette No. 35/1991*.[15] There, the exposure limit for the frequency range from 300 MHz to 30 GHz is 10 µW/cm² (0.1 W/m²).

The Bulgarian legislation for environmental exposure sets two stages of hygienic control of EMF in the surroundings of base stations for mobile communication. The first stage covers check of the technical project documentation with calculation of the safety zone around the base station antennae. The estimated safety zones vary from 20 to 100 m depending on the emitted power, and for most of the base stations, they are approximately 50–70 m. The second stage covers measuring of the EMF values in the proximity of the sources where there is a possibility for people to be exposed.

Health Law, January 2005: final changes published in State Gazette No. 9/2011, where several new points concerning NIR exposures have been implemented:

1. The law obliged the *Minister of Health to manage the national system for analysis, evaluation and control of non-ionizing radiation in urban areas and public buildings.*
2. The list of the *environmental factors* that need to be controlled was enlarged with *non-ionizing radiation in residential, industrial, public buildings, and urban areas.*
3. The law implements in the list of *sites for public purposes* that should be controlled the following sites with sources of NIR:
   a. Emitting facilities that are a part of an electronic communication network, such as base stations and microwave, radio, television transmitters and repeaters, radar and navigation stations, and others
   b. Beauty salons and solariums
   For such objects, the law requires control of the human exposure.
4. The minister can determine by order laboratories with competence on measurements and exposure assessment, as follows:
   a. Accredited laboratories following EN ISO/IEC 17025 and/or EN ISO/IEC 17020
   b. Laboratories without accreditation but started and implemented necessary procedures for accreditation
   c. Laboratories without any accreditation but with evidence that, in conducting analysis for the purpose of control, laboratories implement a system and procedures to ensure the quality of the laboratory activities

20.2.1.2 Occupational Exposure

Concerning *occupational exposure* to RF fields, the Bulgarian Standard Institute withdrawn in 2011 two standards—for RF (60 kHz to 300 MHz) and for microwave (300 MHz to 300 GHz) exposure.

The result is that there are not any requirements for the health and safety of workers exposed to RF and microwave radiation from 2011 up to now.

Finally, in Bulgaria, there is not any legislation for exposure to workers in the frequency range 60 kHz to 300 GHz (exception: magnetic field up to 10 MHz). The legislation for EMF exposure to general population is only for frequencies 30 kHz to 30 GHz, and it is only for stationary communication sources in living environment.

20.2.2 OPTICAL RADIATION

Concerning the optical radiation, Directive 2006/25/EC of the European parliament and of the council of April 5, 2006, on the minimum health and safety requirements regarding the exposure of workers to risks arising from physical agents (artificial optical radiation) (19th individual Directive within the meaning of article 16(1) of Directive 89/391/EEC) was transposed in the national legislation with Ordinance No. 5, Official News No. 49/2010.[14]

In order to put the ordinance in force, a practical guide was developed on the basis of EU— nonbinding guide to good practice for implementing Directive 2006/25/EC *Artificial Optical Radiation*.[1] The national guide contains additional requirements for competence of the personnel performing measurements and exposure assessment, for measurement equipment, protocols of measurement, classification of laser systems, other harmful factors in working environment, a list of photosensitizing agents, etc. The guide to optical radiation ordinance is in a stage of discussion in the ministry of health. So, the ordinance no. 5/2010 for optical radiation practically is still not in force. Efforts in this field should be directed toward implementing the guide and ordinance and training of the control bodies' personnel.

The main problem in the field of UV radiation protection to the general public is the lack of legislation for control on using sources for artificial tanning.

20.3 STUDIED OBJECTS

20.3.1 RF SOURCES

The studied objects are 703 base stations for mobile communication with antennae-type Kathrein, Powerwave, HUAWEI, Tongue, Andrew, with frequencies about 900, 1800, and 2100 MHz, power approximately 20 W (guaranteed by the mobile operator), and different types of mounting of the antennae described earlier.

20.3.2 UV SOURCES

UV sources are used for medical treatment in physiotherapy and dermatology; for water, air, and surface sterilization in scientific laboratories and food industry; for curing of inks in industry; for tanning in solaria.

20.4 EXPOSURE ASSESSMENT

20.4.1 MEASUREMENT EXPOSURE ASSESSMENT OF RF FIELDS

20.4.1.1 Occupational Exposure

Our data of EMF exposure concern personnel working around typical antennae systems and performing routine operations in the close proximity around antennae. Data cover temporary workplaces for adjustment, maintenance, and repair of antennae, mounted on base stations for mobile communication.

The selected sites cover all possible working scenarios for the personnel servicing the base station antennae.

EMF values were measured, and exposure assessment was made for each particular operation. *Energetic loading* of the organism and whole-body specific absorption rate (SAR) were calculated on the basis of the EMF values and time duration of exposure. The scenario, technological operations, and their average duration were received by the mobile operators.

Data presented here do not concern working places on telecommunication masts.

Measuring methods are based on some international and national standards (EN 50384; EN 50413; EN 50499; EN 61566; EN 62311, IEC 62233:2011), also on our own experience. Some of the methods used are standardized by the ministry of health for such measurements. For exposure assessment, *spot* and *scenario method* have been applied.

The exposure and risk assessment were based on two parameters: time-weighted average (TWA) and energetic values calculated by the technological process and duties.

TWA

$$\text{TWA} = \frac{\sum_i S_i \cdot T_i}{T},$$

where

S_i is the power density measured for i process

T_i is the time duration of the i work process

Energetic parameter (energetic loading of the body)

$$W_s = S \cdot T_i,$$

where

S, W/m^2, is the power density measured

T_i is the time duration of the i work process

T is the total time duration of exposure for the working shift

Measurements were made around all studied base stations. Only for selected typical sites, energetic parameters were evaluated, and scenario method was applied. Measuring device was chosen with probe sensitive to electrical component of EMF. Data were collected using Narda EMR-21C software, installed on laptop. In cases when the height of antenna mounting allows performing measurements from the rooftop level, the equipment was mounted on tripod.

Measurements were performed considering the typical work positions of base station service workers. The rms (averaged over 6 min) and maximum power densities were measured. Energetic loading was calculated with respect to the mean time for performing the earlier-mentioned work tasks.

Risk assessment of exposure to EMF is performed according to ILO recommendations.[17] The priority of measures to be taken by the employer was proposed on the basis of the calculated *risk rating*.

The evaluation of EMF levels was performed at the following stages:

1. Measurement of EMF for various procedures performed by the personnel and evaluation of the compliance with the exposure limits
2. Evaluation of the energetic loading for various processes, activities, and shifts and its compliance with the ICNIRP guidelines

$$\sum \frac{S_i \cdot T_i}{S_{\lim} T} \leq 1,$$

where

S_i is the measured power density for the i—technological operation multiplied by the time duration of the exposure for the same operation T_i

$S_{\lim}$ is the reference level from ICNIRP guidelines

T is the total time duration of exposure for the working shift

3. Calculation of SAR, W/kg

Measurement equipment: Power meter *NARDA*, model EMR-21C of *NARDA Safety Test Solutions*; frequency range, 100 kHz to 3 GHz; electric field probe, isotropic response, nonselective. Measurements were made in wide frequency range—nonselective method.

Exposure assessment: on ICNIRP guidelines for occupational exposure; ICNIRP guidelines and Ordinance No. 9 (national legislation) for environmental exposure. Limit values for RF according to ICNIRP guidelines are presented in Table 20.1.

20.4.1.2 Exposure Assessment in Living Environment

As mentioned earlier, data cover 703 base stations for mobile communication with frequencies 900, 1800, and 2100 MHz (GSM/CDMA and UMTS). Most of the antennae (348) are mounted on telecommunication masts, situated in nonurban areas, other 334 on flat rooftops, 11 on slope roofs, and 10 set on building's facades. All measurements are performed in nominal loading regime of the base stations, which corresponds to the power about 20 W, guaranteed by the mobile operator. The method is determined by the type of antennae mounting.

In the case of *flat rooftops* mounting, depending on the orientation of the emitting surface of the antenna toward the roof space, the measurements are made at different distances:

- Behind the antennae panel (when the antenna is directed *outward* of the building)
- Aside the emitting surface (within antennae pattern)
- In front of the antennae panel

Power density values at 2 and 5 m are presented and compared here.

Hot spots on the roofs have been identified by standard measurement method (the same used for occupational exposure). They are characterized as areas with relatively higher RF field values and are associated with the presence of conductive objects/surfaces near the antennae, creating a complex field configuration. EMF in this case is characterized with very rapid alterations of its intensity when changing the distance to the source.

In the case of antennae mounted on *telecommunication masts*, measurements are made at the ground level at different distances from the antennae projections. The considered telecommunication masts are with a standard height—27 m.

TABLE 20.1

Limit Values for RF Fields according to ICNIRP Guidelines 1998

Frequency (MHz)	Power Density S, W/m² Occupations/General Public	Average Time t, min	SAR W/kg, Occupation/General Public
900	22.5/4.5	6	0.4/0.08
1800	45/9	6	0.4/0.08
2100	50/10	6	0.4/0.08

TABLE 20.2

Exposure Limit Values for RF Fields, according to Ordinance No.5/Directive 2006/25/EC

Index	Wavelength, nm	Exposure Limit Value, J/m²
1	180–400	$H_{eff} = 30$
	(UVA, UVB, and UVC)	Daily value 8 h
2	315–400	$H_{UVA} = 10^4$
	(UVA)	Daily value 8 h

Specifics of the *slope roofs*–mounted antennae require different approach: measurements in the subroof space at 5–6 m from the antennae, in residential premises and halls between them, and in front of the building itself.

On *facade*-mounting type, measuring spots are determined depending on the possibility of general public access and exposure.

20.4.2 MEASUREMENT AND EXPOSURE ASSESSMENT OF SOURCES OF UV RADIATION

A detailed work task analysis was performed before starting measurements, which included examination of working pattern and places where the exposure occurs. The measurements are carried out at places where equipment setup, exposure control, and other activities related to the source are performed by the personnel. Such approach allows correct personal exposure. During the measurements, the source is in typical placement in the chamber or working premise. Measurements are carried out both on levels of the exposed parts of the body—eyes, neck, and hands corresponding to the heights 1.60, 1.40, and 0.90 m from the floor—and on source level at different distances for the case of direct exposure (worst case). This scenario is possible mainly for medical applications of UV sources. The received results are compared to the exposure limit values corresponding to the directive (Ordinance No. 5), and as the measurements are not spectral, it is assumed that all of the lamps emission is at 270 nm, and $S(\lambda)$ has a maximum value of 1.

When the measurements are carried out at different levels above the ground, maximal permissible exposure time calculations are based on the maximum of the three measured values. The results are presented at typical distances from the source.

Measurement equipment: Photometer/radiometer IL1400A with the following configurations of detectors and input optics—SEL 400/WBS320/QNDS3/W, SEL 240/UVB-1/W, and SEL 033/UVA/W for measurements in UVA+B, UVB, and UVA ranges, respectively.

The exposure limit values according to Ordinance No.5/Directive 2006/25/EC are presented in Table 20.2.

20.5 RESULTS AND DISCUSSION

20.5.1 ENVIRONMENTAL EXPOSURE

The measured values depend on the type of antennae mounting and the possible access to the antennae. The four main types of antennae mounting are considered. The measurement results depending on the type of mounting in percent compared to the national legislation and ICNIRP guidelines are presented in Table 20.3.

The presented results show that in the majority of cases of rooftop mounting of base station antennae, measured values are not in compliance with the Bulgarian limits. At a few meters away from the emitting surface of antenna, there are areas where the exposure limits are exceeded.

TABLE 20.3

Results of Measurements of EMF Values Depending on the Type of Mounting of the Antenna in Percent to the Limits of National Legislation and ICNIRP Guidelines

Type of Mounting	% of National Legislation[a]	% of ICNIRP Guidelines[a]
Flat rooftops	1500	33
Telecommunication masts	20	0.4
Building's facades	80	2
Slope roofs	30	0.6

[a] For the maximal values.

These areas are not considered as a risk for general public because only incidental stay of people is possible there. It is not the same for occupational exposure, but other limits should be applied there.

All average values are in compliance with ICNIRP limits for general population.

Results from measurements and evaluation of antennae emissions of the earlier-cited number of base stations are presented here.

The measurement results are presented by mounting place, determined by mounting type of base station antennae—on poles with different height on the rooftop of buildings, on building facade, on slope roofs, and telecommunication masts.

20.5.1.1 Measurements on Rooftops

Depending on the orientation of the emitting surface of the antenna toward the roof space, measurements are made in all accessible areas around the antenna panel. The data for two distances—2 and 5 m—are presented later. All measurements are made on 1.8 m height above the rooftop level.

Data of measurements *in front the antennae panels*

Figures 20.1 and 20.2 present charts containing data distribution and medians for 2 and 5 m from the antenna panel. Figure 20.3 presents the mean values of power density at two different distances from the antenna.

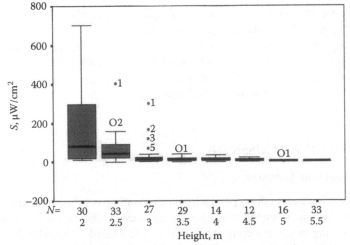

FIGURE 20.1 Data distribution and medians of EMF levels at a distance 2 m from the emitting surface of the antenna panel (on *x*-axis, the number of measurements for every height of antenna mounting is recorded).

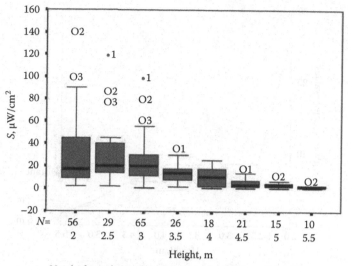

FIGURE 20.2 Data distribution and medians of EMF levels at a distance 5 m from the emitting surface of the antenna panel (on *x*-axis, the number of measurements for every height of antenna mounting is recorded).

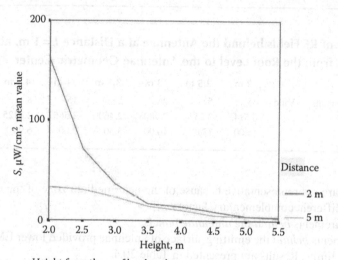

FIGURE 20.3 Mean values of power density distribution at two different distances from the emitting surfaces of the antenna panel.

At measurements in front of the antennae, the EMF power density values are logically the highest ones at small mounting height vs. the level of the roof.

Data of measurements *aside the antennae emitting surface*

When measuring *aside* the antenna, higher values are obtained when

* A part of the antenna is turned to the roof
* The mounting height of the antenna is small
* There are conductive surfaces near the antenna

Figure 20.4 presents the mean values of power density at two different distances from the antenna.

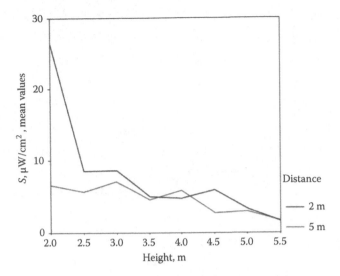

FIGURE 20.4 Mean values of power density distribution at two different distances aside of the antenna panel.

TABLE 20.4

Measurements of RF Fields behind the Antennae at a Distance $L=1$ m, at Different Heights (h, m), from the Roof Level to the Antennae Geometric Center

		2 m	2.5 m	3 m	3.5 m	4 m	4.5 m	5 m	5.5 m
Number of measurements	Valid	35	57	26	25	15	8	12	8
Mean value		3.5143	3.2544	2.9808	2.4600	3.5667	2.8125	1.9167	1.4375
Maximum value		15.00	12.00	10.00	8.50	7.00	6.00	5.00	3.00

These results are not representative because of the near/medium zone of the measurements and the influence of different complementary factors.

Data of measurements *behind the antennae panels*

The measurements *behind* the emitting surface of antennae provided lower EMF values, in most cases, within the limits. Results are presented in Table 20.4.

20.5.1.2 Measurements on Antennae Mounted on Slope Roofs

The antennae are mounted on poles situated at hard to access roofs (usually tiled). The measurements were made in penthouses (in the sub-roof space) at 5–6 m from the antennae, in residential premises and halls between them, and in front of the building itself. The results of the measurements in penthouses are from 0.5 to 2.0 µW/cm², and in front of the building—up to 2.5 µW/cm².

20.5.1.3 Measurements on Antennae Mounted on Facades

At mounting antennae on facades, there are two cases setting different sites of measurement for the determination of the possibilities for irradiation of the population:

1. *Near the antenna, there are no windows of residential premises.* In this case, the measurements are made behind and over the emitting antennae. Values from 0.5 to 2.0 µW/cm² were recorded, that is, below the limits set in Bulgaria.

2. *Near windows of residential premises.* In this case, measurements were made in the premises near the antenna and by the nearest window. The maximal values from such measurements are as follows:

   a. 4.0 µW/cm² in the premises
   b. 8.0 µW/cm² by the open windows of the premises

Differences in the measuring values depend on the distance of the window to the source.

20.5.1.4 Antennae of Base Stations Mounted on Telecommunication Masts

In this case, the base stations are usually situated on uninhabited territories—mainly in mountains, agricultural lands, and near to roads. The mounting height is standard 27 m from the ground level. The results are presented in Table 20.5.

In most cases, the power density values measured near the towers are low (under 2 µW/cm²). Single higher values (exceeding 10 µW/cm²) are found; they are detected to be due to induced currents in the measuring equipment by high-voltage power lines in the area. They are excluded by the discussed and statistically proceeded values.

In single cases, the towers with antennae are situated in courtyards of residential buildings. In these cases, measurements were made in the living premises of the adjacent houses. There values significantly lower than the exposure limits for residential areas were recorded—0.5–1.0 µW/cm².

20.5.2 Occupational Exposure

20.5.2.1 RF Fields

Operation working positions of the personnel around GSM/CDMA and UMTS antennas and the approximate average duration of exposure for every operation are presented in Table 20.6. Data are presented in percent compared with the reference levels proposed by ICNIRP. The average power density measured at temporary workplaces for adjustment, maintenance, and repair operations shows compliance with the ICNIRP guidelines. Maximal values exceed the reference levels as could be seen in Figures 20.1 and 20.2 and also in Figures 20.5 and 20.6 (at the close proximity of the antenna). The average values compared to the ICNIRP guidelines are presented in Table 20.7. The measurement points correspond to the location of the subjects at work with respect to the antennae. The power densities on each operation position, for GSM/CDMA and UMTS antennae, are presented in Figures 20.5 and 20.6.

Calculated *whole-body SAR values* for CDMA antennae are in the range 0.001–0.242 W/kg, depending on the type of operation (the body position toward antennae). The highest value corresponds to operations performed in front of emitting antennae. In the case of UMTS, the highest value is 0.062 W/kg for the same position.

Measured values, compared to ICNIRP guidelines (reference levels), are not in compliance for operations performed in front of the antennae. Nevertheless high measured EMF values, the whole-body SAR values correspond to the basic restriction for occupational exposure. This compliance

---

TABLE 20.5

Measurements of RF Fields around Grid Towers at Different Distances, *L*, m

L (m)		15	20	25	30	40	50
Number of measurements	Valid	173	240	76	99	28	18
Mean value		1.1845	1.3243	1.3036	1.5624	2.0179	1.4500
Maximum value		4.00	4.70	7.00	7.50	8.00	5.00

TABLE 20.6

Operation Types and the Approximate Average Exposure Duration for Every Operation

No.	Operation Position	Operation Duration[a]
	GSM/CDMA antennae	
1	Changing the position of the antenna	30 min
	Single Antennae	
	Operation in a close proximity of antennae	
2	Behind antennae	2 h
3	In front of antennae	2 h
4	Close to the lower edge of a single antenna	2 h
5	Near emitting antennae (distance up to 2 m)	2 h
6	Along the feeder line	4 h
	Antennae mounted on a pole (wrist type)	
	Operation in a close proximity of more than one antenna	
7	In the area among antennae, on a pole	6 h
8	Between side edges of antennae	6 h
9	Operation close to the lower edge of antennae	6 h
10	Operation on new antennae	6 h
	UMTS antennae	
	Single antenna	
	Operation in a close proximity of antennae	
11	Operation behind antennae	2 h
12	Operation in front of antennae	2 h
13	Operation close to the lower edge of antennae	2 h
14	Operation on new antennae	2 h
15	Operation along the feeder line	4 h

[a] Data from the operator.

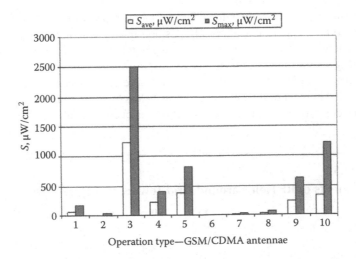

FIGURE 20.5 Power densities on each operation position, for GSM/CDMA antennae.

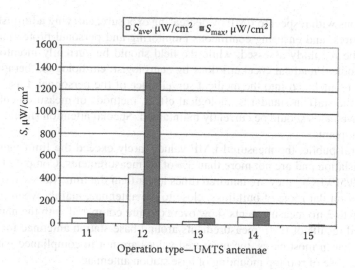

FIGURE 20.6   Power densities on each operation position, for UMTS antennae.

TABLE 20.7

Average EMF Values Compared to the ICNIRP Guidelines

Working Position of the Personnel	% of ICNIRP Guidelines
Close proximity (up to 2 m from the emitting surface of the antennae)	54
Between side edges of antennae	4

doesn't mean that exposure to particular operations is not significant. Therefore, employers should take the following measures in order to protect the workers' health:

- Technological operations in the close proximity of antennae should be performed after switching off emitters.
- The use of personal protective equipment (clothes, glasses, etc.) when it is not possible to switch antennae off.
- When protective equipment is not used, combinations of technological operations should be made respecting the following limitation of the exposure (parameters have been introduced earlier):

$$\sum \frac{S_i \cdot T_i}{S_{\lim} T} \leq 1.$$

Some additional not less important recommendations for the operators' safety work are the following:

- Workers should use personal dosimeters, which provide data about the cumulated dose.
- Periodical training of the staff about the risks of EMF and protection should be performed.

As a result from this investigation, a high risk for EMF exposure is found for workers mounting and maintaining base stations for mobile communication. Measures should be taken to improve the working conditions with respect to EMF by decreasing exposure, carrying administrative and orga-nizational measures, and ensuring continuous monitoring and personal protection. The energetic loading should be regularly assessed, while the field should be currently monitored by personal dosimeters. Periodical medical checkups done by neurologist, cardiologist, therapist, and ophthal-mologist should be included into the medical surveillance of the personnel. Last but not least, the qualification of the staff in standards, biological effect, methods of measurement, and protection against microwave fields should be currently maintained. Special attention should be paid to work-ers with active implants.

For the general public, the measured EMF values rarely exceed the limit values according to the national legislation and are not more than 3% of all measurements. Compared with EC recom-mendation (1999/519/EC),[2] they are hundred times lower than the limit values and could not create health risk. Only on flat roofs of buildings, close to the antennae, mainly when they are mounted at low height (up to 2 m) measurements show overexposure compared with the national legislation.

The presented results of EMF measurements around base station antennae for mobile commu-nications show that in most cases, the measured values are not in compliance with the Bulgarian limits in force in case of rooftop mounting of base station antennae.

The experience from previous measurements in Bulgaria and similar ones in other countries shows that at a few meters away from the emitting surface of antenna, there are areas where the *exposure limits are exceeded*. Usually, this is due to the following reasons:

- Mounting of the base station antennae *at low poles* (small height vs. the rooftop level)
- Geometry creating conditions for *hot spots*
- Inadequate adjusting of the *irradiation cone* vs. the building—carrier and neighboring buildings
- A presence of *background from other sources*, emitting in the area of base station antennae in similar frequency bands

The question in this case is whether the high EMF values can generally be considered as unconfor-mity with the limits as most often the cases refer to roofs of dwellings where population's exposure is only incidental or associated with repair activities. And if there is unconformity, what measures should be undertaken? At established unconformities, the global as well as Bulgarian practice is to recommend the placing of *fences and precautionary measures* for the restriction of population access to places where the EMF is above the limits. In many cases, it is recommended to mount the antennae on higher poles and placing signs and labels restricting the access to that roof area where the exposure limits are exceeded. Such actions are efficient at observing the limitations, that is, population exposure to above-standard EMF levels is avoided.

Another important issue is the EMF exposure from base station antennae and additional EMF sources in the area on the operators maintaining and servicing the stations.

20.5.2.2 Occupational Exposure to UV Radiation

For the case of direct exposure, here, we present data of measurements of irradiance at different distances from the source for two phototherapy units.

The results of irradiance measurements depending on the distance of emitting surface of source for phototherapy unit, UV 208T (1.10), are presented in Figure 20.7.

The results of irradiance measurements depending on the distance of emitting surface of source for phototherapy unit, UV 409T (2.11), are presented in Figure 20.8.

The results of irradiance measurements depending on the distance from emitting surfaces of sources used for the sterilization of the air, Silvania (2.2) and UV-FAN 25H-JON (2.3), are pre-sented in Figures 20.9 and 20.10.

As it could be seen from the graphs, the irradiance values decrease fast with the distance. Nevertheless in almost all space of the chambers, the irradiance values are relatively high.

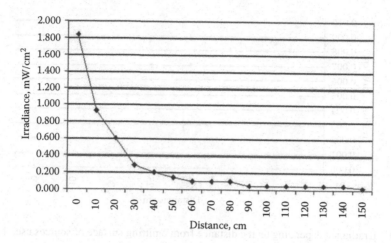

FIGURE 20.7 Irradiance depending on the distance of emitting surface of source for phototherapy unit—UV 208T.

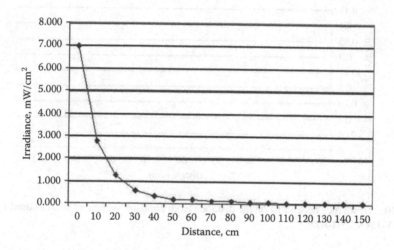

FIGURE 20.8 Irradiance depending on the distance of emitting surface of source for phototherapy unit—UV 409T.

Results of calculated maximal permissible time of exposure for different sources are presented in Table 20.8.

20.5.2.2.1 UV Sources in Medicine

Our study has shown that UV exposure of medical personnel may occur during setup of equipment, placing the patients and control of exposure.

Depending on the height of mounting of the source, applied procedure, respectively, maximum irradiance was measured at different levels: at the level of the neck or arms, and in the case of higher-mounted source, maximum values are at eye level. There are cases with low mounting of the source, where irradiance has background values on eye level.

It was found that around UV sources, often, nonshielded large-area reflective surfaces (mirrors, windows, metal parts of other equipment) could be discovered that may pose additional risk for overexposure of the patients and the staff.

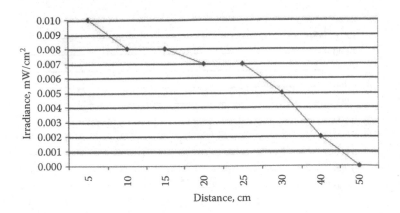

FIGURE 20.9   Irradiance depending on the distance from emitting surface of sources used for sterilization of the air—Silvania.

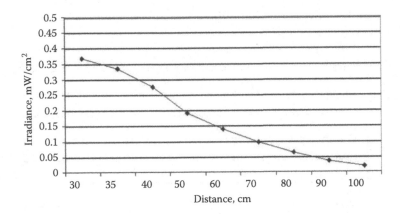

FIGURE 20.10   Irradiance depending on the distance from emitting surface of sources used for sterilization of the air—UV-FAN 25H-JON.

The results of performed measurements of UV irradiance on working places and at different distances from sources used in dermatology and physiotherapy show that exposure limit values, set by Directive 2006/25/EC (Ordinance No. 5/2010), are exceeded in the chambers where the procedures are performed.

20.5.2.2.2   UV Sources Used for Sterilization

UV radiation was not registered outside the systems for sterilization for air and water sterilization, because they are encapsulated and there were no leakages established in the housing of the sources.

Concerning the other sources used in scientific laboratories, exposure limit values [3,4] are exceeded. Depending on their design features, there is a difference in registered exposure levels. The levels are lower for the lamp with barred cover; nevertheless, they exceed exposure limit values for unlimited stay. With this type of sources, it is possible to avoid exposure by remote switching on/off the sources when there is no staff in the premises and restricting access. There are cases, however, where it is necessary for the sources to be switched on during the work of personnel. In these cases, the exposure could be minimized (by installation at higher place, outside the field of view, etc.). When the bactericidal lamps are placed in housing with forced ventilation system, proper mounting could help to avoid the exposure to personnel.

TABLE 20.8

Maximal Permissible Time of Exposure for Different UV Sources

No.	Source	Distance/Place of Measurement	Maximal Permissible Exposure, t
		UV Sources in Medicine	
1.1.	Quartz/sunlamp *Sun 250-1*	At 20 cm to the source	9 s
1.2.	Quartz/sunlamp *Sun 250-2*	At 20 cm to the source	21 s
1.3.	Quartz/sunlamp *Sun*	At 20 cm to the source	6 s
1.4.	*Dinamed* source	At 20 cm to the source	14 s
1.5.	*Sun 400* without tube	At 20 cm to the source	8 s
1.6.	*Sun 400* with tube	At 20 cm to the source	1 min 26 s
1.7.	Portative quartz lamp *Sun*	At 50 cm to the source	9 s
1.8.	PSORILUX 5050	At 20 cm to the source	5 s
1.9.	PUVA 200	At 20 cm to the source	6 min 35 s
1.10.	Phototherapy unit 208 T	At 50 cm to the source	21 s
1.11.	Phototherapy unit 409 T	At 50 cm to the source	16 s
1.12.	UWE SUNSTREAM BRONZARIUM	At the edge of the equipment	1 min 13 s
		UV Sources Used for Sterilization	
2.1.	Bactericidal lamp, Philips	At 50 cm to the source	Not permissible
2.2.	Bactericidal lamp, Silvania	At 50 cm to the source	8 s
2.3.	Bactericidal lamp UV-FAN 25H-JON	At 30 cm to the source	10 min
2.4.	System for water sterilization: DUV 2-75	At 20 cm to the source	Unlimited
2.5.	System for air sterilization: AR-UV-170-1	At 20 cm to the source	Unlimited
		UV Sources in Industry	
3.1.	Printing machine KASE KLP6	At working places	7 s to 50 min
3.2.	Printing machine KASE 6000 CNC	At 20 cm to the source	1 min 7 s to 16 min 40 s
		UV Sources Used for Tanning	
4.1.	Solarium *Prestige 48 C*	At working place	5 min

20.5.2.2.3   UV Sources in Industry

Measurements performed in working environment in industry show exceeded limit values at working places. The maximal values of irradiance are measured primarily at the level of the arms and, in some cases, also at eye level of the workers. Due to continuous working process, exposure could not be controlled by restriction of duration of work in vicinity of the equipment.

20.5.2.2.4   UV Sources Used for Tanning

For the case of solarium, exposure limit values exceeding were registered at the slit of the enclosure of the solarium and to a distance of approximately 1.5 m from the device. So, it is evident that there is a risk of overexposure to personnel working in the same premise where the equipment is situated.

Due to the fact that the calculated maximal permissible exposure times are very small for the most of the working places with UV sources, it is clear that the risk could not be controlled through limiting time duration of exposure during the working shift/day. If results are interpreted only regarding the EU Directive (Ordinance No. 5), it would mean prohibited stay around a lot of sources of UV radiation in the working environment. In many cases, that would be unacceptable. In those cases, taking steps for risk management according to the guide to the directive is the only decision.

Directive/ordinance does not give the answer for cases where overexposure could not be avoided or stay around UV sources should be prohibited.

Lets' do a parallel with MRI equipment where the exposure limits for EMFs are exceeded at working places. There, the problem is solved by specific derogation for this particular source in Directive 2013/35/EC.[4] In this case, it means that exposure may exceed limit values provided that several conditions are met: all technical and/or organizational measures have been applied; the circumstances duly justify exceeding the limit values; the characteristics of the workplace, work equipment, or work practices have been taken into account; and the employer demonstrates that workers are still protected against adverse health effects and against safety risks, including by ensuring that the instructions for safe use provided by the manufacturer are followed.

Such risk management practice without any derogation could be applied to working places with sources of UV radiation.

In conclusion, the medical staff in physiotherapy and dermatology, personnel in scientific laboratories and industry are working at high occupational risk due to UV sources. The only exceptions are sources that are encapsulated.

The risk for the staff is determined not only by the type of source and the radiation spectrum but also by the mode of organization of work and the workplace and the use of personal protective equipment and protective measures.

The risk can be minimized by performing control, training of the personnel, and implementation of appropriate measures for the protection and use of personal protective equipment. Specific risk management approach for ensuring health and safety of workers should be applied at places where the exposure levels are particularly high and permissible time of exposure is small.

20.6 RISK POLICY: LEVEL OF PUBLIC CONCERN

20.6.1 ELECTROMAGNETIC FIELDS

The number of complaints in Bulgaria decreases in the last years (after 2010). Our explanation of this process shows different reasons: more adequate information for the people, more clear messages from the scientists, and hard work of the responsible specialists in the field of risk communication. The tiredness among the activists and the appearance of other deeper problems as poorness and unemployment are also good explanation of the decrease in the interest but with smaller importance.

The analysis shows that for the period from 1994 to 2007, a total of 33 complaints from citizens have been received. Since 2008, a sharp increase in the number of complaints up to the early 2010 can be found. Only for this short period (2 years), the complaints raised to 214. As we mentioned earlier, after 2010, the tendency is for sharp decrease of the complaints—several per year. Now we expect slight increase in the public concern because of the coming elections this year. The risk analysis shows the following:

1. *Results* of exposure assessment show that in *no more than 3% of the places of concern exposure exceeds the limits*. Analysis of these cases shows that they are due to the following reasons:
   a. *New building in the area* of the cone of radiation (inside the safety zone)
   b. *In places of illegal base station*, sometimes working for more than 5 years
   c. *Presence of reflecting surfaces* near the windows of the house, which increase the exposure part of the apartment

   It is important to be mentioned that *overexposure* concerns only the Bulgarian legislation, which is conservative—the limit value for microwaves is 10 $\mu W/cm^2$. Compared with the limits recommended by EC (ICNIRP), it is 45–100 times more stringent for the frequencies used by the GSM/UMTS technology.

2. The analysis of the reasons for a temporary increase in the number of complaints concerning problem identification shows the following[10]:

    a. Interference of laymen in measurements or incorrect interpretation of results of *measurements* and scientific publications

    b. *Accreditation* of measurement laboratories without any competence, knowledge, and experience in the field

    c. Political interests

    d. *Measurements with household, not professional types of devices,* that are not appropriate for the purpose of quality measurements, in order to receive *suitable* results

    e. *Publications* in *media arising the fears among the population,* for example, EMF and cancer; discussion of only *positive* results concerning adverse effects and neglecting the quality in science, and others.

In addition, there is a lack of good legislation to regulate obligations of the industry for ensuring health for the general public as well as citizens' rights to make use of EMF sources, and also for receiving information regarding the exposure. There are also some financial interests for a part of the citizens (rent relations with the operators) and for companies considering ensuring general public health as a business.

Changes in the health law have been developed, and they are in the final stage of discussion, prepared for implementation by the parliament. Those changes follow the requirements for legislation in the field of EMF published in *Model Act* and *Model Regulations,* WHO,[6,13] mainly concerning the environmental exposure. A new chapter "Non-ionizing radiation" is included in section 2: "Activities for public health protection." There, most of the requirements proposed by WHO, ICNIRP, Council recommendations 1999/519/EC have been implemented. Some new points (and limits) have been implemented at the time of discussion with general population, industry, specialists, as follows:

The sector minister that should provide evaluation of the EMF exposure and to take care about possible adverse effects on the general population is the minister of health.

Two differentiated zones of exposure to the population are defined in the following manner:

First zone (short-term stay) sets regions where only short-term human stay is possible: hard-to-access areas and slope roofs of residential buildings.

Second zone (access area) sets regions where temporary and/or periodical human stay is possible, such as agricultural lands, accessible roofs of residential buildings, residence areas, streets, and electric transport.

Exposure limits proposed are for frequency range from static fields (0 Hz) to 300 GHz, concerning all kind of sources of radiation. The limits proposed in CR 1999/519/EC (ICNIRP guidelines) are implemented in the following way:

For the first zone the ICNIRP exposure limits – basic restriction and reference levels for the frequency range >0 Hz to 300 GHz are accepted.

For the second zone the strategy is to keep the actual exposure limits that are in use in Bulgaria. It means frequencies from 30 kHz to 30 GHz. For the others the exposure limits correspond to ICNIRP Guidelines.

There is a legal procedure for implementing sources of radiation emitting in residential areas. It includes calculation of safety zone around the source corresponding to the accepted limits and spot measurements for the control of the emission. The chief health inspector gives the permission for construction or mounting the equipment.

The law requires to create and to support database consisting the sources of radiation, the levels of exposure, and all other documents concerning the permission for use. This information should be open for public access. Sanctions for the industry are recorded in case of overexposure or illegal construction of sites.

The law implements the *agency* proposed in the *Model legislation...*—the National Centre of Public Health and Analysis (NCPHA) belonging to the Ministry of Health—and gives it the responsibility for developing databases, training of the control bodies, and developing methods for exposure and risk assessment of EMF exposure. Control bodies are the regional health inspections belonging to the ministry of health.

New issue is the prohibition of advertisement of mobile phones in different media concerning children and adolescents up to age of 18 and also the use of such people in advertisements.

It is very important that the Ministry of Health should introduce an *Ordinance* including exposure limits, methods of measurement and dosimetry, methods of evaluation of the exposure, some precautionary measures, in a few months after implementing the proposed changes in the Health Law.

Finally, the law defines *non-ionizing radiation* for the first time including static electric and magnetic fields, extremely low-frequency fields, RF, microwaves, and optical radiation up to UV radiation with a wavelength to 180 nm.

In the health law, there are texts dealing with the use of precautionary principle for cases of possible human exposure to environmental factors.

Unfortunately, there are many requirements that should be included in our national legislation as the need of periodic control of EMF exposures, requirements for the competence of the control bodies and the specialists doing measurements, for real-time data about EMF exposure to the population, etc.

For the working conditions, the ICNIRP guidelines will be implemented according to the limits and deadlines for the implementation of the European Directive 2013/35/EC.

20.6.2 OPTICAL RADIATION

Regarding the optical radiation in order to implement the existing legislation into force, the following actions are needed:

1. To finalize the work on practical guide and make it available to interested bodies
2. To organize training on different levels: for health and safety authorities, for control bodies, and laboratories performing measurements
3. To set a control over the competence of laboratories performing measurements

The new regulatory developments concerning UV radiation in our country are related to the establishment of national policy and regulations concerning protection of the general public when using UV sources of artificial tanning. The future regulatory document should include at least the following issues: requirements for devices: spectrum, maintenance, safety precautions; electrical safety, protection from UV radiation; control over qualification and training of the consultants providing UV exposure services, so that they could provide accurate and adequate information to their customers; requirements for periodical control of level of UV radiation; requirements for premises and the installation of equipment; cases when UV exposure should be discouraged; introducing informed consent form for the use of sources of artificial tanning.

Some of these activities should follow the same procedure for radiofrequency fields as well, moreover that Member states will introduce Practical Guide for applying the Directive 2013/35/EC.

There are principal differences in the developing legislation for RF human exposure between the countries: different criteria and rationale for exposure limits, terminology, definitions of adverse and biological effect, long term and short term, nonthermal/athermal effects, safety factors, etc.

This led to a criticism to the proposed exposure limits by ICNIRP. The new recommendations of WHO and EC for legislation in Europe and other countries concerning NIR health and safety include exposure limits that are very debatable. ICNIRP guidelines propose exposure limits for EMFs in the frequency range up to 300 GHz based on short-term exposures and on thermal effects. Still many specialists in Russia, Italy, Poland, Switzerland, United Kingdom, China, Belgium, and also Bulgaria working in the field of developing exposure limits have criticized to the proposed ICNIRP guidelines philosophy. This led to big differences in the exposure limits.

We mentioned in 2002[5] several questions concerning the possibility of developing an international framework for standard and policy in the field of NIR protection. Some of them concern with the standard harmonization and others with terminology, rationale for developing exposure limits, long-term and low-level effects, and precautionary approach in policy.

The main conclusion of the discussion at that time was the need for developing methodological guide[9] consisting of terminological glossary for NIR exposure, guide for developing exposure limits, and guide for good quality of science. This is now the idea of SCENIHR (2013) for developing methodological guide and criteria for good science.

We hope that the new directive is good consensus for the protection of workers exposed to EMF in different occupations. The environmental approach also needs improvement; moreover, there we find low levels of exposure and wide range of emitting sources. Thus, the implementation of more stringent limits in some European countries could aid the harmonization process.

Our communication strategy includes all those parties involving in the process, as[11] follows:

- Mobile operators: during measurements, implementing base stations, discussing legislation, arranging regional meetings with people, administration
- Ombudsman: solving cases of concern of individuals
- Citizens: individual approach and in small groups; NGOs; spreading of brochures; regional meetings; hot telephone; website
- Government: ministry of health; NCPHA; control inspections (started to be active from about 2 years)
- Municipalities: in implementation of new objects, in resolving complaints jointly with the control inspections, ministry of health, and NCPHA
- Private companies: doing measurements of EMF levels
- NGOs: management of the process jointly with the government and industry
- Universities: education and training of specialists in the field of measurement, exposure and risk assessment, also for the control bodies
- *Hypersensitivity*: individual approach, case by case

All the activities are performed by NGO with the participation of NCPHA, control bodies, and ministry of health.

The following activities needs improvement (in national level): implementation of the changes in the health law and the ordinance, training of specialists, increasing competence of the control bodies, and implementation of periodical control and electronic systems for monitoring and reporting the EMF exposures.

REFERENCES

1. A non-binding guide to the artificial optical radiation, Directive 2006/25/EC, EC, catalog N: KE-30-10-384-EN-C.
2. Council Recommendation of 12 July 1999 on the limitation of exposure of the general public to electromagnetic fields (0 Hz to 300 GHz), 1999/519/EC.
3. Directive 2006/25/EC of the European parliament and of the council of 5 April 2006 on the minimum health and safety requirements regarding the exposure of workers to risks arising from physical agents (artificial optical radiation) (19th individual Directive within the meaning of article 16(1) of Directive 89/391/EEC).

4. Directive 2013/35/EU of 26 June 2013 on the minimum health and safety requirements regarding the exposure of workers to the risks arising from physical agents (electromagnetic fields) (20th individual Directive within the meaning of Article 16(1) of Directive 9/391/EEC) and repealing Directive 2004/40/EC.

5. EOARD Project: "Criteria for Standards in the Field of Radiofrequency Radiation in some East European Countries. Standard Harmonization Worldwide." Contract No. F61775, Project Leader M. Israel, 2002.

6. Framework for developing health-based EMF standards, WHO, Geneva, Switzerland, ISBN 9789241594332.

7. Guidelines for limiting exposure to time varying electric, magnetic and electromagnetic fields (up to 300 GHz), ICNIRP, *Health Physics*, April 1998, 74(4).

8. IARC. *Solar and Ultraviolet Radiation*, IARC Monographs on the Evaluation of Carcinogenic risks to Humans, Vol. 100, Part D: Radiation. Lyon, France: International Agency for Research on Cancer, 2012.

9. Israel, M. and P. Tchobanov, The need of methodological guide for developing standards, *Journal of Environmental Protection and Ecology*, 2005; 6(2):441–446.

10. Israel, M., V. Zaryabova, M. Ivanova, T. Shalamanova, and P. Ivanova. *Electromagnetic Fields (EMF) Risk and Policy in Bulgaria*, September 28–October 2, 2008, Città del Mare, Terrasini, Palermo.

11. Israel, M. and V. Zaryabova. National program for training in risk perception, risk communication and risk management as a policy of precautionary approach, *WHO Meeting on EMF Biological Effects and Standards Harmonization in Asia and Oceania*, October 22–24, 2001, Seoul, Korea, pp. 89–90.

12. Mild, K.H. et al. Occupational RF exposure from base stations antennae on roof-tops and buildings, *Workshop on Base Stations and Wireless Networks*, June 15–16, 2005, WHO, Geneva, Switzerland.

13. World Health Organization, *Model Legislation for Electromagnetic Fields Protection*, Geneva, Switzerland, 2006, ISBN 9789241594325.

14. Ordinance No. 5 on the minimum health and safety requirements regarding the exposure of workers to risks arising from exposure to artificial optical radiation, SG No.49/2010.

15. Ordinance No. 9 of 14 March 1991. TLVs for electromagnetic radiation in residential areas and for determining safety zones around electromagnetic sources, SG No. 35/1991.

16. Vecchia, P., M. Hietanen, B.E. Stuck Emilie van Deventer, and S. Niu (eds.). *Protecting Workers from Ultraviolet Radiation*, International Commission on Non-Ionizing Radiation Protection. In Collaboration with International Labour Organization World Health Organization, ICNIRP 14/2007, ISBN 978-3-934994-07-2.

17. Stranks, J.W. *The Handbook of Health and Safety Practice*, Ed. Pitman, 3rd edn, 1996.

21 Environmental and Safety Aspects of the Use of EMF in Medical Environment

Jolanta Karpowicz

CONTENTS

ELECTROMAGNETIC QUANTITIES AND CORRESPONDING
SI UNITS (SYMBOL—QUANTITY—UNIT)

H	Magnetic field strength, ampere per meter (A/m)
B	Magnetic flux density, tesla (T)
E	Electric field strength, volt per meter (V/m)
U	Voltage, volt (V)
I	Current, ampere (A)
σ	Conductivity, siemens per meter (S/m)
λ	Wavelength, meter (m)
ε	Permittivity, farad per meter (F/m)
f	Frequency, hertz (Hz)
t	Time, second (s)
SAR	Specific energy absorption rate, watt per kilogram (W/kg)

Submultiple and Multiple Units Applicable in the Environmental EMF Discussion

Prefix to Unit	Symbol	Submultiple or Multiple Meaning	
Nano	n	$\times 10^{-9}$	($\times$0.000,000,001)
Micro	μ (micro)	$\times 10^{-6}$	($\times$0.000,001)
Milli	m	$\times 10^{-3}$	($\times$0.001)
—	—	$\times 10^{0}$	($\times$1)
Kilo	k	$\times 10^{3}$	($\times$1,000)
Mega	M	$\times 10^{6}$	($\times$1,000,000)
Giga	G	$\times 10^{9}$	($\times$1,000,000,000)

21.1 INTRODUCTION

The electrodynamic effects of exposure to electromagnetic fields (EMFs) are used in medicine for therapeutic or diagnostic purposes. The devices used for such applications may emit EMF significantly stronger than typical environmental exposure and cause intentional exposure to patients undergoing treatment, as well as unintentional exposure to physiotherapists, surgeons, or radiographers, patients not undergoing treatment, and anyone visiting a medical center, as well as any electronic equipment present nearby, including medical implants (Hansson Mild et al. 2009, ICNIRP 1998).

The effects of electromagnetic exposure of each of these subjects may influence the human body and electronic devices and may cause various health or safety hazards, such as excessive thermal effect, electrostimulation of tissues, projectile ferromagnetic objects, or electronic device malfunctions. Safety risks caused by EMF depend mainly on the field frequency, strength, and spatial distribution, as well as the characteristic and configuration of the exposed subjects (ICNIRP 1998, Karpowicz and Gryz 2007, Reilly 1998).

Usually, the main attention is focused on the safety of patients undergoing medical procedure, for example, the use of electronic implants by patients was identified as a contraindication for various physiotherapeutic treatments. However, the attention is also needed for the proper identification of other mentioned EMF-related hazards in the medical environment near to any applicators emitting EMF during the patient's treatment. The consecutive sections discuss these topics without considering the safety of a patient who is the subject of EMF treatment.

21.2 ELECTRODYNAMIC EFFECTS OF EMF EXPOSURE

The health and safety hazards caused by EMF exposure are related to the results of direct electrodynamic interaction between the electromagnetic energy and exposed objects (the human body as well as any other object). The EMFs are characterized by the electric field strength (E) and magnetic field strength (H). The direct interaction of EMF with the human body causes the direct effects of EMF exposure, whereas indirect exposure effects also involve interactions with other objects that, as a result of EMF interaction, have an electric potential different from that of the body.

21.2.1 DIRECT EXPOSURE EFFECTS: INDUCED ELECTRIC FIELD AND ELECTRIC CURRENT, ELECTROMAGNETIC ENERGY ABSORPTION

The dominant mechanisms of electrodynamic interaction between EMF and any electrically conductive object include (IARC 2002, 2013, ICNIRP 1998, Reilly 1998)

- Electric fields induced in an object exposed to the time-varying electric fields, which cause a flow of electric current, the formation of electric dipoles, and the reorientation of electric dipoles already present, for example, in human tissue
- Circulating electric fields and electric currents, recognized as the eddy currents, induced in an object exposed to time-varying magnetic field

The relative magnitudes of different exposure effects depend on the ratio between electric and magnetic field strengths, a spatial distribution of conductivity (σ) and permittivity (ε) of exposed object—which depend on the frequency (f) of the applied EMF and the material composition of the object (e.g., vary with the type of body tissue)—and depend on the conditions of exposure, that is, on the size and shape of the object/body and on the object's/body's position in the field and on affecting EMF distribution in space and time (ICNIRP 1998). The frequency of EMF plays a dominant role in the EMF mechanism of coupling with the exposed object. For example, the magnitudes of the induced current density are proportional to the radius of the conductive loop in the exposed object, the electrical conductivity of the object (i.e., of the tissue), and the rate of change in magnitude of the magnetic flux density (i.e., the frequency composition of magnetic field variability over time). The special case of exposure to a time-varying magnetic field occurs when an object (e.g., a human body) is changing the location against the static magnetic field (SMF) source, which creates a variability over time of the magnetic field affecting an individual.

The spatial distribution of EMF energy absorbed in the human body depends to a large extent on the frequency—in low-frequency EMF, in practice, only the magnetic field may penetrate the body, and in the kHz- and low-MHz-frequency EMF, the internal absorption increases. In the frequency of the range 40–100 MHz, the highest energy deposition in the human body occurs because of the phenomenon of the whole-body resonance absorption (because the electric dimensions of the body, depending on the posture and grounding conditions, may fit with the length (λ) of the electromagnetic wave—3.0–7.5 m in that frequency range). At higher frequencies, the EMF deposition in the body becomes spatially nonuniform, and above a few GHz, energy absorption occurs primarily at the body's surface (IARC 2013, ICNIRP 1998).

21.2.2 INDIRECT EXPOSURE EFFECTS: CONTACT CURRENTS, TRANSIENT DISCHARGES, FERROMAGNETIC FLYING OBJECTS, MEDICAL IMPLANTS

The indirect hazards of EMF exposure concerning the individuals in the proximity of conductive structures include the following (IARC 2002, 2013, ICNIRP 1998, Reilly 1998):

- Contact currents (*I*) when anyone comes into contact with an exposed object at a different electric potential (i.e., when either the body or the object is charged by absorbed EMF energy) or when the human body and objects exposed to the magnetic field create a conductive loop—the volume of contact current depends not only on the structure of exposed objects but also on the frequency and polarization of the EMF and is elevated when the object is in resonance with the electromagnetic wave.
- Transient discharges (sparks) when an individual comes in close proximity to a conducting object exposed to a strong static or time-varying electric field.
- Direct coupling of EMF energy with medical electronic devices—body worn or implanted (active implantable medical devices [AIMDs])—or contact currents/sparks affecting an individual creating a flow of electric currents in AIMD.

Moreover, a spatially heterogeneous SMF (where so-called gradient dB/dx is not equal to zero) exerts translational force on all ferromagnetic objects and rotational force on elongated ones. The magnet may pull the object as it moves toward it with increasing speed along the surface where these forces exceed frictional resistance (depending on the shape and mass of an object as well as its surface and the type of ground it stands on) or turned the object on its long axis according to the polarity of SMF (similar to a compass needle). When gravitational forces are also balanced, such a pulled object may levitate with rapidly increasing speed toward the magnet and become to be projectile object called also flying object (ICNIRP 2009, Karpowicz and Gryz 2013b).

21.3 EMF-RELATED HAZARDS

These interactions of EMF with electrically conductive objects may cause various effects when a high-level EMF influence the human body—recognized as short-term effects occurring during (or immediately after) exposure. The dominant biophysical effects, the most important in the sense of prevention of health and safety hazards, are determined mostly by the frequency of EMF.

21.3.1 Electrostimulation

The electric field induced in the human body by exposure to low-frequency EMF causes the cell membrane's polarization, which may result in the alteration of the cellular membrane's natural resting potential and a depolarization of nerve and muscle membranes. It can lead to their excitation (recognized as electrostimulation). The human body reactions caused by electrostimulation by EMF with a frequency lower than 100 kHz include aversive or painful stimulation of sensory or motor neurons, muscle excitation, excitation of neurons or the direct alteration of synaptic activity within the brain, and cardiac excitation (ICNIRP 2010, Reilly 1998). A special case of electrostimulation caused by exposure to time-varying EMF (with a frequency lower than 10 Hz) is the exposure during movements of the body or eyes near the SMF source. The symptoms reported by individuals affected by such exposure—during rapid movements in SMF exceeding 1 T and even during slow movements in stronger fields—include vertigo, difficulty with balance and hand–eye coordination, nausea, headaches, numbness and tingling, phosphenes, and unusual taste sensations (Jokela and Saunders 2011, Wilén and de Vocht 2011). The levels of thresholds of EMF exposure causing particular effects of electrostimulation are frequency dependent, for example, it is in the millitesla range in the case of magnetic fields with a power frequency of 50/60 Hz.

Exposure to SMF also causes magnetohydrodynamic effects in moving charges in fluids, which may influence the blood flow at many-tesla level of exposure (ICNIRP 2009, Jeheson et al. 1988, Keltner et al. 1990, Kinouchi et al. 1996, Tenforde 1992).

When workers are involved in potentially hazardous activities, this perception of electrostimulation in the body caused by the influence of EMF may cause work imperfections or accidents, and danger to the workers themselves, along with patients or bystanders. The prevention of adverse effects associated with rapid movements and electrostimulation perception within strong SMF is required in lower fields than the prevention of blood flow changes caused by magnetohydrodynamic effects (ICNIRP 2009, WHO 2006).

21.3.2 THERMAL EFFECTS

The exposure to EMF with a frequency exceeding 100 kHz does not cause electrostimulation, which significantly reduces the ability of individuals to recognize dangerous levels of environmental EMF and dangerous thermal effects of absorbing electromagnetic energy in the body. The thermal effects of EMF exposure become more significant than electrostimulation when the frequency of EMF exceeds 100 kHz, but they are dominant in the MHz (localized radiofrequency [RF] heating) and GHz (surface microwave heating) frequency ranges (ICNIRP 1998).

Thermal effects in the human body may be caused not only by direct EMF influence and the absorption of electromagnetic energy by tissues, but also because of contact or induced current flow when individual is touching an object exposed to EMF. The level of thermal effects is determined by EMF exposure patterns, as well as the geometrical and electrical structures of the environment. In the MHz frequency range, thermal effects in the human body causing a pain perception or tissue burns (inside the body or at the surface) may be created either when an individual touches an ungrounded metal object charged by the EMF influence or when a charged ungrounded individual (e.g., by insulating shoes) touches a grounded metal object. For example, the direct thermal results of EMF exposure need attention in the vicinity of sources emitting continuous EMF from the MHz frequency range at a level of several hundred volts per meter, when a contact current needs attention already in that EMF at the level of dozens of volts per meter-depend on the frequency and environment configuration (Hill and Walsh 1985, ICNIRP 1998, Jokela et al. 1994, Korniewicz 1995).

The thermal effects and contact current perception may initiate hazardous situations and accidents involving workers active in the proximity of exposed structures and may also cause the hazards of RF burns in patients.

21.3.3 ELECTRONIC DEVICE MALFUNCTIONS

The function of sensitive electronic devices (such as AIMD or medical diagnostic devices) may be influenced by the induced voltages and currents that occur in electrically conductive structures affected by EMF, spark discharges, and the direct influence of SMF. The kind of changes in the function of exposed devices, and even the risk of their damage, depends on the magnitude and frequency of EMF, along with the geometrical and electrical structures of conductive structures receiving EMF energy. Any uncontrolled interaction between electronic devices and the human body may create health and safety hazards, especially in highly sensitive relations between humans and any AIMD (such as cardiac pacemakers, implantable cardioverter defibrillators, cochlear implants, implantable neurostimulators, implantable infusion pumps, or other types of implantable medical devices), which is supporting a vital function of individual (Carranza et al. 2011, Hocking and Hansson Mild 2008).

Observations of AIMD and other medical device dysfunctions caused by EMF exposure reported in the research papers include, for example, the destruction of cochlear implants during the surgery of the user by an electrosurgical unit (electrosurgical unit [ESU]; the influence of EMF of 0.3–1 MHz frequency), hearing distorted sound when the cochlear implant user is passing near or through an antitheft gate (the influence of EMF of 20 kHz to 20 MHz frequency) and metal detection gates (the influence of EMF not exceeding a kHz frequency range), an overdose of insulin by infusion pumps affected by mobile phones or RF identification devices (influence of EMF of 0.8–2.4 GHz), changes in the operating rate of cardiac implants; spontaneous uncontrolled trigger of implanted defibrillators and dysfunctions of neurostimulators, accidentally setting on an alarm in a medical device, disrupting the sound or image generated by the device, and stopping or resetting or disrupting the transmission between a terminal and its central unit (Calcagnini et al. 2006, Carranza et al. 2011, Christe 2009, Pantchenko et al. 2011, Tognola et al. 2007). A dysfunction of electronic devices that include magnetic memory or battery may also be caused by SMF exposure.

The sensitivity of particular electronic devices to the EMF influence varies significantly by the type and the mode of operation of devices, which require individual evaluation of hazards to a user exposed to EMF. It needs attention that electronic devices may be much more sensitive to EMF exposure than healthy people and that compliance with safety guidelines referring to exposure to workers and the general public may not preclude the EMF interference with AIMD.

21.3.4 FLYING OBJECTS

The pull of ferromagnetic objects toward an SMF source (magnet or electromagnet) constitutes the most crucial hazard for human health and safety and even hazards for the integrity of the device containing magnet itself. In an SMF near the strong magnets (producing SMF of the tesla range), various ferromagnetic objects (such as paperclips, keys, tools, oxygen tanks, furniture) may be moved or even levitate toward the magnet due to the force exerted by the SMF, behaving as bullets flying into the SMF source. Therefore, a striking by a moving ferromagnetic object may not only cause damage to the device containing the SMF source, and to other devices present nearby, but also lead to serious bodily harm, or even death, of anybody who finds themselves between the magnet and the moving object (Chaljub et al. 2001, Chen 2001, Duck 2010, ICNIRP 2009, MRI safety, WHO 2006). Such hazards are recognized as ballistic hazards, projectile hazards, or *flying objects*. The susceptibility of an object to SMF depends on its magnetic parameters, determined mainly by the chemical composition, production method, and processing during construction, so it may vary in time and requires an individual assessment (Karpowicz and Gryz 2013b). Problems caused by *flying objects* may happen in SMF exceeding several millitesla and do not concern objects made out of diamagnetic and paramagnetic materials (Table 21.1).

21.3.5 DELAYED HEALTH HAZARDS RELATED TO OVEREXPOSURE OR CHRONIC EXPOSURE

These short-term effects of EMF exposure and health or safety hazards caused by them are well investigated by the laboratory and field studies. On the contrary, the discussion regarding delayed health hazards, which may be linked to the results of chronic (many-years) exposure or the delayed results of overexposure (single or repetitive), still remains controversial—but they are not excluded. The studies on delayed health effects are much more complex and take longer (even over many decades). The possible mechanisms of such chronic/delayed EMF interactions with the human body are discussed, among other things, in connection with hypotheses related to increased cases of cancer diseases, reproductive or immune system effects, nervous system effects, etc., among exposed

TABLE 21.1

Examples of Ferromagnetics, Paramagnetics, and Diamagnetics

	Ferromagnetics	Paramagnetics	Diamagnetics
Material	Pure iron	Air	Gold
	Technical iron	Aluminum	Silver
	Nickel	Titanium	Copper
		Vacuum	Water

Source: Karpowicz, J. and Gryz, K., *Pol. J. Radiol.*, 78, 31, 2013b.

Note: Stainless steel objects may have ferromagnetic properties or not, and they need individual assessment of potential hazards near SMF source.

population (Bortkiewicz et al. 2012, European Parliament 2013, Hansson Mild et al. 2013, Prato et al. 2010, Vesselinowa 2013, WHO 2006). At present, the limited evidence provided by studies regarding health hazards caused by EMF exposure is summarized, for example, by the International Agency of Research on Cancer (IARC), which decided to evaluate the extremely low-frequency magnetic fields and the RF EMFs to be possibly carcinogenic to humans (Group 2B),* but static electric and magnetic fields and extremely low-frequency electric fields are not classifiable as to their carcinogenicity to humans (Group 3)—the review of examined evidence and rationale for IARC evaluations have been published in monographs No. 80 and 102 (IARC 2002, 2013). As long as such mechanisms of delayed EMF effects cannot be dismissed as being irrelevant, progress in research should be monitored and prevention for EMF exposure limitation should be applied, especially for individuals exposed almost every day like workers.

21.4 MEDICAL APPLICATORS EMITTING RELATIVELY STRONG EMF

The electrodynamic effects of EMF energy absorption may be used to assist the intentional biomedical influence on the human body in therapeutic or diagnostic interactions, such as posttraumatic rehabilitation, degenerative disease treatment, pain reduction, medical imaging involving the magnetic resonance phenomenon, or soft tissue surgery involving a localized EMF energy deposition (Hansson Mild et al. 2009, ICNIRP 1998, Robertson et al. 2006). These medical applications need a relatively strong electromagnetic influence on the patient's body during the use of short-wave diathermy devices, microwave diathermy devices, long-wave diathermy devices, ultrasound therapy units (USTUs), magnetotherapeutic applicators (MTAs), electrosurgery units (ESUs), and magnetic resonance imaging (MRI) scanners. The main principles of the function of particular devices are presented in further sections.

21.4.1 SHORT-WAVE DIATHERMY DEVICES

Physiotherapeutic devices recognized as short-wave diathermy devices (SWDDs) are widely used, for example, for pain reduction caused by the localized thermal effects of EMF exposure (Figure 21.1a). The SWDD therapy involves the use of capacitive (two dipole electrodes) or inductive (single emitting coil) applicators, energized by regulated electric signals delivered by supplying cables from a generator with an output power in the range from several watts to several hundreds of watts (up to over 1 kW in the pulsed-modulated [PM] mode of use) (Gryz and Karpowicz 2014, Kalliomäki et al. 1982, Robertson et al. 2006, Sarwer and Farrow 2013). The capacitive applicators are usually supplied by a continuous wave (CW) signal, whereas the inductive applicators are rather supplied by a PM signal. The most popular frequency of electric signals supplying SWDD applicators is 27.12 MHz (in Europe), but other frequencies also may be used, for example, 13.56 and 40.78 MHz (depending on the administrative decisions in particular countries on the permitted use of electromagnetic spectrum for medical applications).

21.4.2 MICROWAVE DIATHERMY DEVICES

The area of use of physiotherapeutic devices recognized as microwave diathermy devices (MWDDs) is similar to that of SWDD (Figure 21.1b). MWDD therapy involves the use of a single applicator (an emitting dipole electrode) for localized treatment, powered by a regulated CW or PM electric

---

* In IARC methodology of evaluation of the carcinogenicity of the agent to humans, the following definitions were applied (IARC 2002, 2013):

Group 2B—there is limited evidence of carcinogenicity in humans and less than sufficient evidence of carcinogenicity in experimental animals, or there is inadequate evidence of carcinogenicity in humans, but there is sufficient evidence of carcinogenicity in experimental animals.

Group 3—the evidence of carcinogenicity is inadequate in humans and inadequate or limited in experimental animals.

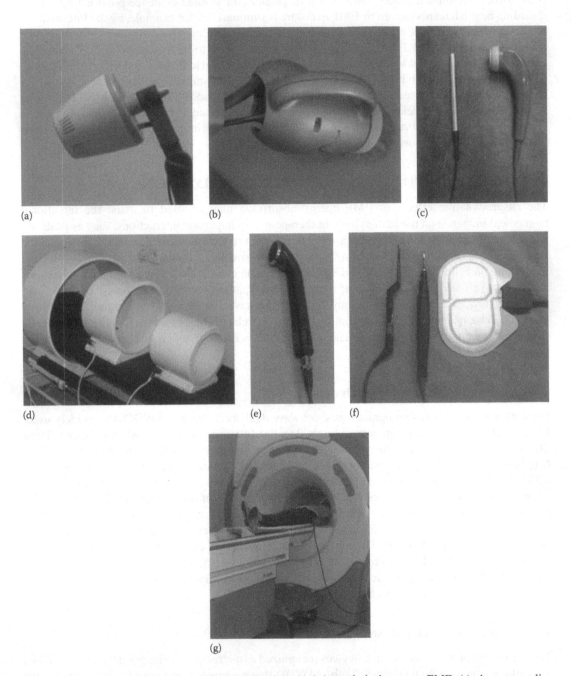

FIGURE 21.1 Examples of the medical applications emitting relatively strong EMF: (a) short-wave diathermy device (SWDD) inductive applicator; (b) microwave diathermy device (MWDD) applicator; (c) long-wave diathermy device (LWDD) applicators; (d) magnetotherapeutic applicators (MTAs); (e) ultrasound therapy unit (USTU) applicator; (f) electrodes of electrosurgery unit (ESU); (g) magnetic resonance imaging scanner (MRI). (Photo courtesy: author's collection.)

signal delivered by coaxial cable from a generator with an output power in the range from several watts to several hundreds of watts (up to over 1 kW in the PM mode of use) (Robertson et al. 2006, Sarwer and Farrow 2013). The most popular frequencies of electric signals supplying MWDD applicators are 434, 915, and 2450 MHz.

21.4.3 LONG-WAVE DIATHERMY DEVICES

Another type of diathermy device uses a capacitive current flowing between electrodes—the passive one grasped by the patient and the active one held by the operator during the treatment. Such devices are called long-wave diathermy devices (LWDDs) because of the use of signals with a frequency of approx. 1 MHz (Figure 21.1c). The LWDD electrodes are powered by a regulated CW or PM electric signal delivered by cables from a generator with an output power of the order of 1 W at 100 Ω load (Gryz and Karpowicz 2014, Robertson et al. 2006).

21.4.4 MAGNETOTHERAPEUTIC APPLICATORS

The most popular MTAs are constituted by many-turn spool coils with diameters that fit the dimensions of the human body under treatment—20–60 cm (Figure 21.1d). Such applicators allow the magnetic field to penetrate into the patient's body, and in the case of time-varying magnetic field application, it induces electric field and current in patient. MTAs are supplied by low-voltage and low-current CW or modulated electric signal of a regulated level, frequency, and shape (e.g., sinusoidal or rectangular)—alternating or rectified (Karpowicz et al. 2009, Robertson et al. 2006). As a result, the magnetic field—alternating or alternating combined with an SMF component—is applied for physiotherapy. The most commonly applied magnetic field has a frequency not exceeding 100 Hz and a magnetic flux density not exceeding 20 mT, though applicators emitting a magnetic field in the kHz frequency range may also be used.

21.4.5 ULTRASOUND THERAPY UNITS

The use of ultrasound energy absorption in tissues for physiotherapeutic treatment involves piezoelectric converters built in the applicators of USTUs, and they are supplied by a regulated CW or PM high electric voltage with a frequency in the range of 1–3 MHz (Figure 21.1e). As a result, ultrasound therapy also involves localized exposure to EMF of that frequency, which is present near the applicator and cable from a generator with an output power ranging from several watts to over a dozen (Gryz and Karpowicz 2014).

21.4.6 ELECTROSURGERY UNITS

ESUs, recognized also as electric knifes, are used for soft tissue cutting (CUT mode) or blood coagulation (COAG mode) during surgical treatments (traditional or laparoscopic), in order to reduce blood loss (Figure 21.1f). Such applications involve tissue heating, caused by the flow of an electric capacitive current (which passes through the air and tissues) between active and passive electrodes, supplied by an RF (usually 0.3–2.0 MHz) at high voltage (exceeding 200 V—in the range of 200–12,000 V peak-to-peak in various applications) delivered by cables from an ESU generator (de Marco and Magi 2006, Eggleston and Von Maltzahn 2000, Karpowicz et al. 2013a, Liljestrand et al. 2003, Schneider and Abatti 2008, Wilén 2010). In order to reduce leakage currents from cables and the active electrode, both the handle of the active electrode held by surgeons and the cables are insulated. The output power of the generator suits various modes of using the ESU—from several watts up to a couple of hundred (approximately 200) watts when using a unipolar active electrode, but up to significantly less than 100 W when using a bipolar electrode. In addition, different time waveforms are used—in CUT mode, usually pure sinusoidal or quasisinusoidal CWs supply the electrode, whereas in COAG modes, amplitude/PM waves are used (with a crest factor U_{peak}/U_{RMS} of 2–20).

21.4.7 MAGNETIC RESONANCE IMAGING SCANNERS

Among medical imaging diagnostics technique is MRI, performed in devices recognized as MRI scanners (Figure 21.1g). Images of the internal structure of a patient's tissues can be created when a patient is affected by SMF and time-varying EMFs. At present, the most commonly used MRI scanners worldwide are equipped with superconductive magnets (active permanently—i.e., 24 h/day) with the main field in the magnet's bore of magnetic flux density 1.5 or 3 T (Hansson Mild et al. 2013, Karpowicz and Gryz 2013a). The examined patient lies on the MRI table, located in the bore of the magnet housing, and is exposed to PM RF EMF (with a frequency of approx. 63 MHz in a 1.5 T magnet or approx. 126 MHz in a 3 T magnet) and pulsed magnetic *gradient* fields (with a frequency spectrum of approx. 0.5–10 kHz) generated in sequences of merged RF and gradient pulses by diagnostic and gradient coils supplied by generators with an output power ranging from several kilowatts to several dozen kilowatts. The magnet of the MRI scanner is located in an electromagnetically shielded chamber (MRI treatment room).

21.5 CHARACTERISTICS OF EMF HAZARDS NEAR MEDICAL APPLICATORS

The electromagnetic environmental impact of devices and installations emitting EMF is usually characterized by the distribution in space and in time of the electric field strength (E) and magnetic field strength (H). Electric and magnetic field strengths are indirect measures of EMF hazards arising from thermal effects and electrostimulation in exposed humans, caused by electric fields and currents induced in tissues, as well as from interference in an electronic device operation caused by electromagnetic influence or hazards from projectile/flying ferromagnetic objects (Hansson Mild et al. 2009). They may be evaluated by measurements or virtual investigations (computer simulations in numerical models of EMF sources and environment).

The degree of exposure effects in the body is also directly evaluated using parameters known as the specific energy absorption rate (SAR) and the strength of the induced electric field (E_{in}) calculated using virtual human body models (phantoms) (Zradziński 2013). The results of such virtual investigations can be used not only for visualizing or verifying the biophysical results of a patient's EMF exposure but also for assessing the results of unintentional exposure in the body of healthcare personnel or other bystanders present near an active medical device. SAR measurements can be performed only under laboratory conditions.

The electric current in the limbs of anyone who is present near an EMF source or is touching objects exposed to EMF allows an indirect evaluation of the hazards related to the strongest localized thermal effects in limbs (as an alternative to the assessment of local *SAR* in the limbs) (Karpowicz and Gryz 2010). It can be evaluated by measurements or virtual investigations.

The EMF measurements summarized in this section were carried out near medical devices of internationally recognized brands under routine conditions of use and settings. The broadband meters, equipped with isotropic probes for measuring the root-mean-square (RMS) values of electric and magnetic field strengths, were used. The current flowing into the arms of the individuals touching EMF RF applicators, cables, or generator housings was measured using a meter showing the RMS value of electric current in the limb. The sensitivity of used devices enabled a reliable assessment of the environmental impact near the medical devices, and the measurement uncertainty did not exceed 20%, in accordance with relevant requirements (Hansson Mild et al. 2009). The measurement devices were also tested in an accredited calibration laboratory CIOP-PIB (accreditation certificate from Polish Centre for Accreditation No AP 061). The measurements of electric and magnetic field strengths are a routine test of the environmental impact of devices used in the workplace, whereas limb currents are measured rather for research purposes.

21.5.1　EMF Hazards near Physiotherapeutic Devices

The unintentional dispersion of EMF from all the physiotherapeutic applicators (of SWDD, LWDD, MWDD, USTU, and MTA) and leakage from cables and generators supplying RF applicators (of SWDD and LWDD) during a patient's treatment cause a common EMF exposure nearby—of not only the patient undergoing physiotherapeutic treatment (intentionally) but also of other individuals and objects present nearby (unintentionally) (Gryz and Karpowicz 2014, Karpowicz and Gryz 2009, 2013b). The level of EMF rapidly drops as the distance from the applicators increases (Figure 21.2, Tables 21.2 and 21.3).

Regular physiotherapeutic treatments do not usually involve an operator assisting the patient. All workers employed in the treatment center (including operators of physiotherapeutic devices, nurses, physicians, masseurs, and administrative personnel) may stay away from the active physiotherapeutic applicators—at a distance depending on the spatial organization of the treatment room and other adjacent rooms, even several meters away—where EMF may be counted as negligible. But if the treatment center is not organized properly, where, for example, the administrative personnel stay on the opposite side of the treatment box, this distance may be even shorter than 1 m. Such a situation may cause a high level of worker exposure because popular lightweight nonconductive walls (e.g., plaster-board modules) do not screen RF EMF and low-frequency magnetic fields. The space where patients are waiting for any treatment (EMF-related or other) and where patients are undergoing other treatments (unrelated to the EMF use) may also be located in the vicinity of active applicators—which may cause EMF exposure to the workers doing such treatments, for example, massage, as well as patients who are not intentionally exposed to EMF. Near to active applicators, even in adjoining rooms, secondary EMF sources may be created by metal objects emitting induced EMF, such as furniture, water supply or a central heating installation. The EMFs near the secondary sources are usually many times weaker than those from the primary ones, but in the vicinity of strong EMF primary sources (e.g., SWDD), they may also influence sensitive subjects (e.g., AIMD). The operators of physiotherapeutic devices may also be exposed to EMF because of modifications to the settings of the treatment parameters or the location of the applicator by the patient's body while carrying out the treatment. Sometimes, a healthcare worker or patient's attendant (e.g., a relative) also has to stay near disabled patients throughout the treatment.

The resulting EMF exposure at a small distance from active devices may take place over the whole duration of treatment (with an order of 10–20 min/patient) or while passing by (a fraction of minute only). Exposure in the meaning of limb current load (flow in the body) is caused only by touching active applicators, cables, or generator housings of SWDD, LWDD, and USTU (Table 21.4).

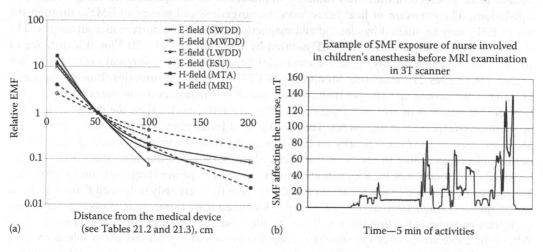

(a) Distance from the medical device (see Tables 21.2 and 21.3), cm

(b) Time—5 min of activities

FIGURE 21.2 EMF near medical devices: (a) normalized distribution of EMF emitted by MTA, SWDD, ESU, and MRI; (b) movement-related variability of SMF exposure of healthcare worker attending MRI patient (Gryz and Karpowicz 2014, Karpowicz et al. 2009, 2011, 2013a,b, Karpowicz and Gryz 2013a,c).

TABLE 21.2

RMS Value of Magnetic Field Strength near Medical Devices (Gryz and Karpowicz 2014, Karpowicz et al. 2009, 2011, 2013a,b, Karpowicz and Gryz 2013a,c)

Place of Measurements— Distance from the Medical Devices (cm)	Electric Field Strength E (V/m)						
	Kind of Device						
	SWDD (27 MHz)	MWDD (2450 MHz)	LWDD (500 kHz)	MTA (50 Hz)	USTU (500 kHz)	ESU (500 kHz)	MRI (1.5 T and 3 T)
10	760–1160	90–250	170–750	<s	4–10	160–320	<s
50	85–110	30–100	20–60	<s	<2	13–27	<s
100	12–30	7–50	5–20	<s	<1	<3	<s
200	2–15	4–20	<2	<s	<s	<s	<s

Notes:

<s—not exceeding the sensitivity of measurement device (0.4–1 V/m).

SWDD, short-wave diathermy device/distance from applicators and supplying cables; MWDD, microwave diathermy device/distance from applicators; LWDD, long-wave diathermy device/distance from applicators and supplying cables; MTA, magnetotherapeutic applicators/distance from applicators; USTU, ultrasound therapy unit/distance from applicators and supplying cables; ESU, electrosurgery unit/distance from electrodes and supplying cables; MRI, magnetic resonance imaging scanner/distance from the edge of magnets bore.

Electric field near housing of supplying generators of SWDD, MWDD, LWDD, MTA, USTU, and ESU does not exceed ICNIRP's general public exposure limits, except in the case of its dysfunction in electromagnetic shielding (e.g., when it is from insulating plastic material or it is mechanically damaged).

Electric field near physiotherapeutic applicators for direct (resistive coupling) application of electric currents and their supplying cables and generators does not exceed sensitivity of measurement devices (0.4–1 V/m).

21.5.2 EMF HAZARDS NEAR ELECTROSURGERY UNITS

The use of high voltage for supplying surgical electrodes causes the emission of EMF, mainly from the active electrode and cables (3–5 m long) connecting both electrodes with ESU generator. EMF emitted by ESU may be high enough that the various measures of worker exposure needed to be assessed in order to evaluate compliance with occupational safety guidelines or legislations. The exposure of healthcare workers, surgeons, and nurses to EMFs, through the use of ESU, may be studied by electric and magnetic fields or limb current measurements. The electric field is dominating in the EMF emitted by ESU (up to 150–650 V/m at a distance of 10 cm from the cable supplying active electrode) because usually a surgical current does not exceed 1 A (Tables 21.2 and 21.3). Monitoring of ESU work during surgeries shown their use in the range from several up to longer than 50% of time over starting several minutes of surgery, but later on not longer than 10% of surgery time—significantly shorter than the duration declared by workers (Karpowicz et al. 2013a). The electric field induced in each electrically conductive object exposed to the E-field nearby an ESU causes the flow of capacitive currents and heating in this object—including the body of the worker. It must be noted that almost all materials are better electric conductors for RF currents than in the case of power frequency ones (50/60 Hz). The spatial distribution of healthcare workers' exposure to a variously polarized E-field depends on the spatial configuration of the ESU cables—localized hand exposure is always present when a surgeon or nurse is holding the handle of the electrode or the supplying cable (Figure 21.2). Where the cable hangs over the shoulder along the torso and leg, almost the whole body may be exposed. However, the most typical position of workers involved in surgical treatment is the first one—where the hand receives the highest EMF exposure and capacitive current (Karpowicz et al. 2013a,e).

TABLE 21.3

RMS Value of Magnetic Field Strength near Medical Devices (Gryz and Karpowicz 2014, Karpowicz et al. 2009, 2011, 2013a,b, Karpowicz and Gryz 2013a,c)

Place of Measurements— Distance from the Medical Devices (cm)	Magnetic Field Strength H						
	Kind of Device						
	SWDD (27 MHz) (A/m)	MWDD (2450 MHz) (A/m)	LWDD (500 kHz) (A/m)	MTA (50 Hz) (mT)	USTU (500 kHz) (A/m)	ESU (500 kHz) (A/m)	MRI (1.5 T and 3 T) (mT)
10	0.7–3.0	0.3–0.7	0.3–0.4	0.45–1.1	0.02–0.05	0.1–0.4	470–1200
50	0.05–0.3	0.1–0.3	0.02–0.1	0.04–0.05	<s	<s	120–290
100	0.02–0.1	0.02–0.15	<s	<0.008	<s	<s	25–60
200	<s	<s	<s	<0.002	<s	<s	3–7

Notes:

<s—not exceeding the sensitivity of measurement device (0.02 A/m).

SWDD, short-wave diathermy device/distance from applicators and supplying cables; MWDD, microwave diathermy device/distance from applicators; LWDD, long-wave diathermy device/distance from applicators and supplying cables; MTA, magnetotherapeutic applicators/distance from applicators; USTU, ultrasound therapy unit/ distance from applicators and supplying cables; ESU, electrosurgery unit/distance from electrodes and supplying cables; MRI, magnetic resonance imaging scanner/distance from the edge of magnets bore.

Magnetic field near housing of supplying generators of SWDD, MWDD, LWDD, MTA, USTU, and ESU does not exceed the sensitivity of measurement device (0.02 A/m).

Magnetic field near physiotherapeutic applicators for direct (resistive coupling) application of electric currents and their supplying cables and generators does not exceed the sensitivity of measurement devices (0.02 A/m).

Electric capacitive current in an exposed worker may be caused by direct interaction between the EMF and the body (known as *induced current*). It may also be caused by an indirect interaction of the EMF with electrically conductive objects, usually metal, along with current flow when touching such objects (known as *contact current*). Both are of special concern in limbs. These currents produce thermal effects in the human body and may lead to increased temperature and thermal damage in tissue (called RF burns) or pain sensations in the contact area where nerve excitation occurs (called electric shock). Measured values of upper limb current varied significantly (in the range from 1 to 58 mA in the case of holding the active electrode handle and 1 to 100 mA in the case of holding the cable supplying electrode) (Table 21.4). Statistically significant differences were found between the old-fashioned and the modern subsets of investigated ESU. Further studies are needed to assess the scale of possible localized overexposure inside the fingers and palm and the possible health consequences of such exposure effects.

21.5.3 EMF HAZARDS NEAR MRI SCANNERS

During examination, usually only a patient in the MRI chamber is being exposed to SMF and RF and gradient EMF. Healthcare personnel remain with a patient only in special cases, and such complex exposure was discussed elsewhere (Karpowicz and Gryz 2006, 2013a, Karpowicz et al. 2007). However, it needs to be underlined that electronic devices that will remain in the MRI chamber need sufficient immunity against all components of EMF used during the examination. The computer console used to control diagnostic process and data acquisition is usually situated out of the MRI chamber, away from the EMF emitted by the MRI scanner. Consequently, during the course of a patient's examination, the healthcare personnel, who are usually present in front of the computer console, are not affected by EMF.

TABLE 21.4

RMS Value of Upper Limb Current While Touching Active Medical Devices
(Gryz and Karpowicz 2014, Karpowicz and Gryz 2013c, Karpowicz et al. 2013b)

	Electric Current (mA)			
	Kind of Device			
Measurement Conditions	SWDD	LWDD	USTU	ESU
When touching the control panel	3–120	2–4	<s	<s
When touching the treatment electrode of SWDD, LWDD, or USTU/the handle of active electrode of ESU	23–560	5–6	<s	1–58 (6±9)
When touching cable supplying electrodes	13–780	4–6	<s	1–100 (8±13)

Notes:

<s—not exceeding the sensitivity of measurement device (1 mA).

SWDD, short-wave diathermy device/distance from applicators and supplying cables; LWDD, long-wave diathermy device/distance from applicators and supplying cables; USTU, ultrasound therapy unit/ distance from applicators and supplying cables; ESU, electrosurgery unit/distance from electrodes and supplying cables.

Magnetic field near housing of supplying generators of SWDD, MWDD, LWDD, MTA, USTU, and ESU does not exceed the sensitivity of measurement device (0.02 A/m).

(Arithmetic mean ± Standard deviation), based on over 250 measurements.

During the other activities of healthcare personnel, a time-varying EMF is not emitted. Therefore, the highest attention is needed for SMF exposure, which occurs during any activity near the MRI magnet—active permanently, 24 h/day, and every day. The spatial distribution of SMF near the magnet is highly heterogeneous; the strongest field exists in front of the bore of the magnet where the patient is moved for scanning (Figure 21.2, Table 21.3). This is why healthcare personnel attending patients between examinations (radiographers) are always exposed to SMF and may receive more than several minutes of exposure to strong SMF with rapid changes in time caused by movement (Figure 21.2) (Karpowicz 2009, Karpowicz et al. 2011, Karpowicz and Gryz 2013a). The most typical tasks during such activities include attending patients before and after the examination while the patient accesses the MRI chamber and lies down on the MRI table or gets off from the MRI table; positioning the diagnostic coils on the MRI table or the patient body; plugging-in/ unplugging the coils cables into the supply socket; and positioning the patient's table in the magnet bore using the manual console on the magnet's housing. A significant percentage of MRI examinations require the administration of pharmaceuticals (e.g., contrast), which are to be applied to a patient during the examination (in the majority of cases, manually by a nurse), but in the pause in time-varying EMF emission. The nurse administering contrast has to approach the patient's limb, for example, and is also exposed to SMF—but the level and duration of exposure depend on the distance from the MRI magnet, which is defined by the work organization.

The level of SMF exposure by the MRI magnet is defined by distance, that is, by the work organization. The maximum SMF affecting the worker is the main field used for exposure in diagnostics (mentioned 1.5 or 3 T), but usually routine worker exposure does not exceed 10% of the main field (Karpowicz and Gryz 2013a). In front of the bore of MRI magnets, SMF reaches the level of 1 T, where rapid movement may cause the mentioned symptoms disturbing the worker's activities (Table 21.3). SMF is permanently switched on—SMF at a level of several mT, in which the hazard of *flying objects* is also permanently present in the main part of the MRI chamber (Table 21.4, Figure 21.2).

Because the MRI magnet is continuously active, anybody who is present nearby is always affected by SMF and is in danger from movement-related symptoms (e.g., loss of balance), *flying objects* (e.g., metallic tools), or AIMD dysfunctions (e.g., cardiac pacemakers)—for example, cleaners, technicians, guards, firefighters, and relatives attending patients.

21.6   EMF ASSESSMENT CRITERIA

Because of the discussed adverse effects of the EMF exposure on humans and electronic devices, various requirements and recommendations for the assessment and mitigation of exposure have been established, with respect to the protection of electronic devices, the general public, and workers. Those requirements do not apply to the influence of EMF on patients undergoing medical procedures—for example, according to the European Directive,

> medical devices must be designed and manufactured in such a way that, when used under the conditions and for the purposes intended, they will not compromise the clinical condition or the safety of patients, or the safety and health of users or, where applicable, other persons, provided that any risks which may be associated with their use constitute acceptable risks when weighed against the benefits to the patient and are compatible with a high level of protection of health and safety

Council of EU (1993)

The requirements regarding EMF exposure are summarized in this section based on the international and European rules, but it needs to be pointed out that in this area, the national legislations are binding and in particular countries requirements may differ from the presented general international rules.

21.6.1   LIMITATIONS TO GENERAL PUBLIC AND WORKERS' EXPOSURE

The minimum recommendations regarding limitations of exposure to EMF in the European Union are based on the guidelines provided by the International Commission on Non-Ionizing Radiation Protection (ICNIRP) developed in order to protect against the effects of short-term exposure (ICNIRP 1998). The frequency-dependent limitation of general public exposure to EMF is determined by the noncompulsory recommendation of European Council, whereas a binding European Directive 2013/35/EU of European Parliament and Council provided minimum requirements regarding exposure limits for workers (Council of EU 1999, European Parliament 2013). In all European Union member countries, a national labor law regarding protection against EMF exposure in the workplace should be harmonized with the directive by 2016. This control over the level of environmental EMF exposure is based on the limits of electric field strength (E), magnetic field strength (H), and limb-induced/contact currents (I), which may be measured near the EMF sources, where individuals are present because of any purposes (Table 21.5). The assessment of EMF hazards based on the measurement of the capacitive currents flowing in the limbs is used for the evaluation of localized thermal effects or pain electric shock perception (ICNIRP 1998)—the electrical current with a frequency from 0.1 to 110 MHz passing through the wrist, regardless of the conditions under which it occurs, should not exceed 40–100 mA.

Many countries decided also to implement additional rules on applying the protection system, covering also the health hazards caused by chronic exposure (mentioned delayed effects). These consist, for example, in limiting the duration of exposure exceeding that permissible for the general public or in applying lower exposure limits for general public exposure than that provided by ICNIRP (Karpowicz et al. 2006).

TABLE 21.5

Requirements regarding the Protection against Undesirable Impact of Environmental EMF of Frequencies Emitted by Typical Medical Devices, according to ICNIRP Guidelines Published in 1998, Referred to in European Council Recommendation Published in 1999 (ICNIRP 1998)

Parameter Characterizing Field/ Frequency/(Kind of Device)	General Public Exposure Limits		Occupational Exposure Limits	
	Electric Field Strength E (V/m)	Magnetic Field Strength H (A/m)	Electric Field Strength E (V/m)	Magnetic Field Strength H (A/m)
Static magnetic field (used in MRI scanners and MTA)	—	32,000 (40 mT)	—	160,000 (200 mT)
50 Hz (magnetic field used in MTA)	—	80 (100 μT)	—	400 (500 μT)
500 kHz (EMF used in ESU, LWDD, and USTU)	87	0.73	610	1.6
27 MHz (EMF used in SWDD)	28	0.073	61	0.16
2450 MHz (EMF used in MWDD)	61	0.16	137	0.36

Notes:

- It needs to be pointed out that over the years 2009–2014, ICNIRP partly revised guidelines, and European Directive 2013/35/EU on workers EMF exposure limitation has been published, but the administrative procedures regarding the EMF emission and immunity testing, involved among others in the process of so-called *CE mark* labeling of an electric devices being available at European market, are still based on the old ICNIRP limits mentioned in the table. To analyze earlier-mentioned newer documents, see the ICNIRP webpage (http://www.icnirp.de) and European law web multilingual (in each European language) service (http://eur-lex.europa.eu).

- It needs to be pointed out that ICNIRP's limits of exposure of general public and workers to EMF of frequency from the range 100 kHz to 10 GHz refer to electric and magnetic field strengths averaged over 6 min, whereas when AIMD immunity to EMF interaction is tested, EMF is not averaged over time.

- In accordance with the IEC EN 60601-1-2 standard, medical devices should be resistant to interference caused by an electric field with a strength of 3 V/m within the frequency range from 80 MHz to 2.5 GHz (IEC/CENELEC 2007). The required immunity level for life-supporting devices is higher: 10 V/m.

21.6.2 Limits regarding the EMF Impact on Medical Devices (Including AIMD)

The immunity of electronic medical devices supporting the human organism (including AIMD) to electromagnetic interference has to be tested and to comply with the technical requirements regarding *electromagnetic compatibility* (EMC). The EMF levels used for testing AIMD immunity according to the provisions of European Standards are lower than the ICNIRP general public exposure limit (Table 21.5) (IEC EN 60601-1-2:2007). As a result, in AIMD exposed to EMF nearing the ICNIRP limit, some interference may cause malfunctions, which is a concern to any user—patients, medical personnel, and bystanders.

That analog watches, credit cards, magnetic tapes, computer disks, precise electronic devices, etc., may be adversely affected by exposure to SMF exceeding 1 mT also needs attention (ICNIRP 2009, WHO 2006). Precautions to prevent hazards from flying metallic objects should be taken in SMF exceeding 3 mT.

21.6.3 SAFETY RULES REGARDING AIMD USERS' EXPOSURE TO EMF

European Standard IEC EN 60601-1-2, concerning general safety requirements for medical life-supporting devices, which is defining the protocol for testing the AIMD immunity to EMF exposure, does not refer to EMF of frequencies generated by the most of physiotherapeutic devices—the immunity of medical devices is tested in radiated EMF from the frequency range 80 to 2500 MHz, at the 10 V/m level of E-field, which covers common environmental exposures to EMF emitted by radio and television broadcasting and wireless communication system facilities. However, the manufacturers of SWDD, LWDD, MTA, and MRI recommend that, for safety reasons, patients with AIMD be excluded from the treatment. Similar recommendations can be found in publications setting out the rules for physiotherapeutic procedures and MRI diagnostics (CENELEC 2002, Robertson et al. 2006).

European Directive 2013/35/EU on the protection of workers against health and safety hazards caused by EMF exposure stresses the need to assess any direct and indirect hazards from electromagnetic interaction on workers, with special attention to the users of AIMD (such as cardiac pacemakers or infusion pumps). AIMDs intended for the European Union market have to be compliant with the European Standard EN50527-1, which requires that in EMF not exceeding the reference level for the general public given by ICNIRP requirements and Council Recommendation regarding EMF of 0–300 GHz frequency, then AIMD dysfunctions caused by EMF may not be expected by users (CENELEC 2010). In case of exposure exceeding such limits, some AIMDs may also function properly, but it needs individual assessment.

According to the requirements provided by ICNIRP, European EMF Directive 2013/35/EU, and Council Recommendation 1999/519/EC regarding the procedure of assessing direct hazards of exposure to RF EMF (hazards caused by thermal effects), E-field and H-field values are averaged over any 6 min period. However, according to the opinion given by European Standard No EN50527-1:2010, the electromagnetic immunity of AIMDs depends on the EMF values nonaveraged over time. It means that, in the case of PM EMF, the rules of an assessment of direct and indirect effects are different. Also the metrological properties of EMF measurement devices, calibrated in the sinusoidal time-varying EMF, may be different in the case of measurement of CW and PM EMF.

Therefore, it may be assumed that, in order to assess whether hazards for AIMD users exist, the limits of general public exposure, as recommended by ICNIRP, may be used (Table 21.5). In consequences based on international recommendations, the contraindications for the EMF exposure, which is exceeding the general public limits, include the use of AIMDs, such as pacemakers and infusion pumps. Similarly, such exposure is not recommended for the general public, pregnant women, and young workers (Karpowicz et al. 2006). Additionally, the most common recommendations concerning the impact of the EMFs are as follows: cardiac pacemaker users should avoid exposure to SMF exceeding 0.5 mT, magnetic fields of power frequency (50/60 Hz) exceeding 0.1 mT, and electric fields of power frequency exceeding 1 kV/m, but users of ferromagnetic implants or electrically activated devices (other than cardiac pacemakers) may be affected by fields exceeding several millisteslas. It should be emphasized that because of various technologies used by manufacturers of AIMD, their immunity for EMF interactions and hazards for users needs individual evaluation with respect to the pattern of EMF exposure to particular users.

21.7 CONCLUSIONS REGARDING THE SCALE OF EMF HAZARDS NEAR VARIOUS MEDICAL APPLICATIONS

The parameters of EMF emitted by selected medical applications emitting relatively strong fields (such as physiotherapeutic devices, electrosurgery devices, MTAs, and MRI scanners), considered in the context of frequency, level of magnetic and electric components, and spatial distribution, as well as exposure effects caused by such fields (such as capacitive RF electric currents [induced or contact], movement-related induced currents, or ferromagnetic *flying objects*), determine the type and level of EMF hazards in various medical centers. The prevention measures need to be suitable for the various types of hazards caused by EMF and considered with respect to requirements regarding the protection of electronic devices, including medical implants, against electromagnetic interference and protection against the health hazards caused by electromagnetic exposure in members of the general public (anyone who is present nearby but not undergoing treatment), as well as in healthcare personnel.

The strongest environmental electromagnetic hazards occur near short-wave diathermy and electrosurgery devices, as well as near MRI scanners (up to the distance of 200–300 cm), and to a lesser degree near long-wave diathermy devices and magnetotherapeutic applicators (up to 100 cm), but it was not found near USTUs and physiotherapeutic applicators equipped with the direct (resistive coupling) application of electric currents. A capacitive electric current exceeding many times the international recommendations regarding protection against excessive thermal effects was measured in the wrists of personnel while touching electrodes and cables of active short-wave diathermy devices. *Flying objects* hazards (which may even cause lethal accident) were identified up to approx. 200 cm from the magnets of the most popular MRI scanners (1.5 or 3 T). Movement-related potentially hazardous influence on personnel moving rapidly near the magnets of such scanners was identified at a distance less than 50 cm from the edge of the bore of the scanner's magnet.

Many factors influence the results of EMF exposure in the AIMD and its user, for example, the type and model of AIMD, the settings of its operation, and the duration and spatial distribution of the user's exposure. As a result, a particular level of EMF exposure may not be dangerous for all users of AIMD, and an individual analysis of EMF hazards for each AIMD user is advised (European Parliament 2013, Tiikkaja et al. 2013). This analysis may involve supporting data from the AIMD manufacturer. The individual sensitivity of the AIMD is determined mainly by its technology and design, and many modern AIMDs from leading manufacturers are less sensitive to EMF interaction than EN 50527-1:2010 (CENELEC 2010) indications, but older devices may still be in use and the safety of their users needs attention.

Workers and other AIMD users should be aware that their presence close to SWDD, LWDD, MWDD, MTA, ESU, and MRI is not recommended because of a possible implant dysfunction. Warning signs are recommended to indicate a location of such hazards, but do not disturb the privacy of interested AIMD users (Hocking and Hansson Mild 2008, Tiikkaja et al. 2013). The location of medical devices emitting EMF in the electromagnetically shielded boxes may also be recommended to reduce the spatial impact of EMF emitted by those devices.

It should also be emphasized that as long as various delayed EMF effects cannot be dismissed, progress in the research on the health effects of EMF exposure should be monitored, and prevention involving EMF exposure limitation should be applied in the use of technologies causing such exposure, including medical ones.

ACKNOWLEDGMENTS

The results of research carried out within the scope of the National Program "Improvement of safety and working conditions" in 2011–2014 were presented: within the scope of state services by the Ministry of Labour and Social Policy and within the scope of research and development by the Ministry of Science and Higher Education/National Centre for Research and Development. The Central Institute for Labour Protection—National Research Institute is the program's main coordinator.

Summarized EMF measurement results have been collected with the help of Dr. Krzysztof Gryz, Dr. Patryk Zradziński, and Mr. Wiesław Leszko from the Laboratory of Electromagnetic Hazards of Central Institute for Labour Protection (CIOP-PIB), Warszawa, Poland.

ABBREVIATIONS

AIMDs	Active implantable medical devices
CENELEC	European Committee for Electrotechnical Standardization
COAG	Coagulating
CUT	Cutting
CW	Continuous wave
EC	European Commission
ELF	Extremely low frequency
EMC	Electromagnetic compatibility
EMF	Electric, magnetic, or electromagnetic fields and radiation
EP	European Parliament
ESU	Electrosurgery unit
EU	European Union
IARC	International Agency of Research on Cancer
ICNIRP	International Commission on Non-Ionizing Radiation Protection
LWDD	Long-wave diathermy device
MRI	Magnetic resonance imaging
MTA	Magnetotherapeutic applicator
MW	Microwave
MWDD	Microwave diathermy device
PM	Pulsed-modulated
RF	Radiofrequency
RMS	Root-mean-square
SMF	Static magnetic field
SWDD	Short-wave diathermy device
USTU	Ultrasound therapy unit

REFERENCES

Bortkiewicz A, Gadzicka E, Szynczak W, Zmyślony M (2012) Heart rate variability (HRV) analysis in radio and TV broadcasting stations workers. *International Journal of Occupational Medicine and Environmental Health*, 25(4):446–455.

Calcagnini G, Floris M, Censi F, Cianfanelli P, Scavino G, Bartolini P (2006) Electromagnetic interference with infusion pumps from GSM mobile phones. *Health Physics*, 90(4):357–360.

Carranza N, Febles V, Hernandez JA, Bardasano JL, Monteagudo JL, Fernandez de Aldecoa JC, Ramos V (2011) Patient safety and electromagnetic protection: A review. *Health Physics*, 5(100):530–541.

Chaljub G, Kramer LA, Johnson RF et al. (2001) Projectile cylinder accidents resulting from the presence of ferromagnetic nitrous oxide or oxygen tanks in the MR suite. *American Journal of Roentgenology*, 177(1):27–30.

Chen DW (2001) Boy, 6, dies of skull injury during M.R.I. *NY Times*, July 31:Sec. B:1,5.

Christe B (2009) Evaluation of current literature to determine the potential effects of radio frequency identification on technology used in diabetes care. *Journal of Diabetes Science and Technology*, 3:331–335.

Council of the European Union (EU) (1993) Council Directive 93/42/EEC of 14 June 1993 concerning medical devices. *Official Journal of the European Communities* 169, 12/07/1993, 0001–0043.

Council of the European Union (EU) (1999) Recommendation of 12 July 1999 on the limitation of exposure of the general public to electromagnetic fields (0 Hz to 300 GHz), 1999/519/EC. *Official Journal of the European Communities*, L 199/59.

De Marco M, Magi S (2006) Evaluation of stray radiofrequency radiation emitted by electrosurgical devices. *Physics in Medicine and Biology*, 51:3347–3358.

Duck B (2010) MRI accident earlier this year kills service engineer who was Sucked in and pinned to the unit— FDA investigation at Barbara Duck at Medical Quack Healthcare Blogger, http://ducknetweb.blogspot.com/2010/06/mri-accident-earlier-this-year-kills.html (accessed May 2, 2014).

Eggleston JL, Von Maltzahn WW (2000) Electrosurgical devices, Chapter 81. In: *The Biomedical Engineering Handbook*, 2nd edn., JD Bronzino (ed.), Boca Raton, FL: CRC Press LLC, pp. 81-1–81-10.

European Committee for Electrotechnical Standardization (CENELEC) (2002) European Standards EN 60601-2-33:2002. Medical electrical equipment. Part 2: Particular requirements for the safety of magnetic resonance equipment for medical diagnosis. Brussels, Belgium.

European Committee for Electrotechnical Standardization (CENELEC) (2010) European Standard EN50527-1:2010. Procedure for the assessment of the exposure to electromagnetic fields of workers bearing active implantable medical devices—Part 1: General. Brussels, Belgium.

European Parliament (EP) and the Council (2013) Directive 2013/35/EU of the European Parliament and of the Council of 26 June 2013 on the minimum health and safety requirements regarding the exposure of workers to the risks arising from physical agents (electromagnetic fields) (20th individual Directive within the meaning of Article 16(1) of Directive 89/391/EEC). *Official Journal of the European Union*, 29.6.2013, L 179/1–21.

Gryz K, Karpowicz J (2014) Environmental aspects of the use of radiofrequency electromagnetic fields in physiotherapeutic treatment. *Annals of the National Institute of Public Health—National Institute of Hygiene*, 65(1):55–61.

Hansson Mild K, Alanko T, Decat G, Falsaperla R, Gryz K, Hietanen M, Karpowicz J, Rossi P, Sandström M (2009) Exposure of workers to electromagnetic fields. A review of open questions on exposure assessment techniques. *International Journal of Occupational Safety and Ergonomics (JOSE)*, 15(1):3–33.

Hansson Mild K, Hand J, Hietanen M, Gowland P, Karpowicz J, Keevil S, Lagroye I, van Rongen E, Scarfi MR, Wilén J (2013) Exposure classification of MRI workers in epidemiological studies. *Bioelectromagnetics*, 34:81–84.

Hill DA, Walsh JA (1985) Radio-frequency current through the feet of a grounded human. *IEEE Transactions on Electromagnetic Compatibility*, EMC-27, 1:18–23.

Hocking B, Hansson Mild K (2008) Guidance note: Risk management of workers with medical electronic devices and metallic implants in electromagnetic fields. *International Journal of Occupational Safety and Ergonomics (JOSE)*, 14(2):217–222.

International Agency for Research on Cancer (IARC) (2002) Non-ionizing radiation, part 1: Static and Extremely Low-Frequency (ELF) Electric and Magnetic Fields. Lyon, France: The WHO/IARC, IARC Monographs, Vol. 80.

International Agency for Research on Cancer (IARC) (2013) Non-ionizing radiation, part 2: Radiofrequency electromagnetic fields, Lyon, France: The WHO/IARC, IARC Monographs Vol. 102.

International Commission on Non-Ionizing Radiation Protection (ICNIRP) (1998) Guidelines for limiting exposure to time-varying electric, magnetic, and electromagnetic fields (up to 300 GHz). *Health Physics*, 4(74):494–522.

International Commission on Non-Ionizing Radiation Protection (ICNIRP) (2009) Guidelines on limits of exposure to static magnetic fields. *Health Physics*, 96:504–514.

International Commission on Non-Ionizing Radiation Protection (ICNIRP) (2010) Guidelines for limiting exposure to time-varying electric and magnetic fields (1 Hz to 100 kHz). *Health Physics*, 99(6):818–836.

International Electrotechnical Committee (IEC)/European Committee for Electrotechnical Standardization (CENELEC) (2007) European Standard IEC EN 60601-1-2:2007. Medical electrical equipment—Part 1–2: General requirements for safety. Collateral standard: Electromagnetic compatibility—Requirements and tests.

Jeheson P, Duboc D, Lavergne T et al. (1988) Change in human cardiac rhythm induced by 2-T static magnetic field. *Radiology*, 166:227–230.

Jokela K, Puranen L, Gandhi OP (1994) Radio frequency currents induced in the human body for medium-frequency/high-frequency broadcast antennas. *Health Physics*, 66(3):237–244.

Jokela K, Saunders RD (2011) Physiologic and dosimetric considerations for limiting electric fields induced in the body by movements in a static magnetic field. *Health Physics*, 100:641–653.

Kalliomäki P-L, Hietanen M, Kalliomäki K, Koistinen O, Valtonen E (1982) Measurements of electric and magnetic stray fields produced by various electrodes of 27-MHz diathermy equipment. *Radio Science*, 17(5 S):29–34.

Karpowicz J (2009) Assessment of induced currents hazards among health care workers operating magnetic resonance scanners—Towards new standardization approach, *Second International Symposium on Applied Sciences in Biomedical and Communication Technologies*, Bratysława, Slovakia, November 21–27, 2009, Piscataway, NJ: IEEE.

Karpowicz J, Gryz K (2006) Health risk assessment of occupational exposure to a magnetic field from magnetic resonance imaging devices. *International Journal of Occupational Safety and Ergonomics* (*JOSE*), 12(2):155–167.

Karpowicz J, Gryz K (2007) Practical aspects of occupational EMF exposure assessment. *Environmentalist*, 27:525–531.

Karpowicz J, Gryz K (2010) Electromagnetic hazards in the workplace. In: *Handbook of Occupational Safety and Health*, D. Koradecka (ed.), CRC Press, Taylor & Francis Group, Boca Raton, FL, pp. 199–218.

Karpowicz J, Gryz K (2013a) The pattern of exposure to static magnetic field of nurses involved in activities related to contrast administration to patients diagnosed in 1.5T MRI scanners. *Electromagnetic in Biology and Medicine*, 32(2):182–191.

Karpowicz J, Gryz K (2013b) Experimental evaluation of ballistic hazards in imaging diagnostic center. *Polish Journal of Radiology*, 78(2):31–37. http://www.polradiol.com/search/form (accessed May 5, 2014)

Karpowicz J, Gryz K (2013c) An assessment of hazards caused by electromagnetic interaction on humans present near short-wave physiotherapeutic devices of various types including hazards for users of electronic active implantable medical devices (AIMD). *BioMed Research International*, Article ID 150143, 8pp., http://dx.doi.org/10.1155/2013/150143 http://www.hindawi.com/journals/bmri/2013/150143/

Karpowicz J, Gryz K, Leszko W, Zradziński P (2013a) Objectivized evaluation of surgeons exposure to radiofrequency electromagnetic fields—In the context of exposure duration and Polish and new international requirements regarding workers protection. *Medycyna Pracy*, 64(4):487–501.

Karpowicz J, Gryz K, Leszko W, Zradziński P (2013b) An assessment of limb current in surgeons and nurses using electrosurgical units. In: *The Bioelectromagnetics Society BioEM2013*, Abstract Collection, June 10–14, 2013, Thessaloniki, Greece, pp. 71–73.

Karpowicz J, Gryz K, Politański P, Zmyślony M (2011) Exposure to static magnetic field and health hazards during operation of magnetic resonance scanners. *Medycyna Pracy*, 62(3): 309–321 (Polish).

Karpowicz J, Gryz K, Zradziński P (2009) Pola elektromagnetyczne w otoczeniu urządzeń fizykoterapeutycznych—aplikatory do terapii zmiennym polem magnetycznym. *Inżynieria Biomedyczna Acta Bio-Optica et Infiormatica Medica*, 15(1):60–63 (Polish).

Karpowicz J, Hietanen M, Gryz K (2006) EU directive, ICNIRP guidelines and polish legislation on electromagnetic fields. *International Journal of Occupational Safety and Ergonomics* (*JOSE*), 12(2): 125–136.

Karpowicz J, Hietanen M, Gryz K (2007) Occupational risk from static magnetic fields of MRI scanners. *Environmentalist*, 27(4):533–538.

Keltner JR, Ross MS, Brakeman PR et al. (1990) Magnetohydrodynamics of blood flow. *Magnetic Resonance in Medicine*, 16(1):139–149.

Kinouchi Y, Yamaguschi H, Tenforde TS (1996) Theoretical analysis of magnetic field interactions with aortic blood flow. *Bioelectromagnetics*, 17:21–32.

Korniewicz H (1995) The first resonance of a grounded human being exposed to electric fields. *IEEE Transactions on Electromagnetic Compatibility*, 2(37):295–299.

Liljestrand B, Sandstrom M, Hansson Mild K (2003) RF exposure during use of electrosurgical units. *Electromagnetic, Biology and Medicine*, 2(22):127–132.

MRI safety com—Your information resource for MRI safety, bioeffects and patient management. http://www.MRIsafety.com (accessed May 2, 2014)

Pantchenko OS, Seidman SJ, Guag JW, Witters DW, Sponberg CL (2011) Electromagnetic compatibility of implantable neurostimulators to RFID emitters. *BioMedical Engineering* (On Line), 10:50.

Prato F, Thomas AW, Legros A et al. (2010) MRI safety not scientifically proven. *Science*, 328:568–569.

Reilly PJ (1998) *Applied Bioelectricity. From Electrical Stimulation to Electropathology*. New York: Springer-Verlag.

Robertson V, Ward A, Low J, Reed A (2006) *Electrotherapy. Explained Principles and Practice*. Edinburgh, New York: Butterworth-Heinemann, Elsevier.

Sarwer SG, Farrow A (2013) Assessment of physiotherapists' occupational exposure to radiofrequency electromagnetic fields from shortwave and microwave diathermy devices: A literature review. *Journal of Occupational and Environmental Hygiene*, 10:312–327.

Schneider B, Abatti PJ (2008) Electrical characteristics of the sparks produced by electrosurgical devices. *IEEE Transactions on Biomedical Engineering*, 55:589–593.

Tiikkaja M, Hietanen M, Alanko T, Lindholm H (2013) *Working in Electromagnetic Fields with a Cardiac Pacemaker*. Helsinki, Finland: Finish Institute of Occupational Health.

Tenforde TS (1992) Interaction mechanisms and biological effects of static magnetic fields. *Automedica*, 14:271–293.

Tognola G, Parazzini M, Sibella F, Paglialonga A, Ravazzani P (2007) Electromagnetic interference and cochlear implants. *Annali dell'Istituto Superiore di* Sanità, 43(3):241–247.

Vesselinowa L (2013) Biosomatic effects of the electromagnetic fields on view of the physiotherapy personnel health. *Electromagnetic Biology and Medicine*, 32(2):192–199.

Wilén J (2010) Exposure assessment of electromagnetic fields near electrosurgical units. *Bioelectromagnetics*, 31:513–518.

Wilén J, de Vocht JF (2011) Health complaints among nurses working near MRI scanners—A descriptive pilot study. *European Journal of Radiology*, 80(2):510–513.

World Health Organization (WHO) (2006) *Environmental Health Criteria 232, Static Fields*, Geneva, Switzerland: World Health Organization.

Zradziński P (2013) The properties of human body phantoms used in calculations of electromagnetic fields exposure by wireless communication handsets or hand operated industrial devices. *Electromagnetic Biology and Medicine*, 32(2):192–199.

22 Long-Term, Low-Intensity, Heterogeneous Electromagnetic Fields
Influence on Physiotherapy Personnel Morbidity Profile

Lyubina Vesselinova

CONTENTS

22.1 INTRODUCTION

Technical and technological development enables widespread use of artificial sources emitting electromagnetic fields (EMFs) in areas other than specialized practice. In the last 20 years, the major contribution to this technology has come from the worldwide network of mobile and satellite communications, which have dramatically increased the high-frequency (HF) component of the evolutionary electromagnetic background, which has thus far been unrecognized. In addition, nuclear accidents and changes in ozone concentration have repeatedly raised the amount of radiation (Bobrakov and Kartashev 2001, Markov and Hazlewood 2009), which has set biological objects in a new electro-ecological situation. This complicated radiation environment poses the problem of reducing the exposure load, especially for professional groups at risk, with a priority to closely follow the fundamental maxima underlying the Precautionary Principle: "better safe than sorry" (Communication on Precautionary Principle EC 2000, Goldsmith 1997). The need for restricting access as well as shielding both devices and staff as a protective measure is one of the key points of the Benevento Resolution 2006 (Recent Research on EMF and Health Risks 2007). The personnel in physiotherapy were traditionally exposed to electromagnetic radiation (EMR) till after 1900, when HF currents were included as a routine therapeutic factor; however, this topic still remains outside research interest. When the first book on magnetotherapy was published in Bulgaria by Professor Nentcho Todorov in 1982, magnetotherapy became yet another basic tool in physiotherapy. Actually, the palette of treatment modalities is much more varied as it comprises fields and currents in different frequency ranges, optic radiation, ultrasound, magnet, and laser. Namely, this wide range of physical factors with possible applications in healing techniques creates the need for serious hazard assessment concerning the health of personnel.

22.2 PROBLEM ANALYZING

22.2.1 RISK GROUPS

Today, the high-tech social lifestyle has globalized the problem of overexposure, forming temporary at-risk groups based on their background characteristics: *residential* (living in the vicinity of power lines, base stations, and broadcasting towers) and *professional* (workers in industry, electronics and communications, base station operators, technicians, engineers, etc.). For professionals, the level of risk depends on the distance from the source, the amplitude modulation, and the synergy of biological, chemical, and physical components of the microclimate. Hence, the occupational security and health (OSH) classification in the Final Technical Report on Occupational MF Exposure 2008 (EMF-NET MT-2 D49) identifies three professional subgroups:

1. *Under high exposure*: Operators of base stations, radars, power-lines, electrosurgical knives, diathermy units, where the work conditions cannot be changed and personnel need precaution and individual risk assessment. Special attention should be paid to the most vulnerable workers in the EMF group—pregnant women, very young workers, and those with active or passive metal implants.
2. *Under possible risk*: Some technological branches. Warning signs and inscriptions are mandatory. In this subgroup, qualification of generator operators is mandatory and the assessment must be carried out individually.
3. *Nonexposed*: Professional group (similar to the public group). The exposure levels are within the standards and warning is only required for people with pacemakers.

The occupational features of electrophysiotherapy assign professionals in this field to the first risk group, independent of the fact that exposures in physiotherapy are predominantly of low intensity and that there is no predominance of the contact with high-frequency therapy (HFTh) sources.

22.2.2 "TAKE AND RUN" AND "TAKE AND HOLD" PRINCIPLE OF THE EMF IN PHYSIOTHERAPY

In fact, all electrotherapy devices generate an EMF. In therapy, it provokes or accelerates processes of healing and reparation in the radiated tissue. The physical characteristics of EMFs—electric and magnetic components, wavelength, penetration depth, processes of refraction, reflection, polarization and oscillation, and superposition—and their cumulative ability determine their potential as powerful therapeutic agents for the patients termed by us ("*take and run*" principle) with anti-inflammatory, anti-swelling, pain relief, trophic improvement, chondroprotection, and biostimulating or biosuppressing (dose-dependent) effects (Aleksiev 2013, Basset 1994, Becker and Marino 1982, Irving 2007, Kositsky et al. 2001, Markov 2010, 2011, Rasoulia et al. 2012, Vesselinova and Kovatchev 2013) for maximal functional recovery and, at the same time, by dispersed long-term professional influence termed by us ("*take and hold*" principle), as potentially adverse bioactive factors. Usually the prescription in the rehabilitation program is complex and combines electrotherapy and natural factors. Monotherapy prescriptions are rare, but more often they don't exceed three or four factors in a prescription. Prescriptions for different factors are implemented in sections of the corresponding fields—electro/light therapy, kinezitherapy, hydrotherapy, thermotherapy, laser therapy, inhalation therapy. Prescribed electrotherapy procedures have been performed by physiotherapists in the electro/light therapy section (ELTS), which is normally a big room divided into cubicles. By medical standards, each cubicle should have one source of artificial physical factors for low or intermediate frequency currents, iontophoresis, ultrasound, light therapy (ultraviolet (UV) or visible light), and magnetotherapy. In most cases, generators of HF currents are also placed in the same big room in shielded compartments. This way the medical devices emitting different frequency ranges—50 Hz, 150 kHz, 1 MHz, 3 MHz, 27.12 MHz, and 2450 MHz—as well as sources of pulsing magnetotherapy and optic radiation (for UV and infrared light therapy) can work simultaneously (Israel and Tschobanoff 2006, Vesselinova 2013a). This, however, generates a specific electromagnetic environment in the physiotherapy–electrotherapy sections that personnel face daily (Vesselinova 2012a). The overexposure from apparatuses for HF treatment up to a distance of 1 m is actually absorbed mainly by the staff (Karabetsos et al. 2010, Karpowicz and Gryz 2013, Markov 2008, Traykov and Israel 1994), and this has been practically proven with a test-lamp for EMR (Vesselinova 2012a). Chronic exposure can unlock different dose- and time-dependent adaptive mechanisms or disadaptive events, which may manifest as complications, disturbances, or disorders that may or may not be reversible. Contradictions in bio effects due to therapeutic applications and professional long-term absorption are presented in Table 22.1.

In the long history of practical and scientific physiotherapy, very few papers have paid attention to working conditions and the impact of physical factors on the personnel. In the scientific literature,

TABLE 22.1
Contradiction of the Physiotherapy EMF on Health by Short-Term *"Take and Run"* and Long-Term *"Take and Hold"* Applications

Therapy Application (Short-Term *"Take and Run"*)	As Professional Background Component (Long-Term *"Take and Hold"*)
Short application duration—5 ÷ 30 min	Long time burden—more than the half or the entire working time (4 ÷ 7 h)
Short term—3 ÷ 20 days (max. 3 or 4 courses yearly)	Long-term—the entire working experience
Local/regional (commonly peripheral)	Whole-body contact or at least the central body sagittal line by application of the procedure
Dose-limitated characteristics (athermal, olygothermal, or thermal, measured in W per cm² or in mA (mV)	Dose unknown burden—indirectly by specific sense occurrence (the measurement concerns the density of power)
Precise determination of the physical factors demanded of the health status (as usual 3 or 4 max. by course)	Many kinds of physical factors with accidental prevalence (depending on performed daily prescriptions)
Focused fields	Dispersed fields with or without superposition
Frequency known characteristics	Stochastic frequency changes and combinations
In a changed biological medium of disturbance, disease, or convalescence	On a principally healthy body

the situation is the same. In the last 4 years, there appears to be a tendency to enhance research focusing on the improvement of the power of density measurement and assessment in physiotherapy. The complicated and stochastically changing frequency of diverse EMF characteristics of the work background in the low-intensity range distinguishes physiotherapy personnel from other professional groups in risky electromagnetic professions that have been recognized and studied so far among the so-called electrical professions in industry and communications, and therefore needs to be studied independently and in depth.

22.3 LOW-INTENSITY EMF

The interaction of an EMF with biological structures is currently interpreted by the concepts of two main mechanisms: thermal and nonthermal. The difference in these perceptions between Western and East European analysts is based on the assumption of the EMF's ability to induce biological effects with or without overheating and tissue shock (Markov 2006, Michaelson 1974, Pilla and Markov 1994, Ueno 1996, Williams 2009). Data from the last 20 years show the increasingly harmful effect of nonthermal EMFs through important biological effects that have a specific meaning—the ability to accumulate repeated weak signals, which usually do not provoke momentary recognizable feelings or biological response (Beal and Fagin 1995, Nordin et al. 2011, Tolgskaya et al. 1987). Currently, Ishido et al. (2001) have found paradoxical dependence in nonthermal EMFs—a stronger effect of low-dose fields, which poses even greater threat to human health, as described by other authors as well (Adair 1991, Adey 1983). It is an axiom that the human body is "transparent" in static and low-frequency magnetic fields and the permeability of the electric field depends on the conductivity of the medium (Becker and Marino 1982, Habash 2008). The main active component for personnel in physiotherapy is the dispersed low-intensity EMF. The absorbed EMR in the work background and the inductive currents thereby provoked in the physiotherapy personnel's bodies may unlock different biological answers depending on the condition of the operator and the anatomical area of the body that is involved (Vesselinova 2012a, 2013c,d).

22.4 ELECTROMAGNETIC BACKGROUND IN PHYSIOTHERAPY

Apparatuses are placed in the electrotherapy sections for electrotherapy (therapeutic application of electric currents with low, intermediate, and high frequencies, in permanent or alternating modality), magnetotherapy, ultrasound therapy, and light therapy (optic radiation sources for UV and visible light). Hence, these are in fact ELTS. The organization demands that the personnel in every section perform the prescribed procedures with the physical factors located in the same section as their duty place. So, the physiotherapists, the patients, and the visiting physiatrists are under the influence of nonhomogeneous by spectral and density characteristics of the EMF, which are strongly dependent on the prescribed physical factors and modalities. The common sources of UV, visible, and infrared light, and extreme HFTh (mm, cm, and dm waves) in the electrotherapy sections within the physiotherapy unit together with their energy characteristics (wavelength and frequency) have been presented by Kositsky et al. (2001).

The first data of overexposure in electrotherapy sections of physiotherapy wards were found more than 50 years ago mainly by researchers from the former Soviet Union and Bulgaria. In the publications of Dimitrova (1965) and Todorov et al. (1965) a correlation was found between building materials in the working environment, shielding of the sources, and the dosimetric characteristics, resulting in the manifestation of certain symptoms: menstrual disturbances, headaches, tiredness, abnormalities in the white blood cells. These findings served as an indirect indicator of the harmful potential of EMR. This fact was already reflected in the legislation documents concerning the organization of physiotherapy departments, which recommended shielding the compartments with radiofrequency (RF) generators. However, almost 30 years later, in 1994, the Bulgarian physicists Traykov and Israel (1994) described areas of overexposure in their measurements of work environmental factors through risk evaluation which are valid for all structures of physical medicine. The authors emphasized that the values of the measured intensity of the electric field at a distance of 50 cm from the electrodes of the ultrahigh-frequency (UHF) therapy generator (27.12 MHz) are between three and seven times higher than the maximum allowed 200 V/m.

Manifestation of skin irritation and burning sensation in the upper limbs has been recorded by our own clinical experience during common application of HFTh. The ICNIRP report (2001) summarized the experience of different authors (Stuchly et al. [1982] and Veit and Bernhardt [1984]) who have identified the risk areas of high-level exposure to personnel at a distance of less than about 1.5–2 m for UHF apparatuses (27.12 MHz) or 1 m for apparatuses using microwaves (433 MHz and 2.45 GHz). In recent works, Miclaus et al. (2010), Karabetsos et al. (2010), and Karpowitcz and Gryz (2013) have confirmed the same overexposure zones near the electrodes and cables. The typical exposure distances and burden levels for patients and personnel involved with shortwave diathermy (27, 12 MHz) and microwave treatment (433 and 2450 MHz) have been assessed and compared with magnetic resonance imaging (MRI) examination (42–300 MHz) by Vecchia et al. (2009) in the ICNIRP report. In their review of the literature on a 20-year period of RF EMF power density measurement, Shah and Farrow (2013) clearly show and recommend revision of the distance for sources of continuous shortwave diathermy (CSWD) tonot less than 2 m and at least 1.5 m for sources of pulsing shortwave diathermy (PSWD).

22.5 RECOGNIZED BIOLOGICAL EFFECTS ON PHYSIOTHERAPY PERSONNEL

The human body is a complex biological system whose proper functioning depends on the proper functioning of its main systems—nervous, endocrine, immune, and vascular circulatory.

These systems are in a dynamic equilibrium, which is dependent on different internal (acid–base balance, tissues hydration, hormonal balance, enzyme system activity, intact metabolism) and external factors such as radiation (ionizing and nonionizing), chemical pollution (in the air, water, and soil), stress, viruses, and microbial agents. In the case of prolonged stimuli that may exceed the body's adaptive capacity, a certain pathology, disturbance, or disease, reversible or nonreversible, may develop depending on the intensity, frequency, and persistency. Work in physiotherapy EMF environments for months or years may cause some interference with the personnel's health.

- *Radio-sickness* (*microwave sickness*): First described in the 1960s by Sadchikova and Glotova (1973)—fatigue, weakness, sleep disorders, irritability, and vegetative nervous system disturbances. Thcy defined three clinical phases of progression: (1) asthenic (vagothonia, artherial hypotension, bradycardia); (2) astheno-vegetative: significant manifested asthenia, sympathicothonia with vascular dystonia, and hypertension; (3) hypothalamic: the pathological manifestation persists, paroxysmal sympatho-adrenal crises, ischemic heart attacks, and ophthalmological problems. Later Firstenberg (2001) confirmed this sickness as "microwave sickness" and described the so-called *additional syndrome*—nonspecific compliances: respiratory (bronchitis, sinusitis, influenza-like symptoms), *oppression*, eyeball pressure, soreness in the throat, perspirations, "flying" body pain, unspecified pelvic pain, leg and foot pain, epistaxis, digestive problems, skin irritability, tinnitus, tooth pain with metallic taste.
- *Circulatory disturbances*: These are described mainly as hypertonia due to irritation of the sympatho-adrenal system with an increase in stress hormonal secretion (Israel et al. 2007, Vangelova et al. 2007).
- *Reproductive outcomes*: The reports of reproductive outcomes are very limited but consequences are serious enough not to be dismissed. Dimitrova (1965) described the majority of young (30–39 years old) female personnel suffering from hypermenorhoea and menstrual cycle shortening.
- *Miscarriages*: In the early studies, this is one of the most worrying findings. Because of the complexity of such outcomes, no secure factor relation has been established. Reported

cases of pregnancy ending in miscarriage among physiotherapists are very controversial in terms of possible physical factors being the reason. Some of them occur between 6 and 10 weeks of gestation, and others during the first trimester. Shortwave diathermy, ultrasound, and transcutaneous electric nerve stimulation (TENS), in combination or separately, are most likely factors that are attributed to miscarriages (Cromie et al. 2002, Dimitrova 1965, Lindbom and Taskinen 2000, Taskinen et al. 1990). Magnavita and Fileni (1994) excluded therapeutical ultrasound as a cause of overexposure due to the fast airborne signal desertion.

- *Malformations*: These have the same uncertainties as miscarriages. The assumed cases of malformation in offspring, partly ending with habitual abortions and partly with the impossibility of excluding suspicious physical factors as a reason, present arguments for the analysts to continue to raise this issue as a likely consequence (Cromie et al. 2002). Previous studies, however, had "higher than expected" congenital abnormality findings (the estimated percentage being 13% versus about 3% in the general community).

- *Hematological deviations*: Very few papers on changes in blood tissue have been published. Leucopenia with lymphocytosis and anemia has been found studied by Dimitrova in 1965, but it must be noted that the requirement for shielding compartments for HFTh did not exist at that time.

- *Infections and mycoses*: These are more often related to hydrotherapy procedures where personnel come into contact with water, chemicals, or contaminants. In particular, skin irritation or dermatitis has been related to different kinds of disinfectant use (Cromie 2002). Standford et al. (1995, by Cromie et al. 2002) reported that physiotherapists' professional group is among the "high-risk occupations for hepatitis B" and Von Guttenberg and Spickett (2009) announced on the other hand that in Australia, physiotherapists in hospitals are significantly exposed to body fluids, causing high risk levels compared to other physiotherapy facilities, especially those practicing acupuncture.

- *Low-back pain*: This can be caused by musculoskeletal disorders and injuries associated mainly with lifting and bad posture due to the execution of different physical methods, for example, massage. The authors considered the need of guidelines to follow a right ergonomic algorithm to reduce injuries, as well as the musculoskeletal load and vertebral complications in all regions of the spine (Cromie et al. 2001, 2002).

- *Rare nonspecific complaints*: Asthma (nonsmoking-related) (Liss et al. 2003); eye, nose, and throat irritations (Tarlo et al. 2004); however, no significant statistics have been obtained for physiotherapists in comparison studies with radiographers. Stress (burnout syndrome) is observed as one reason for taking leave by young physiotherapists in Australia with an inpatient caseload compared with physiotherapists working with outpatients (Lindsay et al. 2008). Wernicky and Karoly (1995, by Cromie at al. 2002) have found cases of hearing loss.

22.6 FIRST CAUSE-RELATED COMPLEX MORBIDITY STUDY

A total of 267 men and women from 30 physiotherapy units were studied (Vesselinova 2013b) over a period of 3 years (2004–2007) under professional conditions to establish specific types of pathology and to find their correlated causes , by means of a complex original survey card (COSC, created and validated in 1999–2001) on somatic and neurobehavioral health deviations. The information was coded and processed anonymously according to the ethical principles and legislations for protecting personal information. The combined retroprospective cohort indirect survey with the option of anonymity among physiotherapy personnel in native work conditions was chosen. Some of the questions were orientated to estimate the role of the HF component or another environmental factor as an aggravating factor. Another focus was the assessment of the work algorithm's ergonomics and the need to assemble a prevention program (Vesselinova 2013b).

Group structure: The following data were obtained from 30 nationwide physical therapy facilities of the Republic of Bulgaria: specialists in physical and rehabilitation medicine (physiatrists) (13%), physiotherapists and kinezitherapists from electro/light therapeutic sections (68%) and from the kinezitherapy section (15%), hospital attendants (4%). The COSC, which focuses on the so-called target somatic body systems and behavioral deviations recognized in the literature for the EMF as well as on some additional specific aspects (myoma, osteoporosis, photosensibilization, menopause), has 102 questions, divided into two sections. The first section contained 82 questions comprising 9 topics: (I) q. 1–13—personal data and general work and health status information; (II) q. 14–21—duty obligations and contact with EMF sources; (III) q. 22–31—screening checkups, skin diseases, musculoskeletal system and blood vessels' disorders, peripheral nervous system status, ophthalmology diseases; (IV) q. 32–54—urinary and genital system/female/ concerning menstruation, pregnancies and births, gynecological disorders, operations, menopause; (V) q. 55–58—neoplasms, localization, diagnostic history; (VI) q. 59–67—cardiovascular problems, immunological status, dermatophotosensibility, endocrinal disturbances, semen analysis deviations/male/, offset retrospection; (VII) q. 68–70—somatological status; (VIII) q. 71–80—personal attitude and recommendations; (IX) q. 81 and 82—height and weight. The second section of the COSC is focused on some behavioral and psycho-emotional reactions. It includes 20 questions about occurrence of headache, tremor, acrocyanosis, sudden perspiration, tachycardia, breathlessness, faintness, fainting fits, unmotivated changes in mood, tenseness, irritation, ungrounded anxiety, hot waves, acids, change in libido, sleep problems, satisfaction, overall psychological adjustment, feelings of euphoria or of capsulation, and harmful habits (Vesselinova 2012a,b).

22.6.1 EMF Propagation in Closed Space by More Than One Emitting Source

As we underlined while analyzing the specific background, the ELTS are overloaded with unhomogeneous and stochastically changing EM exposures from simultaneously emitting apparatuses at different ranges and modalities. For a better understanding of this complicated situation as shown in Figures 22.1 through 22.4, we strive to show the exposure burden surrounding workers and patients in every ELTS. A specially orientated experiment with Specific Anthropomorphic Mannequin SAM1 and SAM2 was conducted. *Design*: The two phantoms, SAM1 and SAM2, are placed in a closed space with metal walls with tree active antennas emitting at 900 MHz, EM-RF (mobile communication zone) in the usual phone position (on ear), and the third at 6 cm for the nose of SAM1. On the 2D images after emission is activated we clearly see the generated EM power field, which

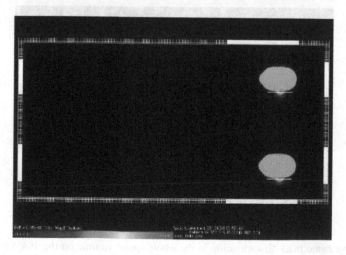

FIGURE 22.1 EMF in closed-space sources activating initializations (158×10^{-12} s).

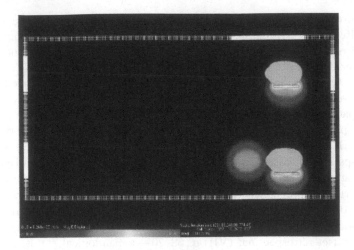

FIGURE 22.2   The field on the 543×10^{-12} s of sources activating.

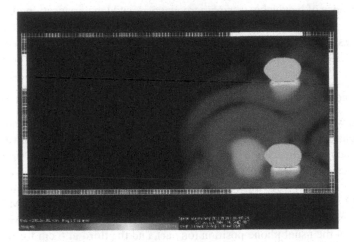

FIGURE 22.3   The propagated electrical field on the 2084×10^{-12} s of emission activating.

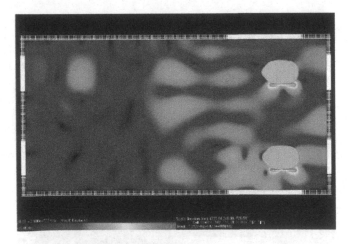

FIGURE 22.4   The same field dissemination in the whole space volume on the 18455×10^{-12} s of emission activating.

is virtually almost instantly dispersed, with overlapping zones, signal reverberation from the walls, and electrical field point confluences, which monotonically decreased with distance from the source (Figures 22.1 through 22.4). The space overload dependence on emission duration and the short time needed to reach this phenomenon are perfectly recognizable. One more very important visual detail that has been shown is that, practically, the electro/light section is not unexposed or free of EMF load zone (Figure 22.4).

22.6.2 Evaluation of the COSC Results

For the evaluation of the results, data were statistically analyzed by SPSS 13.0 (for Windows). The critical level of significance is $\alpha = 0.05$.

22.6.3 Demographic Section of the Study Group

From the 267 examined respondents, 240 were women (90.15%) and 27 men (9.85%). The average age of the researched personnel was 45.3 years (between 22 and 71 years), and demographic section of the group studied is presented in Table 22.2.

The data show that the average length of work experience was 13.8 years, with a significant number of employees having worked for more than 10 years ($p < 0.05$). The temporary absences from work for different reasons (maternity, acute infections, operations) did not exceed 17.98% of the studied occupational group and could not reflect the continuity of the staff in the work environment. The average duration of work in the ELTS was 5.8 h, which exceeded the statutory law in Bulgaria of up to 4 h risk-free stay in such an environment. The results were reliable and allowed for correlation to be conducted.

The sources emitting EMF with which the personnel from ELTS are in contact and are exposed to are presented in Table 22.3.

In 82.45% of the facilities, the sources of HFTh are placed in shielded compartments; therefore, in the interpretations this indicator is accepted as available. From the data presented, the objective results could not describe the contact with sources of HFTh during extended work as dominant, with 47.53% negative responses ($p > 0.05$).

The general somatic morbidity and the subnosology analysis (percentage of the main system nosology) in the group investigated have been presented in a previous publication (Vesselinova 2013a). General morbidity has indicated some unexpected events such as periodontitis to be the leading focus and photosensibilization and osteoporosis as being significantly present. Periodontitis, cardiovascular disturbances, especially arterial hypertension, allergic manifestations, photosensibilization, and musculoskeletal disorders, especially osteoporosis, seem to be the common clinical manifestations of physiotherapy personnel (equal to or over 30%). Using the confidence interval

TABLE 22.2

Presentation of Demographic Section Studied

Assessed Index	Validation	Average Mean
Age (years)	263	45.3 (22 ÷ 71)
Work experience extent (years)	256	13.8
Duration of stay of personnel in the ELTS (hours)	203	5.8 (1 ÷ 8)
Absence from work (months)	48	15.7 (2 ÷ 48)
Height (cm)	236	165.27 (155 ÷ 190)
Wight (kg)	237	65.71 (48 ÷ 115)

TABLE 22.3

Apparatuses for Electro/Light Therapy That Personnel Are in Contact with Simultaneously

| | Personnel in Contact | |
Therapeutic Source	Total	Percentage
Magnet	193	72.28
Ultrasound	193	72.25
Optical light (visible, UV light)	183	68.52
Radar	178	66.67
High frequency	176	65.87
D'Arsonval	137	51.34
Monochromatic light (laser)	62	23.22

(CI) method, the relative rate for developing these events in the general active physiotherapy population was checked (95% confidence in optimistic [<] and pessimistic [>] ranges). Anxiously pessimistic ranges (over 30%) were found for periodontitis (47.9%); cardiovascular disorders (47.5%); allergic manifestations (46.7%); photosensibilization (39.8%); skin diseases (37.0%), among which were herpes simplex (51.0%) and erythema (35.6%); musculo skeletal disorders (35.5%), among which was osteoporosis (54.6%), whose optimistic ranges were also high (32.9%) (Table 22.4).

The huge percentage of periodontitis and cases with low mineral bone density opens the question of the EMF's interaction mechanism and its possible interrelationship with manifestation. Only two aspects of interaction could be proposed: osteoblast suppression and osteoclast stimulation, and probably vitamin D–induced insufficiency due to malabsorption or blockage of absorption processes.

The EMF's influence on the circulatory system is not a surprising finding and its effects on the studied group were expected. Hypertension (45.05%) can be referred as a positive finding among the results for increased stress-hormonal levels by EMF radiation, which is suspected as a deleterious factor on the cardiovascular system as described in many studies (Bortkiewicz et al. 1995, Hillman 2005, Vangelova et al. 2006, Vecchia et al. 2009). Bulgarian research among a group of 52 physiotherapists reported similar effects, with arterial hypertension and metabolite changes (Israel et al. 2007).

Hypotensive effects are another significant aspect of vessel microcirculatory reaction after EMF exposure. Our results showing 33.33% with hypothonia, 11.71% with dystonia, and 22.7% with ischemic strokes support the manifestation of this process, described in experiments as a first response mainly to low-frequency fields due to vasodilatation (Traikov et al. 2005) or to chronic exposures to EMF (Grigoriev et al. 2010).

It is interesting to note the distribution and the variations of morbidity in the exposure criteria of the subgroups. They are formed to observe how occupational nonionizing radiation burden can interfere with health.

22.6.4 Exposed Subgroups

Three exposed subgroups (EsGs) were formed according to the duty obligations of daily work in the ELTS: **Ist EsG ≤ 4 h** ($n = 51\% - 19.1\%$), **IInd EsG > 4 h** (whole working day, $n = 153\% - 57.3\%$), and **IIIrd EsG = 0 h** (personnel without electro/light contact, $n = 63\% - 23.6\%$). This way the real exposure effects (complaint manifestation and further pathology) could be observed. As it is clearly evident, the group mainly affected is the second subgroup of personnel, whose location of duty is in the ELTS (Table 22.4).

TABLE 22.4
General Morbidity, Expositional Subgroups (ESG) Distribution, Confidence Intervals and Statistical Significance

General Morbidity	Percentage (n=267)	Confidence Interval		ESG Predilection and Statistical Significance		
				Ist EsG	IInd EsG	IIIrd EsG
				≤4 h	>4 h	0 h
				(%; n=51)	(%; n=153)	(%; n=63)
Periodontitis	41.95	36.0	47.9	45.1	45.8	30.2
Cardiovascular	41.57	35.7	47.5			
Arterial hypertension	45.05			27.45	20.91	6.3
Hypotonia	33.33			11.8	17.0 (p<0.0001)	7.9
Dystonia	11.71			13.7	3.3	1.6
Stenocardia	5.41					
Allergic manifestations	40.82	34.9	46.7	19.6	28.8	9.5
Urticaria	55.0					
Severe	45.0					
Photosensibilization	34.08	28.4	39.8	29.4	36.6	31.7
Skin diseases	31.50	25.9	37.0	13.7	22.9	12.7
Herpes symplex	40.48	30.0	51.0	11.8	14.4	9.5
Erythema	26.19	16.8	35.6	5.9	11.1	3.2
Staphyloccal infection	14.28			—	5.2	6.3
Musculoskeletal disorders	30.00	24.5	35.5	—	—	—
Osteoporosis	43.75	32.9	54.6	13.7	15.0 (p<0.0001)	7.9
Degenerative	32.5			19.6	9.2	3.2
Fractures	21.3			3.9	8.5 (p<0.0001)	3.2
Excretory system	18.35	13.7	23.0			
Kidney diseases	67.3			21.6	10.5	9.5
Other	32.7			3.9	6.5	6.3
Ophthalmology	17.98	13.4	22.6	—	—	
Inflammations	81.3			9.8	19.6 (p<0.0001)	6.3
Retinopathy	18.8			3.9	3.3	3.2
Endocrinal	13.48	9.4	17.6	MD	MD	MD
Cerebrovascular	8.24	4.9	11.5	—	—	
Ins.haemorragicus	45.5			2.0	5.2 (p=0.010)	1.6
Transitory disturbances	31.8			0	3.9 (p=0.010)	0
Ins.ischaemicus	22.7			3.9	2	1.6
Peripheral nervous system	6.00	3.1	8.8			
Hypestesy	37.5			1.96	2.6	1.6
Neuropathy	37.5			3.9	2.6	0
Hyperestesy	25.0			1.96	0.65	3.2

The analysis of the results leads to the conclusion that the first clinical manifestations of any disturbances take place almost after 1 year of work (CI 13.1%–22.8%) and most of these are reversible. In female physiotherapists and physiatrists, often the first signs of interference in the body are menstrual disturbances (menorrhagia) in the first 6 months of working in specific occupational EMF conditions in physiotherapy (29.7%). The percentage of miscarriages, which were recorded in 28.26% of female respondents, goes outside the referent 10%–15% of the general community. However, it is too difficult to suspect EMF work conditions as the main cause to this finding on this basis alone; therefore, a correlation with the possible causes was made. Two factors—predominant contact with HF sources ($p=0.036$) and the second EsG's daily work in the ELTS ($p=0.025$)—were statistically significant for the interruption of pregnancy due to miscarriage. The problem is that in most cases the first week of pregnancy are "silent" and no precaution is therefore taken. The data obtained from offspring's malformations are difficult to typify by systems because of missing or incorrectly filled out data in the COSC. Hence, to avoid hyperdiagnosis the rate was calculated on the total count of female staff ($n=240$), and not only on the pregnancies took place ($n=152$). The 12 self-reported cases are a serious warning and need further clarification. The medical personnel in the fertility age should be paid special attention and this pathology has to be set in the observed professional pathologies.

The question about the neogenesis potential of the EMF is still controversial in the literature because of the impossibility of setting aside other possible carcinogens nowadays. The confirmatory findings for the prevalence of breast cancer cases among physiotherapy personnel over the other types not only cause anxiety but could also lead to possible disturbances in hormonal balance. The second significant cancer pathology, which we established as the second most prevalent, was ovarial cancer, which led to the same correlation. As a whole, the possible neogenesis unlocking in the second EsG has been statistically confirmed ($p=0.033$). The established tendency for breast cancer predominance among neoplasms (10 cases out of 18 with neoplasms per whole group) confirmed this EMF predilection burden, as reported in subject-related researches (Coogan et al. 1996, Demers et al. 1991, Sellman 2007). Disturbances and diseases of the female reproductive system in the general group by EsG are presented in Table 22.5.

From data obtained from 38.2% of respondents who were smokers, a separate cross-tabulation of the main disease, whose development could have originated from smoking as an accepted universal noxious factor, was made and showed independent smoking manifestation ($p > 0.05$).

The results on neurobehavioral changes are very interesting. Some consistency is noted in the statistical significance. Sleep disturbances, irritability, hot waves, headaches, and sudden perspiration at work are evidenced for second EsG for sure by "p value" examination ($p < 0.0001$). This not only confirms the radio-sickness symptoms' manifestation in the physiotherapy professional group, but also questions the severity of consequences that are obviously caused by

---

TABLE 22.5

Disturbances and Diseases of the Female Reproductive System, Confidence Intervals, Expositional Subgroups (ESG), Distribution with Statistical Significance

			ESG Predilection and Statistical Significance		
			Ist EsG	IInd EsG	IIIrd EsG
			≤ 4 h	> 4 h	0 h
Reproductive System Disturbances	Percentage ($n=240$)	Confidence Interval	(%, $n=51$)	(%, $n=153$)	(%, $n=63$)
Dysmenorrhoea/as menstrual disturbance/	20.0	17.1 25.4	19.6 ($p<0.0001$)	14.4	3.2
Myoma uteri	15.83	11.2 20.5	15.7	17.0	6.3
Hysterectomy	10.0	6.2 13.8	9.8	9.2	4.8
Miscarriages	28.26% ($n=92$)	19.1 37.5	($p=0.001$)	($p=0.001$)	
Offspring's malformations	5%	2.2 7.8	2.0	4.6	6.3
Neoplasms	7.5	4.2 10.8	3.9	7.2 ($p=0.033$)	7.9
Ca gl.mamme	55.6		2.0	4.6	3.2
Ca ovarii	27.8		3.9	2.0	0
Mel. malignum	11.1		0	1.3	0

TABLE 22.6

Neurobehavioral Disturbances and Compliances, Confidence Intervals, Expositional Subgroups (ESG) Distribution with Statistical Significance

Neurobehavioral Disturbances	Percentage ($n=267$)	ESG Predilection and Statistical Significance		
		Ist EsG ≤ 4 h (%, $n=51$)	IInd EsG > 4 h (%, $n=153$)	IIIrd EsG 0 h (%, $n=63$)
Haedache	42.7	23.5	36.6 ($p<0.0001$)	17
Faintness				
At the end of the working day	40.4	43.1	43.8	30.2
During work hours	13.1	15.7	16.3	3.2
Heart palpitations	34.1	47.1	33.3	25.4
Sleep disturbance				
In quality	31.5	25.5	37.9 ($p<0.0001$)	17.5
In the duration	21.0	21.6	22.2 ($p<0.0001$)	17.5
At falling asleep	18.4	15.7	20.9 ($p<0.0001$)	14.3
Nonmotivated mood changes		9.8	19.0	9.5
At the end of the working day	26.6			
During work hours	13.5			
Irritability				
At the end of the working day	26.6	23.5	30.7 ($p<0.0001$)	19.0
During work hours	13.5	5.9	18.3 ($p<0.0001$)	7.9
Perspiration	25.5	25.5	29.4	15.9
Suddenly at work	13.9		($p<0.0001$)	
Breathlessness	13.1	19.6	11.1	12.7
Tremor	9.4	5.9	13.1	3.2
Acrocyanosis	3.4	2	4.6	1.6
Hot waves	22.09	19.6	26.2 ($p<0.0001$)	14.3
Libido changes	7.1	11.8	7.2	3.2
Agitation	40.07	5.9	19.6	11.1

body stress and the disturbed ability for relaxation and recovery. The neurobehavioral disturbances provide conclusive evidence of the existence of radio-sickness symptoms among this EMF professional group. These are very important reactions to an exteriorly applied irritant (low-intensity EMF) (Table 22.6).

22.6.5 MORBIDITY DISTRIBUTION IN THE EXPOSURE SUBGROUPS ASSUMPTION

22.6.5.1 Ist EsG

In this subgroup, which works with EMF sources irregularly and less than 4 h per working day, a prevalent tendency of the association with the following *diseases* is shown: disturbance in the menstrual cycle (termed as dysmenorrhea in the questionaire), leyomyoma uteri development, malignancy of ovarian cancer, cardiovascular disorders—hypertension and dystonia, degenerative diseases of the musculoskeletal system, and kidney diseases; and from the *neurobehavioral interferences*—heart palpitations, breathlessness, seizures, heartburns, and changes in libido. Due to the small number of reported cases of cerebrovascular disorders, peripheral nervous system, and retinopathy (less than 10% of the studied professional population), we could only express an assumption of the risk of an increased likelihood of developing ischemic cerebrovascular disorders, peripheral affecting of the nervous system of the neuropathy type, and retinopathy.

22.6.5.2 IInd EsG

In this group, with the highest level of exposure burden, a prevalence of 79% of the studied indices is established, associated with the manifestation of skin diseases (herpes and erythema), allergy, diseases of the musculoskeletal system (bone fractures and osteoporosis), inflammations of the peripheral urinary system, meno- and metrorrhagia, eye inflammatory diseases, breast cancer (and because of limited cases, a tendency to melanoma malignum), circulatory disorders displaying hypotension, photosensibilization, headache, sudden perspiration, fatigue (tiredness), tremor, acrocyanosis, unmotivated mood changes, increased irritability, agitation, feeling of euphoria, and sleeping problems. Menopause in this EsG covers the largest percentage of women and the additional data processing in this direction shows a shortening of its age at occurrence: average age 49.5 years (between ages of 48 and 51 years). We hypothesize and relate this shortness as an attempt of the body to protect itself from elevated estrogen levels, indirect indications of what we found in the established morbidity in the investigated group.

Due to the small number of reported cases of cerebrovascular disorders and disorders of the peripheral nervous system (less than 10% in all three EsG for each nosology) we could only express an assumption of the risk of an increased likelihood of developing transient and hemorrhagic cerebrovascular disorders, which has shown a great degree of importance and peripheral hypoesthesia. Statistical significance for this group shows that inflammatory eye diseases have a prevalence of $p < 0.0001$, independent of the limited self-reported cases.

22.6.5.3 IIIrd EsG

This group works outside the ELTS. The main concern is the appearance of a bias of migraine headache and stomach acids at work. The number of cases of periodontitis, photosensibilization, and retinopathy among this group are close to the findings in the two other EsGs, which can be explained because of the diffused effects of the EMF in the far zone as well as its dissemination in closed spaces, as was shown in the designed experiment with phantoms SAM1 and SAM2 (Figures 22.1 through 22.4). Due to the small number of reported cases (less than 10% of the studied population) we could only express an assumption of the risk of an increased likelihood of developing staphylococcal infections and hyperesthesia, maybe attributed to the nature of the work of this group in which personnel come into direct contact with organic materials (patient's body).

22.6.5.4 Manifestation of Periodontal Disease Is Significant for All Exposed Subgroups

Analysis of these detailed data shows that all physiotherapy staff are affected to various degrees and emphasizes the necessity of taking specific measures for prevention to reduce the overexposure risk among professional groups in an EMF environment.

22.6.6 EMF–MORBIDITY CORRELATIONS

By ANOVA Fisher's exact test calculation, the EMF–morbidity correlation was studied. The caseload of the ELTS is statistically significant for causing the development of breast cancer ($p = 0.006$) and provoking carcinogenesis in general ($p = 0.044$). The prevalent contact with sources of HFTh is the cause for the manifestation of photosensibilization ($p = 0.013$).

22.6.6.1 Cumulative Effect

Around 80% of personnel have over 10 years of extended work experience. Many authors specify this index as one of the crucial factors for reviewing the provoked pathology due to chronic electrostress (Kositsky et al. 2001, Sadchikova and Glotova 1973). The main affected indices in our study that have major significance for health are as follows: in general—enhancement of carcinogenesis potential ($p=0.025$) and predisposition to ovarian cancer development ($p=0.027$); for the IIIrd EsG—endocrine disorders ($p=0.036$), leomyoma uteri ($p=0.036$), and anxiety ($p=0.024$). On the cumulation index (length of service >10 years), the logistic regression analysis conclusively shows significant increase in the risk for development of erythema by about 11 times. Skin problems such as erythema, hypersensitivity, and herpes simplex were first described by videodisplay workers in Norway, and later many researches obtain such confirmatory results from residents or office staff (Arnetz et al. 1997, Johansson and Liu 1995, Leitgeb and Schrottner 2003). Data obtained through our research show 40.82% for allergies and a non-negligible percentage of physiotherapists with herpes simplex (14.2%), which can be explained as the EMF's influence on immunity (Grigoriev et al. 2010, Kositsky et al. 2001).

22.7 DISCUSSION

To the best of our knowledge, this first not only for Bulgaria complex research studying causal correlations shows that adverse effects caused by chronical low-intensity disparate EMFs on physiotherapy personnel health are well recognizable. The evidenced relation to specific work conditions first established disorders such as periodontitis, myoma uteri, photosensibilization, and osteoporosis, as well as the age-shortening of menopause (praecox menopause), which indirectly elucidates hormonal (estrogen) interference as one of the main mechanisms in pathogenesis, supporting the already recognized effect of EMR for breast cancer, confirmed by us, too. This compromise could also have a deeper relation. If we consider observed sleep disturbances in all sleep components (falling asleep, duration and quality of sleep) in the second EsG, it could be suggested that the melatonin balance is compromised. As a result, one can understand melatonin's protective role against elevated estrogen levels. By Fisher's exact criteria, statistically significant causal relations have been evidenced: ELTS caseload, predominant work with sources of HFTh, external radiation cofactor existence, and high BMI ($p<0.05$). Because specific work conditions are not only due to the RF component (it is not always a prevalent and permanent factor in daily specialized practice), the persisting problem enhancing morbidity is the combined placement and simultaneous working of sources at different frequency ranges: low frequency, intermediate frequency, HF, optic radiation, magnetic field, with complete stochastic changes in the work environment (sometimes there is an additional radiating source in the vicinity) which interfere with normal body functioning. The complicated radiation conditions in ELTS demands measures to reduce the overexposure burden and to "separate" the diverse frequency interferences by administrative decisions and appropriate new risk-mitigated work ergonomy. The evidence of the enhanced burden of cumulative and extended work experience elucidates the adversity of long-term, low-intensity, heterogenous NIR (EM) radiation. The established potential of the enhancement of carcinogenesis is concerning enough to plead for immediate action of precaution and prevention. Outside of the boundaries of physiotherapy professional group the established interrelations can play a key-role in understanding and revealing the nonthermal interactions and bio-activeness of chronically exposed, low-intensity, frequency nonhomogenous EMFs which occur in recent real life and are a silent menace with possibly detrimental potential for the living world.

22.8 CONCLUSIONS

In summary, the results of the analyses of personnel health consequences due to factors of physical medicine undoubtedly show that the specific low-intensity EMF background in clinical physiotherapy practice is an evidenced hazard for personnel health. This is especially convincingly presented in the results of the first cause related complex morbidity study. The work exposure situation is similar wherever physical medicine is practiced integratively. The reported pathology and symptoms should to be assessed as a need to assemble special precaution and risk communication without delay. The diseases of the female reproductive system (leiomyoma uteri, miscarriages, cancer, praecox menopause), the circulatory system (hypertonia), and the musculoskeletal system (osteoporosis), with the manifestation of periodontitis as well as the typical "radio-sickness," have important social significance in the authors' knowledge, while other conditions have yet to be confirmed by observations in professional pathology. Mobile phone usage in these sections has to be prohibited not because this might "disturb the generators' proper work," as many facilities underline in their warning, but because it disturbs the exposure climate with adverse effects risk raising for both personnel and patients. Another less obvious important factor that remains to be realized is the improvement in the—staff's "radiation" competence. This will have to be mobilized as a special part of risk communication, which would be a powerful motivating factor to respect both private and public occupational electro "hygiene" and discipline. Personal attitude to precaution and prevention can be a big contribution.

Even though the hazards of long-term exposure to low-intensity EMF for physiotherapy personnel have been clearly revealed. More efforts are needed to secure protection of both personnel and patients by legislation changes of regulations of such conditions. The conclusive scientific evidence of the causal relationship needs to appropriately adapt precaution and prevention. Thereby, the workspace of physiotherapy personnel will become a safe area to successfully practice the noble medical profession.

In a century that has faced big human losses because of local military conflicts, disasters, accidents, and terrorism, every safe and healthy member of the society is a victory against the civilization's monstrous threats of an advanced society so as the catastrophic spread the EMF sources.

ACKNOWLEDGMENTS

For the presentation of the experiment with Specific Anthropomorphic Mannequin SAM1 and SAM2 to define and figure out the propagation of the EMF in closed space, I owe my thanks to Associate Professor Nikolay Atanasov, PhD, Eng. from the College of Telecommunications, Sofia, Bulgaria.

For the investigation of the physiotherapies in Bulgaria, part of my dissertation was realized with the contribution of Associate Professor Marin Marinkev, PhD, President of the Association of Physical Medicine and Rehabilitation in Bulgaria (2001–2010), and of Associate Professor Michel Izrael, PhD, President of the National committee of NIR.

ABBREVIATIONS

COSC	complex original survey card
CSWD	continuous shortwave diathermy
ELTS	electro/light therapy section
EMF	electromagnetic field
EMR	electromagnetic radiation
EsG	exposed subgroup
EU	European Union
HF	high frequency
HFTh	high-frequency therapy
MF	magnetic field
MRI	magnetic resonance imaging
NIR	nonionising radiation
PSWD	pulsing shortwave diathermy
RF	radio frequency
UV	ultraviolet

REFERENCES

Adair RK (1991) Constraints on biological effects of weak extremely-low frequency electromagnetic fields. *Phys. Rev. Ann.* 43, 1039–1048.

Adey WR (1983) Some fundamental aspects of biological effects of extremely low frequency (ELF). In: *Biological Effects and Dosimetry of Nonionizing Electromagnetic Fields*, Grandolfo M, Michaelson, SM, eds. Plenum, New York, pp. 561–580.

Ahlbom A, E Cardis, A Green, M Linet, D Savitz, A Swerdlow (2001) Review of the epidemiologic literature on EMR and health: ICNIRP (International Commission for Non-Ionizing Radiation Protection) standing committee on epidemiology. *Environ. Health Perspect.* 109(Suppl 6), 911–933.

Aleksiev A (2013) Comparative analysis between electro-analgesia with or without iontophoresis in exacerbated low back pain. *Phys. Med. Rehabilit. Health J.* XII (4), 11–18. (Abstract in English.)

Arnetz BB, M Berg (1996) Melatonin and adrenocorticotropic hormone levels in video display unit workers during work and leisure. *J. Occup. Med.* 38, 1108–1110.

Arnetz BB, M Berg, J Arnetz (1997) Mental strain and physical symptoms among employees in modern offices. *Arch. Environ. Health* 52, 63–67.

Basset A (1994) Therapeutical uses of electric and magnetic fields in orthopedics. In: *Biological Effects of Electric and Magnetic Fields*, Carpenter DO, Ayrapetyan S, eds. Academic Press, San Diego, CA, Vol. 2, pp. 13–48.

Beal J, D Fagin (1995) Are you at risk from sick building syndrome? *Family Circle*, April 25, pp. 70–73.

Becker RO, AA Marino (1982) *Electromagnetism and Life*. State University of New York Press, Albany, NY, p. 156.

Bobrakov SN, AG Kartashev (2001) The electromagnetic component of modern urbanized environment. *Rad. Biol. Radioecol. J.* 41(6), 706–711. (in Russian.)

Bortkiewicz A, M Zmyslony, C Palczynski, E Gadzicka, S Szmigielski (1995) Dysregulation of autonomic control of cardiac function in workers at AM broadcasting stations (0.738–1.503 MHz). *Electro. Magnetobiol.* 14(3), 177–191.

Commission of the European communities (2000) Communication from the commission on the precautionary principle, Brussels. http://ec.europa.eu/dgs/health_consumer/library/pub/pub07_en.pdf.

Coogan PF, RW Clapp, PA Newcomb et al. (1996) Occupational exposure to 60-Hertz magnetic fields and risk of breast cancer. *Women Epidemiol.* 7, 459–564.

Cromie JE, VJ Robertson, MO Best (2001) Occupational health and safety in physiotherapy: Guidelines for practice. *Aus. J. Physiother.* 47, 43–51.

Cromie JE, VJ Robertson, MO Best (2002) Occupational health in physiotherapy: General health and reproductive outcomes. *Aus. J. Physiother.* 48, 287–294.

Demers PA, DB Thomas, KA Rosenblatt et al. (1991) Satariano occupational exposure to electromagnetic fields and breast cancer in men. *Am. J. Epidemiol.* 134(4), 340–347.

Dimitrova SV (1965) On hygiene in electro-light offices of physiotherapy departments. *Resorts Physiother. J.* II(1), 31–35. (in Bulgarian.)

Final technical report on Occ EMF exposure. EMF-NET: Effects of the exposure to electromagnetic fields: from science to public health and safer workplace, 2008; D49:44.

Firstenberg A (2001) *Radio Wave Packet Cellular Phone Taskforce*. September 4.

Goldsmith JR (1997) Epidemiologic evidence relevant to radar (microwave) effects. *Environ. Health Perspect.* 105, 1579–1587.

Grigoriev YG, OA Grigoriev, AA Ivanov et al. (2010) Confirmation studies of Soviet research on immunological effects of microwaves: Russian immunology results. *Bioelectromagnetics* 31, 589–602.

Habash RWY (2008) *Bioeffects and Therapeutic Applications of Electromagnetic Energy*. Taylor & Francis Group, Boca Raton, FL.

Hillman D (2005) Exposure to electric and magnetic fields (EMF) linked to neuro-endocrine stress syndrome: Increased cardiovascular disease, diabetes and cancer. *Shocking News* 8, 1–8.

Irving PH (2007) *Physics of the Human Body*. Springer-Verlag, Berlin, Germany, p. 845.

Ishido M, H Nitta, M Kabuto (2001) Magnetic fields (MF) of 50 Hz at 1.2 µT as well as 100 µT cause uncoupling of inhibitory pathways of adenylyl cyclase mediated by melatonin 1a receptor in MF-sensitive MCF-7 cells. *Carcinogenesis* 22(7), 1043–1048.

Israel M, K Vangelova, M Ivanova (2007) Cardiovascular risk under electromagnetic exposure in physiotherapy. *Environmentalist* 27(4), 539–543.

Israel MS, P Tschobanoff (2006) Exposure to non-ionizing radiation of personnel in Physiotherapy. In: *Bioelectromagnetics: Current Concepts*, Ayrapetyan SN, Markov MS, eds. Springer. pp. 367–376.

Johansson O, PY Liu (1995) Electrosensitivity, electrosupersensitivity and screen dermatitis: Preliminary observations from on-going studies in the human skin. In: *Proceedings of the COST 244: Biomedical Effects of Electromagnetic Fields-Workshop on Electromagnetic Hypersensitivity*, Simunic D, ed., Brusse, Belgium, EU/EC (DG XIII), pp. 52–57.

Karabetsos E, M Basiouka, A Zissimopoulos (2010) Occupational exposure of physical therapists to radio frequency radiation-the situation in Greece. Workshop on biological effects of electromagnetic fields. In: *Sixth International Workshop on Biological Effects of Electromagnetic Fields*, October, Bodrum, Turkey.

Karpowicz J, Gryz K (2013) An assessment of hazards caused by electromagnetic interaction on humans present near short-wave physiotherapeutic devices of various types including hazards for users of electronic active implantable medical devices (AIMD). *BioMed Research International*, p. 8. Article ID 150143, http://dx.doi.org/10.1155/2013/150143

Kositsky NN, AI Nizhelska, GV Ponezha (2001) *Influence of High-frequency Electromagnetic Radiation at Non-thermal Intensities on the Human Body* (A review of work by Russian and Ukrainian researchers) No Place To Hide—Newsletter of the Cellular Phone Taskforce Inc., 3(1 Suppl), 1–33.

Leitgeb N, J Schrottner (2003) Electrosensibility and electromagnetic hypersensitivity. *Bioelectromagnetics* 24(6), 387–394.

Lindbom ML, H Taskinen (2000) Reproductive hazards in the work place. In: *Women and Health*, Goldman MB and Hatch MC, eds. Academic Press, San Diego, CA, pp. 463–473.

Lindsay R, L Hanson, M Taylor, H McBurney (2008) Workplace stressors experienced by physiotherapists working in regional public hospitals. *Aust. J. Rural Health* 16(4), 194–200.

Liss GM, SM Tarlo, J Doherty et al. (2003) Physician diagnosed asthma, respiratory symptoms, and associations with workplace tasks among radiographers in Ontario, Canada. *Occup. Environ. Med.* 60(4), 254–261.

Magnavita N, A Fileni (1994) Occupational risk caused by ultrasound in medicine. *Radiol. Med.* 88(1–2), 107–111. PMID:8066232 (abstract in English).

Markov DV (2008) Hygienic evaluation and optimization of the electromagnetic environment of modern physiotherapy offices. PhD degree dissertation, Moscow, Russia. (in Russian.)

Markov MS (2006) Thermal vs. Nonthermal mechanisms of interactions between electromagnetic fields and biological systems. In: *Bioelectromagnetics*, Ayrapetyan SN, Markov MS, eds. Springer, pp. 1–15.

Markov MS (2010) Angiogenesis, magnetic fields and "window" effects. *Cardiology* 117(1), 154–156. DOI 10.1159/000315433.

Markov MS (2011) Nonthermal mechanism of interactions between electromagnetic fields and biological systems: A calmodulin example. *Environmentalist* 31, 114–120. DOI 10.1007/s10669-011-9321-1.

Markov MS, Hazlewood CF (2009) Electromagnetic field dosimetry for clinical application. *Environmentalist* 29, 161–168.

Michaelson SM (1974) Thermal effects of single and repeated exposures to microwaves—A review. In: *Proceedings of Biological Effects and Health Hazards of Microwave Radiation*, Warsaw, Poland, pp. 1–14.

Miclaus S, P Bechet, C Iftode (2010) Near field radiofrequency measurements for occupational exposure assessment by personal exposimeter: Possibilities and limitations. In: *Proceedings of Sixth International Workshop on Biological Effects of Electromagnetic Fields*, October, Bodrum, Turkey, pp. 10–14.

Nordin NAM, JH Leonard, NC Thye (2011) Work-related injuries among physiotherapists in public hospitals— A southeast Asian picture. *Clinics* 66(3), 373–378. DOI 10.1590/S1807-59322011000300002.

Pilla AA, Markov MS (1994) Weak electromagnetic field bioeffects. *Rev. Environ. Health* 10, 155–169.

Rasoulia J, R Lekhraj, NM White, ES Flamm, AA Pilla, B Strauch, D Caspera (2012) Attenuation of interleukin-1beta by pulsed electromagnetic fields after traumatic brain injury. *Neurosci. Lett.* 519, 4–8.

Recent Research on EMF and Health Risks (2007) Fifth Annual Report from SSI: Independent Expert Group on Electromagnetic Fields, SSI, Rapport, 2008:12.

Sadchikova MN and HW Glotova (1973) Clinic, pathogenesis and disease outcomes radiowave. In *Proceedings of the Laboratory of Radio Frequency Electromagnetic Fields of the Institute of Hygiene and Occupational Diseases of the USSR Academy of Medical Sciences*, 4, pp. 43–48. (in Russian.)

Sarwar Shah SG, A Farrow (2013) Assessment of physiotherapists' occupational exposure to radiofrequency electromagnetic fields from shortwave and microwave diathermy devices: A literature review. *J. Occup. Environ. Hygiene* 10(6), 312–327. DOI 10.1080/15459624.2013.782203.

Sellman S (2007) *Electropollution, Hormones and Cancer NEXUS*, August–September, pp. 1–7.

Tarlo SM, GM Liss, JM Greene et al. (2004) Work-attributed symptom clusters (darkroom disease) among radiographers versus physiotherapists: Associations between self-reported exposures and psychosocial stressors. *Am. J. Ind. Med.* 45(6), 513–521.

Taskinen H, P Kyyronen, K Hemminki (1990) Effects of ultrasound, shortwaves, and physical exertion on pregnancy outcome in physiotherapists. *J. Epidemiol. Commun. Health* 44, 196–201.

Todorov T, Sv. Dimitrova, T Dragiev, V Marinov (1965) Proffesional imparement by medical personnel working in the area of UHF and MW EMF of the physiotherapy departments. *Resorts Physiother. J.* II (3), 118–123. (in Bulgarian.)

Tolgskaya MS, ZV Gordon, VV Markov, VS Vorontsov (1987) Change in the secretory activity of the hypothalamus and certain endocrine glands to intermittent and continuous irradiation modalities. Medicine, Moscow, No. 4, pp. 87–90. (in Russian.)

Traikov I, M Israel (1994) Determining the risk of medical personnel servicing equipment for UHF therapy. *Phys. Resort Rehabilit. Med. J.* XXVI(2), 38–40./in Bulgarian/

Traikov L, A Ushiyama, G Lawlor, R Sasaki, K Ohkubo (2005) Subcutaneous Arteriolar Vasomotion Changes during and after ELF-EMF Exposure in mice in vivo. *Environmentalist*, 25, 93–101.

Ueno S ed. (1996) *Biological Effects of Magnetic and Electromagnetic Fields*, 1st edn. Plenum Press, New York.

Vangelova K, C Deyanov, M Izrael (2006) RF exposure adversely affected cardiovascular system. *Int. J. Hyg. Env. Health* 209(2), 133–138.

Vangelova K, M Israel, D Velkova, M Ivanova (2007) Changes in excretion rates of stress hormones in medical staff exposed to electromagnetic radiation. *Environmentalist* 27(4), 551–555.

Vecchia P, R Matthes, G Ziegelberger, J Lin, R Saunders, A Swerdlow (2009) Exposure to high frequency electromagnetic fields, biological effects and health consequences (100 kHz–300 GHz): ICNIRP 16/2009, ISBN 978-3-934994-10-2.

Vesselinova L (2012a) Electromagnetic fields in clinical practice of physical and rehabilitation medicine: A health hazard assessment of personnel. *Environmentalist* 32(2), 249–255. DOI 10.1007/s10669-011-9379-9.

Vesselinova L (2012b) Presentation the complex original survey card for biological effects of the electromagnetic fields on physiotherapy medical personnel assessment. *Phys. Med. Rehabilit. Health J.* XI(2), 20–27. (Abstract in English.)

Vesselinova L (2013a) Biosomatic effects of the electromagnetic fields on view of the physiotherapy personnel health. *EBM J.* 32(2), 192–199. DOI 10.3109/15368378.2013.776429.

Vesselinova L (2013b) The bioeffects of the electromagnetic fields on the personnel in the practice the physical and rehabilitation medicine: A potential for risk mitigation and leading. PhD degree dissertation, MMA, Sofia. (Abstract in English.)

Vesselinova L (2013c) On the risk communication in the wards of physical and rehabilitation medicine. *Phys. Med. Rehabilit. Health J.* XII(2), 22–27. (Abstract in English.)

Vesselinova L (2013d) Analysis of the specific exposition background in physical and rehabilitation facilities. *Phys. Med. Rehabilit. Health J.* XII(3), 10–18. (Abstract in English.)

Vesselinova L, St. Kovatchev (2013) Constant low-intensive magnetic field by complications after radical gynaecologic oncology operations (case report). *Phys. Med. Rehabilit. Health J.* XII(1), 20–24. (Abstract in English.)

Von Guttenberg Y, JA Spickett (2009) Survey of occupational exposure to blood and body fluids in physiotherapists in Western Australia. *Asia Pac. J. Public Health* 21(4), 508–519.

Williams JM (2009). *Biological Effects of Microwaves: Thermal and Nonthermal Mechanisms*. Basic ISP, CA 94064.

23 Nanoelectroporation for Nonthermal Ablation

Richard Nuccitelli

CONTENTS

23.1 INTRODUCTION

In 2001, the world was first introduced to a new kind of interaction between electric fields and biological systems. That is when Schoenbach et al. (2001) first described the effect of intense pulsed electric fields in the nanosecond range (nsPEF) on cells and tissues. The two main distinguishing characteristics of these pulses is their large amplitude and their ability to penetrate past the outer plasma membrane into the interior of the cells and organelles located between the delivering electrodes. Before this, cells had never experienced intracellular electric fields of this duration and magnitude, and there is no doubt that the range of cellular responses to such fields will be quite diverse and complex. Here, I will discuss one such cellular response, the ability to convince the cells exposed to sufficient numbers of such short, high-voltage pulses to initiate a sequence of programmed cell death or apoptosis. We have named this response *nanoelectroablation* and have shown that it involves a truly nonthermal mechanism leading to programmed cell death.

23.1.1 Characteristic Pulse Parameters

23.1.1.1 Rise Time

This ability of pulsed electric fields to penetrate into the cell is mainly dependent on how quickly the pulse appears, which is also known as the rise time. This dependence on rise time can best be understood by discussing the cell's response to an imposed pulsed electric field. The cell can be considered to be a conductor surrounded by a nonconducting lipid bilayer membrane. When a conductor is placed in an electric field, the internal mobile charges will be moved by the electric force in the conductive cytoplasm until they pile up at the nonconductive plasma membrane. With the positive charges at one end and the negative charges at the opposite end of the cell, they will charge up the membrane capacitance to generate a field that is equal and opposite to the imposed one. However, this charge redistribution is not instantaneous, and it typically takes at least 1 µs to occur. If the field can be applied faster than this microsecond, the mobile charges in the cell interior will not yet have redistributed so the applied field will penetrate into the cell and every organelle. So theoretically, any pulse with a rise time faster than 1 µs can penetrate into the cell and exert its effect until the equal and opposite field has been established by the charge redistribution. The presence of an electric field in the cytoplasm means that organelle membranes will be directly exposed to the field, and if it is large enough, this can have very strong effects, leading us to consider the amplitude of the field.

23.1.1.2 Amplitude

The main cellular targets of pulsed fields are the resistive elements, the lipid bilayer membranes that surround the cell and every organelle. These membranes can generally withstand hundreds of millivolts of applied voltage across them. However, once this field exceeds 500 mV, there is sufficient force to drive dipole water molecules into the bilayer to form small, water-filled, continuous defects or pores through the lipid bilayer. Molecular modeling shows that such defects can form within 10 ns when fields on the order of 30 kV/cm are applied (Levine and Vernier, 2010). In order to generate this voltage difference across the two regions of a 10 µm diameter cell that are perpendicular to the field lines, only 1 kV/cm is required. However, in order to generate the same voltage difference across an organelle that is 1/10 the size of the cell, 10 times more voltage is required or at least 10 kV/cm. In practice, we have found that 30 kV/cm is the smallest field that will routinely lead to nanoelectroablation, and this supports the hypothesis that organelle membranes are the critical target for this response. Direct evidence for the nsPEF-induced permeabilization of intracellular organelles, including mitochondria, endoplasmic reticulum, nucleus, and granules, has come from several investigators (Schoenbach et al., 2001; White et al., 2004; Tekle et al., 2005; Chen et al., 2007; Batista et al., 2012; Beebe et al., 2012). By measuring the transport of different-sized molecules through these nanopores, it has been determined that they are 1 nm in diameter and are transient (Pakhomov et al., 2007a,b, 2009; Fernandez et al., 2012; Levine and Vernier, 2012; Romeo et al., 2013).

23.1.1.3 Comparison of Nanosecond and Microsecond Pulsed Field Effects

Electric pulse technology using pulses in the microsecond domain has been applied to living cells since the 1970s and was first used to generate transient increases in the permeability of the cell's plasma membrane in a technique that is called *electroporation* (Neumann and Rosenheck, 1972). The membrane defects generated by these longer pulses are much larger than those generated by nsPEF and have been used to introduce large molecules and even DNA into cells (Neumann et al., 1982; Wong and Neumann, 1982). These longer pulses on the order of 1 kV/cm typically do not penetrate into the cell interior due to their slower rise time, and the pores that form are usually transient and reversible. However, when the field strength is increased to 2–3 kV/cm, irreversible pores are formed (Davalos et al., 2005), and this *irreversible electroporation* (IRE) permanently eliminates the integrity of the plasma membrane and kills cells by necrosis.

However, it is important to note that even this higher IRE field strength is below the 10 kV/cm required to generate water defects in organelle membranes. If the amplitude were increased to this level, the long pulse length would introduce so much energy into the cell that the cell would vaporize.

For example, assume that we apply 30 kV across a load of 60 Ω for 100 μs generating a current of 30 kV/60 Ω = 500 A.

$$\text{Energy delivered} = I \times V \times t = 500 \times 30{,}000 \text{ V} \times 10^{-4} \text{ s} = 1500 \text{ J}$$

Given that the specific heat of water is 4.19 J/g °C, 1500 J would heat a gram of water by 357°C and would therefore vaporize it.

Now consider the energy delivered using 100 ns pulses with this same amplitude:

Energy delivered = $I \times V \times t = 500 \times 30{,}000$ V $\times 10^{-7}$ s = 1.5 J, which would heat a gram of water by 0.357°C. This small temperature increase would be barely noticed by the cells.

Therefore, one major advantage of nanosecond pulses is the ability to deliver large voltage gradients across organelle membranes and introduce nanopores without damaging the cells by hyperthermia.

23.2 TUMOR NANOELECTROABLATION

23.2.1 DISCOVERY

The first paper to report that nanosecond pulses could slow tumor growth came out in 2002 (Beebe et al., 2002). Fibrosarcoma subdermal allograft tumors were treated using two needle electrodes. The tumors treated twice with five to seven pulses (300 ns, 75 kV/cm) were 60% smaller than controls over an 8-day period. During the following years, Dr. Beebe has studied the response of cells treated in vitro with nsPEF and has obtained strong evidence that nsPEF triggers apoptosis in the treated cells (Beebe et al., 2003; Stacey et al., 2003).

23.2.2 NANOELECTROABLATION WITH MULTIPLE TREATMENTS

The next work on tumor nanoelectroablation was published in 2006 using the murine B16 melanoma allograft model system (Nuccitelli et al., 2006). A range of field strengths were explored by sandwiching subdermal tumors between parallel-plate electrodes so that the entire tumor was exposed to a uniform electric field (Figure 23.1). Pulsed fields of 10 kV/cm and 300 ns had no effect on tumor growth compared to controls even when 100 pulses were applied. However, fields of 20 kV/cm and greater inhibited growth with 10 pulses and caused tumor shrinkage with 100 pulses. Complete regression of melanomas was achieved with multiple applications of 100 pulses at 40 kV/cm. In addition, we observed reduced blood flow to the tumor and measured the tumor temperature during pulsing to increase no more than 3°C. This was followed by a long-term study in which a single melanoma tumor in 17 mice was nanoelectroablated, and the mice were monitored for 150 days (Nuccitelli et al., 2009). Complete tumor remission occurred within an average of 47 days, and none of these melanomas recurred during the following 4 months (Figure 23.2). In addition, these pulses were observed to generate small nanopores lasting for minutes in the plasma membrane of exposed cells, as well as increasing intracellular Ca^{2+} and triggering pyknosis, DNA fragmentation, and apoptosis.

23.2.3 NANOELECTROABLATION WITH A SINGLE TREATMENT

While these subdermal allograft tumors could be ablated routinely with multiple treatments, we next wanted to determine if we could find a *single* treatment that would ablate all treated tumors.

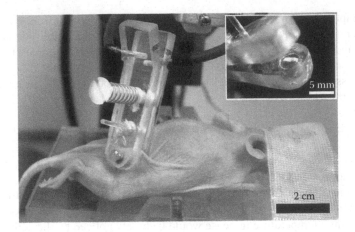

FIGURE 23.1 Photograph of SKH-1 hairless mouse being treated with parallel-plate electrode under isoflurane inhalation anesthesia. Inset: Close-up of one of the plates of parallel-plate electrode showing it recessed by 0.5 mm to allow a space for a conductive agar gel to be placed on it. (Reprinted from Nuccitelli, R. et al., *Biochem. Biophys. Res. Commun.*, 343, 351, 2006. With permission.)

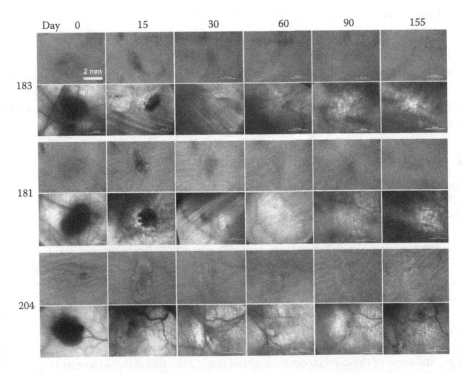

FIGURE 23.2 Photomicrographs of three melanomas taken on the day indicated for each column after nsPEF treatment. Rows are grouped in matched pairs with the top image of the pair being the surface view and the bottom image being the transillumination view of the tumor. All images were taken at the same magnification indicated by the scale bar in the upper left image. The animal number is shown to the left of each grouped pair. Mice 183 and 181 were treated on day 0 with 600 pulses (45 kV/cm, 300 ns long) and again on day 18 with 300 pulses. Mouse 204 was only treated once on day 0 with 300 pulses (40 kV/cm, 90 A, 300 ns), and the tumor exhibited total remission following this single treatment. Red tattoo marks were used to indicate the original tumor location. (Reprinted from Nuccitelli, R. et al., *Int. J. Cancer*, 125, 438, 2009. With permission.)

Using a suction electrode with either two parallel arrays of needles or parallel plates, we treated hundreds of melanoma allograft tumors with 100 ns pulses to determine the minimum number of pulses required to ablate them with a single treatment (Nuccitelli et al., 2010). We found that this varied with the tumor type. For B16 melanoma tumors, 2000 pulses at 30 kV/cm were required, but for subdermal human pancreatic xenograft tumors, 500 pulses were sufficient to ablate with a single treatment (Nuccitelli et al., 2013a). This suggests that the total delivered energy needed to trigger apoptosis in all the tumor cells varies with the tumor type. Indeed, other labs have found this to be the case when applying these pulsed fields to suspensions of cultured tumor cells in a cuvette (Ibey et al., 2010, 2011; Yang et al., 2011).

Allograft tumor models are very useful for identifying the optimal pulse parameters for ablation, but we wanted to determine if *endogenous tumors* would also respond to the pulse parameters identified using allografts. For this, we collaborated with two scientists who had developed transgenic murine tumor models in which tumors appeared spontaneously several months after exposing pups to radiation. The first was Dr. Ervin Epstein's autochthonous basal cell carcinoma (BCC) tumors in Ptch1$^{+/-}$K14-Cre-ER p53 Fl/fl mice. BCCs develop on the skin of these mice and are quite similar to those found in humans. We treated 27 BCCs across 8 mice with either 300 pulses of 300 ns in duration or 2700 pulses of 100 ns duration, all at 30 kV/cm and 5–7 pulses per second (Nuccitelli et al., 2012a). Every treated BCC began to shrink within a day after treatment, and their initial mean volume of 36 ± 5 mm^3 shrunk by $76\% \pm 3\%$ over the ensuing 2 weeks. After 4 weeks, they were 99.8% ablated if the size of the treatment electrode matched the tumor size. Pyknosis of nuclei and DNA fragmentation were detected in the treated tumors.

The second endogenous tumor model we used was Dr. Ed DeFabo's UV-induced melanoma C57/BL6-HGF/SF transgenic mouse. Pups exposed to UV radiation develop visible melanomas 5–6 months later. We treated 27 of these melanomas in 14 mice with 2000 pulses (100 ns, 30 kV/cm) (Nuccitelli et al., 2012b). All treated tumors began to shrink within a day after treatment and gradually disappeared over a period of 12–29 days. We observed pyknosis of nuclei within 1 h of nsPEF treatment and DNA fragmentation by 6 h consistent with the initiation of apoptosis. Serial section histology of the skin regions where the treated melanomas had been located revealed a reduction in melanin density with no residual melanoma.

23.2.4 Immunogenic Apoptosis

One surprising result came out of some concurrent melanoma allograft work. We suspected that nanoelectroablation stimulated an innate immune response to the treated tumor because tumor cell injections made after nanoelectroablation usually led to little or no tumor growth. We used the B16 murine allograft system to show that the secondary tumor shrunk by an average of 70% in immunocompetent mice whose first tumor had been nanoelectroablated. In contrast, lesion excision only slightly inhibited second tumor growth (Nuccitelli et al., 2012b). This difference was not observed in immunodeficient mice. Moreover, we have used immunohistochemistry to identify CD4$^+$ lymphocytes in treated tumors by 19 days following treatment as well as in untreated tumors injected into mice in which a previous tumor had been nanoelectroablated. This is further evidence for an innate immune response. In addition to our work, a recent study from Xinhua Chen in China (2014) indicates that another arm of the immune response, macrophage infiltration, may contribute to ablation as well. If this immune response holds up in humans, it may provide an important advantage for nanoelectroablation therapy over surgical excision of tumors.

23.2.5 Laboratories Conducting Nanoelectroablation Research

This nanoelectroablation technique applied in vivo or in vitro by applying nsPEF has been replicated in seven other laboratories around the world: three in the United States, four in China, and one

TABLE 23.1

Laboratories Conducting Nanoelectroablation Research

United States	Reference
Center for Bioelectrics, Old Dominion University, Norfolk, VA	Beebe et al. (2002)
BioElectroMed Corp., Burlingame, CA	Nuccitelli et al. (2010)
Department of Electrical Engineering-Electrophysics, Viterbi School of Engineering, USC, Los Angeles, CA	Garon et al. (2007)
Cedars-Sinai Medical Center, UCLA School of Medicine, Los Angeles, CA	Yin et al. (2012)
France	
Institut Gustave Roussy, Laboratoire de Vectorologie et Therapeutiques, Paris	Breton and Mir (2012)
Japan	
Global Center of Excellence Program on Pulsed Power Engineering, Kumamoto University, Kumamoto	Shiraishi et al. (2013)
China	
Department of Hepatobiliary and Pancreatic Surgery, First Affiliated Hospital, School of Medicine, Zhejiang Univ. Hangzhou, Zhejiang Province	Chen et al. (2014)
College of Engineering, Peking University, Beijing	Wu et al. (2014)
Laboratory of Obstetrics and Gynecology, Second Affiliated Hospital, Chongqing Medical University, Chongqing	Guo et al. (2014)
Biomedical Ultrasonics/Gynecological lab, West China Second University Hospital, Sichuan University, Chengdu	Zou et al. (2013)

in both France and Japan (Table 23.1). In addition to nanoelectroablation, several other laboratories have been studying a variety of other applications of nsPEF, including platelet activation (Zhang et al., 2008; Hargrave and Li, 2012) and decontamination (Schoenbach et al., 2000; Rieder et al., 2008; Frey et al., 2013).

23.2.6 ENERGY REQUIRED VARIES WITH TUMOR TYPE

Not surprisingly, the energy required to ablate a given tumor varies with the tumor type and delivery system used (Table 23.2). Most of the data have been collected using pulse widths of 100–300 ns, amplitudes of 30–60 kV/cm, and pulse numbers between 30 and 2000. Due to the ease of delivering these pulses to the skin, most of the studies have been limited to treating skin tumors or murine subdermal allograft and xenograft tumors. Melanoma allograft tumors are the most difficult to ablate, requiring 2000 pulses for single-treatment ablation. However, human pancreatic xenograft tumors are ablated with a single treatment of only 500 pulses.

23.2.7 CLINICAL TRIALS

The only human clinical trial that has been conducted so far determined that human BCCs could be ablated with a 78% success rate using only 100 pulses 100 ns long and 30 kV/cm in amplitude (Nuccitelli et al., 2014) (Figure 23.3). This was a small safety clinical trial approved by the IRB as a nonsignificant risk device. Ten lesions on three patients were nanoelectroablated with a treatment time of only a few minutes. Therefore, the first indications are that human skin tumors are easily ablated using nsPEF, and this therapy has the advantages of being fast (100 pulses only take 50 s when applied at 2 pps) and leaving no scar as do most of the standard therapies.

TABLE 23.2
Nanoelectroablation of Tumors or Tissues In Vivo

Tumor or Tissue Type	Host	Pulse Width (ns)	Amplitude (kV/cm)	Lethal Number	References
Murine fibrosarcoma	Mouse	300	60	Not determined	Beebe et al. (2002)
Murine melanoma	Mouse	100, 300	30, 40	2000, 900	Nuccitelli et al. (2006, 2009, 2010, 2012b); Chen et al. (2009)
Human melanoma	Mouse	300	20		Guo et al. (2014)
Murine BCC	Mouse	300, 100	30	2000	Nuccitelli et al. (2012a)
Murine SCC	Mouse	14	30	200	Yin et al. (2012)
Human pancreatic	Mouse	100, 20	30, 43	500, 675	Nuccitelli et al. (2013a); Garon et al. (2007)
Pig liver tissue in vivo	Pig	100	15	30	Long et al. (2011); Beebe et al. (2011)
Murine hepatoma	Mouse	30, 100	68	900	Chen et al. (2012)
Human hepatoma	Mouse	100	40	300	Yin et al. (2014); Chen et al. (2014)
Human BCC	Human	100	30	100	Nuccitelli et al. (2014); Garon et al. (2007)

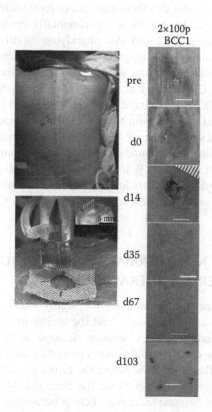

FIGURE 23.3 Human basal cell carcinoma treated using the parallel needle array electrode pictured on the lower left. Since the lesion was larger than the electrode, we treated each half separately with 100 pulses each (30 kV/cm, 100 ns). Images of the lesion taken before and after treatment are shown with the day photographed indicated to the left of each photo. d0 was taken right after treatment and shows some edema. By day 35, after treatment, the lesion was completely eliminated, and by day 103, the only way to know where it was originally located was by using a plastic overlay map made prior to treatment. The white scale bars are 5 mm long.

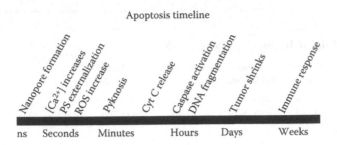

FIGURE 23.4 Approximate timing of the events comprising immunogenic apoptosis following nsPEF application.

23.3 WHAT IS KNOWN ABOUT THE ABLATION MECHANISM?

Much work has been done studying the effects of nsPEF on cells exposed in an electroporation cuvette. This has been reviewed recently (Beebe et al., 2013) so I will not go into extensive detail here. Several cellular changes have been detected following nsPEF application (Figure 23.4). The earliest is the formation of water-filled defects in the plasma membrane as well as intracellular membranes. Molecular modeling indicates that these defects can form within 10 ns after field application (Levine and Vernier, 2010), and this has been experimentally confirmed using the fast voltage-sensitive dye, ANNINE-6 (Frey et al., 2006). Phosphatidylserine diffuses from the inner leaflet of the plasma membrane to the outer leaflet along the water-filled defects that are approximately 1 nm wide (Vernier et al., 2004; Gowrishankar and Weaver, 2006; Pakhomov et al., 2007a). There is then an increase in the cytoplasmic Ca^{2+} due to leaks in both the ER and the plasma membrane (Vernier et al., 2003; White et al., 2004). The Ca^{2+} increase is known to quickly trigger reactive oxygen species generation (Nuccitelli et al., 2013b), which has been linked to the translocation of calreticulin (CRT) from the ER to the cell surface (Krysko et al., 2012). This ecto-CRT has been shown to be an important signal for recruiting the immune system to recognize and destroy similar tumor cells, a process called immunogenic apoptosis. Within an hour or so, DNA fragmentation (Chen et al., 2010) is evident, and nuclear pyknosis occurs followed by the release of cytochrome C (Beebe et al., 2003) from the mitochondria and the appearance of active caspases (Ren et al., 2012) that cut up cellular proteins.

23.4 ADVANTAGES OF NONTHERMAL NANOELECTROABLATION OVER OTHER THERMAL THERAPIES

The main advantages of nanoelectroablation are the highly localized ablation zone, the lack of sensitivity to heat sinks such as blood vessels, and the ability to stimulate an immune response. Thermal therapies such as radiofrequency ablation damage nearby healthy tissues due to the spread of heat and are not very effective at ablating tumors that surround blood vessels because the heat can be carried away by the blood flow before the tumor reaches hyperthermia temperatures. Nanoelectroablation triggers apoptosis in cells via the electric field that is localized to the region between the application electrodes, and there is no heat generated to spread to neighboring tissues. Another extremely important advantage is the stimulation of the immune system by nanoelectroablation. This enhanced immune response has only been observed in mice and rats thus far, but if it also occurs in humans, this will provide a powerful advantage to nanoelectroablation over thermal therapies.

23.5 FUTURE CHALLENGES FOR HUMAN THERAPY

23.5.1 PULSE GENERATORS

There are many different nanosecond pulse generators being used in several laboratories around the world, but none of them have been approved for human use. This must be done before extensive human clinical trials can begin. The human studies completed to date have been restricted to treating skin tumors for which the pulse generator and delivery system are considered to be a nonsignificant risk device. However, once the electrodes are introduced into the body to treat internal tumors, the risk levels increase and a series of FDA requirements come into play involving risk assessment and mitigation as well as electrical safety standards for medical devices. For example, these high-voltage nanosecond pulses generate electromagnetic interference that must be minimized by shielding, and both the pulse generator and delivery device must be made under good manufacturing process regulations.

Pulse amplitude is also a consideration both in generation and in delivery. Once the output voltage exceeds about 20 kV, ionization of gases near the electrode occurs, and this leads to corona generation. In order to reach higher voltages, all circuit components must be insulated or potted to prevent exposure to gases in the air that would lead to coronas.

23.5.2 DELIVERY SYSTEMS

Once the pulse is generated, it must be delivered to the target tumor in the liver, kidney, pancreas, or other organs in a manner that exposes the entire tumor to the pulsed electric field. The main way that this is done for delivering hyperthermia therapies is through a percutaneous approach in which the electrode passes through the abdominal wall and penetrates the tumor where sufficient energy is introduced to raise the tumor temperature to at least 50°C. We have been determining the ablation zone around electrode pairs so that a similar approach might be taken to deliver nsPEF nonthermally. The insulation required on electrode leads to hold off 40 kV dc is 2.5 mm thick so that the cable must be at least 5 mm thick and two such cables required to deliver the pulse will be 1 cm thick. It does not have to be quite as thick for nsPEF as it is for dc, but it is definitely a challenge to design delivery systems for internal tumors with these insulator requirements.

23.5.3 TUMOR SIZE

Human tumor sizes being ablated today range from 1 to 5 cm or more in diameter. The requirement of exposing all the tumor cells to nsPEF of 30 kV/cm is a challenge. Insulation requirements limit us to about 20 kV output voltage so electrodes cannot be more than 0.7 cm apart in order to apply the 30 kV/cm. This would suggest that multiple electrode impalements will be required in order to expose every tumor cell to the appropriate number of pulses.

A second electrode delivery approach that we are implementing is via an echo-endoscope. This instrument is used routinely to image tumors in the pancreas, liver, and kidney from inside the stomach by using an ultrasound transducer at its tip. A 4 ft long aspiration needle fits in the working channel and can be used to penetrate through the stomach wall and into the tumor in order to aspirate a cell sample from tumors in organs located next to the stomach or digestive tract. By placing a delivery electrode into the working channel of the echo-endoscope, the pulses could be delivered to throat and stomach tumors as well as those in the pancreas or other organs imaged by ultrasound once the electrode has penetrated through the duodenal wall and into the imaged tumor. This approach has been successfully demonstrated in the pig and is much less invasive than a laparotomy normally required for treating pancreatic carcinomas.

23.5.4 IMAGING SYSTEMS NEEDED

In order to position the delivery electrodes near the tumor, it is critical to have a method of imaging both the tumor and electrodes. For endoscopy, ultrasound is routinely used, but for percutaneous delivery, CT scans are also popular. Once the CT scan identifies the tumor location, the patient is slid out of the machine and an electrode is fed into the tumor region. The electrode position is then verified by moving the patient back into the scanner with the electrode in place in the tumor.

23.6 CONCLUSION

Much has been learned about this new nanoelectroablation therapy over the past decade. It appears to be very effective in triggering apoptosis in all tumors to which it has been applied, and a single treatment is usually sufficient to ablate the tumors without recurrence. While nearly all of the studies to date have been on mice and rats, the first human clinical trial treating BCC indicated a 78% success rate as well as scarless lesion ablation. The advantages of nanoelectroablation over thermal ablation therapies are significant, particularly if immunogenic apoptosis is triggered in humans as it is in mice.

REFERENCES

Batista, N. T., Wu, Y. H., Gundersen, M. A. et al. (2012). Nanosecond electric pulses cause mitochondrial membrane permeabilization in Jurkat cells. *Bioelectromagnetics 33*, 257–264.

Beebe, S. J., Chen, X., Liu, J. A. et al. (2011). Nanosecond pulsed electric field ablation of hepatocellular carcinoma. *Conf. Proc. IEEE Eng Med. Biol. Soc. 2011*, 6861–6865.

Beebe, S. J., Chen, Y. J., Sain, N. M. et al. (2012). Transient features in nanosecond pulsed electric fields differentially modulate mitochondria and viability. *PLoS ONE 7*, e51349.

Beebe, S. J., Fox, P., Rec, L. J. et al. (2002). Nanosecond pulsed electric field (nsPEF) effects on cells and tissues: Apoptosis induction and tumor growth inhibition. *IEEE Trans. Plasma Sci. 30*, 286–292.

Beebe, S. J., Fox, P. M., Rec, L. J. et al. (2003). Nanosecond, high-intensity pulsed electric fields induce apoptosis in human cells. *FASEB J. 17*, 1493–1495.

Beebe, S. J., Sain, N. M., and Ren, W. (2013). Induction of cell death mechanisms and apoptosis by nanosecond pulsed electric fields (nsPEFs). *Cells 2*, 136–162.

Breton, M. and Mir, L. M. (2012). Microsecond and nanosecond electric pulses in cancer treatments. *Bioelectromagnetics 33*(2):106–123.

Chen, N., Garner, A. L., Chen, G. et al. (2007). Nanosecond electric pulses penetrate the nucleus and enhance speckle formation. *Biochem. Biophys. Res. Commun. 364*, 220–225.

Chen, X., Kolb, J. F., Swanson, R. J. et al. (2010). Apoptosis initiation and angiogenesis inhibition: Melanoma targets for nanosecond pulsed electric fields. *Pigm. Cell Melanoma Res. 23*, 554–563.

Chen, X., Swanson, R. J., Kolb, J. F. et al. (2009). Histopathology of normal skin and melanomas after nanosecond pulsed electric field treatment. *Melanoma Res. 19*, 361–371.

Chen, X., Yin, S., Hu, C. et al. (2014). Comparative study of nanosecond electric fields in vitro and in vivo on hepatocellular carcinoma indicate macrophage infiltration contribute to tumor ablation in vivo. *PLoS ONE 9*, e86421.

Chen, X., Zhuang, J., Kolb, J. F. et al. (2012). Long term survival of mice with hepatocellular carcinoma after pulse power ablation with nanosecond pulsed electric fields. *Technol. Cancer Res. Treat. 11*, 83–93.

Davalos, R. V., Mir, I. L., and Rubinsky, B. (2005). Tissue ablation with irreversible electroporation. *Ann. Biomed. Eng. 33*, 223–231.

Fernandez, M. L., Risk, M., Reigada, R. et al. (2012). Size-controlled nanopores in lipid membranes with stabilizing electric fields. *Biochem. Biophys. Res. Commun. 423*, 325–330.

Frey, W., Gusbeth, C., and Schwartz, T. (2013). Inactivation of *Pseudomonas putida* by pulsed electric field treatment: A study on the correlation of treatment parameters and inactivation efficiency in the short-pulse range. *J. Membr. Biol. 246*, 769–781.

Frey, W., White, J. A., Price, R. O. et al. (2006). Plasma membrane voltage changes during nanosecond pulsed electric field exposure. *Biophys. J. 90*, 3608–3615.

Garon, E. B., Sawcer, D., Vernier, P. T. et al. (2007). In vitro and in vivo evaluation and a case report of intense nanosecond pulsed electric field as a local therapy for human malignancies. *Int. J. Cancer 121*, 675–682.

Gowrishankar, T. R. and Weaver, J. C. (2006). Electrical behavior and pore accumulation in a multicellular model for conventional and supra-electroporation. *Biochem. Biophys. Res. Commun. 349*, 643–653.

Guo, F., Yao, C., Li, C. et al. (2014). In vivo evidences of nanosecond pulsed electric fields for melanoma malignancy treatment on tumor-bearing BALB/c nude mice. *Technol. Cancer Res. Treat 13*, 337–344.

Hargrave, B. and Li, F. (2012). Nanosecond pulse electric field activation of platelet-rich plasma reduces myocardial infarct size and improves left ventricular mechanical function in the rabbit heart. *J. Extra. Corpor. Technol. 44*, 198–204.

Ibey, B. L., Pakhomov, A. G., Gregory, B. W. et al. (2010). Selective cytotoxicity of intense nanosecond-duration electric pulses in mammalian cells. *Biochim. Biophys. Acta 1800*(11), 1210–1219.

Ibey, B. L., Roth, C. C., Pakhomov et al. (2011). Dose-dependent thresholds of 10-ns electric pulse induced plasma membrane disruption and cytotoxicity in multiple cell lines. *PLoS. ONE 6*, e15642.

Krysko, D. V., Garg, A. D., Kaczmarek, A. et al. (2012). Immunogenic cell death and DAMPs in cancer therapy. *Nat. Rev. Cancer. 12*, 860–875.

Levine, Z. A. and Vernier, P. T. (2010). Life cycle of an electropore: Field-dependent and field-independent steps in pore creation and annihilation. *J. Membr. Biol. 236*, 27–36.

Levine, Z. A. and Vernier, P. T. (2012). Calcium and phosphatidylserine inhibit lipid electropore formation and reduce pore lifetime. *J. Membr. Biol. 245*, 599–610.

Long, G., Shires, P. K., Plescia, D. et al. (2011). Targeted tissue ablation with nanosecond pulses. *IEEE Trans. Biomed. Eng. 58*(8), 2161–2167.

Neumann, E. and Rosenheck, K. (1972). Permeability changes induced by electric impulses in vesicular membranes. *J. Membr. Biol. 10*, 279–290.

Neumann, E., Schaefer-Ridder, M., Wang, Y. et al. (1982). Gene transfer into mouse lyoma cells by electroporation in high electric fields. *EMBO J. 1*, 841–845.

Nuccitelli, R., Chen, X., Pakhomov, A. et al. (2009). A new pulsed electric field therapy for melanoma disrupts the tumor's blood supply and causes complete remission without recurrence. *Int. J. Cancer. 125*, 438–445.

Nuccitelli, R., Huynh, J., Lui, K. et al. (2013a). Nanoelectroablation of human pancreatic carcinoma in a murine xenograft model without recurrence. *Int. J. Cancer, 132*, 1933–1939.

Nuccitelli, R., Lui, K., Kreis, M. et al. (2013b). Nanosecond pulsed electric field stimulation of reactive oxygen species in human pancreatic cancer cells is Ca^{2+}-dependent. *Biochem. Biophys. Res. Commun. 435*, 580–585.

Nuccitelli, R., Pliquett, U., Chen, X. et al. (2006). Nanosecond pulsed electric fields cause melanomas to self-destruct. *Biochem. Biophys. Res. Commun. 343*, 351–360.

Nuccitelli, R., Tran, K., Athos, B. et al. (2012a). Nanoelectroablation therapy for murine basal cell carcinoma. *Biochem. Biophys. Res. Commun. 424*, 446–450.

Nuccitelli, R., Tran, K., Lui, K. et al. (2012b). Non-thermal nanoelectroablation of UV-induced murine melanomas stimulates an immune response. *Pigment Cell Melanoma Res. 25*, 618–629.

Nuccitelli, R., Tran, K., Sheikh, S. et al. (2010). Optimized nanosecond pulsed electric field therapy can cause murine malignant melanomas to self-destruct with a single treatment. *Int. J. Cancer 127*, 1727–1736.

Nuccitelli, R., Wood, R., Kreis, M. et al. (2014). First-in-human trial of nanoelectroablation therapy for basal cell carcinoma: Proof of method. *Exp. Dermatol. 23*, 135–137.

Pakhomov, A. G., Bowman, A. M., Ibey, B. L. et al. (2009). Lipid nanopores can form a stable, ion channel-like conduction pathway in cell membrane. *Biochem. Biophys. Res. Commun. 385*, 181–186.

Pakhomov, A. G., Kolb, J. F., White, J. A. et al. (2007a). Long-lasting plasma membrane permeabilization in mammalian cells by nanosecond pulsed electric field (nsPEF). *Bioelectromagnetics 28*, 655–663.

Pakhomov, A. G., Shevin, R., White, J. A. et al. (2007b). Membrane permeabilization and cell damage by ultrashort electric field shocks. *Arch. Biochem. Biophys. 465*, 109–118.

Ren, W., Sain, N. M., and Beebe, S. J. (2012). Nanosecond pulsed electric fields (nsPEFs) activate intrinsic caspase-dependent and caspase-independent cell death in Jurkat cells. *Biochem. Biophys. Res. Commun. 421*, 808–812.

Rieder, A., Schwartz, T., Schon-Holz, K. et al. (2008). Molecular monitoring of inactivation efficiencies of bacteria during pulsed electric field treatment of clinical wastewater. *J. Appl. Microbiol. 105*, 2035–2045.

Romeo, S., Wu, Y. H., Levine, Z. A. et al. (2013). Water influx and cell swelling after nanosecond electropermeabilization. *Biochim. Biophys. Acta. 1828*, 1715–1722.

Schoenbach, K. H., Beebe, S. J., and Buescher, E. S. (2001). Intracellular effect of ultrashort electrical pulses. *Bioelectromagnetics 22*, 440–448.

Schoenbach, K. H., Dobbs, F. C., and Beebe, S. J. (2000). Bacterial decontamination of liquids with pulsed electric fields. *IEEE Trans. Dielect. Elect. Insul. 7*, 637–645.

Shiraishi, E., Hosseini, H., Kang, D. K. et al. (2013). Nanosecond pulsed electric field suppresses development of eyes and germ cells through blocking synthesis of retinoic acid in Medaka (Oryzias latipes). *PLoS ONE 8*, e70670.

Stacey, M., Stickley, J., Fox, P. et al. (2003). Differential effects in cells exposed to ultra-short, high intensity electric fields: Cell survival, DNA damage, and cell cycle analysis. *Mutat. Res. 542*, 65–75.

Tekle, E., Oubrahim, H., Dzekunov, S. M. et al. (2005). Selective field effects on intracellular vacuoles and vesicle membranes with nanosecond electric pulses. *Biophys. J. 89*, 274–284.

Vernier, P. T., Sun, Y., Marcu, L. et al. (2003). Calcium bursts induced by nanosecond electric pulses. *Biochem. Biophys. Res. Commun. 310*, 286–295.

Vernier, P. T., Sun, Y., Marcu, L. et al. (2004). Nanoelectropulse-induced phosphatidylserine translocation. *Biophys. J. 86*, 4040–4048.

White, J. A., Blackmore, P. F., Schoenbach, K. H. et al. (2004). Stimulation of capacitative calcium entry in HL-60 cells by nanosecond pulsed electric fields. *J. Biol. Chem. 279*, 22964–22972.

Wong, T. K. and Neumann, E. (1982). Electric field mediated gene transfer. *Biochem. Biophys. Res. Commun. 107*, 584–587.

Wu, S., Wang, Y., Guo, J. et al. (2014). Nanosecond pulsed electric fields as a novel drug free therapy for breast cancer: An in vivo study. *Cancer Lett. 343*, 268–274.

Yang, W., Wu, Y. H., Yin, D. et al. (2011). Differential sensitivities of malignant and normal skin cells to nanosecond pulsed electric fields. *Technol. Cancer Res. Treat. 10*, 281–286.

Yin, D., Yang, W. G., Weissberg, J. et al. (2012). Cutaneous papilloma and squamous cell carcinoma therapy utilizing nanosecond pulsed electric fields (nsPEF). *PLoS ONE 7*, e43891.

Yin, S., Chen, X., Hu, C. et al. (2014). Nanosecond pulsed electric field (nsPEF) treatment for hepatocellular carcinoma: A novel locoregional ablation decreasing the lung metastasis. *Cancer Lett. 346*(2), 285–291.

Zhang, J., Blackmore, P. F., Hargrave, B. Y. et al. (2008). Nanosecond pulse electric field (nanopulse): A novel non-ligand agonist for platelet activation. *Arch. Biochem. Biophys. 471*, 240–248.

Zou, H., Gan, X., Linghu, L. et al. (2013). Intense nanosecond pulsed electric fields promote cancer cell apoptosis through centrosome-dependent pathway involving reduced level of PLK1. *Eur. Rev. Med. Pharmacol. Sci. 17*, 152–160.

24 Electroporation for Electrochemotherapy and Gene Therapy

Maja Cemazar, Tadej Kotnik, Gregor Sersa, and Damijan Miklavcic

CONTENTS

24.1 ELECTROPORATION: THE PHENOMENON

The theoretical understanding of the phenomenon of electroporation is crucial for planning and optimization of protocols for drug and/or gene delivery. In the last four decades, a number of tentative theoretical descriptions of this phenomenon have been proposed, assuming either deformation of membrane lipids,[1-3] their phase transition,[4] breakdown of interfaces between the lipid domains,[5] or denaturation of membrane proteins.[6] However, each of these former descriptions has serious flaws,[7] and today, there is broad consensus that electroporation is best described as the formation of aqueous pores in the lipid bilayer.[8-11] This also clarifies the prevalent choice of the term *electroporation*, as opposed to the broader term of *electropermeabilization*, which is also applicable to all the alternative explanations of the phenomenon.

The theory of electroporation is largely based on thermodynamics and describes the initial stage of pore formation by penetration of water molecules into the lipid bilayer of the membrane, forming unstable structures termed *water wires* or *water fingers*. This subsequently causes the adjacent lipids to reorient with their polar heads toward these structures, forming metastable aqueous pores.

Both the theory and molecular dynamics simulations suggest that small unstable pores are forming and closing within nanoseconds even in the absence of an external electric field, but an exposure of the membrane to an electric field reduces the energy required for penetration of water into the bilayer. As such exposure starts, the external field infiltrates the membrane so that the membrane field is of the same order of magnitude as the external field, but within less than a microsecond, the external field also causes a polarizing flow of dissociated ions in the media surrounding the membrane, resulting in the gradual buildup (inducement) of transmembrane voltage that amplifies the membrane field by about three orders of magnitude.[12,13]

Exposure of the membrane to an electric field thus increases the probability of pore formation in the membrane's bilayer so that on the average pores form more frequently and with much longer lifetimes than those formed in the absence of the electric field. For transmembrane voltages of hundreds of millivolts, the number of pores and their average lifetime become sufficient for detectable increase in membrane permeability to molecules otherwise unable to cross the membrane.

Metastable aqueous pores in the bilayer are at most several nanometers larger in diameter, which is too small to be observable by optical microscopy, while sample preparation techniques required for electron microscopy of soft matter are too harsh for reliable preservation of metastable formations in the bilayer and often themselves cause pore-like structures in the bilayer. Still there is growing and increasingly convincing indirect support for aqueous pore formation in the form of molecular dynamics simulations. These computational studies largely confirm the theoretically predicted stages of aqueous pore formation, including the strong increase in the rate of pore formation with the increase in the electric field to which the membrane is exposed—first through the direct action of the external field and then augmented by the inducement of transmembrane voltage due to polarization.[13–16]

The characteristics of electroporation and the accompanying phenomena depend on the amplitude and duration of the electric field to which the cells are exposed, and this relation is sketched in Figure 24.1. With low amplitudes and durations of the electric field, there is no detectable effect on the membrane and the transport across it. With moderate amplitudes and durations, electroporation is reversible so that after the exposure ceases, the pores gradually reseal and the cells remain viable. With higher field amplitudes and/or longer durations, electroporation is irreversible, as the transport through the pores—particularly the leakage of intracellular content—is too extensive,

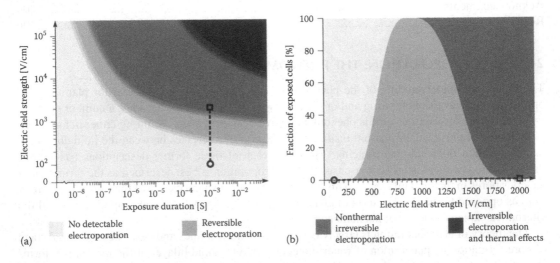

FIGURE 24.1 Electroporation and thermal effects caused by exposure of cells to electric fields. (a) Reversible electroporation, irreversible electroporation, and thermal damage as functions of electric field strength and duration. (Adapted from Bower, M. et al., *J. Surg. Oncol.*, 104, 22, 2011; Yarmush, M.L. et al., *Annu. Rev. Biomed. Eng.*, 16, 295, 2012.) (b) The fractions of non-electroporated, reversibly electroporated, and irreversibly electroporated cells as functions of electric field strength, for a fixed exposure duration of 1 ms (i.e., along the dashed vertical in panel a). Note that the field scale is logarithmic in panel a but linear in panel b, where it covers a much narrower range. (Reprinted from Delemotte, L. and Tarek, M., *J. Membr. Biol.*, 245, 531, 2012. With permission.)

and the resealing is too slow for the cells to recover, resulting in their death. At still stronger and/ or longer exposures, irreversible electroporation becomes accompanied by thermal damage to the cell, as well as to the molecules released from it. Since pore formation is a stochastic process and exposed cells are typically not all identical in size, shape, and orientation, the ranges of no poration, reversible poration, irreversible poration without thermal damage, and irreversible poration with thermal damage partly overlap. The bounds of these four ranges also vary with the type of the cells exposed and by the properties of the medium surrounding the cells. In addition, thermal damage is both organism and molecule dependent, as proteins already start to denature at relatively small temperature increases (at ~43°C–45°C in human cells), DNA melting occurs only above ~70°C, and most lipids and simpler saccharides are not affected even by boiling.

Similarly to pore formation, pore resealing is a stochastic process, but it proceeds on a much longer time scale. Namely, the formation of electropores takes nano- to microseconds, while their resealing—as revealed by the return of the membrane's electric conductivity to its preporation value and by termination of detectable transmembrane transport—is often completed only within seconds or even minutes after the end of the exposure.[20] More detailed measurements reveal that the resealing proceeds in several stages with time constants ranging from micro- and/or milliseconds up to tens of seconds.[21,22] Unfortunately, neither the existing theory nor the experiments can provide a reliable picture of specific events characterizing each of these distinctive stages, while reliable molecular dynamics simulations, even in their most simplified versions (e.g., coarse-grained), cannot yet cover time scales that are extensive.

24.1.1 Induced Transmembrane Voltage and Electroporation

In most applications of electroporation, biological cells to be porated are not brought into direct contact with the electrodes, so that the voltage on the membranes of the exposed cells, termed the induced transmembrane voltage ($\Delta\Psi_m$), represents only a part of the voltage delivered to the electrodes. Unlike with clamped membrane patches, where $\Delta\Psi_m$ is a constant all over the exposed patch, with cells exposed as a whole in a contactless manner, $\Delta\Psi_m$ is position dependent; in spherical cells, its spatial variation is described by the steady-state Schwan equation[23]:

$$\Delta\Psi_m = 1.5ER\cos\theta$$

where

E is the amplitude (strength) of the external electric field
R is the radius of the spherical cell
θ is the angle between the direction of the applied field and the radial line connecting the cell center with the considered point on the membrane

Thus, $\Delta\Psi_m$ is proportional to the applied electric field and the cell radius, and it varies as $\cos\theta$, with extremal values at the two points where the field is perpendicular to the membrane, that is, at $\theta = 0°$ and $\theta = 180°$ (the *poles* of the cell).

The induced transmembrane voltage is typically established within microseconds after the onset of the field. To describe the initial transient behavior, one uses the more general first-order Schwan equation[11]:

$$\Delta\Psi_m = 1.5ER\cos\theta\left(1-\exp\left(\frac{-t}{\tau_m}\right)\right)$$

where τ_m is the time constant of membrane charging (approximately 0.5 μs under physiological conditions).

Induced transmembrane voltage as a function of position and time can also be assessed for non-spherical cells. For cells resembling a regular geometrical body such as a cylinder (e.g., a muscle cell, an axon of a nerve cell), an oblate spheroid (e.g., an erythrocyte), or a prolate spheroid (e.g., a bacillus), this can be done by means of analytical derivation, solving the Laplace equation in a suitable coordinate system with the appropriate boundary conditions.[24–26] For irregularly shaped cells and cells in dense suspensions or clusters, $\Delta\Psi_m$ can be computed only numerically, using either the finite-differences or the finite-elements method; the latter is used more frequently and more efficiently both for irregularly shaped cells[27,28] and for clusters of cells.[29–31]

Experimental alternatives to analytical derivation and numerical computation of $\Delta\Psi_m$ are provided by measurements with microelectrodes and with potentiometric fluorescent dyes. The use of microelectrodes is invasive, characterized by a rather low spatial resolution, and the physical presence of the electrodes distorts the electric field and hence the voltage it induces; these are serious disadvantages. On the other hand, measurements with potentiometric dyes are noninvasive; with no physical disruption of the membrane, they offer higher spatial resolution than microelectrodes, and their presence does not distort the electric field, but such measurements can be taken only on the cells that are visually accessible. Their use in tissues is thus rather limited, but for experiments in vitro, potentiometric dyes, such as di-8-ANEPPS,[32,33] RH292,[34] and ANNINE-6,[35] have become established tools for measurements of $\Delta\Psi_m$, experimental studies of voltage-gated membrane channels, as well as for monitoring of nerve and muscle cell activity. A potentiometric dye incorporates into the lipid bilayer of the membrane, where it starts to fluoresce with a spectrum dependent on the amplitude of the induced voltage. With a suitable setup comprising a pulse laser, a fast sensitive camera, and a system for synchronization of acquisition with field exposure, these dyes also allow to monitor the time course of $\Delta\Psi_m$ with a resolution of microseconds, and even nanoseconds for ANNINE-6.[35]

As the pores in the membrane caused by electroporation are not observable directly with the currently available techniques, electroporation can be detected and studied only indirectly, by assessing its larger-scale manifestations—mainly the changes in electrical or optical properties of the membrane resulting from the formation of pores or transport through them. The changes in electrical properties of the membrane can be measured by patch-clamp techniques, and they show that during electroporation, the electric conductivity of the membrane increases by several orders of magnitude, and its dielectric permittivity is also affected.[36,37] In dense cell suspensions, electroporation can also be monitored by measuring the bulk electric conductivity, which increases significantly if a large fraction of the exposed cells is electroporated.[38,39] A similar approach is also used in tissues and can be augmented by measuring the conductivity and permittivity at several frequencies, typically in the kilohertz range, which allows to distinguish between nonporated, reversibly porated, and irreversibly porated tissues.[40,41]

The bulk optical properties of the membrane, particularly light scattering and absorption, are also affected by the reorientation of lipids around the pores, and measurements of these properties can also be used to assess electroporation.[42] Finally, an even more indirect, and perhaps also the most frequently used method of electroporation assessment, is by means of imaging the transport of molecules that cannot permeate an intact membrane, as described in more detail later.

24.1.2 TRANSPORT ACROSS THE ELECTROPORATED MEMBRANE

Electroporation-mediated transport across the membrane is strongly correlated with the transmembrane voltage induced by the exposure to the electric field, which is in turn proportional to this field.[20,34] This correlation can be demonstrated particularly clearly by combining potentiometric measurements and monitoring transmembrane transport on the same cell.[21] On the tissue level, this same correlation is reflected in the fact that the tissue regions with the highest local electric field are generally also the regions containing the highest fractions of electroporated cells.[22]

The transport of molecules across an electroporated membrane can be characterized by the Nernst–Planck equation[42]

$$\frac{V}{S}\frac{dc}{dt} = -DE\frac{zF}{\rho T}c - D\nabla c$$

where

V is the volume of the cell

S is the surface area of the electroporated part of the membrane

c is the concentration of molecules or ions transported across this part of the membrane

D is the diffusion coefficient for such transport

z is the electric charge of the molecules or ions

E is the local electric field acting on them

F is the Faraday constant

ρ is the gas constant

T is the absolute temperature

The first term on the right-hand side of the Nernst–Planck equation corresponds to the electrophoretic transport driven by the exposure of the cell to the electric field, and the second term to the diffusive transport that persists until either the concentrations of the transported molecules on both sides of the membrane equalize or all the pores reseal.

During an electric pulse, the electric field is the main source of the driving force acting on charged molecules and ions, and the electrophoretic term dominates the right-hand side of the Nernst–Planck equation. As a pulse ceases, so does the electrophoretic transport, with only the diffusive component persisting. Although diffusive transport proceeds at a much slower rate than electrophoretic transport, complete pore resealing takes seconds or even minutes,[42] while pulses used for electroporation last at most several milliseconds. As a consequence, despite the fast initial rate of electrophoretic transport, the total transport of both ions and small molecules through an electroporated membrane is often predominantly diffusive.[43,44] In contrast, electrophoretic transport can contribute crucially in the transport of macromolecules, particularly DNA, across the electroporated membrane.[45,46] Besides electrophoretic transport for macromolecules, also electroporation-enhanced endocytotic transport of plasmid DNA has been demonstrated.[47]

24.2 ELECTROCHEMOTHERAPY: PRECLINICAL IN VITRO AND IN VIVO STUDIES

24.2.1 ELECTROCHEMOTHERAPY: IN VITRO STUDIES

Application of electric pulses to the cells in vitro, aiming to increase cytotoxicity of chemotherapeutic drug bleomycin, was first described by Orlowski et al.[48] Thereafter, several other chemotherapeutic drugs were tested in vitro on cells for potential application in combination with electroporation; among them only cisplatin was the most promising drug. Electroporation of cells increased the cytotoxicity of bleomycin (up to several 1000-fold) and cisplatin (up to 70-fold). The prerequisite for the drug to be effective in combination with electroporation is that they are either hydrophilic or lack transport system in the membrane, since electroporation can facilitate the drug transport through the cell membrane only for poorly or non-permeant molecules.[49-51]

Increased cytotoxicity of cisplatin due to electroporation of cells was demonstrated also in cell lines resistant to cisplatin, however, to a lesser degree than on parental cell line.[52] Furthermore, it was demonstrated that endothelial cells are sensitive to bleomycin and to cisplatin, especially when the drug delivery was increased by electropulsation. These data are important for the explanation of vascular disrupting effect of electrochemotherapy (ECT).[53]

24.2.2 ELECTROCHEMOTHERAPY: IN VIVO STUDIES

Bleomycin and cisplatin were tested in ECT protocol on a number of animal models in vivo. Extensive studies on different animal models with different tumors, either transplantable or spontaneous, were performed. Antitumor effectiveness of ECT was tested on tumors in mice, rats, hamsters, and rabbits. Tumors treated by ECT were either subcutaneous, grew in the muscle, brain, or liver, and were of different types, for example, sarcomas, carcinomas, glioma, or melanoma.[50,54–58] The studies demonstrated that with drug doses that have minimal or no antitumor effectiveness, high (up to 80%) complete responses (CRs) of the ECT-treated tumors were obtained. The drug doses used were so low to have no systemic toxicity. Route of administration was either intravenously (for bleomycin) or intratumorally (bleomycin and cisplatin). The time interval between drug injection and application of electric pulses is important. The prerequisite is that, at the time of the application of electric pulses to the tumor, a sufficient amount of drug is present in the tumor. Therefore, after intravenous drug administration into small laboratory animals (4 mg/kg of cisplatin or 0.5 mg/kg bleomycin), only a few minutes' interval is needed to reach the maximal drug concentration in the tumors. After intratumoral administration (2 mg/cm^3 of cisplatin and 3 mg/cm^3 of bleomycin), this interval is even shorter, and the application of electric pulses has to follow the administration of the drug as soon as possible (within a minute).[58] Some other well-established drugs or drugs in development were also tested in combination with electric pulses for potential increase in effectiveness. The majority of results showed some potential benefit; however, the results of the studies were not as pronounced as for bleomycin or cisplatin; therefore, further studies were not conducted.[59–63]

The application of electric pulses of suitable parameters to the tumors, which led to adequate and sufficient electric field distribution in the tumor to obtain cell electroporation, had no antitumor effectiveness and no systemic side effects.[64] Local side effects were contractions of the muscles underlying the treated area, but these are present only during the application of electric pulses and were tolerable, so in most cases, anesthesia of laboratory animal was not necessary.[65]

24.2.3 ELECTROCHEMOTHERAPY: STUDYING IN VETERINARY ONCOLOGY

In the first veterinary clinical trial, conducted in 1997, 12 cats with spontaneous large soft tissue sarcomas that had relapsed after treatment with conventional therapies were treated with ECT with bleomycin combined with immunotherapy consisting of intratumoral injection of CHO (interleukin-2 [IL-2]) living cells that secreted IL-2, which makes this study substantially different from other studies.[66] In most of the studies on ECT in small animals, cisplatin was used as a chemotherapeutic agent. In these studies, ECT was used as single treatment and not as an adjuvant treatment. It was used for the treatment of dogs, cats, and horses with up to 100% tumor cures.[67–71] Studies using intratumorally injected bleomycin were performed either alone or as an adjuvant treatment to surgery. ECT with bleomycin injected intratumorally was performed in pets with spontaneous tumors of different histological types, and the therapy resulted in good response rate.[71] Comparison of ECT of mastocytoma to surgical excision demonstrated that ECT is equally effective and can represent an alternative to surgery.[72] In the case of adjuvant treatment, ECT proved to be very effective as an adjunct to surgery for the treatment of mast cell tumors and soft tissue sarcoma in dogs and hemangiopericytoma and soft tissue sarcoma in cats.[71] Furthermore, several recent studies evaluated ECT with either bleomycin or cisplatin in cats.[73–77] For example, ECT with bleomycin of superficial squamous cell carcinoma in cats resulted in 82% CR, making ECT as a good alternative option for treatment, especially when other treatment approaches are not acceptable by the owners, owing to their invasiveness, mutilation, or high cost.[77] ECT with cisplatin injected intratumorally was tested in several clinical trials on larger numbers of equine sarcoids. The results of the studies confirmed that ECT with cisplatin is a highly effective treatment with long-lived antitumor effects and good treatment tolerance.[70,71]

24.2.4 MECHANISMS OF ANTITUMOR ACTION OF ELECTROCHEMOTHERAPY

Several mechanisms of antitumor effectiveness of ECT were described. Recently, a lot of studies were devoted to elucidation of vascular targeted action of ECT; therefore, it will be explained in more detail. Nevertheless, the principal mechanism of ECT is increased permeabilization of the membranes of cells in the tumors, leading to increased drug effectiveness by enabling the drug to reach the intracellular targets. In preclinical studies on murine tumors, increase in the uptake of bleomycin and cisplatin in the electroporated tumors was demonstrated compared to those tumors without electroporation.[78,79] Furthermore, twofold increase in cisplatin DNA adducts was determined in electroporated tumors.[79]

Another mechanism involved in the antitumor mechanism of ECT is the *involvement of immune system*. It was demonstrated by the difference in response to ECT of tumors growing in immunocompetent and immunodeficient laboratory mice.[80] The tumors growing in immunodeficient mice did not completely regressed after ECT, while tumor growing in immunocompetence mice did. We also demonstrated the increased activity of T lymphocytes and monocytes in tumor-bearing mice treated with ECT.[81] In addition, due to the massive tumor antigen shedding in the organisms after ECT, systemic immunity can be induced and can be upregulated by additional treatment with biological response modifiers like IL-2, IL-12, GM-CSF, and TNF-α.[82–84]

24.2.5 VASCULAR TARGETED ACTION OF ELECTROPORATION AND ELECTROCHEMOTHERAPY

It was shown in preclinical studies that the application of electric pulses to the tissues induces a transient, but reversible, reduction of blood flow. The first study, using albumin-(Gd-DTPA) contrast-enhanced magnetic resonance imaging, has demonstrated that 30 min after application of electric pulses to SA-1 tumors, tumor blood volume was reduced from 20% in untreated tumors to 0% in electroporated tumors.[85] A pharmacological study with [86]RbCl extraction technique in the same tumor model was also done, demonstrating that significant reduction of tumor perfusion (~30% of control) was observed within 1 h following the application of electric pulses to the tumors, which returned to pretreatment value with 24 h. The degree of tumor blood flow reduction was dependent upon the number and amplitude.[86] In subsequent studies, it was demonstrated that the results obtained with the [86]RbCl extraction technique correlated with the Patent Blue staining technique, which is a much more simple method for measuring tissue perfusion,[87] and with tumor oxygenation, which was measured by the electronic paramagnetic resonance technique.[88]

In vitro studies have shown that application of electric pulses to a monolayer of endothelial cells results in a profound disruption of microfilament and microtubule cytoskeletal networks, resulting in increased permeability of endothelial monolayer.[89] Furthermore, mathematical model demonstrated that endothelial cells in the lining of small tumor blood vessels are exposed to an electric field that can increase their permeability and it's higher than in the surrounding tumor tissue.[90] Changes in endothelial cell shape were observed also in histological analysis 1 h after the application of electric pulses. Endothelial cells turned spherical in shape and became swollen, and the lumen of blood vessels was narrowed.[90] The observed effects of tumor blood flow modification after the application of electric pulses were also observed in normal muscle tissue in mice. Similar effects on leg perfusion, measured by Patent Blue, were observed in mice, with a wide variety of electric pulse amplitudes and pulse durations (10–20,000 µs and 0.1–1.6 kV/cm).[91] Based on all the gathered information on vascular effects of electric pulses in the tumor, a model of the sequence of changes was proposed.[92]

Compared to vascular changes obtained by the application of electric pulses, the changes observed after ECT were more severe, but depending on the type and the dose of chemotherapeutic drug used.[87,88] Studies on ECT with bleomycin as well as with cisplatin have demonstrated that changes, within 2 h in tumor perfusion and oxygenation, are identical to those observed after the application of electric pulses alone. Immediately after the treatment, tumor perfusion was maximally reduced.

Approximately 30 min later, the tumors started to reperfuse in both groups; in the tumors treated by ECT, the reperfusion leveled after ~1 h and stayed at 20% up to 48 h after the treatment, whereas the tumors treated with the application of electric pulses alone continued to reperfuse. If using low dose of chemotherapeutic drug, gradual reperfusion of the tumors occurred, whereas the higher dose of bleomycin resulted in complete shutdown of tumor perfusion and a high percentage of tumor cures (70%).[87,88,90]

In vitro data supported the observed in vivo effects. It was demonstrated that electroporation of human endothelial HMEC-1 cells, even after short-term drug exposure, significantly enhanced the cytotoxicity of bleomycin or cisplatin[93] and resulted in significant disruption of cytoskeletal network of endothelial cells.[94]

Detailed histological analyses of tumors after ECT demonstrated that the same morphological changes in endothelial cells occurred as after the application of electric pulses to the tumors, endothelial cells turned spherical in shape and became swollen, and the lumen of blood vessels was narrowed. However, apoptotic morphological characteristics were found in some vessels 8 h after ECT. Furthermore, blood vessels were stacked with erythrocytes, and extravasation of erythrocytes was also observed. Apoptotic endothelial cells were not observed in the control group or in tumors treated with either electric pulses or bleomycin alone,[90] while intravital microscopy of tumors in dorsal window chamber also confirmed differential effect: tumor blood vessels were more affected than normal blood vessels surrounding the tumor, which has a significant clinical applicability (significance).[95]

24.3 CLINICAL APPLICATIONS OF ELECTROCHEMOTHERAPY

Based on vast preclinical data, ECT soon entered clinical trials. The first clinical study on ECT was published already in 1991 by Mir et al.[96] It has demonstrated the feasibility, safety, and effectiveness of ECT. Soon followed the reports from the group in the United States (Tampa), Slovenia (Ljubljana), France (Toulouse and Reims), and Denmark (Copenhagen) with their own clinical results, confirming the results of the first study.[97–101] The first results were compiled in a mutual paper in 1998, which is still a hallmark of clinical ECT.[102] The development of the field was then marked by the report of the European project called "European Standard Operating Procedures on Electrochemotherapy" (ESOPE). Results from this prospective multicenter study were published in 2006[103] together with the standard operating procedures for ECT using the electric pulse generator CLINIPORATOR (SOP).[104] This was the foundation for wider acceptance of ECT into broader clinical use throughout the Europe. So far, the predominant tumor type was skin metastases of melanoma, along with skin metastases of other tumor types. ECT for skin tumors is predominantly used in palliative intent and also in previously heavily pretreated area (Figure 24.2).

The clinical indications were published in a review paper,[105] along with the compiled results of all published studies till then. Recently, the systematic review and meta-analysis of all clinical data have demonstrated that overall effectiveness of ECT was 84.1% objective responses (ORs), from these 59.4% CRs.[106] Data analysis confirmed that ECT had a significantly ($p < 0.001$) higher effectiveness (by more than 50%) than bleomycin or cisplatin alone. Furthermore, ECT was more efficient in sarcoma than in melanoma or carcinoma tumors. Another recent review and a clinical study suggested that SOP may need refinement since the currently used SOP for ECT may not be suitable for tumors bigger than 3 cm in diameter, but such tumors are suitable for the multiple consecutive ECT treatments.[59,107] In line with these findings, future investigations are needed to focus on the prognostic and predictive markers for the response of the tumors, in order to adjust ECT for the specific tumor type. Several studies are ongoing on superficial tumors, not only on melanoma but also on the treatment of chest wall breast cancer recurrences[108–111] and head and neck cancers,[112] Kaposi sarcomas,[113] and metastatic soft tissue sarcomas.[114] Furthermore, the technology is being adapted also for the treatment of deep-seated tumors, like colorectal tumors, soft tissue sarcomas, and brain, bone, and liver metastases.[117] The first clinical study on liver metastases of colorectal carcinoma has

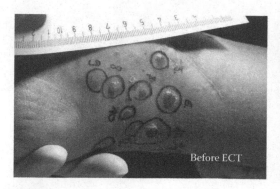
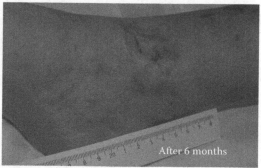

Before ECT

After 6 months

FIGURE 24.2 The antitumor effectiveness of electrochemotherapy with intravenously administered bleomycin in skin melanoma metastases. Electric pulses were delivered by plate electrodes, encompassing the nodules. Two electrochemotherapy sessions were performed. Excellent antitumor and cosmetic effects are visible.

demonstrated feasibility, safety, and effectiveness of ECT.[116] The data indicated again on 84% CR rate of the treated tumors, verified with histology and/or radiology. Specifically, ECT was demonstrated to be effective also in tumors that are close to major hepatic vessels, and not amenable for radiofrequency ablation. This study has set the stage for the use of ECT of other tumors in the liver and also in other organs in the abdomen.

Technology, electric pulse generators, as well as electrodes, were adapted for the treatment of deep-seated tumors. Several different electrode types have been prepared.[117] However, to meet the prerequisite that the whole tumor needs to be covered with sufficient electric field in order to provide good clinical response, for deep-seated tumors bigger that 2 cm, which are being treated with the placement of individual electrodes, treatment planning is recommended.[117] It is similar to the treatment plan that is prepared for radiation therapy,[118] providing the amplitudes of electric pulses that need to be delivered between the pairs of electrodes.[119] The plan that is based on the segmentation of the target tumor with safety margins is then by numerical modeling prepared for the specific tumor, with the placement of the electrodes and the treatment parameters to enable whole coverage of the tumor with the sufficiently high electric field.[120,121]

24.4 PRECLINICAL AND CLINICAL APPLICATION OF GENE ELECTROTRANSFER: GENE THERAPY

Another application of electroporation in biomedicine is gene electrotransfer—electrogene therapy. It can be used either for DNA vaccination against infectious diseases or for the treatment of various diseases, such as cancer, where therapies either are targeted directly to tumor cells or aim to increase the immune response of the organism against cancer cells. In vivo gene delivery using electroporation was first performed in the 1990s,[122] and since then, a number of different types of tissues have been successfully transfected using this approach (for instance, tumors, skeletal muscle, skin, and liver).[123,124] Transfection efficiency of electrotransfer is still low compared to viral vectors; yet its advantages, mostly lack of pathogenicity and immunogenicity, make it a promising new method.

Gene therapy can be performed using two different approaches. The first one is ex vivo gene therapy, where cells, including stem cells, are removed from patient, transfected in vitro with the plasmid or viral vector, selected, amplified, and then reinjected back into the patient. The other approach is in vivo gene therapy, where exogenous DNA is delivered directly into host's target tissue, for example, locally to tumor or peritumorally and for systemic release of the therapeutic molecule into skeletal muscle depending on the type of therapeutic molecules and intent of treatment (Figure 24.3).

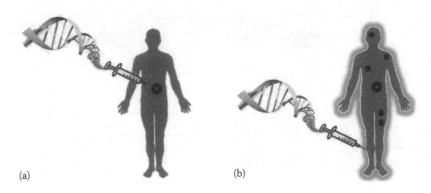

(a)   (b)

FIGURE 24.3 (a) Local approach to cancer gene therapy. Injection of plasmid DNA directly into the tumor. (b) Systemic approach to cancer gene therapy by injection of plasmid DNA into the muscle, which then produces therapeutic protein that is distributed throughout the body reaching distant tumors.

Gene electrotransfer of therapeutic genes into tumors facilitates local intratumoral production of therapeutic proteins, enabling sufficient therapeutic concentration and thus therapeutic outcome. This is especially important in case of cytokines, where high systemic concentrations are associated with severe toxicity.[125] Gene electrotransfer can be used as a single therapy or in combination with other modalities for cancer treatment, such as standard treatment options surgery and radiotherapy, but also for example electrochemotherapy.[126,127]

The first evaluation of intratumoral electrogene therapy for cancer treatment was performed on murine melanoma tumor model in 1999 by Niu et al.[128] Since then, not only a variety of therapeutic genes, mostly encoding cytokines, but also tumor suppressor proteins, siRNA molecules against various targets, etc., have been tested in numerous animal tumor models, for example, melanoma, squamous cell carcinoma, sarcoma, and hepatocellular carcinoma.[129,130] Results of preclinical studies indicate that intratumoral therapeutic gene electrotransfer enables efficient transgene expression with sufficient production of therapeutic proteins, which can lead to pronounced antitumor effect on treated tumor (e.g., suppression of tumor growth, partial or complete reduction of tumor nodule), and even induces long-term antitumor immunity in treated animals.[84,131] Interestingly, some of the studies reported that even control plasmid without therapeutic gene can result in CR of the tumors, especially melanoma B16 tumor model. It was demonstrated that the underlying mechanism for this result is multifactorial, including direct toxicity of DNA, selection of electric pulses parameters, and induction of immunity.[132]

Some of the most significant antitumor effect to date in cancer gene therapy have been achieved with the employment of active nonspecific immunotherapy, that is, the use of cytokines. Gene electrotransfer of genes, encoding different cytokines, has already shown promising results in preclinical trials on different animal tumor models. Cytokine genes, which showed the most potential for cancer therapy, are IL-2, IL-12, IL-18, interferon (IFN) α, and GM-CSF.[129,133] Currently, the most advanced therapy is using IL-12, which plays important role in the induction of cellular immune response through stimulation of T-lymphocyte differentiation and production of IFN-γ and activation of natural killer cells.[134] Antitumor effect of IL-12 gene electrotransfer has already been established in various tumor models, for example, melanoma, lymphoma, squamous cell carcinoma, urinary bladder carcinoma, mammary adenocarcinoma, and hepatocellular carcinoma.[133] Results of preclinical studies show that besides regression of tumor at primary and distant sites, electrogene therapy with IL-12 also promotes induction of long-term antitumor memory and therapeutic immunity, suppresses metastatic spread, and increases survival time of experimental animals.[130] Gene therapy with IL-12 was successfully combined also with other therapies, such as ECT and radiotherapy resulting in potentiated effect[126,127,135–137] (Figure 24.4).

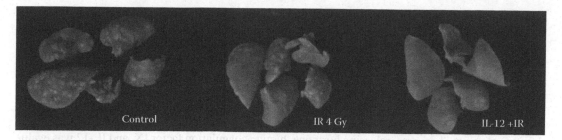

FIGURE 24.4 Effect of combined IL-12 gene electrotransfer and radiotherapy (IR) on lung metastases. Combined therapy resulted in complete eradication of metastases. (From Heller, L. et al., *Cancer Gene Ther.*, 20, 695, 2013.)

Recently, clinical studies performed in patients with melanoma, as well as in veterinary patients, show great promise for further development of this therapy.[135,136] In human clinical study, 24 patients with malignant melanoma subcutaneous metastases were treated three times. The response to therapy was observed in treated as well as in distant nontreated tumor nodules. In 53% of patients, a systemic response was observed resulting in either stable disease or an OR. The major adverse side effect was transient pain after the application of electric pulses. In posttreatment biopsies, tumor necrosis and immune cell infiltration were observed. This first human clinical trial with IL-12 electrogene therapy in metastatic melanoma proved that this therapy is safe and effective.[138] In veterinary oncology, eight dogs with mastocytoma were treated with IL-12 gene electrotransfer. A good local antitumor effect with significant reduction of treated tumors' size, ranging from 15% to 83% (mean 52%) of the initial tumor volume, was obtained. Additionally, a change in the histological structure of treated nodules was seen as a reduction in the number of malignant mast cells and inflammatory cell infiltration of treated tumors. Furthermore, systemic release of IL-12 and IFN-γ in treated dogs was detected, without any noticeable local or systemic side effects.[139] Again, the data suggest that intratumoral IL-12 electrogene therapy could be used for controlling local as well as systemic disease.

On preclinical level, gene electrotransfer to tumors was also employed in suicide gene therapy of cancer. The concept of suicide gene therapy is intratumoral transfer of a prodrug-activating gene, which selectively (intratumorally) activates otherwise nontoxic drugs. The most often used strategy in suicide gene therapy is the delivery of gene, encoding herpes simplex virus thymidine kinase (HSV-TK) and prodrug ganciclovir (GCV). HSV-TK activates GCV, which blocks extensions of DNA strands, leading to cell death by apoptosis. Results of several studies show that electroporation-based HSV-TK/GCV gene therapy may provide potentially effective gene therapy for cancer.[140–143]

Another approach in cancer gene therapy, which is currently being widely investigated, is based on the inhibition of angiogenesis of tumors. The basic concept of antiangiogenic gene therapy is the transfection of cells with genes, encoding inhibitors of tumor angiogenesis. Electrotransfer of plasmids encoding antiangiogenic factors (angiostatin and endostatin) was demonstrated to be effective in the inhibition of tumor growth and metastatic spread of different tumors.[144–146] Recently, we showed that the RNA interference approach, using siRNA molecule against endoglin, which is a coreceptor of transforming growth factor β and is upregulated in activated endothelial cells, also resulted in vascular targeted effect in mammary tumors.[130]

Besides tumors, skeletal muscle is an attractive target tissue for the delivery of therapeutic genes, since it is usually a large mass of well-vascularized and easily accessible tissue with high capacity for the synthesis of proteins, which can be secreted either locally or systemically.[147] Furthermore, transfection efficiency in muscle is very high compared to other tissues, especially tumors.[147] Owing to the postmitotic status and slow turnover of skeletal muscle fibers, which ensures that transfected DNA isn't readily lost, it is possible to achieve long-term expression of exogenous DNA, which can last up to 1 year.[147,148] This is due to the dynamics of naked DNA transfer since plasmid does not integrate into the genome of transfected cell, and thus the duration of exogenous DNA expression in part depends on the lack of cell division. In contrast to muscle cells, in tissues, where cell turnover is much higher, for example, tumors, plasmid DNA is rapidly lost from the cells.[147,149]

Gene electrotransfer into skeletal muscle can be applied for the treatment of various muscle diseases, for local secretion of angiogenic or neurotrophic factors, or for systemic secretion of different therapeutic proteins, such as erythropoietin, coagulation factors, cytokines, and monoclonal antibodies.[147,150–152] In cancer gene therapy, gene electrotransfer of plasmid DNA encoding IL-12, IL-24, and antiangiogenic factors was evaluated with encouraging results. In clinical studies, intramuscular delivery of growth hormone–releasing hormone, human coagulation factor IX, and IL-12 was evaluated.[153–155] Results of our study indicate that in canine cancer patients, intramuscular IL-12 EGT is a safe procedure, which can result in systemic shedding of hIL-12 and possibly trigger IFN-γ response in treated patients, leading to prolonged disease-free period and survival of treated animals.[155]

24.5 PERSPECTIVES

Electroporation-based biomedical applications, such as ECT and gene electrotransfer, are the most advanced of its applications. Very likely, electroporation will find its place also in vaccination, in the treatment of cancer, and in the delivery of drugs to and through the skin, for local and systemic treatment of diseases other than cancer.

ECT is now on the verge to enter into standard of care in many European Oncology Centers. Its application has spread; experiences are being gained, and further profiling of ECT have begun. The next steps are in the translation of this technology into the treatment of deep-seated tumors. Furthermore, many possibilities exist to combine ECT with other local or systemic treatments, either to potentiate the local effect, that is, radiotherapy, or to augment the systemic response by adjuvant immunotherapy. This will, again, take time, but will broaden current clinical indications of ECT.

In the future for gene therapy, some crucial questions need to be resolved, such as how optimization of treatment protocols for different tumor types should be performed, with respect to defining optimal plasmid dose, number of treatment repetitions, optimal route of administration (intratumoral, peritumoral/intradermal, intramuscular), and effect of combination with other treatment protocols (e.g., local intratumoral plasmid delivery or ECT) in order to achieve effective long-term antitumor effect in cancer patients.

ACKNOWLEDGMENTS

The authors acknowledge support for their work through various grants from the Slovenian Research Agency and the European Commission within its Framework Programs, in particular within projects Cliniporator and ESOPE. Research was conducted in the scope of the EBAM European Associated Laboratory (LEA). This manuscript is a result of the networking efforts of the COST Action TD1104 (http://www.electroporation.net).

REFERENCES

1. Michael DH, O'Neill ME. 1970. Electrohydrodynamic instability in plane layers of fluid. *Journal of Fluid Mechanics* 41: 571–580.
2. Crowley JM. 1973. Electrical breakdown of bimolecular lipid membranes as an electromechanical instability. *Biophysical Journal* 13: 711–724.
3. Steinchen A, Gallez D, Sanfeld A. 1982. A viscoelastic approach to the hydrodynamic stability of membranes. *Journal of Colloid and Interface Science* 85: 5–15.
4. Sugár IP. 1979. A theory of the electric field-induced phase transition of phospholipid bilayers. *Biochimica et Biophysica Acta* 556: 72–85
5. Cruzeiro-Hansson L, Mouritsen OG. 1988. Passive ion permeability of lipid membranes modelled via lipid-domain interfacial area. *Biochimica et Biophysica Acta* 944: 63–72.

6. Tsong TY. 1991. Electroporation of cell membranes. *Biophysical Journal* 60: 297–306.
7. Weaver JC, Chizmadzhev YA. 1996. Theory of electroporation: A review. *Bioelectrochemistry and Bioenergetics* 41: 135–160.
8. Spugnini EP, Arancia G, Porrello A et al. 2007. Ultrastructural modifications of cell membranes induced by electroporation on melanoma xenografts. *Microscopy Research and Technique* 70: 1041–1050.
9. Freeman SA, Wang MA, Weaver JC. 1994. Theory of electroporation of planar bilayer membranes: Predictions of the aqueous area, change in capacitance, and pore-pore separation. *Biophysical Journal* 67: 42–56.
10. Kotnik T, Kramar P, Pucihar G, Miklavcic D, Tarek M. 2012. Cell membrane electroporation—Part 1: The phenomenon. *IEEE Electrical Insulation Magazine* 28: 14–23.
11. Kotnik T, Miklavčič D, Slivnik T. 1998. Time course of transmembrane voltage induced by time-varying electric fields—A method for theoretical analysis and its application. *Bioelectrochemistry and Bioenergetics* 45: 3–16.
12. Kotnik T, Miklavčič D. 2006. Theoretical evaluation of voltage inducement on internal membranes of biological cells exposed to electric fields. *Biophysical Journal* 90: 480–491.
13. Leontiadou H, Mark AE, Marrink SJ. 2004. Molecular dynamics simulations of hydrophilic pores in lipid bilayers. *Biophysical Journal* 86: 2156–2164.
14. Tarek M. 2005. Membrane electroporation: A molecular dynamics simulation. *Biophysical Journal* 88: 4045–4053.
15. Böckmann RA, De Groot BL, Kakorin S, Neumann E, Grubmüller H. 2008. Kinetics, statistics, and energetics of lipid membrane electroporation studied by molecular dynamics simulations. *Biophysical Journal* 95: 1837–1850.
16. Dev SB, Rabussay DP, Widera G, Hofmann GA. 2000. Medical applications of electroporation. *IEEE Transactions on Plasma Science* 28: 206–223.
17. Bower M, Sherwood L, Li Y, Martin R. 2011. Irreversible electroporation of the pancreas: Definitive local therapy without systemic effects. *Journal of Surgical Oncology* 104: 22–28.
18. Yarmush ML, Golberg A, Serša G, Kotnik T, Miklavčič D. 2014. Electroporation-based technologies for medicine: principles, applications, and challenges. *Annual Review of Biomedical Engineering* 16: 295–320.
19. Delemotte L, Tarek M. 2012. Molecular dynamics simulations of lipid membrane electroporation. *Journal of Membrane Biology* 245: 531–543.
20. Gabriel B, Teissié J. 1999. Time courses of mammalian cell electropermeabilization observed by millisecond imaging of membrane property changes during the pulse. *Biophysical Journal* 76: 2158–2165.
21. Kotnik T, Pucihar G, Miklavčič D. 2010. Induced transmembrane voltage and its correlation with electroporation-mediated molecular transport. *Journal of Membrane Biology* 236: 3–13.
22. Miklavčič D, Šemrov D, Mekid H, Mir LM. 2000. A validated model of in vivo electric field distribution in tissues for electrochemotherapy and for DNA electrotransfer for gene therapy. *Biochimica et Biophysica Acta* 1532: 73–83.
23. Pauly H, Schwan HP. 1959. Über die Impedanz einer Suspension von kugelformigen Teilchen mit einer Schale. *Zeitschrift für Naturforschung B* 14: 125–131.
24. Bernhard J, Pauly H. 1973. Generation of potential differences across membranes of ellipsoidal cells in an alternating electrical field. *Biophysik* 10: 89–98.
25. Kotnik T, Miklavčič D. 2000. Analytical description of transmembrane voltage induced by electric fields on spheroidal cells. *Biophysical Journal* 79: 670–679.
26. Gimsa J, Wachner D. 2001. Analytical description of the transmembrane voltage induced on arbitrarily oriented ellipsoidal and cylindrical cells. *Biophysical Journal* 81: 1888–1896.
27. Pucihar G, Kotnik T, Valič B, Miklavčič D. 2006. Numerical determination of transmembrane voltage induced on irregularly shaped cells. *Annals of Biomedical Engineering* 34: 642–652.
28. Pucihar G, Miklavčič D, Kotnik T. 2009. A time-dependent numerical model of transmembrane voltage inducement and electroporation of irregularly shaped cells. *IEEE Transactions on Biomedical Engineering* 56: 1491–1501.
29. Susil R, Šemrov D, Miklavčič D. 1998. Electric field induced transmembrane potential depends on cell density and organization. *Electro- and Magnetobiology* 17: 391–399.
30. Pavlin M, Pavšelj N, Miklavčič D. 2002. Dependence of induced transmembrane potential on cell density, arrangement, and cell position inside a cell system. *IEEE Transactions on Biomedical Engineering* 49: 605–612.
31. Ying W, Henriquez CS. 2007. Hybrid finite element method for describing the electrical response of biological cells to applied fields. *IEEE Transactions on Biomedical Engineering* 54: 611–620.

32. Gross D, Loew LM, Webb W. 1986. Optical imaging of cell membrane potential changes induced by applied electric fields. *Biophysical Journal* 50: 339–348.

33. Pucihar G, Kotnik T, Miklavčič D. 2009. Measuring the induced membrane voltage with di-8-ANEPPS. *Journal of Visual Experiments* 33: 1659.

34. Hibino M, Itoh H, Kinosita K. 1993. Time courses of cell electroporation as revealed by submicrosecond imaging of transmembrane potential. *Biophysical Journal* 64: 1789–1800.

35. Frey W, White JA, Price RO, Blackmore PF, Joshi RP, Nuccitelli RL, Beebe SJ, Schoenbach HK, Kolb JF. 2006. Plasma membrane voltage changes during nanosecond pulsed electric field exposure. *Biophysical Journal* 90: 3608–3615.

36. Benz R, Conti F. 1981. Reversible electrical breakdown of squid giant axon membrane. *Biochimica et Biophysica Acta* 645: 115–123.

37. Rytssen F, Farre C, Brennan C, Weber SG, Nolkrantz K, Jardemark K, Chiu DT, Orwar O. 2000. Characterization of single-cell electroporation by using patch-clamp and fluorescence microscopy. *Biophysical Journal* 79: 1993–2001.

38. Kinosita K, Tsong TY. 1979. Voltage-induced conductance in human erythrocyte membranes. *Biochimica et Biophysica Acta* 554: 479–497.

39. Pavlin M, Leben V, Miklavčič D. 2007. Electroporation in dense cell suspensions—Theoretical and experimental analysis of ion diffusion and cell permeabilization. *Biochimica et Biophysica Acta* 1770: 12–23.

40. Pliquett U, Prausnitz MR. 2000. Electrical impedance spectroscopy for rapid and noninvasive analysis of skin electroporation. *Methods in Molecular Medicine* 37: 377–406.

41. Ivorra A, Rubinsky B. 2007. In vivo electrical impedance measurements during and after electroporation of rat liver. *Bioelectrochemistry* 70: 287–295.

42. Pucihar G, Kotnik T, Miklavčič D, Teissié J. 2008. Kinetics of transmembrane transport of small molecules into electropermeabilized cells. *Biophysical Journal* 95: 2837–2848.

43. Rols MP, Teissié J. 1990. Electropermeabilization of mammalian cells: Quantitative analysis of the phenomenon. *Biophysical Journal* 58: 1089–1098.

44. Puc M, Kotnik T, Mir LM, Miklavčič D. 2003. Quantitative model of small molecules uptake after in vitro cell electropermeabilization. *Bioelectrochemistry* 60: 1–10.

45. Pavlin M, Flisar K, Kanduŝer M. 2010. The role of electrophoresis in gene electrotransfer. *Journal of Membrane Biology* 236: 75–79.

46. Escoffre JM, Portet T, Favard C, Teissié J, Dean DS, Rols MP. 2011. Electromediated formation of DNA complexes with cell membranes and its consequences for gene delivery. *Biochimica et Biophysica Acta* 1808: 1538–1543.

47. Rosazza C, Buntz A, Riess T, Woll D, Zumbusch A, Rols MP. 2013. Intracellular tracking of single-plasmid DNA particles after delivery by electroporation. *Molecular Therapy* 21: 2217–2226.

48. Orlowski S, Belehradek J Jr, Paoletti C, Mir LM. 1988. Transient electropermeabilization of cells in culture. Increase in cytotoxicity of anticancer drugs. *Biochemical Pharmacology* 37: 4727–4733.

49. Mir LM. 2006. Bases and rationale of the electrochemotherapy. *European Journal of Cancer Supplements* 4: 38–44.

50. Sersa G, Cemazar M, Miklavcic D, Mir LM. 1994. Electrochemotherapy: Variable anti-tumor effect on different tumor models. *Bioelectrochemistry and Bioenergitics* 35: 23–27.

51. Gehl J, Skovsgaard T, Mir LM. 1998. Enhancement of cytotoxicity by electropermeabilization: An improved method for screening drugs. *Anti-Cancer Drugs* 9: 319–325.

52. Cemazar M, Sersa G, Miklavcic D. 1998. Electrochemotherapy with cisplatin in treatment of tumor cells resistant to cisplatin. *Anticancer Research* 18: 4463–4466.

53. Cemazar M, Parkins CS, Holder AL et al. 2001. Electroporation of human microvascular endothelial cells: Evidence for anti-vascular mechanism of electrochemotherapy. *British Journal of Cancer* 84: 556–570.

54. Okino M, Mohri H. 1987. Effects of a high-voltage electrical impulse and an anticancer drug on in vivo growing tumors. *Japanese Journal of Cancer Research* 78: 1319–1321.

55. Mir LM, Orlowski S, Belehradek J Jr, Paoletti C. 1991. Electrochemotherapy potentiation of antitumor effect of bleomycin by local electric pulses. *European Journal of Cancer* 27: 68–72.

56. Salford LG, Persson BRR, Brun A, Ceberg CP, Kongstad PCH, Mir LM. 1993. A new brain tumor therapy combining bleomycin with in vivo electropermeabilization. *Biochemical and Biophysical Research Communications* 194: 938–943.

57. Heller R, Jaroszeski M, Leo-Messina J, Perrot R, Van Voorhis N, Reintgen D, Gilbert R. 1995. Treatment of B16 mouse melanoma with the combination of electropermeabilization and chemotherapy. *Bioelectrochemistry and Bioenergetics* 36: 83–87.
58. Sersa G. 2000. Electrochemotherapy: Animal work review. In: Jaroszeski MJ, Heller R, Gilbert R, (eds.). *Electrochemotherapy, Electrogenetherapy, and Transdermal Drug Delivery: Electrically Mediated Delivery of Molecules to Cells.* Totowa, NJ: Humana Press, pp. 119–136.
59. Miklavcic D, Mali B, Kos B, Heller R, Sersa G. 2014. Electrochemotherapy: From the drawing board into medical practice. *BioMedical Engineering Online* 13: 29. DOI:10.1186/1475–925X-13–29.
60. Bicek A, Turel I, Kanduser M, Miklavcic D. 2007. Combined therapy of the antimetastatic compound NAMI-A and electroporation on B16F1 tumour cells in vitro. *Bioelectrochemistry* 71: 113–117.
61. Cemazar M, Pipan Z, Grabner S, Bukovec N, Sersa G. 2006. Cytotoxicity of different platinum (II) analogues to human tumour cell lines in vitro and murine tumour in vivo alone or combined with electroporation. *Anticancer Research* 26: 1997–2002.
62. Frandsen SK, Gissel H, Hojman P, Eriksen J, Gehl J. 2014. Calcium electroporation in three cell lines: A comparison of bleomycin and calcium, calcium compounds, and pulsing conditions. *Biochimica et Biophysica Acta* 1840(3): 1204–1208.
63. Hudej R, Turel I, Kanduser M et al. 2010. The influence of electroporation on cytotoxicity of anticancer ruthenium(III) complex KP1339 in vitro and in vivo. *Anticancer Research* 30: 2055–2063.
64. Miklavcic D, Beravs K, Semrov D et al. 1998. The importance of electric field distribution for effective in vivo electroporation of tissues. *Biophysical Journal* 74: 2152–2158.
65. Miklavcic D, Pucihar G, Pavlovec M et al. 2005. The effect of high frequency electric pulses on muscle contractions and antitumor efficiency in vivo for a potential use in clinical electrochemotherapy. *Bioelectrochemistry* 65: 121–128.
66. Mir LM, Devauchelle P, Quintin-Colonna F et al. 1997. First clinical trial of cat soft-tissue sarcomas treatment by electrochemotherapy. *British Journal of Cancer* 76: 1617–1622.
67. Tozon N, Sersa G, Cemazar M. 2001. Electrochemotherapy: Potentiation of local antitumour effectiveness of cisplatin in dogs and cats. *Anticancer Research* 21: 2483–2488.
68 Pavlica Z, Petelin M, Nemec A et al. 2006. Treatment of feline lingual squamous cell carcinoma using electrochemotherapy—A case report. *Proceedings of the 15th European Congress of Veterinary Dentistry*, Cambridge, England, pp. 19–22.
69. Tamzali Y, Teissie J, Rols MP. 2001. Cutaneous tumor treatment by electrochemotherapy: Preliminary clinical results in horse sarcoids. *Revue de Medicine Veterinaire* 152: 605–609.
70. Rols MP, Tamzali Y, Teissie J. 2002. Electrochemotherapy of horses. A preliminary clinical report. *Bioelectrochemistry* 1–2: 101–105.
71. Cemazar M, Tamzali Y, Sersa G et al. 2008. Electrochemotherapy in veterinary oncology. *Journal of Veterinary Internal Medicine* 22: 826–231.
72. Kodre V, Cemazar M, Pecar J, Sersa G, Cor A, Tozon N. 2009. Electrochemotherapy compared to surgery for treatment of canine mast cell tumours. *In Vivo* 23: 55–62.
73. Spugnini EP, Di Tosto G, Salemme S, Pecchia L, Fanciulli M, Baldi A. 2013. Electrochemotherapy for the treatment of recurring aponeurotic fibromatosis in a dog. *The Canadian Veterinary Journal* 54: 606–609.
74. Spugnini EP, Fanciulli M, Citro G, Baldi A. 2012. Preclinical models in electrochemotherapy: The role of veterinary patients. *Future Oncology* 8: 829–837.
75. Spugnini EP, Filipponi M, Romani L et al. 2010. Electrochemotherapy treatment for bilateral pleomorphic rhabdomyosarcoma in a cat. *Journal of Small Animal Practice* 51: 330–332.
76. Spugnini EP, Renaud SM, Buglioni S et al. 2011. Electrochemotherapy with cisplatin enhances local control after surgical ablation of fibrosarcoma in cats: An approach to improve the therapeutic index of highly toxic chemotherapy drugs. *Journal of Translational Medicine* 9: 152.
77. Tozon N, Pavlin D, Sersa G, Dolinsek T, Cemazar M. 2014. Electrochemotherapy with intravenous bleomycin injection: An observational study in superficial squamous cell carcinoma in cats. *Journal of Feline Medicine and Surgery* 16: 291–299.
78. Belehradek J Jr., Orlowski S, Ramirez LH et al. 1994. Electropermeabilization of cells and tissues assessed by the quantitative and qualitative electroloading of bleomycin. *Biochimica et Biophysica Acta* 1190: 155–163.
79. Cemazar M, Miklavcic D, Scancar J et al. 1999. Increased platinum accumulation in SA-1 tumour cells after in vivo electrochemotherapy with cisplatin. *British Journal of Cancer* 79: 1386–1391.

80. Sersa G, Miklavcic D, Cemazar M et al. 1997. Electrochemotherapy with CDDP on LPB sarcoma: Comparison of the anti-tumor effectiveness in immunocompetent and immunodeficient mice. *Bioelectrochemistry and Bioenergetics* 43: 279–283.
81. Sersa G, Kotnik V, Cemazar M, Miklavcic D, Kotnik A. 1996. Electrochemotherapy with bleomycin in SA-1 tumor-bearing mice—Natural resistance and immune responsiveness. *Anti-Cancer Drugs* 7: 785–791.
82. Mir LM, Roth C, Orlowski S et al. 1995. Systemic antitumor effects of electrochemotherapy combined with histoincompatible cells secreting interleukin 2. *Journal of Immunotherapy* 17: 30–38.
83. Sersa G, Cemazar M, Menart V, Gaberc-Porekar V, Miklavcic D. 1997. Antitumor effectiveness of electrochemotherapy is increased by TNF-α on SA-1 tumors in mice. *Cancer Letters* 116: 85–92.
84. Heller L, Pottinger C, Jaroszeski MJ, Gilbert R, Heller R. 2000. In vivo electroporation of plasmids encoding GM-CSF or interleukin-2 into existing B16 melanoma combined with electrochemotherapy inducing long-term antitumour immunity. *Melanoma Research* 10: 577–583.
85. Sersa G, Beravs K, Cemazar M, Miklavcic D, Demsar F. 1998. Contrast enhanced MRI assessment of tumor blood volume after application of electric pulses. *Electro- and Magnetobiology* 17: 299–306.
86. Sersa G, Cemazar M, Parkins CS, Chaplin DJ. 1999. Tumour blood flow changes induced by application of electric pulses. *European Journal of Cancer* 35: 672–677.
87. Sersa G, Cemazar M, Miklavcic D, Chaplin DJ. 1999. Tumor blood modifying effect of electrochemotherapy with bleomycin. *Anticancer Research* 19: 4017–4022.
88. Sersa G, Krzic M, Sentjurc M, Ivanusa T, Beravs K, Kotnik V, Coer A, Swartz HM, Cemazar M. 2002. Reduced blood flow and oxygenation in SA-1 tumours after electrochemotherapy with cisplatin. *British Journal of Cancer* 87: 1047–1054.
89. Kanthou C, Kranjc S, Sersa G, Tozer G, Zupanic A, Cemazar M. 2006. The endothelial cytoskeleton as a target of electroporation based therapies. *Molecular Cancer Therapeutics* 5: 3145–3152.
90. Sersa G, Jarm T, Kotnik T et al. 2008. Vascular disrupting action of electroporation and electrochemotherapy with bleomycin in murine sarcoma. *British Journal of Cancer* 98: 388–398.
91. Gehl J, Skovsgaard T, Mir LM. 2002. Vascular reactions to in vivo electroporation: Characterization and consequences for drug and gene delivery. *Biochimica et Biophysica Acta* 1569: 51–58.
92. Jarm T, Cemazar M, Miklavcic D, Sersa G. 2010. Antivascular effects of electrochemotherapy: Implications in treatment of bleeding metastases. *Expert Review of Anticancer Therapy* 10: 729–746.
93. Cemazar M, Parkins CS, Chaplin DJ, Tozer GM, Sersa G. 2001. Electroporation of human microvascular endothelial cells: Evidence of an anti-vascular mechanism of electrochemotherapy. *British Journal of Cancer* 84: 565–570.
94. Meulenberg CJW, Todorovic V, Cemazar M. 2012. Differential cellular effects of electroporation and electrochemotherapy in monolayers of human microvascular endothelial cells. *Plos One* 7(12): e52713.
95. Markelc B, Sersa G, Cemazar M. 2013. Differential mechanisms associated with vascular disrupting action of electrochemotherapy: Intravital microscopy on the level of single normal and tumor blood vessels. *Plos One* 8(3): e59557.
96. Mir LM, Belehradek M, Domenge C, Orlowski S, Poddevin B, Belehradek J Jr., Schwaab G, Luboinski B, Paoletti C. 1991. Electrochemotherapy, a new antitumor treatment: First clinical trial. *Comptes Rendus Academic Science III* 313: 613–618.
97. Heller R. 2995. Treatment of cutaneous nodules using electrochemotherapy. *The Journal of the Florida Medical Association* 82: 147–150.
98. Rudolf Z, Stabuc B, Cemazar M, Miklavcic D, Vodovnik L, Sersa G. 1995. Electrochemotherapy with bleomycin: The first clinical experience in malignant melanoma patients. *Radiology and Oncology* 29: 229–235.
99. Sersa G, Stabuc B, Cemazar M, Miklavcic D, Rudolf Z. 2000. Electrochemotherapy with cisplatin: Clinical experience in malignant melanoma patients. *Clinical Cancer Research* 6: 863–867.
100. Rols MP, Bachaud JM, Giraud P, Chevreau C, Roche H, Teissie J. 2000. Electrochemotherapy of cutaneous metastases in malignant melanoma. *Melanoma Research* 10: 468–474.
101. Gehl J, Geertsen P. 2000. Efficient palliation of hemorrhaging malignant melanoma skin metastases by electrochemotherapy. *Melanoma Research* 10: 585–589.
102. Mir LM, Glass LF, Sersa G et al. 1998. Effective treatment of cutaneous and subcutaneous malignant tumours by electrochemotherapy. *British Journal of Cancer* 77: 2336–2342.
103. Marty M, Sersa G, Garbay JR et al. 2006. Electrochemotherapy—An easy, highly effective and safe treatment of cutaneous and subcutaneous metastases: Results of ESOPE (European Standard Operating Procedures of Electrochemotherapy) study. *European Journal of Cancer Supplements* 4: 3–13.

104. Mir LM, Gehl J, Sersa G et al. 2006. Standard operating procedures of the electrochemotherapy: Instructions for the use of bleomycin or cisplatin administered either systemically or locally and electric pulses delivered by the Cliniporator™ by means of invasive or non-invasive electrodes. *European Journal of Cancer Supplements* 4: 14–25.
105. Sersa G, Miklavcic D, Cemazar M, Rudolf Z, Pucihar G, Snoj M. 2008. Electrochemotherapy in treatment of tumours. *EJSO* 34: 232–240.
106. Mali B, Jarm T, Snoj M, Sersa G, Miklavcic D. 2013. Antitumor effectiveness of electrochemotherapy: A systematic review and meta-analysis. *EJSO* 39: 4–16.
107. Mali B, Miklavcic D, Campana LG et al. 2013. Tumor size and effectiveness of electrochemotherapy. *Radiology and Oncology* 47: 32–41.
108. Sersa G, Cufer T, Paulin SM, Cemazar M, Snoj M. 2012. Electrochemotherapy of chest wall breast cancer recurrence. *Cancer Treatment Reviews* 38: 379–386.
109. Campana LG, Valpione S, Falci C et al. 2012. The activity and safety of electrochemotherapy in persistent chest wall recurrence from breast cancer after mastectomy: A phase-II study. *Breast Cancer Research and Treatment* 134: 1169–1178.
110. Campana LG, Galuppo S, Valpione S et al. 2014. Bleomycin electrochemotherapy in elderly metastatic breast cancer patients: Clinical outcome and management considerations. *Journal of Cancer Research and Clinical Oncology* 140: 1557–1565.
111. Matthiessen LW, Johannesen HH, Hendel HW, Moss T, Kamby C, Gehl J. 2012. Electrochemotherapy for large cutaneous recurrence of breast cancer: A phase II clinical trial. *Acta Oncologica* 51: 713–721.
112. Gargiulo M, Papa A, Capasso P, Moio M, Cubicciotti E, Parascandolo S. 2012. Electrochemotherapy for non-melanoma head and neck cancers: Clinical outcomes in 25 patients. *Annals of Surgery* 255: 1158–1164.
113. Di monta G, Caraco C, Benedetto L et al. 2014. Electrochemotherapy as a new standard of care treatment for cutaneous Kaposi's sarcoma. *EJSO* 40: 61–66.
114. Campana LG. 2014. Electrochemotherapy treatment of locally advanced and metastatic soft tissue sarcomas: Results of a non comparative phase II study. *World Journal of Surgery* 38: 813–822.
115. Miklavcic D, Sersa G, Brecelj E, Gehl J, Soden D, Bianchi G, Ruggieri P, Rossi CR, Campana LG, Jarm T. 2012. Electrochemotherapy: Technological advancements for efficient electroporation-based treatment of internal tumors. *Medical and Biological Engineering and Computing* 50: 1213–1225.
116. Edhemovic I, Brecelj E, Gasljevic G et al. 2014. Intraoperative electrochemotherapy of colorectal liver metastases. *Journal of Surgical Oncology* 110: 320–327.
117. Pavliha D, Kos B, Marčan M, Županič A, Serša G, Miklavčič D. 2013. Planning of electroporation-based treatments using web-based treatment-planning software. *Journal of Membrane Biology* 246: 833–842.
118. Pavliha D, Kos B, Županič A, Marčan M, Serša G, Miklavčič D. 2012. Patient-specific treatment planning of electrochemotherapy: Procedure design and possible pitfalls. *Bioelectrochemistry* 87: 265–273.
119. Edhemović I, Gadžijev EM, Brecelj E et al. 2011. Electrochemotherapy: A new technological approach in treatment of metastases in the liver. *Technology in Cancer Research and Treatment* 10: 475–485.
120. Miklavčič D, Snoj M, Županič A, Kos B, Čemažar M, Kropivnik M, Bračko M, Pečnik T, Gadžijev E, Serša G. 2010. Towards treatment planning and treatment of deep-seated solid tumors by electrochemotherapy. *Biomedical Engineering Online* 9: 10.
121. Kos B, Županič A, Kotnik T, Snoj M, Serša G, Miklavčič D. 2010. Robustness of treatment planning for electrochemotherapy of deep-seated tumors. *Journal of Membrane Biology* 236: 147–153.
122. Titomirov AV, Sukharev S, Kistanova E. 1991. In vivo electroporation and stable transformation of skin cells of newborn mice by plasmid DNA. *Biochimica et Biophysica Acta* 1088: 131–134.
123. Chabot S, Rosazza C, Golzio M, Zumbusch A, Teissié J, Rols MP. 2013. Nucleic acids electro-transfer: From bench to bedside. *Current Drug Metabolism* 14: 300–308.
124. Mir LM. 2014. Electroporation-based gene therapy: Recent evolution in the mechanism description and technology developments. *Methods in Molecular Biology* 1121: 3–23.
125. Leonard JP, Sherman ML, Fisher GL et al. 1997. Effects of single-dose interleukin-12 exposure on interleukin-12-associated toxicity and interferon-gamma production. *Blood* 90: 2541–2548.
126. Sedlar A, Dolinsek T, Markelc B et al. 2012. Potentiation of electrochemotherapy by intramuscular IL-12 gene electrotransfer in murine sarcoma and carcinoma with different immunogenicity. *Radiology and Oncology* 46: 302–311.
127. Sedlar A, Kranjc S, Dolinsek T, Cemazar M, Coer A, Sersa G. 2013. Radiosensitizing effect of intratumoral interleukin-12 gene electrotransfer in murine sarcoma. *BMC Cancer* 13: 38.

128. Niu GL, Heller R, Catlett-Falcone R et al. 1999. Gene therapy with dominant-negative Stat3 suppresses growth of the murine melanoma B16 tumor in vivo. *Cancer Research* 59: 5059–5063.

129. Andre F, Mir LM. 2004. DNA electrotransfer: Its principles and an updated review of its therapeutic applications. *Gene Therapy* 11 (Suppl 1): S33–S42.

130. Dolinsek T, Markelc B, Sersa G et al. 2013. Multiple delivery of siRNA against endoglin into murine mammary adenocarcinoma prevents angiogenesis and delays tumor growth. *Plos One* 8(3): e58723.

131. Li S, Zhang X, Xia X. 2002. Regression of tumor growth and induction of long-term antitumor memory by interleukin 12 electro-gene therapy. *Journal of National Cancer Institute* 94: 762–768.

132. Heller L, Todorovic V, Cemazar M. 2013. Electrotransfer of single-stranded or double-stranded DNA induces complete regression of palpable B16.F10 mouse melanomas. *Cancer Gene Therapy* 20: 695–700.

133. Cemazar M, Jarm T, Sersa G. 2010. Cancer electrogene therapy with interleukin-12. *Current Gene Therapy* 10: 300–311.

134. Trinchieri G. 2003. Interleukin-12 and the regulation of innate resistance and adaptive immunity. *Nature Review Immunology* 3: 133–146.

135. Tevz G, Kranjc S, Cemazar M et al. 2009. Controlled systemic release of interleukin-12 after gene electrotransfer to muscle for cancer gene therapy alone or in combination with ionizing radiation in murine sarcomas. *Journal of Gene Medicine* 11: 1125–1137.

136. Kishida T, Asada H, Itokawa Y et al. 2003. Electrochemo-gene therapy of cancer: Intratumoral delivery of interleukin-12 gene and bleomycin synergistically induced therapeutic immunity and suppressed subcutaneous and metastatic melanomas in mice. *Molecular Therapy* 8: 738–745.

137. Torrero MN, Henk WG, Li SL. 2006. Regression of high-grade malignancy in mice by bleomycin and interleukin-12 electrochemogenetherapy. *Clinical Cancer Research* 12: 257–263.

138. Daud AI, DeConti RC, Andrews S et al. 2008. Phase I trial of interleukin-12 plasmid electroporation in patients with metastatic melanoma. *Journal of Clinical Oncology* 26: 5896–5903.

139. Pavlin D, Cemazar M, Coer A, Sersa G, Pogacnik A, Tozon N. 2011. Electrogene therapy with interleukin-12 in canine mast cell tumors. *Radiology and Oncology* 45: 31–39.

140. Tamura T, Sakata T. 2003. Application of in vivo electroporation to cancer gene therapy. *Current Gene Therapy* 3: 59–64.

141. Goto T, Nishi T, Kobayashi O et al. 2004. Combination electro-gene therapy using herpes virus thymidine kinase and interleukin-12 expression plasmids is highly efficient against murine carcinomas in vivo. *Molecular Therapy* 10: 929–937.

142. Shibata MA, Horiguchi T, Morimoto J, Otsuki Y. 2003. Massive apoptotic cell death in chemically induced rat urinary bladder carcinomas following in situ HSVtk electrogene transfer. *Journal of Gene Medicine* 5: 219–231.

143. Shibata MA, Horiguchi T, Morimoto J, Otsuki Y. 2002. Suppression of murine mammary carcinoma growth and metastasis by HSVtk/GCV gene therapy using in vivo electroporation. *Cancer Gene Therapy* 9: 16–27.

144. Cichon T, Jamrozy L, Glogowska J, Missol-Kolka E, Szala S. 2002. Electrotransfer of gene encoding endostatin into normal and neoplastic mouse tissues: Inhibition of primary tumor growth and metastatic spread. *Cancer Gene Therapy* 9: 771–777.

145. Uesato M, Gunji Y, Tomonaga T et al. 2004. Synergistic antitumor effect of antiangiogenic factor genes on colon 26 produced by low voltage electroporation. *Cancer Gene Therapy* 11: 625–632.

146. Weiss JM, Shivakumar R, Feller S et al. 2004. Rapid, in vivo, evaluation of antiangiogenic and antineoplastic gene products by nonviral transfection of tumor cells. *Cancer Gene Therapy* 11: 346–353.

147. McMahon JM, Wells DJ. 2004. Electroporation for gene transfer to skeletal muscles: Current status. *BioDrugs* 18: 155–165.

148. Mir LM, Bureau MF, Gehl J et al. 1999. High-efficiency gene transfer into skeletal muscle mediated by electric pulses. *Proceedings of the National Academy of Sciences USA* 96: 4262–4267.

149. Chiarella P, Fazio VM, Signori E. 2013. Electroporation in DNA vaccination protocols against cancer. *Current Drug Metabolism* 14: 291–299.

150. Lefesvre P, Attema J, van Bekkum D. 2002. A comparison of efficacy and toxicity between electroporation and adenoviral gene transfer. *BMC Molecular Biology* 3: 12.

151. Rubenstrunk A, Mahfoudy A, Scherman D. 2004. Delivery of electric pulses for DNA electrotransfer to mouse muscle does not induce the expression of stress related genes. *Cell Biology and Toxicology* 20: 25–31.

152. Perez N, Bigey P, Scherman D et al. 2004. Regulatable systemic production of monoclonal antibodies by in vivo muscle electroporation. *Genetic Vaccines and Therapy* 2: 2–5.

153. Bodles-Brakhop A, Draghia-Akli R. 2008. DNA vaccination and gene therapy: Optimization and delivery for cancer therapy. *Expert Review of Vaccines* 7: 1085–1101.
154. Prud'homme G, Glinka Y, Khan A, Draghia-Akli R. 2006. Electroporation-enhanced nonviral gene transfer for the prevention or treatment of immunological, endocrine and neoplastic diseases. *Current Gene Therapy* 6: 243–273.
155. Cemazar M, Sersa G, Pavlin D, Tozon N. 2011. Intramuscular IL-12 electrogene therapy for treatment of spontaneous canine tumors. In: You Y, (ed.). *Targets in Gene Therapy*. Rijeka, Croatia: InTech, Cop., pp. 299–320.

25 Dirty Electricity within the Intermediate-Frequency Range: The Possible Missing Link Explaining the Increase in Chronic Illness

Magda Havas

CONTENTS

25.1 INTRODUCTION

Most of the research on the potentially harmful effects of nonionizing radiation falls within two parts of the electromagnetic spectrum: extremely low-frequency-electromagnetic fields (ELF-EMF) and radiofrequency radiation (RFR). Very little attention has been paid to intermediate frequencies (IFs) that fall within the frequency range of 300 Hz to 10 MHz according to the World Health Organization (2005) despite the fact that IFs are now ubiquitous in our environment (Graham, 2000).

According to the WHO (2005), common sources of IF are found in the following settings:

General public: Domestic induction cookers, proximity readers, electronic article surveillance systems and other antitheft devices, computer monitors, and television sets.

Industry: Dielectric heater sealers, induction and plasma heaters, broadcast and communications transmitters.

Hospitals: MRI systems, electromagnetic nerve stimulators, electrosurgical units, and other devices for medical treatment.

Military: Power units, submarine communication transmitters, and high-frequency (HF) transmitters.

Also according to the WHO (2005), "The scientific evidence is not convincing that adverse health effects occur from exposure to IF fields normally found in the living and working environment." This chapter will attempt to refute that conclusion.

Major advances in environmental toxicology are often preceded by technology that enables us to monitor a particular toxicant in our environment and that provides a numerical value related to exposure. According to the WHO (2005), the degree and type of EMF exposure currently encountered in occupational and domestic settings need to be better characterized.

In 2003, Dr. Martin Graham, professor emeritus at UC Berkeley, designed the microsurge meter that enabled measurements of a form of electrical pollution that had been virtually ignored by the scientific community. The microsurge meter is a portable device that measures—as the name implies—small changes in power surges or transients that are associated with poor power quality or dirty electricity flowing along electrical wires (Graham, 2003).

Dr. Martin Graham, with the help of Dave Stetzer—a power quality expert in Wisconsin—also designed specially tuned capacitors (Graham/Stetzer [GS] filters) to short out the HF voltage transients (HFVTs) and thus improve power quality (Graham, 2002). These two devices, the microsurge meter and the GS filter, have enabled scientists to conduct research on a relatively small frequency range (4–150 kHz) that appears to be important biologically (Figure 25.1).

To date, studies have shown an association between poor power quality and asthma, cancer, diabetes, multiple sclerosis (MS), and symptoms of electrohypersensitivity (EHS) (Havas, 2006, 2008; Havas et al., 2004; Milham 2014; Milham and Morgan, 2008). The rest of this chapter will summarize that research and will include information on what constitutes dirty electricity as used in this chapter, definition of the GS unit, and the effectiveness of the GS filters from an exposure perspective.

25.1.1 Compact Fluorescent Bulbs Produce Dirty Electricity

One of the simplest ways to demonstrate the generation of dirty electricity and its reduction is with light bulbs. Many countries have mandated the use of energy-efficient light bulbs, and one of the most common types available is compact fluorescent light (CFL) bulbs. Most CFLs generate HFVTs within the IF range that contribute to poor power quality. Some also generate UV radiation, and various federal health authorities have issued warnings to not sit within 30 cm of these bulbs for more than three consecutive hours (Health Canada, 2014). These authorities do not mention poor power quality produced by the bulbs.

We measured the waveforms and spectra of various types of light bulbs in a remote region to eliminate background levels of dirty electricity (Havas et al., unpublished). Both waveforms and spectra were measured using a two-channel 196 Fluke Scopemeter and Flukeview Software version 3.0. The spectral distribution documents harmonics of the fundamental 60 Hz frequency up to 10 kHz. Two 10:1 shielded probes were used. Channel A (dark gray) measured the voltage through air, 0.5 m from the test lamp. Channel B (light gray) measured voltage on the electrical wire. The channel B probe was connected to the ac power line through a Graham ubiquitous filter, which removes the fundamental 60 Hz frequency (Graham, 2000).

Dirty electricity on electrical wires was measured with a microsurge II meter, which measures changing voltage as a function of time (dV/dt, expressed as GS units) for the frequency range 4–150 kHz and with an accuracy of +/–5% (Graham, 2003).

A GS filter was used to reduce the dirty electricity on the ac power line. The GS filter is a power line filter that plugs into an electrical outlet and shorts out high frequencies on the circuit in the

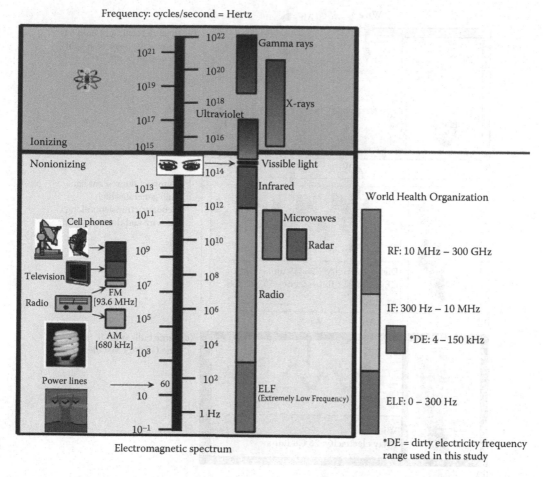

FIGURE 25.1 Electromagnetic spectrum with WHO classification for radio (RF) intermediate (IF) and extremely low (ELF) frequency. Showing range of dirty electricity used in this study.

4–150 kHz range. Power quality with and without light bulbs turned on and with and without GS filters was tested.

Figure 25.2 shows the waveform through the air (sinusoidal wave) and along the wire for (a) an incandescent light bulb, (b) a CFL bulb, and (c) the same CFL with one GS filter installed. The waveform shown in Figure 25.2a is the same whether the incandescent light bulb is on or off because incandescent light bulbs do not generate dirty electricity.

The CFL bulb in Figure 25.2b shows a disturbed waveform with HF transients flowing along the wire (light gray) and through the air (dark gray). The dirty electricity increased from a background of 65–298 GS units when this light bulb was turned on.

One GS filter was able to smooth out the waveform and reduce dirty electricity to 26 GS units. It also reduced the HF transients flowing through the air (Figure 25.2c).

Another example of how the GS filter shorts out transients flowing along the wires and through the air is provided in Figure 25.3, which shows both the waveform and the spectra for a CFL bulb with and without one filter installed. In this example, one filter reduced dirty electricity from 1449 to 24 GS units and significantly reduced the high voltage transients along the wire and through the air as shown by the spectral analysis.

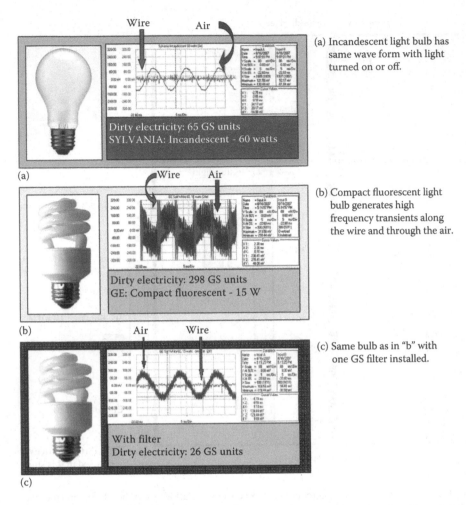

FIGURE 25.2 Waveform for incandescent (a) and compact fluorescent light bulb without (b) and with (c) one GS filter. The voltage traces were measured with a Fluke 196 Scopemeter using Flukeview software. Channel A (dark gray) was connected to a 10:1 probe 0.5 m from the test lamp. Channel B (light gray) was connected to the AC power line through the Graham Ubiquitous filter, which removes the 60 cycle. All test equipment was operated on internal batteries. The power quality reading on the microsurge meter was (a) 65 GS units, (b) 298 GS units, and (c) 26 GS units.

25.1.2 GS Unit as a Measure of Dirty Electricity

The GS unit is named after the two inventers of the GS filters, Martin Graham and David Stetzer. It is a measure of the energy associated with HFVTs and is sensitive to both the number of surges and the intensity of the surges per unit time as shown in Figure 25.4. The microsurge II meter has a maximum range of 1999 GS units, and according to Head State Sanitary Physician of the Republic Kazakhstan, values of electrical pollution should not exceed 50 GS units (HSSP, 2003).

We tested the response of the microsurge II meter under controlled laboratory conditions to changing frequency and changing voltage using a frequency generator and a voltage generator, respectively (Havas and Frederick, 2009). The results are shown in Figure 25.4. Response to both frequency and voltage is linear up to 1800 GS units and then begins to taper off as voltage increases (Figure 25.4a and b). The higher the GS unit, the greater the frequency at a particular voltage and the greater the voltage at a specific frequency.

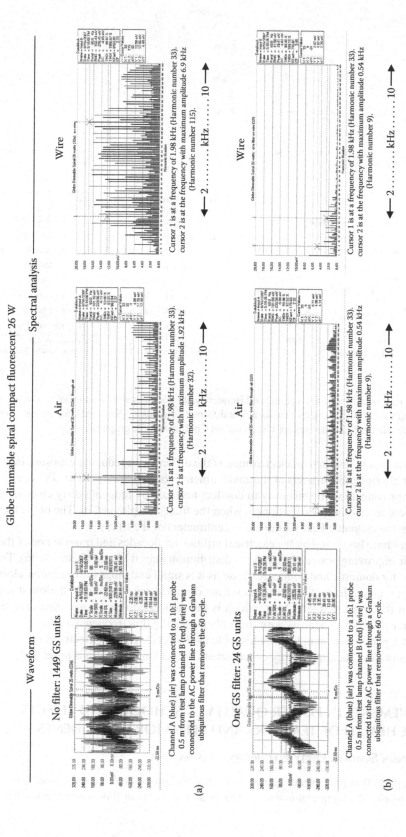

FIGURE 25.3 Waveform and spectra for compact fluorescent light bulb with and without GS filter. The voltage traces were measured with a Fluke 196 Scopemeter using Flukeview software. Channel A (dark gray) was connected to a 10:1 probe 0.5 m from the test lamp. Channel B (light gray) was connected to the AC power line through the Graham Ubiquitous filter, which removes the 60 cycle. All test equipment was operated on internal batteries. The power quality reading on the microsurge meter was (a) 1449 GS units with no filter and (b) 24 GS units with one GS filter installed.

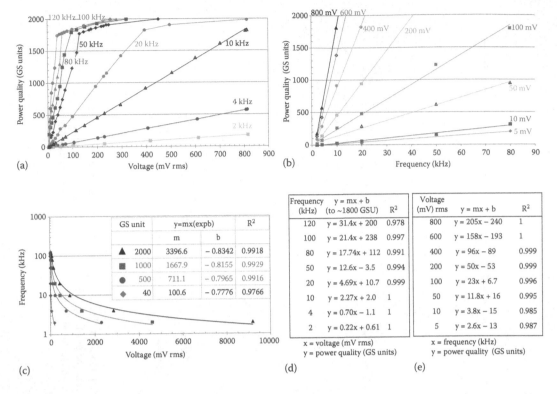

FIGURE 25.4 GS units as a function of voltage and frequency. (a) GS units as a function of different frequencies. (b) GS units as a function of frequency at different voltages. (c) Selected GS units as a function of voltage and frequencies. (d) Coefficients for Graph (a). (e) Coefficients for graph (b).

One of the challenges to research on electrosmog is the difficulty in reducing exposure in a natural setting without altering other aspects of the environment. The inventions of the GS filter and the microsurge meter provide a rare opportunity to conduct research by reducing dirty electricity in a specific environment as no other metric changes when the filters are plugged in. The only change is a small increase in the magnetic field within a few centimeters of the filters.

Power line filters have been used by the electrical utilities for decades and provide one of the first lines of defense for improving power quality along distribution lines (Ontario Hydro, 1996). The GS filter was designed for domestic and commercial use as it is small and plugs into a regular electrical outlet and does not require changes in electrical wiring or an electrician for installation. It reduces frequencies within the range of 4–150 kHz, which is the same frequency range measured by the microsurge II meter providing scientists with the tools for testing this frequency range.

The concept that poor power quality is undesirable as it affects sensitive electrical equipment is not new. The concept that this same energy can adversely affect human health is novel and requires further testing.

25.2 IMPROVEMENTS IN POWER QUALITY IN SCHOOLS AND EFFECTS ON THE HEALTH OF TEACHERS AND BEHAVIOR OF STUDENTS

25.2.1 WISCONSIN SCHOOL, DIRTY ELECTRICITY, AND ASTHMA

One of the first reports that poor power quality is related to human health comes from a school in Wisconsin that had *sick building syndrome*. Teachers and students at this midwestern school had been complaining about a myriad of health symptoms that they believed were associated with the school environment. When the teachers threatened to involve their union, the school administrators took them seriously and began to improve the school environment. Initially—as with many cases of sick building syndrome—they assumed that the health concerns came from either mold or chemicals used in the school. The school went through a thorough cleaning, and when teachers and students returned in September, so did many of their symptoms.

Eventually, a local electrician was called in to determine if the problem was electrical. He measured high levels of IF using a Fluke Scopemeter. As in most schools, banks of fluorescent lights, computers, photocopy machines all contribute to IF that flow along electrical wires contributing to poor power quality or dirty electricity. GS filters were installed throughout the school, and health complaints began to disappear. Of the 37 students with asthma who required inhalers on a daily basis, only 3 needed them and only for exercise-induced asthma after the filters were installed, according to the school nurse (Sbraggia, 2003). These students were still using their inhalers at home but no longer needed them at school suggesting that their asthmatic symptoms were influenced by the environment.

Students had fewer migraine headaches. Both students and staff appeared to have more energy and were less tired. One staff member reported her sense of smell returning, and several who took allergy medication did not have to take medication or to see their doctors because they experienced fewer problems (Sbraggia, 2003).

One of the principals of the school was diagnosed with MS. Her symptoms were so severe that she was unable to remember the names of her students in her class. After the GS filters were installed, her symptoms began to fade. Her energy, her ability to walk properly, and her memory steadily improved to the point that she no longer had symptoms of MS unless she entered an environment where levels of dirty electricity were high.

A few years after the school was filtered, her symptoms began to return at school. By this time, she had also filtered her home and was able to tell if she was in an electromagnetically clean environment. Something at the school had changed. The original contractor was called in, and he discovered that one of the teachers brought a small heater to the school that she kept by her desk. This heater generated HFVTs (dirty electricity). When it was removed, the environment returned to normal and the principal's symptoms disappeared.

25.2.2 Toronto School, Dirty Electricity, and Symptoms of Electrohypersensitivity

I first learned about the health effects of dirty electricity when a mother contacted me about her daughter. This particular girl was attending a new school where regularly she had to leave school by midday and would then sleep all afternoon. I was told the daughter was electrically hypersensitive, and when faulty wiring in their home was fixed and GS filters were installed, her health improved dramatically, as did that of the health of the rest of the family.

The purpose of the call was to invite me to do a study at the school with the GS filters. The parent had permission from the principal to install GS filters at this school to help her daughter, and I was invited to study the effects of these filters.

Even though I was skeptical that there would be dramatic changes after the filters were installed, I designed a single-blind study that consisted of questionnaires for the teachers to complete at the end of each school day related to their health, energy level, mood, and the behavior of students in their last class of the day. Participation was voluntary, and the questionnaire took only a minute or 2 to complete each day. A total of 18 teachers participated on a regular basis, and data were provided for 25 classes. The study was designed to document changes with and without filters. The study ended in March and consisted of 2 weeks without filters (prefilters), 3 weeks with filters, and 1 week without filters (postfilters) to minimize seasonal effects. Teachers were blinded to the purpose of this study and did not relate the questionnaire to the GS filters that were plugged in at the school as this

TABLE 25.1

Changes in Teacher Well-Being and Classroom Behavior at Toronto School with GS Filters

	Teacher Well-Being n = 18 Teachers	Teacher Symptoms n = 16 Symptoms	Student Behavior n = 25 Classrooms
Better (%)	55	88	72
Same (%)	33	6	8
Worse (%)	11	6	20
Net improvement[a] (%)	44	82	52

[a] Net improvement = better − worse.

was done after school hours by one of the custodians and the mother whose daughter was electrically hypersensitive. We also monitored power quality using a Fluke 79 III meter and measured mV as root-mean-squared (rms) up to 20 kHz (Havas et al., 2004). Fifty GS filters were installed that reduced IFs (up to 20 kHz) by 43%. For a school of this size, at least 150 GS filters were required. We did not have a microsurge meter and could not measure values in GS units.

Fifty-five percent of the teachers had improvements in their well-being during the period the GS filters were installed compared with 11% who were worse during that period (Table 25.1). Of the 16 symptoms in the questionnaire, 88% showed improvements as follows: fatigue, frustration, irritability, health, energy, satisfaction, well-being, headache, mood, accomplishment, pain, ability to focus, coughing, and amount of medication taken during this period. There were no changes in asthma as this was not a concern among the teachers, and there was a slight increase in flu-like symptoms when the filters were installed.

Student behavior was also monitored in 25 classes (5 elementary, 3 middle school, and 17 high school classrooms), and they showed significant improvements with elementary having the greatest improvement (100%) and high school showing the least improvement (65%) (Table 25.1). Areas of improvement include the following: fewer students were late for class; it took less time to start class; less classroom noise; less unproductive time during class; students' were better able to focus; more active student participation; and less need to repeat instructions. Time dealing with disruptions remained unchanged. Clearly, in this study, the GS filters were associated with improved teacher well-being and improved student behavior, suggesting that dirty electricity may interfere with education.

25.2.3 MINNESOTA SCHOOLS, DIRTY ELECTRICITY, AND SYMPTOMS OF ELECTROHYPERSENSITIVITY

The results in the Toronto school study were so dramatic that they needed to be replicated. We obtained permission from three schools in Minnesota (elementary, middle, and high school) to replicate the Toronto school study. The three schools were close to each other, and the elementary and middle schools were in the same building so their results are combined. In the Minnesota study, we had access to more than 500 GS filters as well as dummy filters for the sham exposure (hence this was an improved experimental design). The dummy filters were identical to the real filters except they were disconnected internally. Instead of using a Fluke meter, we used the microsurge II meter that extended the range to 150 from the previous 20 kHz. The microsurge meter monitored the same frequency range that the GS filters reduced.

Teachers completed a daily questionnaire regarding their health and the behavior of their students for an 8-week period starting in January and ending in March. The exposure consisted of 2 weeks without filters (pre-filters), 4 weeks with filters, and 2 weeks after filters were removed (post-filters) to minimize seasonal effects. This was a single blind study as teachers were unaware of when the real vs. dummy filters were installed at any one time.

With the installation of 541 GS filters, dirty electricity was reduced by 94% (Havas and Olstad, 2008). The highest levels of dirty electricity before filters were installed exceeded the limit of the microsurge meter at 1999 GS units. After filters were installed, the highest levels were 60 GS units in the elementary school and 150 GS units in both middle and high schools in two areas, one with a large number of photocopy machines and the other was the boiler room.

Of the 44 teachers, 64% were better, 7% were the same, and 30% were worse when the filters were installed leading to a net improvement of 34% (Table 25.2). The symptoms at the top of the list for improvement were as follows: headache, weakness, dry eyes/mouth, facial flushing, depression, dizziness, asthma, itchiness, shortness of breath, and anxiety. Asthma symptoms diminished among teachers in the Minnesota school just as they did among students at the Wisconsin school.

The levels of dirty electricity with and without filters differed for the various classrooms. All teachers improved who taught in classrooms that had levels of dirty electricity above 300 GS units without filters and below 50 GS units with filters providing more evidence that this frequency range is biologically active.

Behavioral traits among students in elementary and middle schools were better for 70% of the traits and for 42% of the classrooms. While the filters were installed, students were more focused, more active, more responsive and there was less time spent unproductively and with disruptions.

In high school, the classroom with the greatest behavioral improvements was the computer lab with a net improvement of 62% of the 13 behavioral traits monitored. The computer lab had 74 filters installed, the most of any classroom, and this reduced the dirty electricity from between 496–1214 GS units to 22 GS units. However, when all the classrooms were combined, there was no net improvement in student behavior with the filters; indeed, there was a slight deterioration. One factor we did not control for was cell phone use, and by the time this study was conducted, many of the high schools students (but not the middle school or elementary students) had their own cells phones and would have been exposed to microwave radiation throughout the study period. Cell phone radiation provides an additional source of electrosmog although at a much higher frequency range.

These three examples of schools that were filtered in Wisconsin, Ontario, and Minnesota demonstrate that dirty electricity can have a significant effect on teacher well-being and student behavior and that poor power quality may adversely influence the learning environment. In both the Toronto and Minnesota studies, younger students showed greater improvement than older students suggesting that age may relate to sensitivity. Clearly, more research is required to better understand the mechanisms and the range of biologically active frequencies involved. School administrators should

TABLE 25.2
Changes in Teacher Well-Being and Classroom Behavior at Three Minnesota Schools with GS Filters

| | Teacher Well-Being | Teacher Symptoms | Student Behavior | |
| | All Schools | All Schools | Elementary/Middle | High School |
	n = 44 Teachers	n = 38 Symptoms	n = 14 Classrooms	n = 17 Classrooms
Better (%)	64	79	50	35
Same (%)	7	8	42	12
Worse (%)	30	13	8	53
Net improvement[a] (%)	34	66	42	−18

consider monitoring and improving power quality as it may very well improve the learning environment for both teachers and students at minimal cost and with potentially enormous benefit.

25.3   CALIFORNIA SCHOOL, DIRTY ELECTRICITY, AND CANCER

Teachers at La Quinta Middle School (LQMS) in La Quinta, California, appeared to be experiencing an abnormally high incidence of different types of cancers. The school opened in 1990, and by 2005, 16 teachers among the 137 who had worked at the school had been diagnosed with 18 different types of cancers, a ratio nearly 3 times the expected number. Nor were the students spared with a dozen cancers detected so far among former students. A couple of them have died (Milham, 2012).

School authorities denied there was a cancer cluster in the school although they did test the air, water, and soil and also tested for radioactivity in the school with nothing significant explaining the cancers. Teachers filed a complaint, and the California Department of Health Services investigated the situation.

Dr. Sam Milham, a retired epidemiologist who worked for State Health Departments in Washington and New York for 40 years, and Lloyd Morgan, a retired electrical engineer, asked the California Department of Health Services to measure power quality in the school. They also obtained the teacher rosters for all teachers who ever taught at the school. By matching the levels of dirty electricity in all classrooms with the teachers and their health records, a disturbing picture began to emerge.

Milham and Morgan (2008) compared the cancer incidence at LQMS with the published California Cancer Registry cancer incidence rates and matched teachers by age, gender, and race. The observed-to-expected risk ratio for all cancers in the school was 2.78. The risk ratios were 13.3 for thyroid cancer, 9.8 for melanoma, and 9.2 for uterine cancer.

Levels of dirty electricity at LQMS were abnormally high compared with homes, offices, and other schools (Milham and Morgan, 2008). Thirteen rooms out of fifty-one tested had dirty electricity readings that exceeded the range of the microsurge meter ($\geq$2000 GS units). The highest cancer risks were for teachers who taught in those 13 classrooms. Risk of cancer from *never* being in a classroom above 2000 GS units was 1.8-fold. This risk increased to 5.1-fold for those who *ever* taught in a classroom above 2000 GS units, and it increased further to 7.1-fold for those teachers who ever taught in a classroom above 2000 GS units and had been employed at the school for more than 10 years (Table 25.3). A single year of employment at this school increased

---

TABLE 25.3

Cancer in Teachers Who Ever Taught in a Classroom with At Least One Overload GS Reading (>2000 GS Units) by the Duration of Employment

Ever in a Room >2000 GS Units	Employed for 10+ Years	Total Teachers	Observed (O)	Expected (E)	(O/E)	Poisson p
Yes	Yes	10	7[a]	0.988	7.1*	0.00007
Yes	No	30	3[a]	0.939	3.2	0.054
Total	—	40	10	1.93	5.1*	0.00003
No	Yes	19	2	1.28	1.6	0.23
No	No	78	6	3.25	1.8	0.063
Total	—	97	8	4.56	1.8*	0.047
Grand total	—	137	18	6.49	2.8*	<0.0001

Source: Milham, S. and Morgan, L.L., *Am. J. Indust. Med.*, 51(8), 579, 2008.

a teacher's cancer risk by 21%. Two rooms had high magnetic fields, and this was not associated with the cancer risks.

The authors concluded that, "The cancer incidence in the teachers at this school is unusually high and is strongly associated with high frequency voltage transients... If our findings are substantiated, high frequency voltage transients are a new and important exposure metric and a possible universal human carcinogen similar to ionizing radiation."

There is only one other study that found a link between IF and cancer. Armstrong et al. (1994) found an association between cumulative exposure to HF transient fields (which they refer to as short-duration pulsed EMFs) and lung cancer (also possibly stomach cancer) among electric utility workers in Quebec. The association was substantial and not explained by smoking or other occupational exposures. However, several factors limit the strength of the evidence for a causal relationship including the lack of precision in their measurements of PEMFs. The technology used to measure the PEMFs included frequencies in the low RF band of the electromagnetic spectrum and cannot be attributed to IF alone. However, this study did report an undeniable association between cancer and electromagnetic frequencies in the IF and low RF ranges of the electromagnetic spectrum.

The fact that IFs are associated with cancer is not surprising. According to the WHO (2005), IF fields act on the body in a way similar to ELF and RF fields, depending on the frequency of the IF field, and the International Agency for Research on Cancer (IARC) has classified both ELF and RF as a Class 2b carcinogen—possibly carcinogenic to humans (IARC, 2002, 2013).

The classification is based primarily on an increased risk of leukemia among children who are exposed to magnetic fields above 3 mG and who live near power lines, transformers, or substations (Wertheimer and Leeper, 1979; Savitz et al., 1988; Feychting and Albom, 1993; Schuz et al., 2001). Additional evidence comes from occupational studies that show an increased risk of brain tumors and breast cancer as well as adult leukemia for those occupationally exposed to levels ranging from 3 to 12 mG. See review by Havas (2000).

RF EMF is also classified as a possible human carcinogen (Class 2b) based primarily on studies of cell phone users and the development of brain tumors and salivary gland tumors (IARC, 2013). Supporting evidence is provided by laboratory studies with rats (Chou et al., 1992) and by epidemiological studies of people who live within 500 m of cell phone base stations (Eger et al., 2004; Wolf and Wolf, 2004; Dode et al., 2011) and within 2 km of broadcast antennas who have a greater risk of developing and dying from cancer (Hocking et al., 1996; Dolk et al., 1997; Michelozzi et al., 1998).

25.4 MULTIPLE SCLEROSIS AND DIRTY ELECTRICITY

The principal at the Wisconsin school who had been diagnosed with MS had significant improvement in her symptoms after the GS filters were installed in her school. So we did a study exclusively working with people who had been diagnosed with MS. The studies we did were not blinded in that we used the real filters and simply asked subjects to record the severity of their symptoms on a daily basis along with information about the weather, presence of stress in their lives, and changes in medication (as some people with MS are sensitive to inclement weather and to stressful situations).

One of the individuals we tested was a 27-year-old male who had been diagnosed with primary progressive MS 2 years earlier. He walked with a cane or had to hold onto the wall or furniture to maintain his balance. He had tremors, was exceptionally tired, and was beginning to have difficulty swallowing. We installed 16 GS filters in his home, and this reduced the dirty electricity from 135–410 GS units to 32–38 GS units. Three days after the filters were installed, his symptoms began to subside. He assumed his body was recovering spontaneously, but he had been diagnosed with progressive MS and not relapsing/remitting MS, so spontaneous recovery was unlikely in his case.

A week after the filters were installed in his home, he had enough energy to go shopping with his father. He did not take his cane because he had not needed it, but after a couple of hours in the store, his symptoms began to reappear, and he had difficulty walking to the car. His tremors began to subside 3 h after arriving home. He has repeated this experience on several occasions, and he now knows that if he goes into an environment with dirty electricity, his MS symptoms will reappear.

The change in his symptoms 2 weeks after the GS filters were installed is provided in Figure 25.5. All of his symptoms improved. This may have included a placebo effect although he told me he did not think that the GS filters would do anything for his symptoms, and when he began to improve, he attributed the improvement to a misdiagnosis of his MS rather than the filters. It wasn't until he spent time away from home in a dirty electrical environment that his symptoms returned and he was able to get relief only after spending several hours at home in a clean electrical environment. I visited him several years later, and his symptoms were still markedly reduced from when I first met him.

His results were both rapid and dramatic. Several other people responded in the same way, and they are videotaped showing gross body movements before and after filter installations (Havas, 2006, 2011). A few subjects did not change during a 6-week period, which was the duration of our testing.

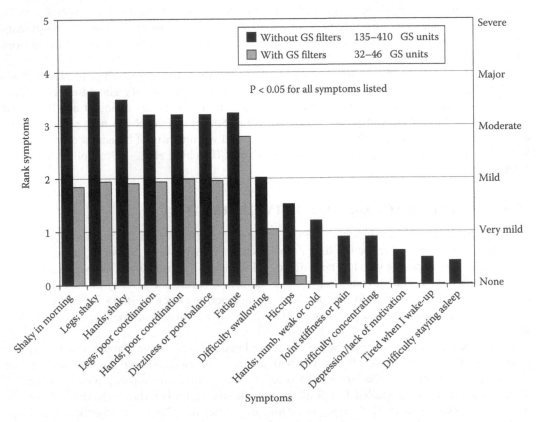

FIGURE 25.5 Symptoms of 27-year old male with primary progressive multiple sclerosis with and without GS filters in his home.

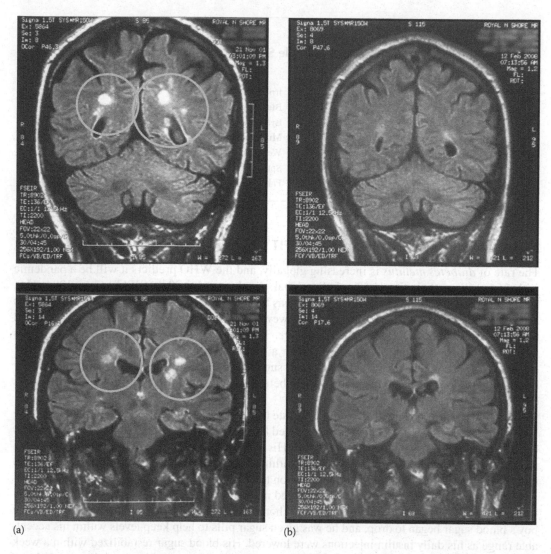

(a)   (b)

FIGURE 25.6   MRI scans for MS patient pre- and 7 years post-GS filters in his home. (a) 2001 Before GS Filters and (b) 2008 After GS Filters.

A subject who had filtered his home and was not exposed to elevated levels of dirty electricity returned to his doctor for an MRI scan of his brain. Figure 25.6 shows considerable improvement with much less sclerosis in his MRI scan after 7 years of having the GS filters installed in his home.

Interestingly, many of the symptoms for MS are similar to those of EHS. See Table 25.4. Three possible explanations come to mind. If the improvements with the GS filters are real and not a placebo effect, it is possible that some individuals may be misdiagnosed with MS rather than EHS. It is also possible that individuals with MS who have a compromised nervous system with demyelization of the nerves in their brain may be more sensitive to exogenous EMFs. A third possibility is that electrosmog may contribute to the onset of MS. Clearly, more research is necessary to determine the mechanisms involved and the role exogenous EMFs play in the onset and exacerbation of both MS and EHS.

TABLE 25.4
Symptoms Common to Both Multiple Sclerosis and Electrohypersensitivity

Central Nervous System	Visual	Sensation
Fatigue	Irritation	Pain
Cognitive impairment	Blurred vision	Numbness
Depression	Eye pain	Pricking
Anxiety	**Musculoskeletal**	Burning
Dizziness/balance	Weakness	**Urinary**
Unstable mood	Spasms	Frequency
Restless leg syndrome	Pain	Urgency

25.5 DIABETES AND DIRTY ELECTRICITY

The rate of *diabetes mellitus* is increasing globally, and the WHO predicts it will be a pandemic with 366 million diabetics (4.5% of the global population) by 2030 (Wild et al., 2004; U.S. Census Bureau, 2005). Much of this increase has been attributed to a rise in obesity due to a less active lifestyle and poor dietary behavior. However, the rising blood sugar may be exacerbated by stress associated with electrosmog exposure. There are now numerous examples of changes in power quality affecting blood sugar within a matter of a few minutes. A decrease in dirty electricity is accompanied by reduced blood sugar levels and an increase in dirty electricity by increased blood sugar levels among prediabetics as well as both type 1 and type 2 diabetics (Havas, 2008).

One example will be provided here to illustrate this. A 12-year-old boy was diagnosed with type 1 diabetes in December 2002 and was hospitalized for high blood sugar levels. Once the blood sugar began to decrease, he was allowed to go home. His parents administered the insulin and reported a steady decrease in blood sugar that stabilized within his acceptable range of 4–11 mmol/L at 70–80 units of daily insulin injections. This stabilization took 2 weeks to accomplish (Figure 25.7).

On January 14, 2003, GS filters were installed in the home to combat symptoms of EHS felt by the mother and her five children, who were all homeschooled. Once the filters were installed, the boy's blood sugar began to drop, and he was given sugar pills to help keep levels within his acceptable range as his daily insulin injections were lowered. His blood sugar restabilized within a week at 35–40 units of daily insulin injections. The only difference during this month-long period was the installation of GS filters. What was even more convincing that blood sugar is affected by dirty electricity is that this boy had a younger sister who had been diabetic since birth and her insulin requirements also decreased during the same period. This requirement for less insulin cannot be explained by the *honey moon* effect because the younger sister had been diabetic for years and she responded in the same way.

What is unique about this situation is that the environment has an influence on blood sugar. Individuals with brittle diabetes, those who are unable to control their blood sugar, may be responding to environmental stressors that alter blood sugar and go unmonitored and thus unnoticed. Due to the large population with diabetes, this needs to be studied further as this study and all the others mentioned in this chapter are *proof of concept* studies and require a larger sample size for testing.

Milham (2014) took this concept one step further and hypothesized that dirty electricity is causing worldwide epidemics of both obesity and diabetes. He tested his hypothesis with island communities and reasoned that diesel generators are used almost universally to provide electricity to small islands and places unreachable by the conventional electric grid. These generators are a major source of dirty electricity. A search of worldwide data banks on prevalence of diabetes,

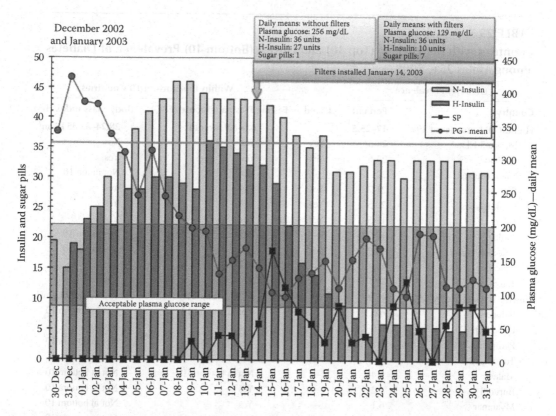

FIGURE 25.7 Mean daily plasma glucose (PG) levels, total daily insulin injections, and daily sugar pills (SP) taken by 12-year old male with type 1 diabetes who was admitted to hospital December 30, 2002 and returned home on January 1, 2003. On January 14, 2003, GS filters were installed in his home to improve power quality. The total amount of insulin and sugar pills to maintain acceptable plasma glucose levels differed significantly before and after GS filters were installed.

fasting plasma glucose (FPG), and body mass index (BMI) identified all the countries in the highest and lowest 10 disease rankings for males age 25+ in 2008. A web search identified the extent of their electrical grids and the sources of their electricity. The highlights are provided in Table 25.5. Island countries represent seven of the top ten countries for prevalence of diabetes and only one of the bottom ten countries.

25.6 CONCLUSIONS AND RECOMMENDATIONS

If poor power quality within the frequency range of 4–150 kHz is as biologically active, as these studies seem to demonstrate, then we have a serious ubiquitous pollutant that is going unnoticed by the scientific community and the regulatory agencies. All of the illnesses with which dirty electricity has been associated are on the rise, including cancer, diabetes, MS, asthma, and EHS. The degree to which dirty electricity affects other illness or disabilities that are also increasing such as attention deficit disorder, attention deficit hyperactivity disorder, Alzheimer's, heart disease, hormonal disruptions, and stress-related illness needs to be tested. Dirty electricity may be the missing link between the extremely low-frequency EMFs and radio and microwave radiation that have been studied for decades. It deserves greater attention from both the scientific community and regulatory bodies especially since meters to measure and devices to reduce poor power quality are readily available.

TABLE 25.5

Countries with the Highest (Top 10) and Lowest (Bottom 10) Prevalence of Diabetes among Males 25+, 2008

	Diabetes Prevalence			Within Top/Bottom 10 Countries	
Country	Percent	Island	Fasting Plasma Glucose (FPG)	Body Mass Index (BMI)
Highest 10 countries of 199	17–25.5		6.0–6.9 mmol/L	28.24–33.85 kg/m²
Marshall Island	25.5	Yes	Yes	Yes
Kiribati	23.6	Yes	Yes	Yes
Saudi Arabia	22	—	Yes	Not in top 10
Samoa	21.2	Yes	Yes	Yes
Cook Islands	20.5	Yes	Yes	Yes
Palau	17.5	Yes	Yes	Yes
Jordan	17.2	—	Yes	Not in top 10
Solomon Islands	17.2	Yes	Yes	Not in top 10
Kuwait	17	—	Yes	Yes
Tonga	17	Yes	Yes	Yes
Lowest 10 countries of 199	<6.6		4.7–5.1 mmol/L	19.88–20.99 kg/m²
Indonesia	6.6	—	Not in bottom 10	Not in bottom 10
Dem. Rep. Congo	6.6	—	Yes	Yes
Philippines	6.5	—	Yes	Not in bottom 10
Timor Leste	6.4	Yes	Yes	Yes
Malawi	6.4	—	Not in bottom 10	Not in bottom 10
Burundi	6.2	—	Yes	Not in bottom 10
Myanmar	6.1	—	Yes	Not in bottom 10
Peru	5.8	—	Yes	Not in bottom 10
Cambodia	4.7	—	Yes	Not in bottom 10

Source: Based on data from Milham, S., *Electromagn. Biol. Med.*, 33(1), 75, 2014.

The health care community needs to be made aware of the sources and consequences of exposure to dirty electricity to help their patients heal. Since reducing exposure (as shown in the school, diabetes, and MS studies) significantly improves health, health insurance agencies should invest in this form of preventative health care to minimize medical expenses and hospitalization costs.

Practicing good electromagnetic hygiene is as important as washing your hands before you eat. Teaching electromagnetic hygiene, especially with the growth of EMF- and RF-generating devices, needs to be a common practice in schools for all grades.

REFERENCES

Armstrong B, G Theriault, P Guenel, J Deadman, M Goldberg, P Heroux. Association between exposure to pulsed electromagnetic fields and cancer in electric utility workers in Quebec, Canada, and France. *Am J Epidemiol* 140(9) (1994): 805–820.

Chou C-K, AW Guy, LL Kunz, RB Johnson, JJ Crowley, JH Krupp. Long-term, low-level microwave irradiation of rats. *Bioelectromagnetics* 13(6) (1992): 469–496.

Dode AC, MMD Leão, F de AF Tejo et al. Mortality by neoplasia and cellular telephone base stations in the Belo Horizonte municipality, Minas Gerais state, Brazil. *Sci Total Environ* 409 (2011): 3649–3665.

Dolk H, G Shaddick, P Walls et al. Cancer incidence near radio and television transmitters in Great Britain. *Am J Epidemiol* 145(1) (1997): 1–9.

Eger H, KU Hagen, B Lucas, P Vogel, H Voit. The influence of being physically near to a cell phone transmission mast on the incidence of cancer (Einfluss der räumlichen Nähe von Mobilfunksendeanlagen auf die Krebsinzidenz). *Umwelt Medizin Gesellschaft* 17 (April 2004: 326–333.).

Feychting M, A Albom. Magnetic fields and cancer in children residing near Swedish high-voltage power lines. *Am J Epidemiol* 136(7) (1993): 467–481.

Graham MH. A ubiquitous pollutant. Memorandum No. UCB/ERL M00/55, Electronics Research Laboratory, College of Engineering, University of California, Berkeley, CA, 3 pp., October 28, 2000.

Graham MH. Mitigation of electrical pollution in the home. Memorandum No. UCB/ERL M02/8, Electronics Research Laboratory, College of Engineering, University of California, Berkeley, CA, 2002.

Graham MH. A microsurge meter for electrical pollution research. Memorandum No. UCB/ERL M03/3, Electronics Research Laboratory, College of Engineering, University of California, Berkeley, CA, February 19, 2003.

Havas M. Biological effects of non-ionizing electromagnetic energy: A critical review of the reports by the US National Research Council and the US National Institute of Environmental Health Sciences as they relate to the broad realm of EMF bioeffects. *Environ Rev* 8 (2000): 173–253.

Havas M. Electromagnetic hypersensitivity: Biological effects of dirty electricity with emphasis on diabetes and multiple sclerosis. *Electromagn Biol Med* 25 (2006): 259–268.

Havas M. Dirty electricity elevates blood sugar among electrically sensitive diabetics and may explain brittle diabetes. *Electromagn Biol Med* 27(2) (2008): 135–146.

Havas, M. Diabetes and electrosensitivity, 2010. http://www.magdahavas.com/diabetes-and-electrosensitivity/. Video: https://www.youtube.com/watch?v=gJcM6RZwyfA.

Havas, M. Multiple sclerosis and electrohypersensitivity, 2011: http://www.magdahavas.com/multiple-sclerosis-and-electrohypersensitivity/ Video: Multiple sclerosis and dirty electricity, http://www.youtube.com/watch?v=xdtIPb3Veuw.

Havas M, R Frederick. Dirty electricity explained. What are GS units? 2009. http://www.magdahavas.com/gs-units-explained/ Video clip at http://www.youtube.com/watch?v=vbebpRvwd8k.

Havas M, M Illiatovitch, C Proctor. Teacher and student response to the removal of dirty electricity by the Graham/Stetzer filter at Willow Wood School in Toronto, Canada. *Third International Workshop on Biological Effects of EMFs*, Kos, Greece, October 4–8, 2004, pp. 311–317.

Havas M, A Olstad. Power quality affects teacher wellbeing and student behavior in three Minnesota Schools. *Sci Total Environ* 402(2–3) (2008): 157–162.

Havas M, D Stetzer, E Kelley, R Frederick, S Symington. Compact fluorescent light bulbs, electromagnetic emissions, and health, unpublished.

Health Canada. The safety of compact fluorescent lamps. Modified on February 24, 2014, 4pp. http://healthy-canadians.gc.ca/consumer-consommation/home-maison/cfl-afc-eng.php.

Hocking B, IR Gordon, HL Grain, GE Hatfield. Cancer incidence and mortality and proximity to TV towers. *Med J Aust* 165 (1996): 601–605.

HSSP. Permissible levels of high-frequency electromagnetic pollutions' voltage in a wire of industrial frequency alternating current. Sanitary-epidemiologic norms. Confirmed by the Order of the Head State Sanitary Physician of the Republic Kazakhstan, November 28, 2003.

IARC. *IARC Monographs on the Evaluation of Carcinogenic Risks to Humans, Non-Ionizing Radiation, Part 1: Static and Extremely Low-Frequency (ELF) Electric and Magnetic Fields*, Vol. 80. World Health Organization, IARC Press, Lyon, France, 2002, 429 pp. http://monographs.iarc.fr/ENG/Monographs/vol80/mono80.pdf.

IARC. *IARC Monographs on the Evaluation of Carcinogenic Risks to Humans, Non-Ionizing Radiation, Part 2: Radiofrequency Electromagnetic Fields*, Vol. 102. World Health Organization, IARC Press, Lyon, France, 2013, 460 pp. http://monographs.iarc.fr/ENG/Monographs/vol102/mono102.pdf.

Michelozzi P, C Ancona, D Fusco, F Forastiere, CA Perucci. Risk of leukemia and residence near a radio transmitter in Italy. *Epidemiology* 9(Suppl.) (1998): 354.

Milham S. *Dirty Electricity: Electrification and the Diseases of Civilization*. iUniverse.com, Bloomington, IN, 2012, 128 pp.

Milham S. Evidence that dirty electricity is causing the worldwide epidemics of obesity and diabetes. *Electromagn Biol Med* 33(1) (2014): 75–78.

Milham S, LL Morgan. A new electromagnetic exposure metric: High frequency voltage transients associated with increased cancer incidence in teachers in a California School. *Am J Indust Med* 51(8) (2008): 579–586.

Ontario Hydro. *Power Quality Reference Guide*, 6th edn. Ontario Hydro, Toronto, Ontario, Canada, 1996, 162 pp. http://wesco.cafe24.com/download/Technique/POWERQ.pdf.

Savitz DA, H Wachtel, FA Barnes, EM John, JG Tvrdik. Case-control study of childhood cancer and exposure to 60 Hz magnetic fields. *Am J Epidemiol* 128 (1988): 21–38.

Sbraggia C. *Changes Noted Since Filters Installed*. Notes from School Nurse, Melrose-Mindoro School, Melrose, WI, 2003. http://www.electricalpollution.com/images/MelMinNurse.jpeg.

Schuz J, J-P Grigat, K Brinkmann, J Michaelis. Residential magnetic fields as a risk factor for childhood acute leukaemia: Results from a German population-based case-control study. *Int J Cancer* 91 (2001): 728–375.

U.S. Census Bureau. Total midyear population for the world: 1950–2050, April 26, 2005. www.census.gov/ipc/www/worldpop.html.

Wertheimer N, E Leeper. Electrical wiring configurations and cancer. *Am J Epidemiol* 109(3) (1979): 273–284.

Wild S, G Roglic, A Green et al. Global prevalence of diabetes. *Diabetes Care* 27 (2004): 1047–1053.

Wolf R, D Wolf. Increased incidence of cancer near a cellphone transmitter station. *Int J Cancer Prev* 1 (2004): 123–128.

World Health Organization. Electromagnetic fields & public health: Intermediate Frequencies (IF). Information Sheet, February 2005. http://www.who.int/peh-emf/publications/facts/intmedfrequencies/en/#.

26 Electromagnetic Fields in the Treatment of Tendon Injury in Human and Veterinarian Medicine

Richard Parker and Marko S. Markov

CONTENTS

26.1 INTRODUCTION

Equine and human tendon injuries or lesions differ. In the equine model, the lesions appear as cavities within the tendon itself, whereas in the human patient, lesions are predominantly found to be edge tears. Equine tendon lesions are found mostly among young horses and gradually diminish in frequency with age. Although a number of options exist today for the treatment of tendon lesions, useful imaging of the tendon cellular structure is limited, fostering the condition whereby the trainer does not have a good opinion of the tendon tissue injury state. Using advanced biophysical and engineering principles, the approach described herein permits the amplification of the unique natural signals from healthy and injured tissues. The derived difference therapy signal is then reintroduced into the specimen to enhance recovery. The analytical design and use of SQUID therapy signals (STS) may dramatically affect the performance of therapeutic devices for both human and animals.

The study referenced herein, although equine, is a pilot for a future human study to be implemented using the same principles. None of the previous therapies used in standard or traditional treatments have demonstrated a useful ability to process the injured tissue into healthy tissue, and few have demonstrated an ability to close tendon lesions in such an efficacious manner.

26.2 REVIEW OF LITERATURE FOR TENDON INJURIES

Thoroughbred (race horse) injuries cover a wide landscape, chiefly caused by activities that are unusual to the animal, such as a sustained run along a track that curves in one direction. The stresses set up by that type of unbalanced weight shift under a focused pounding weight is widely regarded as responsible for many of these injuries, along with the practice of introducing horses so young that they have not yet fully matured.

Performance horse (dressage, polo, hunters, jumpers) injuries, however, are associated with a massive coordinated feat, straining the hind tendons during launch and the forelimb tendons and associated ligaments at landing. These stresses set up tendon and ligaments forces (Figure 26.1), accounting for 46% of performance horse injuries (Tan, 2013).

For thoroughbreds, data show that the superficial digital flexor tendon (SDFT) has the greatest number of injury incidences due to these horses operating close to their physical limits and on the threshold of injury. Injuries to the tendon have fostered a large variety of treatment options. However, these treatment options all address the initial lesion closure but offer little guidance for further maturation of the tissues into a healthy state, including an avoidance of scar tissue in favor of well-formed tendon fibers (Dowling and Dart, 2005; Thorpe et al., 2010; Tan, 2013).

Performance horse specialties, on the other hand, see a larger incidence of hock (rear leg reverse knee) and back injuries. Suspensory ligament (SL) injuries are common among both categories of thoroughbred and performance horses.

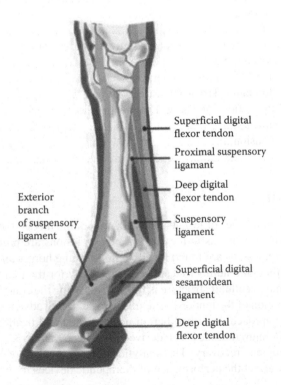

FIGURE 26.1 Suspensory and ligament structure. (HorseJournals.com. http://www.horsejournals.com/ Ligament Injuries: Advances in Diagnosis and Treatment, accessed November 6, 2013.)

The treatment options for both horse classifications focus on cellular or physical techniques. Cellular treatment options (stem cell, platelet-rich plasma [PRP]) have the greatest history, whereas physical treatments such as shock wave (which commenced 1996) and the system known as STS (commenced 2013) have less research trial background.

Table 26.1 lists the various types of injuries and the diagnosis and prognosis for those injuries. It should be noted that a wide variation is observed for proponents of the same treatment option. Industry-sponsored studies *uniformly tend to be positive*, and university-sponsored studies *uniformly tend to be negative*. This gives rise to the question of bias on the part of industry-sponsored studies, making the job of the investigator much more difficult.

It can be seen that among the injuries listed earlier, many have the following general characteristics:

- Limited theoretical understanding of the treatment method
- Poor or slow response to treatment
- High incidence of reinjury
- High cost of treatment

It is apparent that the practicing veterinarian is limited by the state of currently available therapy tools, coupled with a wide disparity in treatment modalities and responses. The many competing therapy types are a serious complication to veterinarians, with wide disagreement on their use and effectiveness. In addition, cellular techniques such as stem cell require weeks for laboratory time and considerable layup (recovery) for a process that delivers mixed results. Cellular modality PRP is frequently employed with shock wave and then repeated several months later to improve outcomes. It is felt by some that the application of both shock wave and PRP will result in a synergistic effect, although there is no solid evidence to support this theory to date. Further research on comparative outcomes is clearly needed to assist the veterinarian and trainer in making an accurate diagnosis and may yet determine the final efficacy judgment of these modalities as developed.

Many injury sites such as the back and the hock require specialized attention. Yet they receive little attention at the present time. The lack of a unified targeted application is reflective of the general weak nature of the therapeutic field and illustrates the need for a more generalized but focused approach.

STS therapy, a relatively new application, shows promise of positively influencing the poor landscape of therapeutic applications. Derived from natural signals, and employing no chemicals or injectable products, this technology is rapidly entering the thoroughbred and performance horse field.

26.3 STS: SQUID DERIVATION OF THE THERAPEUTIC SIGNAL

Activities of cells and tissues generate electrical currents and fields that can be detected on the skin surface, which induces a corresponding magnetic field. However, because these signals are so weak, biologists previously assumed that they could have no physiological significance unless they included cellular depolarization.

Consider the following: if one examines 1 g of tissue, it is seen that it contains 10^9 cells. When arranged in a sphere, 1 g would comprise 10^6 cells on the surface. This density can naturally produce an iron concentration of $10–30$ $\mu g/10^6$ cells (Anastasiadis et al., 2006), which is detectable by the induction of a local magnetic signal arising from the local concentration by ultrasensitive SQUID devices that are capable of detecting 2×10^{-15} T/cm (Anninos et al., 1987; Rose et al., 1987). Operational bandwidths being considered (0.1–40 Hz) allow for sensitivities as low as approximately 1.5×10^{-13} T. Noise levels are then determined by system design and ambient fields in the vicinity of the probe. From the foregoing, pathologic electromagnetic signals should be detectable by means of the SQUID and should not require cellular depolarization for assessment.

TABLE 26.1
Injury Types

Type	Prevalence	Recurrence	Diagnosis	How to Diagnose	Prognosis
Tendon (SDFT) and ligament	46% of sport horse injuries; SDF injuries: 8%–30% thoroughbreds, 90% of National Hunt horses (Williams et al., 2001)	43%–93% after return to full work (Robinson and Sprayberry, 2009)	7–10 days following the injury (Tan, 2013)	Specific region lameness, heat, swelling, pain (Blunden et al., 2006). Monitor throughout rehabilitation.	Most require 9–12 months for optimal healing (UC Davis, 2013). Upon lesion closure, collagen type 3 rather than type 1, being less elastic (Blunden et al., 2006; Duenwald-Kuehl et al., 2012; Kasashima et al., 2004)
Navicular bone	3% of horses with foot pain	Rarely return to work. 17% with DDFT return to full function (Adams and Stashak, 2011)	Lameness on one or both feet	Diagnosis is based on a combination of history, symptoms, nerve blocks, and radiography (Casey, 2013a,b,c).	Poor
DDFT (Deep Digital Flexor Tendon)	Tendonitis 33%, abnormalities 59% (Blunden et al., 2006)	Only 46% recovery (Robinson and Sprayberry, 2009)	7–10 days following the injury (Tan, 2013)	Shape, focal core lesions, mineralization (ultrasound), and marginal tears (Blunden et al., 2006).	Depends on the structure involved, severity, and extent of the initial injury. DDFT poorer prognosis than SFDT (Blunden et al., 2006; Duenwald-Kuehl et al., 2012; Kasashima et al., 2004)
Suspensory ligament proximal (upper) suspensory desmitis (PSD)	The center for Equine Health reported that suspensory desmitis was the most common injury next to colic (1999)	Not published, but recurrence frequently spells the end of a career	7–10 days following the injury (Adams and Stashak, 2011)	Shape, focal core lesions, mineralization (ultrasound), and marginal tears.	Prognosis good with 80% return to work 3–6 months rest; hindlimb 14%–69% return to full athletic function without lameness. Dressage and jumpers PSD 37% and 46% recurrence, racehorses 27%, show hunters 19%, field hunters 18% (Adams and Stashak, 2011; Halper et al., 2006; Herthel et al., 2001)
Bucked shins	70% of young thoroughbred racehorses (Burba, 2009)	Cold water icing and rest 4–6 weeks or condition returns (Tan, 2013)	Heat, pain upon pressure over the area, and swelling is detected (Tan, 2013)	Physical exam using palpation to reveal heat, pain, and tenderness, with or without swelling (Tan, 2013).	Good, with convalescent time of 6–8 months (Daniel, 2009; Tan, 2013).

26.3.1 DEVELOPMENT OF DETECTION METHODS

It is well accepted now that all tissues and organs produce specific magnetic pulsations, also known as biomagnetic fields. Biomagnetic field recordings such as magnetocardiograms and magneto-encephalograms are now supplementing traditional electrical recordings, such as the electrocardiogram and electroencephalogram (it should be noted that magnetograms are several orders of magnitude more sensitive than electrical recordings).

Although most biological magnetic field effects are the result of cellular depolarization, static (or VLF) patterns have been observed as well with maximal amplitude in the range 100–250 nG/cm. For reference, a 100 nG/cm field can be produced by a dipole of strength 1.0 µA cm or an infinite wire carrying 0.5 pA, both at 1.0 cm from the coils—the depth of scalp subcutaneous layer. Magnetic fields produced by naturally occurring steady currents in the body were measured by a specialized magnetic gradiometer in a magnetically shielded room, and a reproducible magnetic field of 0.1 µG/cm over the head and limbs was observed, whereas the torso field was weaker (Cohen et al., 1980). Therefore, we easily see that therapeutic actions are associated with natural biological fields. This proposition was addressed in the 2008 U.S. Patent 7,361,136 B2: *Method and apparatus for generating a therapeutic magnetic field.*

From Figure 26.2, the difference waveform (the mathematical difference between the injury waveform and healthy waveform) contains information about therapeutic recovery representing the biological activity of the body in effecting the injury repair. This signal is applied using an external signal generator that contains the recorded waveform patterns, connected to a *boot* applicator that is a modified Helmholtz coil wrapping around the foreleg (see Figure 26.3).

In the early stages of investigation, this technique shows promise and potential of being a successful therapeutic application as can be seen from a 2013 equine study using this technology in which 24 horses were treated for severe tendon lesions with an excellent positive response (Parker et al., 2014).

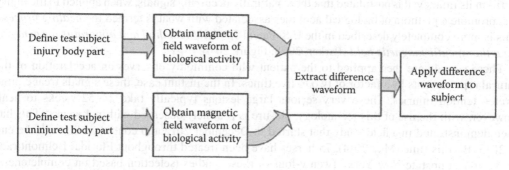

FIGURE 26.2 Derivation of therapeutic waveform.

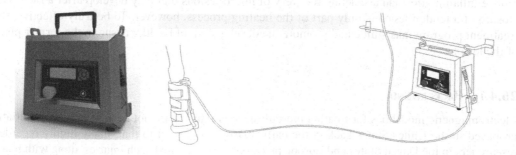

FIGURE 26.3 Equipment setup—*Boot* application and controller connection.

26.4 APPLICATION OF THE MAGNETIC THERAPY SIGNAL

Magnetic and electromagnetic fields (EMFs) are now recognized by twenty-first-century human and equine medicine as real physical entities that can promote healing of various health problems, even when conventional methods had failed, and provide a noninvasive, safe, and easy method to directly treat the site of injury, the source of pain and inflammation, and other types of diseases and pathologies.

Contemporary magnetotherapy began in Japan immediately after World War II by introducing both magnetic and electromagnetic fields in clinical practice. This modality quickly moved to Europe, first in Romania and the former Soviet Union. During the period of 1960–1985, nearly all European countries designed and manufactured their own magnetotherapeutic systems that utilized various waveshapes. Indeed, the first book on magnetotherapy, written by Todorov, summarized the experience of specific pulsating magnetic fields for the treatment of 2700 patients having 33 different pathologies (Todorov, 1982).

It is now commonly accepted that selected weak EMFs are capable of initiating various healing processes from delayed fractures, to pain relief, to multiple sclerosis, and to Parkinson's disease (Rosch and Markov, 2004).

During the recent decades, the use of EMFs for therapy slowly increased in Western medicine. However, these therapeutic modalities have been typically described with little attention to engineering details. In most cases, the devices for magnetotherapy have been designed based upon the intuition of the designing engineer with little or no consideration of the nature of detection and response to these signals from biological tissue that are subject to therapy. For that reason, empirically developed methods did not utilize the full potential of the therapeutic signals.

It is observed that electromagnetic patterns exist that approach the natural signal generated by the body in the course of healing an injury. These signals are a composite determined by experiments and that registered at SQUID scanning installations found in major universities—in our case, SQUID installations at Alexandropoulos, Greece, and Madison, Wisconsin. Although this harvest is still in its infancy, it is postulated that these naturally occurring signals, when applied to the injury site, promote a plethora of biological activities associated with what is termed the *healing process*. This is more completely described in the U.S. Patent 7,361,136: *Method and apparatus for generating a therapeutic magnetic field* (Parker, 2008; Figure 26.4).

These signals are then applied to the patient who commonly observes an acceleration of the natural repair events by a factor of two to five times. In the instant case, these signals treat equine foreleg tendon injuries. These very serious large lesions typically take 24–52 weeks to heal. However, with the use of this technology, closure is routinely observed within 4 weeks. This has been demonstrated in a field study that started in December 2012 and continued through the end of 2013. By this time (May 2014), 75 horses have been treated throughout Florida, Belmont race track, and in upstate New York. Twenty-four of these studies (selection based on completeness) were processed into a compilation. The data show that the effects on horse tendons have been uniformly predictable over 13 months. This system may also be used in a preventative fashion, to reduce inflammation and accelerate recovery of micro-lesions that may develop after a race. The closure of a tendon lesion is only part of the healing process, however. To be truly effective, the treatment program must guide and promote the development of healthy tissue in the former place of the lesion.

26.4.1 Background

Electromagnetic modalities for treating musculoskeletal injuries are not new, having been initially proposed in the United States back in the early 1970s. Subsequent to that, many highly regarded researchers in the United States and Europe proposed means to treat such injuries along with many operational theories.

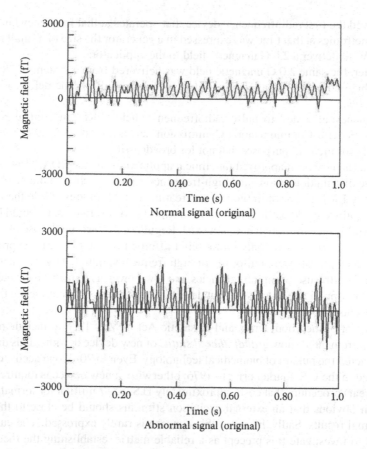

FIGURE 26.4 (a) Normal/(b) abnormal SQUID signals.

The fundamental questions related to the biophysical conditions under which EMF signals could be used by cells in order to modulate cell and tissue functioning remain to be elucidated. The scientific and medical communities still lack the understanding that different magnetic fields applied to different tissues cause different effects.

The medical part of the equation should identify the exact target and the *dose* of EMF that the target needs to receive. Then, physicists and engineers should design the exposure system in such way that the target tissue receives the required magnetic flux density. Particular attention must be paid to the biophysical dosimetry, which should predict which EMF signals could be bioeffective and monitor the corresponding efficiency. This raises the question of using theoretical models and biophysical dosimetry in selection of the appropriate signals and in engineering and clinical application of new EMF therapeutic devices.

In Europe, the concentration initially was on low-frequency applications, that is, <50 Hz. European low-frequency reports of beneficial performance far outweighed the U.S. high-frequency experience. The basic reasons for the prevalence of low-frequency signals are the fact that common electrical supplies are at this frequency range (60 Hz in the United States and 50 Hz in the rest of world). Variations of signals have been implemented in this technology—from sine waves through rectified sinusoidal signals to various pulsed signals. It should be noted that all of these signals were implemented according to engineering limitations, but do not take into account biological and clinical reasons.

The high-frequency modalities used for therapeutic purposes are based upon the first application of a continuous 27.12 MHz signal for deep heating in an attempt to kill cancerous tissues. In 1983, the FDA approved the first radiofrequency device that operates at that frequency but in pulse mode. The status of electronics at that time was expressed in a generator the size of a small refrigerator that required 1000 W to deliver a 2.0 G magnetic field to the applicator.

A decade later, the same 2.0 G magnetic field was delivered from a system of 380 W having the size of a portable computer (not a laptop). Today, technology allows the delivery of 50 G from a laptop-sized device.

As the frequency allocated to pulse radiofrequency field (PRF) in clinical practice utilized 27.12 MHz, the Federal Communication Commission declared that 13.56, 27.12, and 40.68 MHz can be used only for medical purposes, but not for broadcasting.

Two frequencies have been approved for clinical applications by the FDA. This naturally led to a difference in the distribution of low- and high-frequency signals in clinical practice. Thus, the label for low-frequency EMF has been limited to the treatment of nonunions, while the PRF label is for large areas—of pain and edema (inflammation) in superficial soft tissues. It should be noted that in the rest of the world, such limitation does not exist. Regulatory efforts on the use of these signals led to an effective *design freeze* with only minor relief afforded by the FDA 510(k) process. For these reasons, prior developments were uniformly of high frequency, initiated by the then-frozen state of U.S. EMF devices—the use of high frequencies and low power. This factor has been neither previously published nor publically expressed, but device manufacturers cannot escape the logic and the facts of these events.

The amendment of the Food, Drug, and Cosmetic Act in May 1976 to include medical devices fostered the requirement to show *grandfathered status* of new device designs. This dramatically and negatively impacted the pursuit of biomedical technology. Even in 2014, one needs to prove that this modality existed in the U.S. market prior to 1976; otherwise, a new device is required to go through the long (3–5 year) procedure that costs approximately U.S. $5–7 million for formal FDA approval.

It may seem obvious that an external excitation stimulus should be close to the nascent body signal for optimal results. Sadly, however, this concern is rarely expressed. The current subject of research seeks to investigate this precept as a reliable metric, establishing the theory that natural biological frequencies and natural biological signals will be the most therapeutic when applied to musculoskeletal injuries.

26.4.2 EXAMPLE STUDY

This study included treating race horses at Gulfstream (FL), Calder (FL), and Belmont (NY) tracks and included stables in Ocala (FL) and Wellington (FL), New York, and California in a classic field study. That is, the study participants were working trainers who used the equipment in their normal course of business. In this study, ultrasound scans were taken by licensed veterinarians before the treatment and after the tendon appeared from visual examination to close. The images in Figure 26.5 are a transverse view of an SDFT lesion.

You will note that the transverse (left image) shows a marked *hole* in the SDFT tendon structure. The right scans show the beneficial results of the therapy. The equipment is shown as follows.

The study used a research model of the controller capable of delivering a uniform high-fidelity EMF signal up to 40 G from 4 to 50 Hz.

26.5 WHY EQUINE INSTEAD OF A HUMAN STUDY

In evaluating a new device, the number of subjects should be of such a power that statistical significance is assured. In our study of equine subjects, statistical power has been achieved. STS technology, which employs the electromagnetic nature of tendon tissue, finds ready similarities between human and equine tendon tissues when their electromagnetic properties are compared. However, the stresses placed on the equine tendon are much greater than on the human tendon, as equine stresses operate close to the functional limit.

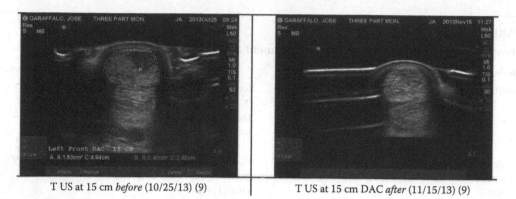

T US at 15 cm *before* (10/25/13) (9) | T US at 15 cm DAC *after* (11/15/13) (9)

FIGURE 26.5 Transverse scan before and after treatment.

Current guidelines for equine tendon therapy permit a greater flexibility in therapy techniques and practices. The human Institutional Review Board protocol and long administrative demands limit experimental application; hence, the equine model develops information that assists the design and format of the human study.

For the equine case, the investigation found the following:

1. The assessment of the damage should optimally take place within 4–7 days after the injury (this metric is the result of consistent observation).
2. The equine tendon, after application of therapy, should show a reduction of swelling, pain, and heat within 3 days (this observation made from 75 controlled cases).
3. The equine tendon, after application of therapy, should show an initial reduction of the tendon lesion from 75% closed to 100% closed within 21 days.

26.6 EVALUATING EQUINE TREATMENT OPTIONS

This is a complicated issue, with many different proponents and little comprehensive comparisons. Table 26.2 helps to identify these options:

In his *British Journal of Sports Medicine* article (Smith and Webbon, 2005) on harnessing stem cells for the treatment of tendon and ligament injuries, Roger Smith states, "a multitude of treatments have been advocated for the management of tendon over-strain injuries, but there is little evidence in any species that any is more effective than a prolonged period of rehabilitation with carefully controlled exercise."

It is seen that most therapies are invasive (10 out of 12), but yet yield little improved performance. As seen in Figure 26.2, the new STS therapy available only since 2013 shows promise of a noninvasive technique that is reliable and efficient in resolving ligament and tendon injuries. Although STS is still in an early form, future studies should yield additional tracking information regarding the reliability and effectiveness of this new technique.

According to numerous references, no new technologies on the horizon exist that will substantially alter the low-current state of equine injury repair. For example, consider the quote from a frequently cited reference, Linda A. Dahlgren, DVM, Ph.D.:

It bears mention that none of the therapies described and none of those that may become available in the foreseeable future are likely to speed the healing process.

Dahlgren (2005)

TABLE 26.2

Treatment Options for Equine Tendon Ligament and Joint Disease

Method	Methodology	Features	Invasive?
Pinfiring	Stimulates tendon repair by counter irritation. Controversial procedure with mixed results. Used for *bucked shins* (Burba and Daniel, 2009).	Not an effective tendon injury treatment (Tan, 2013). Scientists have shown no difference in collagen arrangement and scarring (Hayward and Adams, 2001).	Y
Intralesional corticosteroids (BAPN)	Substitution for surgery.	Discontinued due to steroid-induced demineralization and tissue necrosis in clinical trials.	Y
Stem cell	Stem cells promote the reparative response of injured tissue by differentiating into the target tissue, thus contributing to repairing the defective tissue.	Return to performance rate: 80% compared to typical 30% with half the risk of reinjury with other techniques (Dahlgren, 2005). Sources quote a 48-week rehab with an 18% reinjure rate (Koch et al., 2009). Recovery period unchanged; regeneration is still inferior to native tendon (Reed and Leahy, 2013).	Y
PRP	Platelets contain growth factors released upon activation, which help promote new blood vessel growth, formation of new connective tissue (fibroplasia), and the regrowth of skin.	56% improvement with moderate injury, 64% with severe injury (Mehrjerdi et al., 2008). Application 30 days after the initial injury (Center for Equine Health, 2011; Sutter et al., 2004).	Y
Rest	Called a *natural* form of recovery, involving up to 12 months of farm retirement.	Most tendon injuries require at least 3 months of restricted exercise (e.g., walking in hand or under tack), along with turn out in a paddock (Casey, 2013a,b,c).	N
Extracorporeal shock wave therapy (ESWT)	This technique uses a machine (extracorporeal: from outside the body) to produce high-intensity shock or pressure waves that are directed, often with ultrasound, to a specific site within injured tissue.	Most users report a 50:50 success rate. ESWT seems to be most effective for front-leg suspensory desmitis, but for hind-leg suspensory problems and for navicular disease, it works about half the time (Marcella, 2009).	N
Tissue engineering (Acell)	A *resorbable bioscaffold* that proponents claim generates healthy tissue.	12 months prior to data collection, of the 107 cases, a total of 92%, or 86%, had returned to full work/function (Mitchell, 2009).	Y
Stem cell	Bone-marrow-derived stem cells have been the most studied and currently are favored because of their superior performance. Fat-derived stem cells are surgically harvested from the tail head of the horse. They are implanted via ultrasound.	Reinjury following SDF stem cell therapy in 3 years following repair was 24% (UK 168-horse National Hunt study) compared to 56% in horses getting tendon treatment. Dr. Roger Smith: "This technology (stem cell) has shown encouraging, but not yet proven, efficacy for treating acute tendon lesions in horses" (Frisbie and Smith, 2010; Smith and Webbon, 2005)	Y
IRAP (interleukin receptor antagonist protein)	Cartilage destruction is caused by interleukin-1 (IL-1), whereas IRAP is an antibody acting against IL-1 (House and Morton, 2011).	Repeated delivery of IL-1ra using gene transfer necessitates a better vector, which is currently being researched (Mcilwraith, 2010).	Y

(Continued)

TABLE 26.2 (*CONTINUED*)

Treatment Options for Equine Tendon Ligament and Joint Disease

Method	Methodology	Features	Invasive?
Tildren	Available in Europe. Regulates bone destruction as it reduces the activity of osteoclasts (bone destroyers) and activates osteoblasts (bone producers) (Dowling and Dart, 2005).	Some effectiveness in treating navicular disease, but there is some debate about its use in hock arthritis relief (Dowling and Dart, 2005).	Y
STS (SQUID signal therapy)	SQUID therapy signal—an induction-based system for applying signals derived from superconductive quantum interference device signals to accelerate the healing process (Parker and Markov, 2012).	24 horses were treated by using SQUID-based signals, with complete set of data for 19 subjects. For these animals, statistical data were average healing time D and standard deviation SD are $D \pm SD = (23.5 \pm 1.2)$ days (Parker et al., 2014).	N

26.6.1 EQUINE HOSPITALS

The various U.S. equine hospitals that manage the many types of therapies offered (2014) are shown in Table 26.3. This large variation is indicative of the general disagreement on a preferred method, rather than an endorsement of the technologies.

26.6.2 HUMAN ACHILLES TENDON

The human Achilles tendon (AT) is an excellent counterpart to the equine SDFT as it is the strongest and thickest tendon in the human body, serving as a basic function to connect the soleus and gastrocnemius muscles to the calcaneus bone, allowing plantar flexion about the ankle joint.

The AT also experiences one of the most common tendon lesions, found in 18 out of 100,000 people between 30 and 50 years of age (Hogan et al., 2011). This injury normally appears at the small cross section region 2–6 cm from the calcaneal insertion (Olsson et al., 2011), whereas the insertion area itself is able to withstand higher strains than the rest of the tendon (Bressel and McNair, 2001). In addition, the AT influences the capacity of many human movements. Without surgery, the prognosis for AT recovery is poor, with most patients experiencing only minor improvement after the first year (Magnussen et al., 2011; Maquirriain, 2011). In addition, most are unable to achieve full function 2 years after the injury without surgery, leading to disappointing recovery results (Tan, 2013).

26.7 ULTRASOUND IMAGING AS THE ONLY OBJECTIVE SOURCE OF INFORMATION FOR HEALING PROGRESSION IN THE FIELD

For field use, a portable imaging system is required that is capable of delivering high-quality images sufficiently diagnostic as to permit isolation of lesions or other ligament and tendon lesions consequent to acute or attritional injury. Of all the imaging mediums available (CT, MRI, x-ray, ultrasound), only ultrasound is readily portable and immediate enough to permit field diagnoses. Practical transducers now have a frequency range from 3.0 to 40 MHz with axial and spatial resolution in the hundreds of millimeters (transducer and scanner dependent) (Table 26.4).

TABLE 26.3
Equine Hospitals and Therapies Offered (from Public Records)

# Vets	Hospital Name	State	Website	Shock Wave	PRP	Stem	Laser	IRAP
6	Merritt & Associates Equine Hospital	IL	www.merrittequine.com	X	X	X	X	X
10	Pioneer Equine Hospital	CA	http://www.pioneerequine.com/index.php	X	X	X		X
48	Rood & Riddle Equine Hospital	KY	www.roodandriddle.com			X		
8	Tennessee Equine Hospital	TN	www.tnequinehospital.com	X	X	X	X	X
8	Idaho Equine Hospital	ID	www.idahoequinehospital.com	X	X	X		
19	Peterson & Smith Equine Hospital	FL	http://www.petersonsmith.com/	X		X		X
3	Reata Equine Hospital	TX	www.reataequinehospital.com	X		X	X	
6	Columbia Equine Hospital	OR	columbiaequine.com	X		X		X
6	Tryon Equine Hospital	NC	www.tryonequine.com	X	X	X		X
15	Ocala Equine Hospital	FL	www.ocalaequinehospital.com	X	X			X
9	Brazos Valley Equine Hospital	TX	http://bveh.com/navasota.htm	X	X	X		X
7	Texas Equine Hospital	TX	www.texasequinehospital.com		X	X		X

TABLE 26.4

Sound Wave Propagation through Various Media

Ultrasound Penetration vs. Frequency (MHz)			Velocity of Sound Waves in Media			
			Medium	Velocity (m/s)	Acoustic Impedance (Rayls)	Absorption Coefficient (dB/MHz/cm)
Low resolution	2.0	High penetration				
	3.5					
	5.0		Air	>331	0.0004	12.00
	7.5		Fat	1450	1.38	0.63
	10.0		Soft	1540	1.6	0.94
High resolution	12.0	Low penetration	Bone	4080	7.8	20.00

26.7.1 Application of Ultrasound Properties

Reviewed here is the property that sound waves travel through different media at different speeds, depending on the density of the medium. It is important to recognize this characteristic, as it is the metric that permits us to make tissue-specific diagnosis.

An understanding of the acoustic properties of biological materials is essential for tissue characterization as shown. Higher tissue density causes an increase in acoustic velocity, which permits us to understand tissue composition (Table 26.5).

As sound velocity changes for different media, this property can be used to assess the type of tissue under consideration and render a diagnosis of the injury condition.

26.7.2 Types of Ultrasound Tissue Characterization

There are numerous types of ultrasound characterization models. Each has specific advantages and limitations, and the most promising are early technologies that are yet to make a significant impact for the practitioner (Table 26.6).

The various image forms of these types are shown later.

26.7.2.1 Implications of Tissue-Diagnostic Shock Filters

The promise and ability to perform fiber bundle-level analysis of a tendon or ligament as shown in the last case (shock filters) present the veterinarian with an entire new capability. This device, now found only in research laboratories, may be implemented as an attachment to the

TABLE 26.5

Acoustic Characteristics of Biologic Materials

Materials	ρ (kg/m³)	$\exists$ (kg⁻¹ s⁻¹)	c (m/s)	Z (kg/m²/s)
Units	Volume Mass	Compressibility	Velocity	Acoustic Impedance
Fat	950	508	1440	1.37
Neurons	1030	410	1540	1.59
Kidney	1040	396	1557	1.62
Liver	1060	375–394	1547–1585	1.64–1.68
Spleen	1060	380–389	1556–1575	1.65–1.67
Bone	1380–1810	25–100	2700–4100	3.75–7.40

TABLE 26.6

Types of Ultrasound Tissue Characterization

Technology	Advantage	Disadvantage
B-mode	Portable, inexpensive, popular	Hand-held probe introduces manual positioning errors resulting in inability to differentiate at fiber bundle level.
Sector scan	Inexpensive, popular; used for fetal scans	Not useful for tissue differentiation.
Structure diagnostic	Applied to whole structure for general diagnosis	Not useful for tissue diagnosis at the fiber bundle level.
Tissue diagnostic	Applied to any specific site on tendon	Morphological operation image segmentation, using a shock filter thinning algorithm—digital (Meghoufel et al., 2011).

veterinarian's own high-definition ultrasound probe. Blurry ultrasound images, once processed, may permit the *field assessment* of tissue damage, the extent of a lesion, and the most important characteristic—the ability to determine if the local repair is with scar or healthy tissue (Meghoufel et al., 2011).

> Tendon fibers are inherently dry and do not reflect ultrasound. Tendon interfascicular *linings*, however, are inherently wet and do reflect ultrasound energy, which is why it is possible to differentiate the tendon fiber bundle with a tissue-diagnostic approach (Figures 26.6 and 26.7).

This new imaging capability gives the trainer and veterinarian a diagnostic capability that is 98% equivalent to histology (Meghoufel et al., 2014). This narrow area permits the investigator to identify healthy tissue against scar tissue. This ability has the potential to substantially alter the course of veterinary medicine, as the current design of tendon/ligament treatment centers around identifying and encouraging healthy tissue.

In addition, the technology is predicate and specific and is not subject to the vagaries of statistical methods, as it specifically addresses the characteristics of an individual ultrasound scan layer (as seen in Table 26.7) and not an approximation of those above and below (Table 26.8).

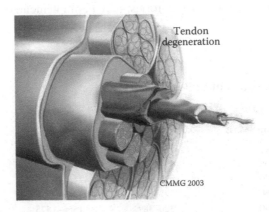

FIGURE 26.6   Tendon degeneration. (Image courtesy of Medical Multimedia Group LLC, www.eOrthopod.com.)

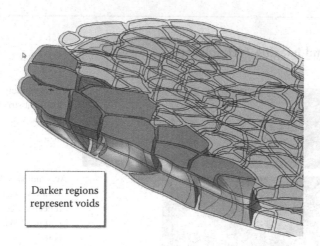

FIGURE 26.7 Tendon layer structure.

26.8 PROMISING DEVELOPMENTS

Of all the therapy technologies offered to date, the most promising for gaining immediate useful results is the STS therapy technology as described in Section 26.3. No other development thus far shows promise of changing the landscape of tendon injury therapy as fast or as dependable (Tables 26.9 and 26.10).

New imaging technologies may significantly impact veterinary medicine. These recent events serve to make the evaluation process more efficient (motorized scanner), add new features such as acoustoelasticity (Echometrix, 2014), and derive new information from the underlying information source (image segmentation). The future is very bright for effective, time-saving, and image-centered equine therapies as new developments are permitting much of the confusion to be replaced with confidence.

Considering only the issue of *restoring healthy tissue*, we review the following promising therapies.

From the reports of study participants, the field responses to the STS therapy appear to be as follows (Figure 26.8).

26.9 SUMMARY AND CONCLUSION

Although this chapter has centered on ligament and tendon injuries, one cannot escape the profound affiliation of these two novel technologies: EMF natural signal stimulation coupled with the computational intensive image segmentation possibilities of performance-enhanced ultrasound. This merge of capabilities suggests the following possible applications in the future:

1. Sports field diagnosis of severe firmament/tendon injuries
   This technology will permit the team physician to make a judgment call *on the field* that ordinarily would require an MRI scan, as imaging capabilities will reveal tendon and ligament tears.
2. Fast assessment of drug protocols
   As the local tissue response is structural in nature, it will be possible to see the effects at the *cellular* level of a live patient.

TABLE 26.7

Types of Ultrasound Imaging Technologies

Technology	Image	Attribute
B-mode		Can accurately reveal tendon and ligament structures in sufficient detail as to be structure diagnostic.
Sector scan		Accomplished by taking a sweep of views with a preset spacing, the object reflections contain sufficient detail to permit a 3D reconstruction as shown here.
Gray scale		B-mode with motorized scanner to provide rigid registration of 600 images along the 12 cm scanner length. Uses first-order analysis to average scans above and below to arrive at a classification. Good analytical capability for whole tendon evaluation. Image shows whole tendon in color longitudinally (van Schei et al., 2003).
Structure diagnostic— texture segmentation	(a) (b)	Discrete packet waveform technology. This tool provides more quantitative and objective information relative to that obtained using visual estimation methods (Kim et al., 1998).
Tissue diagnostic— shock filters with morphological operations		Morphological operation image segmentation, using a shock filter thinning algorithm—a severely complicated digital process to portray the fiber structure of the tendon at a resolution exceeding native ultrasound (Meghoufel et al., 2011).

3. On-site use as a diagnostic tool of soft and dense tissue injuries
   As a local injury is clouded by the associated tissue inflammation, image segmentation techniques, coupled with specific applied processing algorithms, will permit the segregation of tissue injury from bruise effects, thus effectively *wiping away the clouded ultrasound image*.
4. Extraction of specific tissue injuries from background images
   The intelligence of the image processing algorithms will permit the separation of the tissue injury from the normally clouded foreground of soft-tissue impact effects.

TABLE 26.8
Comparison of Ultrasound Scan Technologies

Characteristic	Gray Scale	Sector Scan	Structure Diagnostic	Tissue Diagnostic
Accuracy	Very good (van Schie et al., 2012)	Nominal	Good, as applied to a whole structure (see Table 26.7)	98% (Meghoufel et al., 2014)
Technology	B-mode with motorized scanner	B-mode with motorized scanner	Discrete wavelet packet frame (DWPF—digital) (Kim et al., 1998)	Morphological operation image segmentation, using a shock filter thinning algorithm—digital (Meghoufel et al., 2011)

TABLE 26.9
New Developments in Supporting Equine Therapy

Development	Impact or Observation
Motorized scanner	Available only in association with imaging technology package from UTC61 (UTC Imaging, 2014).
Image segmentation	DWPF (discrete wavelet packet frames) useful for gross approximation, image segmentation using morphological operations only described in research (Meghoufel et al., 2011).
Acoustoelasticity	Only described as a research project but appears to have a commercial offering (Echometrix, 2014).
Tissue engineering	Promises to grow healthy tissue. Limited evaluation due to contaminated study; porcine urinary bladder matrix (UBM) is marketed as an acellular biologic scaffold (Acell Vet; Acell, Columbia, MD) for use in the horse (Badylak et al., 2009).

TABLE 26.10
Therapies That Promise to Restore Healthy Tissue

Therapy	Benefit	Drawback	Invasive?
Tissue engineering (Acell)	A technique that shows promise of tissue regeneration	A 12-month posttreatment analysis showed that of the 107 cases, a total of 92, or 86%, had returned to full work/function. This study is challenged, however (Koch et al., 2009).	Y
STS (SQUID signal therapy)	Applies SQUID signals to an injury (Robinson and Sprayberry, 2009). Demonstrated an average healing time of 23.5 ± 1.2 days with a 4% reinjure rate (Parker and Markov, 2014)	Therapy application is new, having only been available commercially since early 2014.	N

5. Monitoring incision healing processes
   The usefulness of image segmentation as applied to tissue evaluation will permit tissue injuries to be monitored and displayed in real time, and not after a laboratory analysis.
6. Evaluation of the skin damage healing process
   High-frequency (up to 40 MHz) ultrasound probes, coupled with the intelligence of image segmentation processing, will reveal the actual healing processing and display the ordered/disordered nature of the cellular structure beneath the skin, thus permitting much better control of disfiguring scar tissue formation.

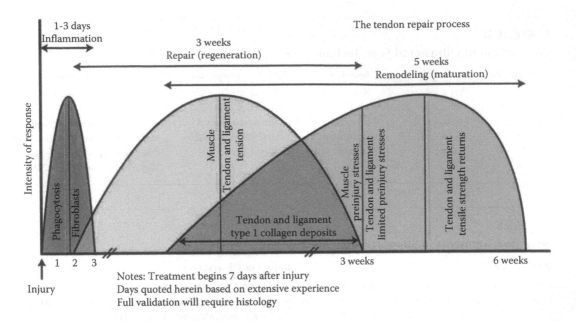

FIGURE 26.8 Field response to treatment.

The two new applications, therapy and imaging, offer profound opportunities in research and application. However, although most other established equine therapy techniques appear by their individual results to have limited usefulness, implementation by a dedicated operator appears to give uniform results unique to that practitioner, suggesting that much of the tool application is skill based, rather than technology based.

To evaluate the significant value of proper therapy application, consider this:

- If a horse can run a mile in 1:40.00 or 100 s over a fast track, his relative value would be about $20,000.
- If this same horse were to lose 1% of his performance or run a mile in 1:41.00 or 101 s over a fast track, his value would likely drop to about $10,000.
- If this same horse were to improve 1% in his performance or run a mile in 1:39.00 or 99 s, his value would likely increase to $40,000.

James M. Casey, DVM, MS Equine Veterinary Service Shenandoah Valley, VA.

The benefits can easily be seen of STS therapy applications and the imaging service to sports medicine for accelerated repair of injuries. In addition, the ability to determine on the field, whether the injury is serious (ligament or tendon) or minor (muscle bruise), will permit the player to resume the game or potentially save his career from a bad call. This type of instant diagnosis that may otherwise require a large imaging service will create an entire new set of opportunities for this technology.

REFERENCES

Anastasiadis P, Anninos P, Kotini A, Koutlaki N. (2006) Differentiation of myomas by means of biomagnetic and doppler findings. *Biomagn Res Technol*, 4:3.

Anninos P, Anogianakis G, Lehnertz G, Pantev C, Hoke M. (1987) Biomagnetic measurements using SQUID. *Int J Neurosci*, 37:149.

Badylak S et al. (2009) Strength over time of a resorbable bioscaffold for body wallrepair in a dog model. *J Surg Research*, 99: 282–287.

Blunden T, Dyson S, Murray R, Schramme M. (2006) Histopathology in horses with chronic palmar foot pain and age-matched controls, part 2: The deep digital flexor tendon. *Equine Vet J*, 38:23–27.

Bressel E, McNair PJ. (2001) Biomechanical behavior of the plantar flexor muscle tendon unit after an Achilles tendon rupture. *Am J Sports Med*, 29(3):321–326.

Burba DVM (2009) Understanding the condition of bucked shins so we can take better care of our horses. *LSU AgCenter*. http://www.lsuagcenter.com.

Daniel B. (2009) The Dilemma of Bucked Shins in the Racehorse, *Equine Health Studies Program*. LSU School of Veterinary Medicine.

Casey J. (2013a) Equine joint disease. http://www.equinehorsevet.com/jmc-news.htm. Accessed September 11, 2002.

Casey J. (2013b) Navicular disease in horses. *Equine Sports Medicine, Dentistry, & Surgery*. www.equine-horsevet.com. Accessed June 27, 2008.

Casey J. (2013c) Tendon injuries (bowed tendon) in horses. www.equinehorsevet.com. Accessed June 27, 2008.

CEH Horse. (2011) *Report Platelet-Rich Plasma: Improving Treatment for Tendon and Ligament Injuries*, Vol. 29(1). Center for Equine Health UC-Davis, Davis, CA.

Cohen D, Yoram P, Cuffin BN, Schmid SJ. (1980) Magnetic fields produced by steady currents in the body. *Proc Natl Acad Sci USA*, 77(3):1447–1451 (*Biophysics*).

Dahlgren LA (2005) Review of treatment options for equine tendon and ligament injuries: What's new and how do they work? In: *Proceedings of the 51st Annual Convention of the American Association of Equine Practitioners*, Seattle, WA.

Dowling B, Dart AJ. (2005) Mechanical and functional properties of the equine superficial digital flexor tendon. *Vet J*, 170(2):184–192.

Duenwald-Kuehl S, Lakes R, Vanderby Jr R. (2012) Strain-induced damage reduces echo intensity changes in tendon during loading. *J Biomech*, 45:1607–1611.

Ellison ME, Duenwald-Kuehl S, Forrest LJ, Vanderby R, Brounts SH (2014) Reproducibility and feasibility of acoustoelastography in the superficial digital flexor tendons of clinically normal horses. *American Journal of Veterinary Research*, 75(6):581–587.

Frisbie DD, Smith RK. (January 2010) Clinical update on the use of mesenchymal stem cells in equine orthopaedics. *Equine Vet J*, 42(1):86–89. doi: 10.2746/042516409X477263.

Gary M, Baxter, G. (2011) *Adams and Stashak's Lameness in Horses*, 6th edn, Wiley-Blackwell.

Halper J, Kim B, Khan A, Yoon J, Mueller E. (2006) Degenerative suspensory ligament desmitis as a systemic disorder characterized by proteoglycan accumulation. *BMC Vet Res*, 2:12. doi:10.1186/1746-6148-2-12.

Hayward M, Adams D. (2001) *The Firing of Horses: A Review for the Animal Welfare Advisory Committee of the Australian Veterinary Association*. AVA Animal Welfare Advisory Committee.

Herthel D. (2001) Enhanced suspensory ligament healing in 100 horses by stem cells and other bone marrow components. In: *AAEP Proceedings, Lameness in:The Athletic Horse*, Vol. 47.

Hogan MV, Bagayoko N, James R, Starnes T, Katz A, Chhabra AB. (2011) Tissue engineering solutions for tendon repair. *J Am Acad Orthop Surg*, 19:134–142.

HorseJournals.com. http://www.horsejournals.com/Ligament Injuries: Advances in Diagnosis and Treatment, accessed November 6, 2013.

House A, Morton A. (2011) *Interleukin-1 Receptor Antagonist Protein (IRAP) Therapy for Equine Osteoarthritis*, UF College of Veterinary Medicine, University of Florida.

Kasashima Y, Takahashi T, Smith RK, Goodship AE, Kuwano A, Ueno T, Hirano S. (2004) Prevalence of superficial digital flexor tendonitis and suspensory desmitis japanese thoroughbred flat racehorses in 1999. *Equine Vet J*, 36(4):346–350.

Kim N, Booth L, Amin V, Lim J, Udpa S. (1998) *Ultrasonic Image Processing For Tendon Injury Evaluation*. Iowa State University, Ames, IA. 0-7803-5073-1/98/.

Koch TG, Berg L, Betts D. (2009) Current and future regenerative medicine—Principles, concepts, and therapeutic use of stem cell therapy and tissue engineering in equine medicine. *Can Vet J*, 50(2):155–165.

Magnussen RA, Glisson RR, Moorman III CT. (2011) Augmentation of Achilles tendon repair with extracellular matrix xenograft: A biomechanical analysis. *Am J Sports Med*, 39(7):1522–1527.

Maquirriain J. (2011) Achilles tendon rupture: Avoiding tendon lengthening during surgical repair and rehabilitation. High Performance National Sports Center (CeNARD) & Nixus Foundation, Buenos Aires, Argentina. *YALE J Biol Med*, 84:289–300.

Marcella K. (2009) *Evaluating Treatment Options for Equine Tendon, Ligament and Joint Disease*. DVM360 Magazine.

Mcilwraith CM. (2010) Management of joint disease in the sport horse Colorado State University, *Proceedings of the 2010 Kentucky Equine Research Nutrition Conference*, Ft. Collins, Colorado.

Meghoufel A, Cloutier G, Crevier-Denoix N, deGuise J. (2011) Tissue characterization of equine tendons with clinical B-scan images using a shock filter thinning algorithm. *IEEE Trans Med Imaging*, 30(3):597–605.

Meghoufel A, Cloutier G, Crevier-Denoix N, de Guise J. (2014) Segmentation of clinical B-scan ultrasound images to predict the equine tendon anatomopathology and to prevent repeat injuries, unpublished.

Mehrjerdi K, Sardari K, Emami M, Movassaghi A, Goli A, Lotfi A, Malekzadeh A. (2008) Efficacy of autologous platelet-rich plasma (PRP) activated by thromboplastin-D on the repair and regeneration of wounds in dogs. *Iran J Vet Surg*, 3(4):9.

Mitchell R. (2009) *Treatment of Tendon and Ligament Injuries with UBM Powder*. Fairfield Equine Associates, Newtown, CT.

Multi-Media Group. (2011) Image courtesy of Medical Multimedia Group LLC. www.eOrthopod.com.

Olsson N, Nilsson-Helander K, Karlsson J, Eriksson BI, Thomée R, Faxén E et al. (2011) Major functional deficits persists 2 years after acute Achilles tendon rupture. *Knee Surg Sports Traumatol Arthrosc*, 19(8):1385–1393.

Parker R. (2008) Method and apparatus for generating a therapeutic magnetic field, Patent 7,361,136.

Parker R, Markov M, Allen J, DVM (2014, submitted) 30-day ligament/tendon lesion closure: A 24-horse case report study, *AAEP Salt Lake City Conference*.

Parker R, Markov M. (2012) Analytical versus empirical design of EMF devices. In: *BEMS Conference*, Halifax, Nova Scotia, Canada, pp. 4–5.

Parker R, Markov M, Allen J. (2014) 30-Day ligament/tendon lesion closure: A 24-horse case report study. In: *AAEP Salt Lake City Conference*, submitted for publication.

Reed SA, Leahy E. (January 2013) Growth and development symposium: Stem cell therapy in equine tendon injury. *Anim Sci*, 91(1):59–65. doi: 10.2527/jas.2012–5736.

Robinson N, Sprayberry K. (2009) *Current Therapy in Equine Medicine*, Vol. 6. Elsevier Health Sciences.

Rosch PJ, Markov MS. (eds.) (2004) Magnetic and electromagnetic field therapy: Basic principles of application for pain relief. In: *Bioelectromagnetic Medicine*. Marcel Dekker, NY, pp. 251–264.

Rose D, Smith P, Sato S. (1987) Magnetoencephalography and epilepsy research. *Science*, 238:329.

Smith R, Webbon PM. (2005) Harnessing the stem cell for the treatment of tendon injuries: Heralding a new dawn? *Br J Sports Med*, 39(9):582–584.

Smith R, Webbon PM. (2005a) Stem cell treatment harnessing the stem cell for the treatment of tendon injuries: Heralding a new dawn? *Br J Sports Med*, 39:582–584. doi:10.1136/bjsm.2005.015834.

Smith R, Webbon PM. (2005b) Harnessing the stem cell for the treatment of tendon injuries: Heralding a new dawn? *Br J Sports Med*, 39:582–584. doi:10.1136/bjsm.015834.

Sutter WW, Kaneps AJ, Bertone AL. (2004) Comparison of hematologic values and transforming growth factor-B and insulin-like growth factor concentrations in platelet concentrates obtained by use of buffy coat and apheresis methods from equine blood. *Am J Vet Res*, 65:924–930.

Tan J. (2013) *A Review and Update on Tendon and Ligament Injuries*. TheHorse.com. Article #32963.

Thorpe CT, Clegg PD, Birch HL. (2010) A review of tendon injury: Why is the equine superficial digital flexor tendon most at risk? *Equine Vet J*, 42(2):174–180. doi: 10.2746/042516409X480395.

Todorov N. (1982) *Magnetotherapy*. Meditzina i Physcultura Publishing House, Sofia, Bulgaria, 106p.

UC Davis Veterinary Hospital, William R. Pritchard Veterinary Medical Teaching Hospital, Diagnostic ultrasound & Musculoskeletal Injuries in Horses. http://www.vetmed.ucdavis.edu/vmth/large_animal/ultrasound/lausbroch.cfm.

van Schie HT, Bakker EM, Jonker AM, van Weeren PR. (2003) Computerized ultrasonographic tissue characterization of equine superficial digital flexor tendons by means of stability quantification of echo patterns in contiguous transverse ultrasonographic images. *Am J Vet Res*, 64(3):366–375.

van Schie HT, Docking S, Daffy J, Praet S, Rosengarten S, Cook J. (2012) Ultrasound tissue characterisation, an innovative technique for injury-prevention and monitoring of tendinopathy. In 2nd *International Scientific Tendinopathy Symposium* (Vancouver, 2012). *Br J Sports Med* 2013, 47:e2. doi:10.1136/bjsports-2013-092459.27.

Index

A

Acetylcholinesterase (AChE), 108
Active implantable medical device (AIMD), 345, 356–358
Age-dependent magnetosensitivity
definition, 217
Na$^+$/Ca^{2+} exchange, 224–225
^{45}Ca^{2+} uptake, 225–227
SMF effect, 226–227
Na$^+$/K$^+$ pump dysfunction, 217
[^{3}H]-ouabain binding, 218–220
tissue dehydration
cell swelling theory, 220–221
Na$^+$/K$^+$ pump, 221
ouabain dose-dependent changes, 221–222
SMF exposure, 223–224
AIMD, *see* Active implantable medical device (AIMD)
Anterior cruciate ligament reconstruction, 289
Antiangiogenic gene therapy, 405

B

Back pain/whiplash syndrome, 283
BBB permeability, *see* Blood–brain barrier (BBB) permeability
B cell receptor (BCR), 155
Binhi's mechanism, 59–60
Bioelectromagnetics
exposure system
animal immobilization, 76
block diagram, 75
counterpoise, 74
envelope detection, 74
low-level effect, 76
orthogonal current generator, 76–77
personal radio radiations, 73–74
transmittance measurements, 74–75
TEM cell
advantages, 70
E-field probe, 71–72
mutual coupling phenomenon, 72–73
schematic diagram, 70
V_2/V_1^+ *vs.* output reflection (Γ_2), 70–71
Biological windows
information transfer, 3
MF/EMF exposure, 2–3
resonance mechanism, 2
static magnetic fields, 3–4
threshold/thermal effect, 3
Biophysical dosimetry, 23, 439
Biophysical stimulation (BS)
bone repair
capacitive system, 254–256
clinical trials, 257
electrokinetic phenomenon, 254

inductive system, 254–256
in vitro studies, 255
mineral apposition rate, 256
nonunion treatment, 258–259
osteonecrosis treatment, 259–260
osteotomies, 258
physical agents, 255
piezoelectric effect, 254
prosthesis, 259
risk fractures, 257–258
spinal fusion, 260
ultrasound system, 254–256
cartilage repair
adenosine receptors, 261
anabolic function, 261
bovine synovial fibroblasts, 261
catabolic function, 261
chondroprotection, 261
dose–response curves, 262
I-ONE therapy (*see* I-ONE therapy)
PEMF effect, 263
surgical treatments, 260
Blood–brain barrier (BBB) permeability, 51, 105, 117
Bone repair
capacitive system, 254–256
clinical trials, 257
electrokinetic phenomenon, 254
inductive system, 254–256
in vitro studies, 255
mineral apposition rate, 256
nonunion treatment, 258–259
osteonecrosis treatment, 259–260
osteotomies, 258
physical agents, 255
piezoelectric effect, 254
prosthesis, 259
risk fractures, 257–258
spinal fusion, 260
ultrasound system, 254–256
Brain and heart muscle tissues, *see* Age-dependent magnetosensitivity
Breast augmentation surgery, 288, 300–301
Breast reconstruction surgery, 304–305
Breast reduction surgery
5/20 regimen, 302–303
inflammatory response, 301–302
NO/cGMP/PDE signaling, 303
Q2 treatment regimen, 302–303
Q4 treatment regimen, 301–303
therapeutic evidence, 301–302
BS, *see* Biophysical stimulation (BS)

C

Calcium-ion efflux
electromagnetic field

453

Printed in the United States
by Baker & Taylor Publisher Services

Printed in the United States
by Baker & Taylor Publisher Services